AF846695

EAI/Springer Innovations in Communication and Computing

Series Editor

Imrich Chlamtac, European Alliance for Innovation, Ghent, Belgium

The impact of information technologies is creating a new world yet not fully understood. The extent and speed of economic, life style and social changes already perceived in everyday life is hard to estimate without understanding the technological driving forces behind it. This series presents contributed volumes featuring the latest research and development in the various information engineering technologies that play a key role in this process. The range of topics, focusing primarily on communications and computing engineering include, but are not limited to, wireless networks; mobile communication; design and learning; gaming; interaction; e-health and pervasive healthcare; energy management; smart grids; internet of things; cognitive radio networks; computation; cloud computing; ubiquitous connectivity, and in mode general smart living, smart cities, Internet of Things and more. The series publishes a combination of expanded papers selected from hosted and sponsored European Alliance for Innovation (EAI) conferences that present cutting edge, global research as well as provide new perspectives on traditional related engineering fields. This content, complemented with open calls for contribution of book titles and individual chapters, together maintain Springer's and EAI's high standards of academic excellence. The audience for the books consists of researchers, industry professionals, advanced level students as well as practitioners in related fields of activity include information and communication specialists, security experts, economists, urban planners, doctors, and in general representatives in all those walks of life affected ad contributing to the information revolution.

Indexing: This series is indexed in Scopus, Ei Compendex, and zbMATH.

About EAI - EAI is a grassroots member organization initiated through cooperation between businesses, public, private and government organizations to address the global challenges of Europe's future competitiveness and link the European Research community with its counterparts around the globe. EAI reaches out to hundreds of thousands of individual subscribers on all continents and collaborates with an institutional member base including Fortune 500 companies, government organizations, and educational institutions, provide a free research and innovation platform. Through its open free membership model EAI promotes a new research and innovation culture based on collaboration, connectivity and recognition of excellence by community.

Anandakumar Haldorai • Babitha Lincy R
Suriya Murugan • Minu Balakrishnan

Artificial Intelligence for Sustainable Development

Springer

EAI
RESEARCH MEETS INNOVATION

Anandakumar Haldorai
Centre for Future Networks
and Digital Twin
Sri Eshwar College of Engineering
Coimbatore, Tamil Nadu, India

Babitha Lincy R
Sri Eshwar College of Engineering
Coimbatore, Tamil Nadu, India

Suriya Murugan
Sri Eshwar College of Engineering
Coimbatore, Tamil Nadu, India

Minu Balakrishnan
Sri Eshwar College of Engineering
Coimbatore, Tamil Nadu, India

ISSN 2522-8595 ISSN 2522-8609 (electronic)
EAI/Springer Innovations in Communication and Computing
ISBN 978-3-031-53971-8 ISBN 978-3-031-53972-5 (eBook)
https://doi.org/10.1007/978-3-031-53972-5

© European Alliance for Innovation 2024
This work is subject to copyright. All rights are solely and exclusively licensed by the Publisher, whether the whole or part of the material is concerned, specifically the rights of translation, reprinting, reuse of illustrations, recitation, broadcasting, reproduction on microfilms or in any other physical way, and transmission or information storage and retrieval, electronic adaptation, computer software, or by similar or dissimilar methodology now known or hereafter developed.
The use of general descriptive names, registered names, trademarks, service marks, etc. in this publication does not imply, even in the absence of a specific statement, that such names are exempt from the relevant protective laws and regulations and therefore free for general use.
The publisher, the authors, and the editors are safe to assume that the advice and information in this book are believed to be true and accurate at the date of publication. Neither the publisher nor the authors or the editors give a warranty, expressed or implied, with respect to the material contained herein or for any errors or omissions that may have been made. The publisher remains neutral with regard to jurisdictional claims in published maps and institutional affiliations.

This Springer imprint is published by the registered company Springer Nature Switzerland AG
The registered company address is: Gewerbestrasse 11, 6330 Cham, Switzerland

Paper in this product is recyclable.

Preface

Welcome to the world of emerging artificial intelligence (AI) and its various applications for sustainable development. This book discusses about the trends, techniques, and applications of AI that create a major effect in various business sectors.

The development of AI technology has created new ways for analyzing and addressing different challenges such as ensuring food security, improving healthcare, and creating sustainable cities. By applying AI, we explore on providing solutions to handle complex issues and create a successful future.

Our major focus is on the application of AI in precision agriculture, where techniques such as crop yield prediction using optimized convolutional neural network (CNN) models are improving farming practices and productivity. Similarly, in smart healthcare, AI-driven systems are applied for early detection and right diagnosis of medical conditions, helping in improved patient care and treatment.

The book also delves into the intersection of AI and urban development for creating more smarter cities. From smart charging approaches for electric vehicles to vision-based solutions for water garbage cleaning, AI is driving innovation and transforming the way we live and interact with our urban environments.

We also discuss the role of AI in enhancing cybersecurity, business modeling, and predictive analytics. By harnessing the power of advanced AI techniques in business, we can gain valuable insights for core development.

As you travel through the pages of this book, I encourage you to approach each chapter with curiosity and an open mind. The book will have valuable inputs whether you are an AI practitioner, a researcher, a policymaker, or who is interested in the future of technology.

I extend my heartfelt gratitude to all the co-authors who have shared their expertise and insights to make this book possible. Their dedication and passion for advancing AI for sustainable development have been truly inspiring.

Let us immerse ourselves in the world of emerging artificial intelligence for sustainable application development and explore the transformative potential of AI in shaping a better tomorrow.

Coimbatore, Tamil Nadu, India

Anandakumar Haldorai
R. Babitha Lincy
Suriya Murugan
Minu Balakrishnan

Contents

About the Authors

Anandakumar Haldorai is Professor and Research Head, Department of Computer Science and Engineering, Sri Eshwar College of Engineering, Coimbatore, Tamil Nadu, India. He has received his Master's in Software Engineering and PhD in Information and Communication Engineering from PSG College of Technology, Coimbatore, under Anna University, Chennai. His research areas include big data, cognitive radio networks, mobile communications, and networking protocols. He has authored more than 158 research papers in reputed International journals and IEEE conferences. He has authored 14 books and many book chapters with reputed publishers such as Springer and IGI. He is Editor in Chief of Keai—Elsevier—IJIN, Associate Editor IEEE Access, Area Editor—EAI Energy Web and served as a reviewer for IEEE, IET, Springer, Inderscience and Elsevier journals. He is also the guest editor of many journals with Elsevier, Springer, Inderscience, etc. He has been the General Chair, Session Chair, and Panelist in several conferences. He is senior member of IEEE, IET, and ACM and Fellow member of EAI research group.

Babitha Lincy R received BTech degree from the Department of Electronics and Communication Engineering, Sri Venkateswara College of Engineering and Technology, Anna University, Tamil Nadu, India, in 2014. She got an ME degree from Vel Tech Multi Tech Engineering College, India, in 2015. She has done research on "Handwritten character recognition system using deep learning approach" at Sri Venkateswara College of Engineering, Anna University, Chennai, India. Currently, she is working as Assistant Professor in Sri Eshwar college of Engineering, Coimbatore. She has filed two patents. She has more than 10 research papers published in SCI, book chapters, and Scopus journals. She attended more than 10 workshops relevant to her research area. Currently she is working as Assistant Professor in the Department of Computer and Communication Engineering at Sri Eshwar College of Engineering, Coimbatore. Besides, she is acting as a Peer Reviewer of journals such as *Current Science* (SAGE), *Concurrency and Computation: Practice and Experience* (Wiley), etc. Her research interests are VLSI circuit design, image processing, and deep learning.

Suriya Murugan has received her BE in Information Technology from Avinashilingam University, Coimbatore, in 2008 and Master's in Computer Science and Engineering from Anna University of Technology, Coimbatore, in 2010. She is pursuing PhD in Information and Communication under Anna University, Chennai. She is having about 12 years of teaching experience and her areas of interest include artificial intelligence, machine learning, and wireless communication. She has authored more than 30 research papers including 4 SCI and 25 Scopus indexed publications with high impact journals like *IEEE Access*, Elsevier, Springer, and Taylor and Francis. She has authored one book with Springer International Publishing and published about 11 book chapters with reputed publishers. She has obtained two patents, and she has been serving as a Reviewer for more than 15 peer journals like *IEEE Access*, Wiley, MDPI, Springer Wireless Networks, Inderscience, Taylor and Francis, MONET, etc. She has also delivered special lectures at FDPs and workshops on machine learning and deep learning at reputed institutions. She is a senior IEEE member.

Minu Balakrishnan obtained BTech in Information Technology and ME in Computer Science and Engineering from Anna University in 2012 and 2017, respectively. She is pursuing PhD in Information and Communication under Anna University, Chennai. She has about 2 years of experience in industry and is currently working as Assistant Professor of Information Technology at Sri Eshwar College of Engineering, Coimbatore. Her research interests include wireless ad-hoc network, image processing, and cognitive radio network.

Part I
Leveraging AI for Precision Agriculture – Significance and Applications

Significance of AI in Smart Agriculture: Methods, Technologies, Trends, and Challenges

1 Introduction

One of the oldest livelihoods of humans is agriculture, according to theory. It offers the primary component of nourishment for humans and other earthly living things. Along with the rapid expansion in global population, food consumption is also rising swiftly. Despite the fact that the production of agriculture is now 17% higher than it was 30 years ago, 821 million people worldwide lack the ability to obtain enough food. A multitude of concepts, including "agriculture 4.0," "digital agriculture," "smart agriculture," and "agri-food 4.0," are used to characterize the application of developing technology in agriculture. These words are motivated by PA and industrial 4.0 concepts. The employment of various technology in conventional farming methods in order to increase agricultural output and sustainability is known as precision agriculture, sometimes it is also known as intelligent agriculture. It has advanced closer to an information strategy whose objective is to acquire as much data as it can about the crop's health in order to carry out a personalized management in line with its needs. Precision agriculture (PA) may be thought of as a large control loop in which farm employees and machinery operate as the actuators to maintain a sustainable and productive output. The farmers are in responsibility of making the necessary corrections in accordance with both production requirements and environmental protection [1]. The transformation that is taking place in agriculture is fueled by the development of science and technology, which is dependent on both the growth in productivity and the constraints of the period.

Figure 1 depicts Agriculture (Agri) Agri 1.0, Agri 2.0, Agri 3.0, and Agri 4.0, the four generations of the agricultural revolution. Low efficiency in operation was agriculture's primary issue throughout the conventional agricultural period (1784–1870), or the nineteenth century, which was ruled by human and animal resources. When we talk about "Agri 1.0," it is a discussion about the farming practices that were used from the beginning of human civilization until the final

© European Alliance for Innovation 2024

A. Haldorai et al., *Artificial Intelligence for Sustainable Development*,
EAI/Springer Innovations in Communication and Computing,
https://doi.org/10.1007/978-3-031-53972-5_1

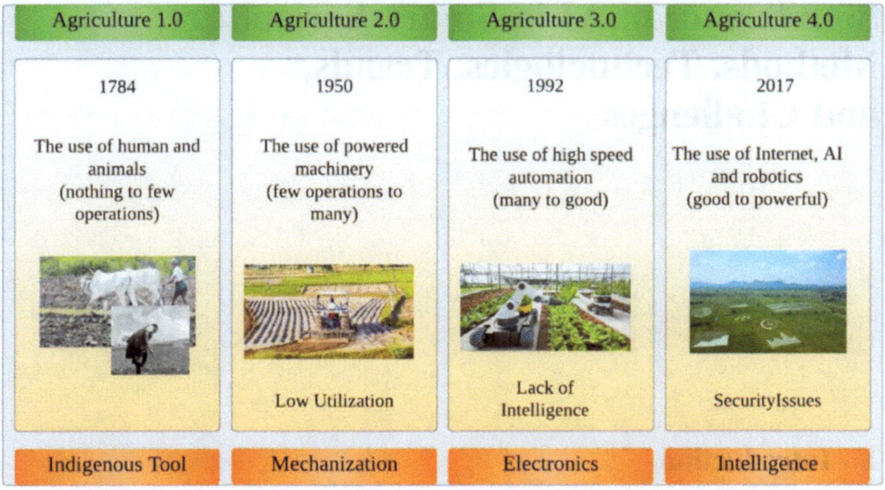

Fig. 1 Agriculture (Agri) Agri 1.0, Agri 2.0, Agri 3.0, and Agri 4.0

decades of the nineteenth century. Farmers often used conventional cultivation equipment like the common plow to loosen the soil for planting seeds and encourage plant development during this time. Agriculture 2.0, the introduction of agricultural machinery for weeding, irrigation, seeding, and harvesting. Agriculture mechanization dramatically enhanced food output while reducing manual labor.

From 1992 to 2017, automated agriculture advanced quickly, but the main issue was a lack of intelligence. This was known as "Agri 3.0." Geothermal, wind, hydropower, solar power, and bioenergy were used in its construction as environmentally friendly renewable energy sources. "Industry 4.0" was followed by the statement "Agri 4.0" [2]. Blockchain technology known as software-defined networking (SDN), artificial intelligence, the Internet of Things (IoT), devices connected to the IoT, 5G communications, drones, fog/edge computing, cloud-based computing, network function virtualization (NFV), smart grids, etc. are some of the emerging technologies that define Industry 4.0 [3]. Agricultural production has tremendous prospects thanks to current information technology, but it also faces significant demands and difficulties in the area of smart agriculture, security, and privacy. For instance, smart agriculture's development objectives include both intellectual and autonomous operations. These traits not only boost production but also raise the danger to equipment and data security.

Using sensors and communication devices, the Smart Agri System continually observes, checks, and gathers information about important factors from the agricultural areas [4]. The important variables include things like pH, moisture, temperature, and water level in the soil. Different types of field-installed sensors must be used to measure these characteristics. The sensor nodes are typically small and have little battery power [5]. To provide a constant and uninterrupted smart agricultural system, energy collecting technologies might be utilized to power the sensor nodes

[6]. Data from the field is collected by the gateway node, which is located at the border of the network, and sent to the cloud for analysis and further processing.

The main contribution of this paper is as follows:

1. In consideration of the smart agricultural ecosystem, comprising crop harvesting, fertilizers or pesticides, irrigation, sowing, and plowing, and animals (livestock), many forms of smart architecture are taken into account.
2. Three standard development methods for precision, facility, and orderly agriculture are all examples of smart agriculture—are used to summarize and categorize its current state of development.
3. Various integration of technologies into agriculture and some security issues has been outlined in this article.

The section of this paper is divided as follows, in Sect. 2, related works are mentioned, follows with Sect. 3 by describing smart architecture in detail. Section 4 has continued with different modes of smart agriculture. In Sect. 5, technology integration with agriculture sector is explained. General security issues in smart agriculture are briefly described in Sect. 6. In Sect. 7, conclusion of this paper is given.

2 Literature Review

Basem Almadani et al. in [6] explains that by 2050, it is predicted that there will be 9 billion people on the planet, which would result in a 70% increase in the global need for food. Agriculture, the industry currently in charge of pollution and consuming about 70% of the world's fresh drinking water, needs to use 41% more water to fulfill the rising demand for food. As the fourth agricultural revolution makes clear, connection is the foundation of farming's digitization.

Sherali Zeadally et al. in [7] explains that IoT is thus the primary enabling technology for the advancement of agriculture. The Internet of Things (IoT) is anticipated to be one of the key technologies that transforms conventional farming and animal raising practices into a new level of intelligence in precision agriculture. The author explored common uses of the cloud computing technologies in agricultural Internet of Things and addressed a straightforward Internet of Things in agriculture scenario with a wireless network.

Arvind Kumar et al. in [8] explains that, aimed to leverage automation and IoT technologies to make agriculture smart. Some examples are intelligent irrigation-integrated decision-making systems, smart GPS-based remote-controlled robotics, and smart warehouse management.

Biswaranjan Acharya et al. in [9] explains that modern and advanced agricultural practices have enhanced production of food and support for the struggle against world hunger, but they have also significantly worsened health catastrophes. As an illustration, using contemporary pesticides lowers antioxidant levels, which can cause neurodegenerative, cardiovascular, pulmonary, renal, endocrine, cancer, and reproductive issues. Researchers have so begun to assess and critique contemporary farming practices.

Abhishek Srivastava et al. in [10] explains that the impact of climate change including the 30% exploitation of natural resources are additional problems that worldwide society and agriculture must address. Despite the evident challenges of the agricultural industry, it has been estimated that 50% of food is wasted along the food-producing processes, from harvesting to final packaging, underlining the need for enhanced management of agriculture that may be supported by evolving technology. Surveying the entire area and carrying out the intended task(s) in sequence is a straightforward strategy for persistent sampling (and/or monitoring).

Joash Mageto et al. in [11] explains that the key disadvantage is that the robot would then thoroughly inspect every sampling place without giving those that would produce a greater benefit or would be more time-critical priority. Orienteering can partially address this issue by identifying routes that maximize the cumulative gain while keeping costs constant. When visiting nearby areas, the robot gives them priority if their combined benefits are larger than those of visiting isolated high-gain locations, provided that no budgetary restrictions are broken.

Anusha Vangala et al. in [12] explains that furthermore, a security technique based on symmetric encryption of data in agricultural sensors and an effective transmission strategy was developed for connection to the internet of intelligent agriculture applications. The security issues faced in rural agriculture are given in Table 1.

Table 1 Security issues in rural agriculture

Sl. no	Parameters	Rural agriculture	Security issues
1	Transport	Bad traffic with farm-land environment	Bad traffic creates a situation of delay in repairing the damaged or stolen facilities
2	Resident	Security consciousness is lacking	Dealing with security threats are harder
3	Communication	Base station is limited	Vulnerable to attacks from false base station
4	Density of facility	Low	Neighbor nodes cannot replace the data of attacked nodes
5	Characteristics of facility	Low cost and low power consumption	In the facility, complex security methods cannot be applied
6	Distribution of facility	Deployed outside	Facilities can be stolen or damaged
7	Ascription of facility	Private facilities for farmers	Private devices contain more personal information
8	Production data	Weak security measures	Vulnerable to unauthorized access
9	Production cycle	Long production cycle	More loss due to the attacks
10	Management system	Immature architecture	Vulnerable to key theft and control system intrusion

Ghadeer Yassin et al. in [13] explains that smart agricultural security challenges have been researched as a full investigation of safety and confidentiality in a smart agricultural ecosystem was presented.

Dmitrii Shadrin et al. in [14] looked at cybersecurity in the context of smart agriculture. Privacy-focused blockchain-based approaches were also recommended in *Sergio Trilles et al.* in [15].

Sina Sontowski et al. in article [16] provided a foundation for a prediction model for cyber assaults on precision agriculture, as well as suggested that anomaly detection be necessary to eliminate erroneous alarms.

3 Smart Agriculture for Smart Farming

In order to enhance information about farming perceptions, quantitative decision-making processes, intelligent control, precise investment, and individualized service, "smart agriculture"refers to an innovative approach to agricultural production which combines modern information technologies with agriculture. These technologies include the Internet, Internet of Things (IoT), big data, cloud computing, and the use of artificial intelligence (AI). Briefly stated, the new mode has been described as an intelligent agricultural approach that combines agriculture with modern information technologies. Smart agriculture is created as a result of the incorporation of computing, actuating, and sensing technology to the conventional agriculture. Many steps of the smart agricultural ecosystem include plowing, planting, crop harvesting, irrigation, fertilizers or the pesticides and animals. The majority of the traditional farming issues, such as drought response, crop optimization, environmental concerns, over or under irrigation, pest controls, and so on, can be resolved by incorporating modern technologies including Internet of Things (IT), Artificial intelligence (AI), drones, remote sensing, and agbots into the various stages of smart agriculture.

The current technological advancements in the agriculture industry have the potential to drastically alter both individual farm's and major businesses' labor processes. Modern farming technologies are being introduced at such a quick rate that they are now accessible not just in developed regions but also in underdeveloped ones. Figure 2 shows the progress agriculture to smart agriculture. The following benefits are often offered to producers by smart farming: speeding up data collection and processing, raising accuracy and precision levels, lowering production costs, requiring less manual labor, increasing crop yield, reducing driver stress, streamlining task recording and reporting, and enhancing sustainability [17].

Fig. 2 Agriculture to smart agriculture

3.1 Smart Irrigation System

One of the primary sources of agricultural production is irrigation. It has been a crucial component of agriculture for ages. It utilizes about 85% of the available sources of freshwater. Water resources have been significantly impacted recently by global misappropriation of resources and climate change. A farming method known as "smart irrigation" aims to use less water overall while applying water management to crops. Smart irrigation systems are typically built using wireless and Internet of Things (IoT) technologies, including WSNs, or wireless sensor networks.

Environmental variables like soil humidity level, atmosphere temperature, and humidity will be discovered where to set the threshold value for pump working. The pump is going to turn off/on as per the water need. In Fig. 3, it shows the gathering of all data that combined with the microcontroller system for measuring and regulating water during the construction of this model. Considering the saving of water while maintaining crop yield was a great accomplishment, it occurred due to sprinkler irrigation and drip irrigation system which is high efficiency system in smart

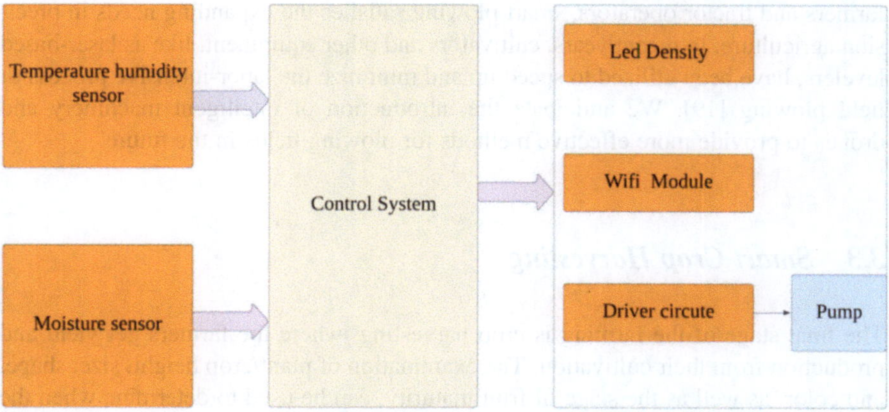

Fig. 3 Microcontroller system for measuring and regulating water

agriculture. By adopting closed-loop irrigation, water use in vertical greenhouse farming is significantly decreased. Contrarily, aquaponics that uses 90% of bone-dry conventional farming is done while hydroponics utilizes 13 times waterless typical soil farming. For intelligent irrigation systems to be capable of continuous tracking data which must be accessible at edge networks for actuation as opposed to being sent to the cloud for additional processing and analysis. In recent years, a brand-new, highly promising method known as Radio Frequency (RF) Energy Harvesting (EH) has evolved that may be used in wireless sensor networking [18]. It can be used as a substitute method of energy delivery in low-power electronic devices, such as wireless and Internet of Things sensors. Mobile communication networks are one of the main ambient sources for energy harvesting in outdoor settings.

3.2 *Smart Plowing System*

Before we start to sow our desired seeds and crops, the preparation of soil is very important task. The analysis of the soil can help a farmer to understand the state of the field's nutrients and several choices can be taken at various soil plowing stages and management of crops. There are some tool kits that assist farmers to understand the soil parameters like moisture, texture, and soil stress. This information can prescribe to get a better solution for deterioration of soil and crops. This will lead to the mitigation of compaction, degradation, salinization, and acidification issues, by this the usage of fertilizer can be avoided. The plowing of the soil is to enhance the field for sowing the seeds. When the fields get plowed, the soil gets loosen and helps the plant roots to pass through it easily. In traditional time, the cutting blades and hoe with animals were used in field. Later on, the farmers started to use machinery like tractors. In addition to providing a wide range of advantages for farmers and tractor operators, smart plowing satisfies the expanding needs in precision agriculture. In recent years, cultivators and other equipment, like as laser-based levelers, have been utilized to speed up and minimize the labor-intensive process of field plowing [19]. We anticipate the introduction of intelligent machinery and drones to provide more effective methods for plowing fields in the future.

3.3 Smart Crop Harvesting

The final stage of the farming is crop harvesting, where the farmers get yield and production from their cultivation. The examination of plant/crop height, size, shape, and color, as well as the stage of fruit maturity, can be used to determine when the crop should be harvested. In addition to optimizing crop quantity, quality, and yield, harvesting on the right calendar also makes it possible to create an effective strategy for collecting and transporting the harvests. To get the most money out of their crops, farmers need to be aware of when these plants are actually ready for harvest. On the other hand, year-round production is possible in greenhouses (vertical, hydroponic, and aquaponic). The ecosystem may be rebuilt to develop and harvest any particular crop.

To help farmers using a smartphone application to acquire real-time visual harvesting data, which is then uploaded to the cloud, the authors created a yield monitoring device. Fruit development measurements can be very helpful in estimating predicted output volume and the quality yield. Robots are also being utilized in fruit picking with the following goals:

- Robots and driverless tractors are gradually taking the place of human-driven machinery in the seeding industry.
- to assess the fruit's maturity.
- A robot with picking arms used to harvest ripe fruit and vegetables is a common use for robots on farms, especially given the present manpower crisis.
- to delicately pluck the fruit without injuring.

Strategies for processing visual or image data and remote sensing are typically employed to determine the freshness of fruits and crops [20]. Using vision, motion, and the pest detection sensors installed, the pests and rodents in fields, as well as greenhouses, are inspected and tracked on the ground or on drones that caught to protect the crops and boost productivity.

3.4 Smart Sowing System

Due to enhanced plant establishment and correct input application, studies in various regions of the nation have revealed that seeding devices implemented in rainfed areas have boosted agricultural yields by 10–20% above conventional techniques of seeding. Traditionally, an instrument in the shape of a funnel is used to plant seeds. With the aid of the tool's pointed ends, the seeds are inserted into the soil through the funnel. In addition to these conventional techniques, seed dropping, drilling, and broadcasting are also used. All of these techniques involve physical labor, are time-consuming, and demand more seeds. We can exactly determine the spacing in between the sowed seeds and the depth of the soil where the seeds need to be sown by utilizing smart agriculture, particularly vision-based technology. Additionally, the global positioning system (GPS) is a technology that has been applied to the spreading of seeds. Because of the regulated climate, sowing may occur in greenhouses yearly. Using IoT sensors, including temperature of atmosphere, humidity, and moisture, seeds are planted in open seed bedding trays. Sowing on beds with pebbles or gravel is possible with hydroponics and aquaponics. Compared to regular agricultural planting, such situations require much fewer seeds. However, just a few crops can still be cultivated in hydroponic and aquaponic systems.

3.5 *Smart Live Stock*

Although raising cattle may seem like a pastime to some country dwellers, it is nevertheless a significant economic activity. Livestock is a significant source of food in many regions of the world, including milk, eggs, and meat. The demand for food is anticipated to rise significantly along with population and economic development. Therefore, the sector's goal must be to meet future demand, guarantee constant quality, and do it sustainably without endangering the environment or depleting the world's resources. Additionally, in certain areas, farmers utilize the draught horses to carry themselves and to till their crops. Such activities have steadily decreased as a result of agricultural mechanization. Additionally, farmers have started creating biomass gas and fertilizing their fields using animal waste and dung.

Figure 4 shows how IoT, cloud, and AI technologies may use the cloud to connect farmers with farms in faraway locations to biological data on livestock and environmental information gathered through sensors. Information gathered by IoT sensors is sent to and stored on an information server in a database in the cloud in large amounts [21]. Depending on the information, the circumstances for cattle reproduction are either automatically or manually regulated. As an example, AI can automatically switch on the fan or air conditioner if biological sensors see a rise in body temperature. Farmers may rapidly take care of the cattle if AI alerts them to a potential ailment based on all available data, such as an increase in body temperature and a reduction in activity and it filters the data and provides only valuable data to the farmer. Animal behavior is observed using a variety of cattle and also livestock sensing equipment. These sensors can be inserted inside the animals or fastened to their collars. Farmers may keep tabs on the behavior, activities, and overall wellness of their livestock by wearing a sensor device.

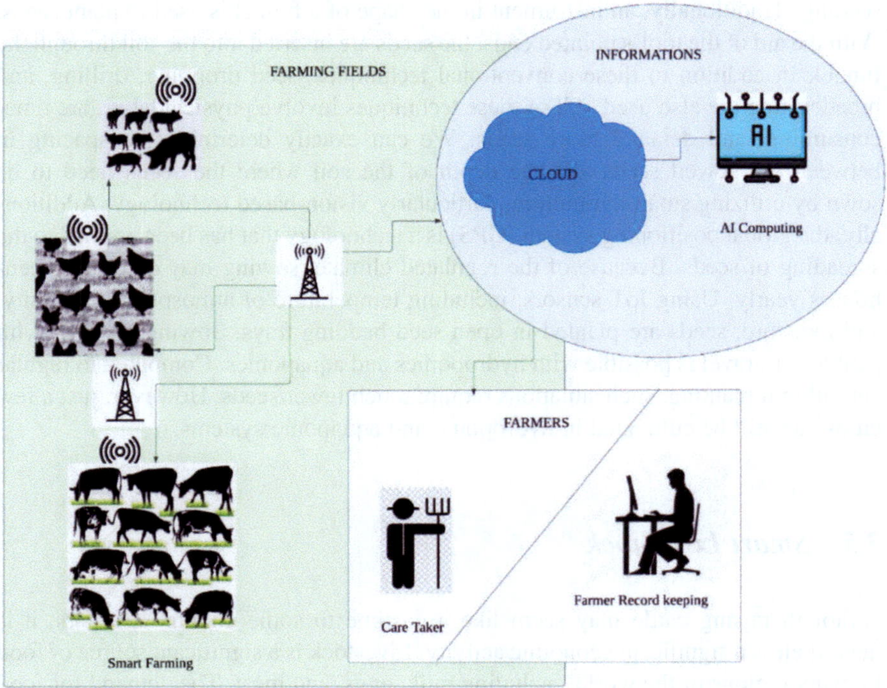

Fig. 4 IoT, cloud, and AI technologies in farms

4 Agri 4.0 and Smart Agricultural Methods

The present generations must assume the burden of utilizing available resources carefully and implementing affordable, sustainable farming practices to meet the expanding demands of the world as the human population explodes and sets new records every year. The UN Food and Agriculture Organization predicts that in order to fulfill the population demand that will occur in 2050, we would need to increase agriculture production globally by 70% during the ensuing decades. The current global agricultural development agenda has the shared objective of smart agriculture. These modalities significantly advance production management, industrial operations, and unmanned agricultural growth.

4.1 Agriculture 4.0

The idea of industry 4.0 gave rise to the more comprehensive Agriculture 4.0 (Agri 4.0) concept in 2011. Agri 4.0 describes the use of technology to connect an organization's value chain with its consumers and other stakeholders [22]. In this context, the term "Agri 4.0" refers to the use of technology for communication and information, including Big Data, Analytics, as well as Machine Learning, to analyze the diversity of data in order to utilize it for handling developments in the agricultural environment. Automated grain selection, machine learning algorithms for managing water resources, artificial intelligence for analyzing soil, and highly developed systems for recognizing and tracking pests and diseases are just a few examples of advanced technology that is directly linked to Agri 4.0.

4.2 Precision Agriculture

To achieve the highest possible yield and the greatest possible economic benefits, as well as safeguarding the agricultural ecological system and the agricultural natural resources, the amount, quality, and time of application of water, fertilizer, seeds, pesticides, and other agricultural inputs should be carefully and accurately optimized to correspond with the environmental circumstances of each operation unit on the field and the temporal-spatial deviations of crop yield. With precision agriculture, sometimes referred to as smart farming, plants are precisely recognized using state-of-the-art technology and then provided the care they require. Several technical techniques are used to do this, including a global positioning system sensor methods, ICT, and robotics.

Different kinds of data in precision agriculture, data from GPS, GIS, RS, and AI techniques are utilized to infer the plant development process and suggest management of crop production through an expert decision system. Making decisions concerning operations at the crops level as well as strategic ones at the farm level may be aided by technology [23]. This makes it possible to optimize output and work on more sustainably grown crops. Contrary to conventional agriculture, precision agriculture enables actions that can be determined per square meter or even per plant, as opposed to having to identify the appropriate action for each individual field.

Fig. 5 shows the functions that are given by precision agriculture. By using precision agriculture, the ideal circumstances for crop development may be maintained. This is done by managing the pesticide dose, increasing irrigation efficiency, enhancing the ecological environment of farmland, and analyzing the regulations of crop growth.

4.3 Facility Agriculture

The industrial-style agricultural production method known as facility agriculture aims to produce crops of high quality and yield. This form of agriculture, which is notable for its creation of modern and advanced agriculture, has been characterized through high labor demand, money, and different innovations and technologies.

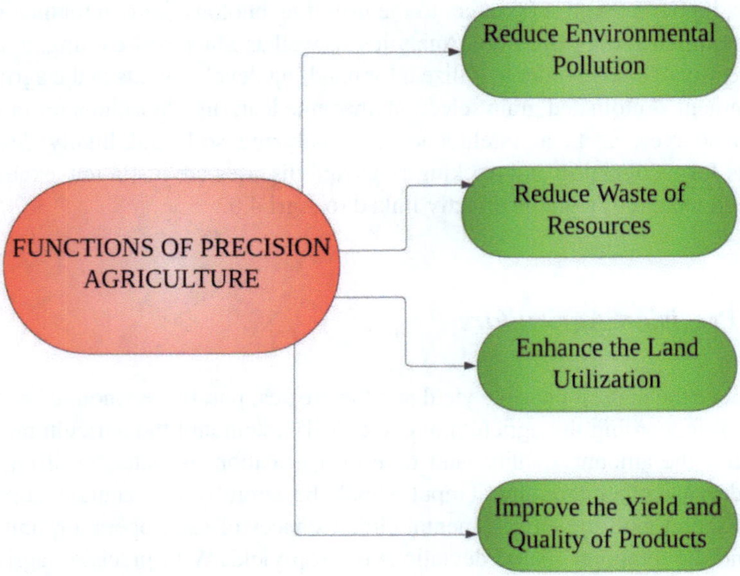

Fig. 5 Agricultural function

Facility agriculture uses engineering technology to create artificial environments that are good for crop development and facility protection, as well as to remove environmental constraints on agricultural output and boost the effectiveness of automated processes. Facilities are used in facility agriculture to modify the internal environment element, such as planting of crops, its breeding, and other agricultural production areas, and to create favorable conditions to enhance the quality of agricultural as well as livestock production and increase agricultural production. In order to guarantee that crops develop under the most optimum circumstances, the facility's components, such as temperature, air pressure, relative humidity, light intensity, and fertilizer application rate, are continuously monitored and managed [24]. The intelligent greenhouse is the most prevalent example. Different environmental factors, various sensor equipment, and particular computer-controlled programmers are some of the distinctions across various studies, such as greenhouse, aquaculture, plant factory, poultry, and cattle breeding. The functions of facility agriculture are shown in Fig. 6.

Facility agriculture, which combines both agricultural and industry, will be the future of agriculture and is already a major sector in industrialized nations like the United States, the Netherlands, and Japan. Facility agriculture is susceptible to control system infiltration and unauthorized access because of the traits of centralized administration and automation management.

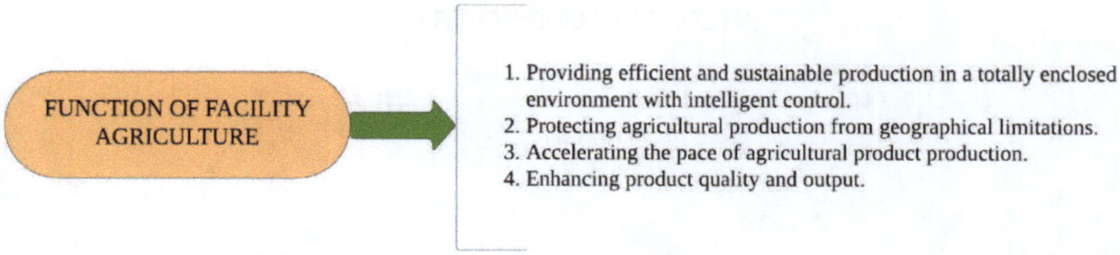

Fig. 6 Facility agriculture

4.4 Order Agriculture

Information island transactions within agricultural products, agriculture goods' quality and security are subject to covert risks and weak agricultural foundations all imply that advanced and modern agricultural production technologies are insufficient and only contribute to value creation when they are combined with enabling market opportunities [25]. Order agriculture, which may decrease planting and breeding risks and prevent blind production, has therefore developed a viable business model by outsourcing the production need for agricultural goods in advance. In addition, the supply chain for agricultural products uses a traceability system to manage agricultural products and promote food security by giving customers the right information about agricultural outputs.

5 Technologies in Smart Agriculture: An Overview

The modern agriculture industry is simultaneously changing in many different directions. However, its main objective is to use agricultural technology to increase crop yields through improved management and planning. Modern agricultural technology supports more productive and environmentally friendly farming practices, which benefit farmers in the agribusiness sector. The viability of agriculture is influenced by both tried-and-true methods like crop rotation and cutting-edge techniques like tracking field production with equipment, satellite pictures, or specialized farming software. Modern agriculture frequently makes use of robots, moisture and temperature sensors, aerial photography, and GPS technologies. These cutting-edge tools, robotic systems, and precision methods of agriculture may help businesses become more prosperous, productive, safe, and less harmful to the environment.

Figure 7 illustrates how many different technologies and computer networks are included in the recommended smart agricultural design. The suggested smart agricultural architecture is made up of the sensor layer, the link to the communication layer, the edge/fog layer, and the cloud layer [26]. Each layer of the various smart agricultural regions consists of a number of technologies that must be successfully integrated. It is no longer necessary for farmers to apply water, fertilizers, and

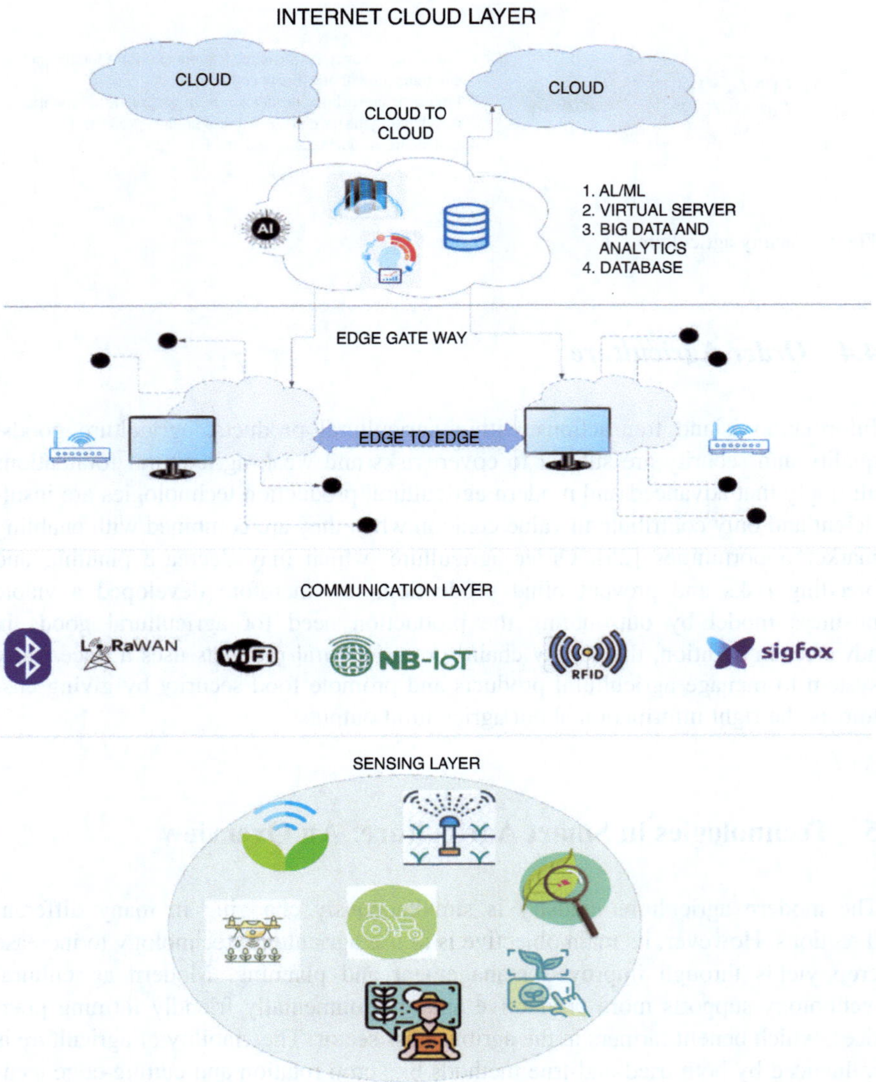

Fig. 7 Smart agricultural design

pesticides uniformly throughout whole fields. Instead, they could use the very bare minimum required, concentrate on extremely specific areas, or even treat other plants differently. Benefits comprise:

- Crop output can be increased.
- Arid, fertilizer or pesticides used.
- It contributes to cheap food prices.
- A diminished impact on natural ecosystems.

- Reduced chemical release into groundwater and rivers.
- Improved workplace safety.

5.1 *Internet Cloud Layer for Smart Agriculture*

The cost-effective pooling with resources is made feasible by cloud computing, which provides services like Infrastructure as a Service, also known as IaaS, Platform as a Service (PaaS), as well as Software as a Service (SaaS). The sensor and edge/fog layer transfers the vast amounts of data generated by smart agriculture to the internet layer or cloud layer for further processing and analysis. Big data analysis in the smart farming sector provides an in-depth understanding of the data obtained from many sources, such as choosing the best crop protection strategy, by processing data from weather conditions, irrigation, and plant/crop sensors. An artificial intelligence (AI)-based system that assesses plant development using recurrent neural networks (RNNs). Deep learning (DL) techniques are frequently employed in precision farming to help farmers effectively detect pests and insects and observed their crops. Additionally, image processing also enables data processing technology to more accurately regulate irrigation, estimate rainfall, and choose the right field fertilizers. The authors in [27] used a Convolutional Neural Network (R-CNN) model on Mask region so several photos to assess the level of tomato ripeness and avoid premature picking.

5.2 *Edge Gate Way Smart Agriculture*

With the use of shared resources near to the network, edge computing technologies are essential to solving various agricultural demands and the requirements of various agricultural applications such as latency, throughput, etc. The edge/fog computing infrastructure aids in carrying out agricultural operations, many of which demand additional computation or precision. Remote users are made more convenient by fog computing by giving them access to the essential agricultural-sensed data for analysis. The authors in [28] suggested an edge-based, IoT-integrated solution for smart agriculture. Blockchain technology can provide a verifiable, private, and unaltered archive for disseminating revolutionary information to the farming community and other significant stakeholders. It will be simpler to track the origin of food with blockchain implementation at the edge or in the fog layers, creating trustworthy food supply chains. Additionally, using smart contracts makes it possible for farmers and stakeholders to get compensation on time in the event that data on the blockchain change.

5.3 Communication Layer for Smart Agriculture

In order to enable real-time processing and operations, such as controlled irrigation and crop monitoring, and livestock raising, as well as deploy weather stations and greenhouses, the IoT ecosystem links various Internet-connected gadgets and sensors. Agricultural services are managed and controlled in large part by IoT technologies. In IoT-based smart agriculture, wireless communication technologies are crucial for disseminating the gathered data to the appropriate parties. For the conveyance of information, the communication network needs to be trustworthy, precise, and secure. Some of the technologies for wireless communication used in smart agriculture are (1) Bluetooth, (2) LORA, (3) WiFi, (4) Sigfox, (5) NB-IoT, (6) RFID.

5.4 Internet of Things (IoT) in Smart Agriculture

The IoT, a major technology in Industry 4.0, has changed many aspects of our daily lives by fostering a smart, connected society. Several use cases include connected autos, smart homes, and the industrial internet. As a result, it is believed that it would help boost the agriculture sector. Some of the IoT's agricultural applications include precision farming, animal monitoring, smart greenhouses, fisheries management, and weather tracking. Agricultural Internet of Things is a crucial innovation in smart agriculture that enables autonomous processing, analysis, and access to measure environmental elements, crop development, and agricultural production processes [29]. IoT architecture may be broken down into perception, transmission, and application layers in accordance with the concept of IoT. IoT technology allows for remote access to data from mobile devices as well as the monitoring of both plants and animals. Sensors and equipment enable farmers to forecast crop levels and assess the weather. In greater numbers than ever, the Internet of Things (IoT) has an impact in water harvesting, monitoring, and regulating the flow amount, figuring out how much water crops require, when to provide it, and water conservation. Sensors and cloud connectivity through the gateway can remotely check the condition and water delivery based on the needs of the soil and plants. To address nutritional shortfalls, pests, and diseases, farmers cannot physically monitor and check every plant; nonetheless, IoT technology is still useful and has assisted farmers in reaching a new milestone in modern agriculture.

Fundamentals of IoT Applications in Agriculture

- Sustainability: Due to sturdy pressure by economic and competition worldwide, the subject of sustainability has gained significant importance.
- Scalability: The results should be scalable due to the ranging from modest to larger farms. The strategy ought to be progressively built up while expending less money on evaluation and placement.

- Affordability: Farming capacity to succeed depends on price; thus, prices should be reasonable with substantial help. Platforms, goods, equipment, and facilities that are standardized may be priced reasonably.
- Robust Model: The diversity, complexity, spatiotemporal variability, and uncertainty of the appropriate sorts of the sector's defining qualities include harvests and facilities.

5.5 Artificial Intelligence in Smart Agriculture

One of the most significant technologies nowadays is artificial intelligence which has technical developments across many sectors, including agriculture, banking, education, and robots. It is fundamentally changing the agricultural business and has a significant impact on it. AI shields the agriculture sector from a variety of difficulties, such as food safety, population growth, climate change, and employment issues. Modern and advanced agriculture has evolved to a new and advanced level as a result of AI. Crop production, continuous tracking, gathering, processing, and marketing have all been enhanced by artificial intelligence. To identify a variety of crucial criteria, including weed proof of identity, yield identification, quality of crops, and many more, several high-tech computer-based approaches have been created. Figure 8 illustrates how AI is used. Similar to conventional agricultural practices, farmers would face a variety of challenges. AI is commonly used in this area to address these issues.

Predictive analytics has significant advantages for the agricultural industry. It aids farmers in overcoming their biggest obstacles, such as determining the best time to plant and harvest a crop and gauging market demand and price expectations. AI-powered machinery can also monitor the weather, assess crop quality, suggest fertilizer applications, and assess the soil's and crops' overall health. All of these benefits of AI in agriculture allow farmers to conduct successful farming and make better decisions. Farmers can make informed decisions at every step of farming because of the real-time insights that AI offers them. With this wise choice, there will be less product and chemical loss and more effective use of both time and money. This problem of farming automation can be resolved by AI. Farmers can complete tasks without adding additional workers because of AI and automation. When compared to human agricultural workers, AI-powered machinery and equipment are significantly faster and more precise. Even though using AI in agriculture has many benefits, most people are unaware of its use in products and equipment. AI producers should first give farmers the basic tools they need to solve the issues, and as they get acclimated to them, more advanced technology. Furthermore, it is crucial to understand the nature of the aforementioned agricultural problems and how to solve them before using AI. However, most AI researchers are generally not very knowledgeable in the field of professional agriculture.

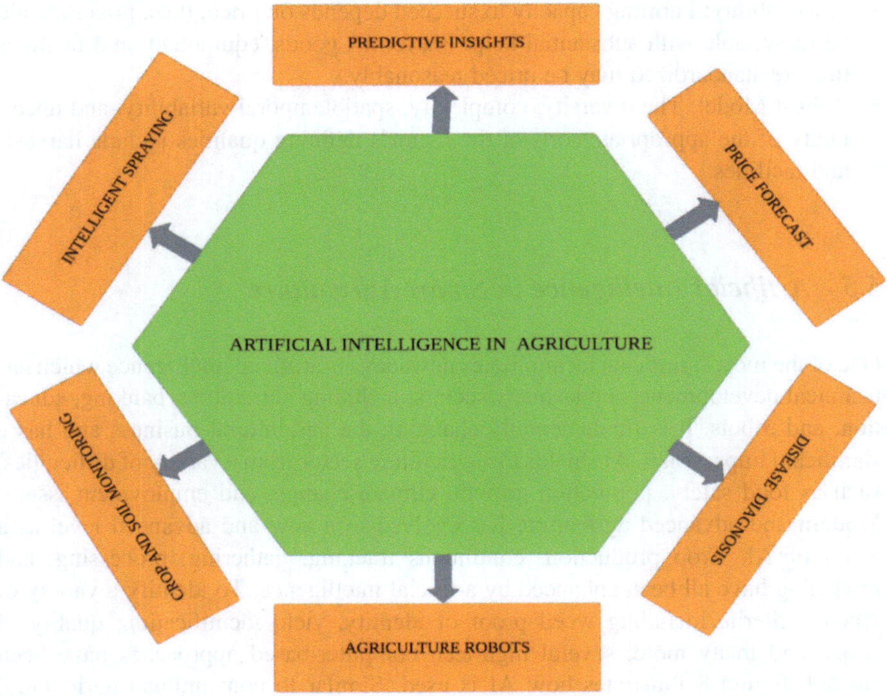

Fig. 8 AI in agriculture

5.6 *Big Data Analytics in Smart Agriculture*

Big data and AI will significantly improve sustainable agriculture and provide several benefits. With the advent of Big Data, the revolution in technology has also pounded on the entrances of the agriculture industry. The process of gathering and analyzing big data is essential for increasing agricultural output, but it also helps reduce the possibility of a worldwide food crisis. Big Data has the potential to completely transform the agriculture industry. Big data in the agricultural sector is entirely dependent on using information, technology, and analytics to provide farmers with meaningful information. Big data may be used to accumulate information about the agriculture sector or can be useful for any particular sector. Big data analytics suggest the best crop for agricultural practitioners to choose based on market demand. The system tracks and maps the crops that have the matching demand, preventing farmers from overharvesting the crops that have the anticipated demand.

Big data is also essential for managing the agri-food supply chain to address issues with food safety, an unbalanced supply, in addition to the loss of food and waste. For instance, offered a data-driven agriculture solution for managing sustainability in the supply chain for byproducts. In [29], a thorough examination of the effects of big data analytics on the supply chain for agricultural products was provided from a variety of angles, including functional, economic, environmental, social, commercial, and technological effects.

6 Security Challenges in Smart Agriculture

The farming industry includes a large amount of risk. Earnings from agriculture may shift significantly due to the unpredictability of the weather, yields, pricing, governmental regulations, international markets, and other factors that affect farming. Making choices among potential remedies to lessen the potential financial effects of these uncertainties is known as risk management. For farmers, protection is becoming a significant concern. Although there is little chance that your farm will be attacked, you need at least take into account your susceptibility to crimes like the following:

- Theft of ammonia that is anhydrous for the creation of methamphetamine.
- Theft of agricultural machinery or chemicals.
- Whether it be the intentional opening of a chemical tank valve, well poisoning, or arson.
- Unsecured gear and equipment used in criminal mischief.
- Destroying trapped animals, items, or goods.
- Bioengineered plants are destroyed.
- Discharge or introduction of a dangerous animal or diseased plant on purpose.
- Broader vandalism.

These cutting-edge technologies have found extensive use in Industry 4.0, and it is simple to replicate their use in agricultural contexts. Since hundreds of devices are being deployed in an open field, the primary problem of creating future technology is not being implemented in agriculture 4.0. Instead, it is the promise of privacy and security. In smart agriculture, physical and cyberattacks are both highly sought-after. Cybersecurity is the defense of data against intrusions that jeopardize its integrity, confidentiality, availability, and dependability. These online dangers can affect people, processes, and/or technology, and can come from a variety of sources. Foreign governments, organized criminal organizations, insiders, and issue-driven organizations are only a few examples of actors [30].

Understanding and accounting for cyber potential and weaknesses within the agricultural sector are necessary for successful security solutions, which include proactive monitoring to thwart and avoid threats. Making ensuring that all interested parties are knowledgeable of and educated to counter common dangers and contribute to increased resilience is an important technique. Additionally, effective access control—also known as access restriction—is one method for preventing unauthorized access, which is frequently the result of ransomware assaults. IoT and smart agriculture apps and equipment are becoming widely adopted in the agricultural industry. Because of this, it has been planned that smart agriculture

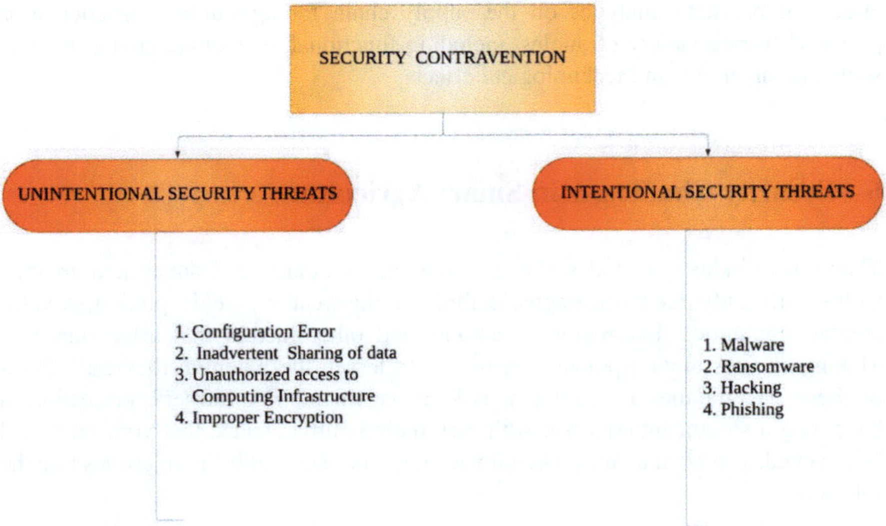

Fig. 9 Security threats in farming

would manage a sizable amount of sensitive data generated by numerous sensors installed in the fields and greenhouses. As we deploy smart agricultural systems in agricultural fields and greenhouses, we must ascertain and assess the security needs and privacy problems for various forms of data created by various smart agricultural applications. For there to be transparency in the ecosystem of smart agriculture, blockchain technology must be employed for contracts and farmer-to-customer communication. Security assaults may lead to security breaches, which may be harmful. Security failures are categorized in Fig. 9. A farm system is susceptible to deliberate or inadvertent security vulnerabilities. For instance, hacking is the deliberate attempt by a person to breach a system to take advantage of vulnerabilities in a network. Malware is another illustration of a planned attack. A software Program known as malware is made to disrupt systems or networks or to access data without authorization. Ransomware is a common type of virus that targets businesses and cutting-edge data processing technology. A device's files might be encrypted by ransomware, preventing users from accessing their data.

In order to unlock the encrypted data, the attackers often present a ransom demand on the computer screen. Another prevalent cybersecurity fraud that can damage farms' digital infrastructure is phishing. A hacker can collect sensitive information by using phishing to send fraudulent emails, calls, or text messages. Phishing can target a single person or a wider number of receivers, and the perpetrator starts a conversation that appears to be with a frequent customer of a company or a familiar contact. However, many security lapses can also result from inadvertent behaviors. The term accidental risks include unintentionally disclosing sensitive information, transmitting it without the necessary encryption, and getting unauthorized access to computing equipment owing to incomplete configurations. Farmers should take information security seriously as a crucial need. To safeguard agricultural systems from possible cyberattacks, farmers should follow security best practices. Despite the fact that some instances of sensitive data leakage in a farm may be the result of farmworkers' carelessness or insufficient action, a breach of data may occasionally be the deliberate act of farm employees [31]. Insider assaults are what these instances are known as. Because insider attackers have authorized access to the system, proactive attack prevention is essential [32]. The danger of insider assaults can be reduced by limiting people who have administrative access to the network, also known as access privilege, and by restricting farmworkers' access to agricultural system components irrelevant to their work.

To prevent malware and ransomware, agricultural equipment should always have the most recent software installed. It is advisable to set computers, tablets, mobile phones, and other digital devices to automatically update software since having up-to-date software reduces the chance of security risks. Farmers should receive guidance from Agriculture Technology Providers (ATPs) on how to set up and upgrade security software products. In order to maintain the systems and the entire network updated at all times, it is also crucial to deploy trustworthy firewalls on them. Using security rules as well as detectors based on machine learning, a firewall is an instance of network security either hardware or software that can monitor traffic from and to the network and filter out suspicious activities. Blockchain can guarantee transaction privacy, ownership rights, and give farmers more control over their data by making it possible for a reliable identity management system. The usage of a password-free authentication system is a reliable technique to confirm a user's identity and grant them safe access to the farm's computer infrastructure. They have shown to be compatible and more user-friendly than passwords.

7 Conclusion

Smart agriculture, which combines agriculture with information technology, is emerging as a trend in agricultural growth as a result of the quickly growing state of contemporary technology. Understanding the intricate agricultural ecosystems is now more crucial than ever because of the growing production challenges in agriculture. Drones, IoT, automation, and AI-assisted precision farming are also big data and are anticipated to be very important in various areas of agriculture. In this article, the survey has done in different sections like methods, technologies, trends, and security challenges even though more survey is needed for each and every section especially security threats. The security of smart agriculture is very important because of integration of various technologies.

References

1. Y. Liu, X. Ma, L. Shu, G. P. Hancke, and A. M. Abu-Mahfouz, From Industry 4.0 to Agriculture 4.0: Current Status, Enabling Technologies, and Research Challenges, IEEE Transactions on Industrial Informatics, vol. 17, no. 6, pp. 4322–4334, Jun. 2021, https://doi.org/10.1109/tii.2020.3003910.
2. S. Koul, Machine learning and deep learning in agriculture, Smart Agriculture, pp. 1–19, Jan. 2021, https://doi.org/10.1201/b22627-1.
3. R. Taghizadeh-Mehrjardi, K. Nabiollahi, L. Rasoli, R. Kerry, and T. Scholten, Land Suitability Assessment and Agricultural Production Sustainability Using Machine Learning Models," Agronomy, vol. 10, no. 4, p. 573, Apr. 2020, https://doi.org/10.3390/agronomy10040573.
4. M. Doshi and A. Varghese, Smart agriculture using renewable energy and AI-powered IoT, AI, Edge and IoT-based Smart Agriculture, pp. 205–225, 2022, https://doi.org/10.1016/b978-0-12-823694-9.00028-1.
5. D. Siddharth, D. K. Saini, and A. Kumar, Precision Agriculture With Technologies for Smart Farming Towards Agriculture 5.0, Unmanned Aerial Vehicles for Internet of Things (IoT), pp. 247–276, Jul. 2021, https://doi.org/10.1002/9781119769170.ch14.
6. B. Almadani and S. M. Mostafa, IIoT Based Multimodal Communication Model for Agriculture and Agro-Industries, IEEE Access, vol. 9, pp. 10070–10088, 2021, https://doi.org/10.1109/access.2021.3050391.
7. S. Zeadally, F. K. Shaikh, A. Talpur, and Q. Z. Sheng, Design architectures for energy harvesting in the Internet of Things, Renewable and Sustainable Energy Reviews, vol. 128, p. 109901, Aug. 2020, https://doi.org/10.1016/j.rser.2020.109901.
8. A. Kumar, M. Rani, Aishwarya, and P. Kumar, Drone Technology in Sustainable Agriculture: The Future of Farming Is Precision Agriculture and Mapping," Agriculture, Livestock Production and Aquaculture, pp. 3–12, 2022, https://doi.org/10.1007/978-3-030-93262-6_1.
9. B. Acharya, K. Garikapati, A. Yarlagadda, and S. Dash, Internet of things (IoT) and data analytics in smart agriculture: Benefits and challenges, AI, Edge and IoT-based Smart Agriculture, pp. 3–16, 2022, https://doi.org/10.1016/b978-0-12-823694-9.00013-x.
10. A. Srivastava, D. K. Das, and R. Kumar, Monitoring of Soil Parameters and Controlling of Soil Moisture through IoT based Smart Agriculture, 2020 IEEE Students Conference on Engineering & Systems (SCES), Jul. 2020, https://doi.org/10.1109/sces50439.2020.9236764.
11. J. Mageto, Big Data Analytics in Sustainable Supply Chain Management: A Focus on Manufacturing Supply Chains, Sustainability, vol. 13, no. 13, p. 7101, Jun. 2021, https://doi.org/10.3390/su13137101.
12. A. Vangala, A. K. Das, N. Kumar, and M. Alazab, Smart Secure Sensing for IoT-Based Agriculture: Blockchain Perspective, IEEE Sensors Journal, vol. 21, no. 16, pp. 17591–17607, Aug. 2021, https://doi.org/10.1109/jsen.2020.3012294.
13. G. Yassin and L. Ramaswamy, "Effective & Efficient Access Control in Smart Farms: Opportunities, Challenges & Potential Approaches," Proceedings of the 8th International Conference on Information Systems Security and Privacy, 2022, https://doi.org/10.5220/0010873000003120.
14. D. Shadrin, A. Menshchikov, A. Somov, G. Bornemann, J. Hauslage, and M. Fedorov, Enabling Precision Agriculture Through Embedded Sensing With Artificial Intelligence, IEEE Transactions on Instrumentation and Measurement, vol. 69, no. 7, pp. 4103–4113, Jul. 2020, https://doi.org/10.1109/tim.2019.2947125.
15. S. Trilles, A. González-Pérez, and J. Huerta, An IoT Platform Based on Microservices and Serverless Paradigms for Smart Farming Purposes, Sensors, vol. 20, no. 8, p. 2418, Apr. 2020, https://doi.org/10.3390/s20082418.
16. S. Sontowski et al., Cyber Attacks on Smart Farming Infrastructure, 2020 IEEE 6th International Conference on Collaboration and Internet Computing (CIC), Atlanta, GA, USA, 2020, pp. 135–143, https://doi.org/10.1109/CIC50333.2020.00025.

17. Murali Dhar M S, Kishore Kumar A, Rajakumar B, Poonguzhali P K, Hemakesavulu O and Mahaveerakannan R, Implementation of the Internet of Things for early Floods in Agricultural Land using Dimensionality Reduction Technique and Ensemble ML, Journal of Machine and Computing, vol.3, no.4, pp. 591–600, October 2023. https://doi.org/10.53759/7669/jmc202303050.
18. A. Sagheer, M. Mohammed, K. Riad, and M. Alhajhoj, A Cloud-Based IoT Platform for Precision Control of Soilless Greenhouse Cultivation, Sensors, vol. 21, no. 1, p. 223, Dec. 2020, https://doi.org/10.3390/s21010223.
19. G. Gokilakrishnan, V. M, A. Dhanamurugan, A. Bhasha, R. Subbiah, and A. H, "A Review of Applications, Enabling Technologies, Growth Challenges and Solutions for IoT/IIoT," 2023 9th International Conference on Advanced Computing and Communication Systems (ICACCS), Mar. 2023, https://doi.org/10.1109/icaccs57279.2023.10112825.
20. A. J. Hati and R. Ranjan Singh, Towards Smart Agriculture: A Deep Learning based Phenotyping Scheme for Leaf Counting, 2020 International Conference on Smart Technologies in Computing, Electrical and Electronics (ICSTCEE), Oct. 2020, https://doi.org/10.1109/icstcee49637.2020.9277402.
21. M. Lezoche, J. E. Hernandez, M. del M. E. Alemany Díaz, H. Panetto, and J. Kacprzyk, Agri-food 4.0: A survey of the supply chains and technologies for the future agriculture, Computers in Industry, vol. 117, p. 103187, May 2020, https://doi.org/10.1016/j.compind.2020.103187.
22. F. M. Enescu et al., Implementing Blockchain Technology in Irrigation Systems That Integrate Photovoltaic Energy Generation Systems, Sustainability, vol. 12, no. 4, p. 1540, Feb. 2020, https://doi.org/10.3390/su12041540.
23. S. V. Shinde, R. Shastri, A. K. Dwivedi, A. Haldorai, V. Sahni, and B. Adusumalli, Multi sensor data and temporal image fusion cross validation technique for Agri yield monitoring system, Sep. 2021, https://doi.org/10.21203/rs.3.rs-943821/v1.
24. J. R. Lamichhane and E. Soltani, Sowing and seedbed management methods to improve establishment and yield of maize, rice and wheat across drought-prone regions: A review, Journal of Agriculture and Food Research, vol. 2, p. 100089, Dec. 2020, https://doi.org/10.1016/j.jafr.2020.100089.
25. J. Feng and X. Hu, An IoT-based Hierarchical Control Method for Greenhouse Seedling Production, Procedia Computer Science, vol. 192, pp. 1954–1963, 2021, https://doi.org/10.1016/j.procs.2021.08.201.
26. A Smart IoT-Based Model to Improve the Agriculture Industry by Sensor Mobile Computing (SMC), International Journal of Nanotechnology and Nanomedicine, vol. 8, no. 1, May 2023, https://doi.org/10.33140/ijnn.08.01.03.
27. C. Zheng, A. Abd-Elrahman, and V. Whitaker, Remote Sensing and Machine Learning in Crop Phenotyping and Management, with an Emphasis on Applications in Strawberry Farming, Remote Sensing, vol. 13, no. 3, p. 531, Feb. 2021, https://doi.org/10.3390/rs13030531.
28. Ali-Khusein and Urquhart, Present and Future Applications of Robotics and Automations in Agriculture, Journal of Robotics Spectrum, vol.1, pp. 047–055, 2023. https://doi.org/10.53759/9852/JRS202301005.
29. A. Rehman, T. Saba, M. Kashif, S. M. Fati, S. A. Bahaj, and H. Chaudhry, A Revisit of Internet of Things Technologies for Monitoring and Control Strategies in Smart Agriculture, Agronomy, vol. 12, no. 1, p. 127, Jan. 2022, https://doi.org/10.3390/agronomy12010127.
30. A. Galieni, N. D'Ascenzo, F. Stagnari, G. Pagnani, Q. Xie, and M. Pisante, Past and Future of Plant Stress Detection: An Overview From Remote Sensing to Positron Emission Tomography, Frontiers in Plant Science, vol. 11, Jan. 2021, https://doi.org/10.3389/fpls.2020.609155.
31. S. Ghatrehsamani, T. Wade, and Y. Ampatzidis, The adoption of precision agriculture technologies by Florida growers: a comparison of 2005 and 2018 survey data, Acta Horticulturae, no. 1279, pp. 311–316, Jun. 2020, https://doi.org/10.17660/actahortic.2020.1279.44.
32. F. Abbas, H. Afzaal, A. A. Farooque, and S. Tang, Crop Yield Prediction through Proximal Sensing and Machine Learning Algorithms, Agronomy, vol. 10, no. 7, p. 1046, Jul. 2020, https://doi.org/10.3390/agronomy10071046.

Crop Yield Prediction Using Optimized Convolutional Neural Network Model Based on Environmental and Phenological Data

1 Introduction

In the real world, farming is carried out in routine ways. It is made even more irregular by the fact that the majority of our farmers lack the necessary understanding [1]. The predictions, which occasionally are incorrect, are the basis of a significant amount of farming and agricultural activity. Farmers frequently commit suicide as a result of having to endure significant losses. The significance of factors such as soil moisture, soil quality, air quality, and irrigation cannot be undermined, as they play a crucial role in the growth of crops. In many developing nations, agriculture serves as a primary source of income. The realm of modern agriculture is continuously expanding with advancements in farming techniques and agricultural innovations [2, 3]. Farmers face challenges in adapting to the evolving demands of the environment, businesses, consumers, and other stakeholders. Some of the difficulties encountered by farmers include: (i) Adapting to climate change resulting from soil erosion and industrial emissions. (ii) Mitigating reduced crop growth caused by nutrient deficiencies in the soil, which can be attributed to inadequate levels of essential minerals like potassium, nitrogen, and phosphorus. (iii) Avoiding the pitfalls of consistently cultivating the same crops without exploring new varieties. Indiscriminate application of fertilizers without considering their quality or quantity is also a common mistake made by farmers. It is crucial to address these challenges and implement sustainable practices to ensure the long-term viability and success of agricultural endeavors.

As a result, this article has developed a novel concept for smart farming and crop prediction. The proposed idea is to digitalize farming operations so that farmers can accurately assess crop needs and predict crops. This idea will undoubtedly help them grow their company and make it more profitable [4]. Our project's implementation heavily relies on raising understanding among farmers, which we think will be simple given all of its benefits. Farmers benefit from the numerous tools and

© European Alliance for Innovation 2024

A. Haldorai et al., *Artificial Intelligence for Sustainable Development*,
EAI/Springer Innovations in Communication and Computing,
https://doi.org/10.1007/978-3-031-53972-5_2

approaches developed for them by working with top public and commercial organizations. With the use of machine learning and deep learning algorithms, the power of artificial intelligence is also being realized in agriculture [5–7]. Although several models have been suggested using traditional methods, the accuracy of the predictions still needs to be increased.

The field of study involving artificial intelligence, machine learning, and deep learning algorithms has attracted the attention of numerous authors, institutions, and organizations. Within this domain, machine learning and deep learning are concepts derived from the broader field of artificial intelligence. These methodologies have led to the development of various methods and models aimed at supporting farmers [8]. Compared to traditional statistical methods, these approaches exhibit superior performance. Deep learning, in particular, is a cutting-edge technique employed to tackle complex problems and achieve optimal outcomes. Its primary objective is to build models based on statistical principles to comprehend a given dataset. It is essential to provide an illustration and algorithmic explanation of a completed deep learning model to enhance understanding. Classification accuracy is often used as an evaluation metric in many applications of deep learning, such as credit card fraud detection, analysis of biomedical data, and suggestion support systems for forecasting [9–11]. In general, deep learning proves beneficial in addressing various complex decision-making tasks. By leveraging historical lessons and observations, it becomes possible to identify and resolve issues. Therefore, the proposed methodology also incorporates a unique approach inspired by nature, utilizing an optimizer. The integration of these advanced techniques holds great potential in addressing the challenges faced by farmers and can contribute to the development of effective solutions in the agricultural sector.

A hybrid strategy is created to improve prediction accuracy in the suggested model by fusing MLO methods with optimization and deep learning algorithms. CNN is utilized as a deep learning technique in the proposed model. MLO is one of the nature-inspired optimizers (NIO). It is true that enhancing the performance of NIOs is a very challenging subject and that learning strategy, topological structure, and parameter tweaking techniques all have significant effects [12]. We make some preliminary conclusions about this study to give NIOA selection and improvement a referential framework. More than 120 metaheuristic algorithms and their variations have been proposed during the past 30 years to efficiently and effectively address optimization issues in AI. Since, the NIOAs have made enormous strides and have been widely and successfully applied to various application domains, the novel MLO combined with CNN in the planned work.

Through collecting real-time crop status data and disseminating that information to farmers, this project aims to incorporate the most recent technology into the agricultural industry and improve crop productivity. In this essay, we describe a crop cultivation planning issue that aims to maximize predicted overall return while keeping an eye on the investment budget and probable losses. Therefore, a novel deep learning model was to be used in this paper. Based on the comprehensive fitness evaluation approach, we adopt an effective evolutionary algorithm, MLO, for the challenge. MLO evolves the solutions to increase fitness in the three related but separate metrics simultaneously. The following is a summary of this paper's significant contributions:

- We outline an automatic crop prediction problem with CNN model that employs uncertain factors to define them.
- To tackle the optimization problem of deep learning model effectively, we suggest an altered MLO method.
- Through several test cases, we validate the suggested strategy.

The remaining sections of the essay are structured as follows. The related work is reviewed in Sect. 2, the crop prediction planning system model is presented in Sect. 3, the adapted LO algorithm is described in Sect. 4, the training model is explained in Sect. 5, the detailed experimentation presented in Sect. 6, the test results are shown in Sect. 7, and the discussion closes in Sect. 8.

2 Related Works

This section explores prior research on hybrid machine learning techniques employed for crop yield prediction. Researchers have sought to improve the accuracy of hybrid crop yield forecasting systems by combining different machine learning methods. U. Muthaiah, S. Markkandeyan, and Y. Seetha [12] proposed a classification model and hybrid feature selection approach to enhance crop performance using mango and maize datasets. The researchers employed the Particle Swarm Optimization-Support Vector Machine (PSO-SVM) classification approach to select the most relevant features for the Mango and Maize datasets. Each optimization solution is viewed as a particle in the Particle Swarm Optimization approach, and the authors used PSO-SVM to pinpoint the key features. The authors used the PSO-SVM method and a fuzzy decision tree in their experiments on these datasets, which led to the development of new rules and enhanced feature selection. These results emphasize the potential of combining several strategies to improve accuracy and feature selection in agricultural forecasting systems and forward the investigation of hybrid machine learning techniques for crop yield prediction.

A hybrid approach integrating machine learning and deep learning approaches was presented to anticipate the best crop production in 2021 in a study by Sandhya Tarar and Sonal Agarwal [13]. Recurrent neural networks and long short-term memory (LSTM) were used for deep learning to handle the dataset, while the support vector machine (SVM) method was used for machine learning. The most appropriate crops for cultivation on smaller sections of land were determined using this hybrid model's estimation and evaluation of many parameters. To reach the best level of accuracy, the authors investigated a number of techniques. The authors achieved a 93% accuracy rate by using the random forest (RF) and artificial neural network (ANN) algorithms on various crops. A mathematical programming model was presented for yearly crop planning in a different work by Lin et al. [14]. The model attempted to maximize total profit while minimizing the overall irrigation water utilized for multiple cropping. It did this by optimizing the allocation of land area for producing dryland and wetland crops. The researchers also suggested a condensed swarm optimization method that improved the particle swarm optimization (PSO) algorithm by combining four probabilities to properly update solutions. By merging machine learning, deep learning, and optimization methodologies, these research findings increase crop planning and yield optimization. The suggested models and algorithms offer insightful information and practical strategies for raising crop productivity and resource efficiency in agricultural systems. Alemany et al. [15] have devised a series of mathematical programming models to facilitate the scheduling of tomato planting and harvest within multifarmer supply chains, considering uncertain decision-making scenarios. Their approach involves integrating individual solutions from each farmer across various scenarios to achieve a comprehensive supply that meets the market's demand. Notably, their research focuses on employing evolutionary algorithms to tackle crop planning problems at a large scale, involving numerous plots of land and diverse crop types, while accounting for uncertain factors.

Numerous solutions have been put up in various studies to address the issues relating to the impact of changing weather conditions on crops. According to the research done by Safieh et al., climate change affects crop water needs as well as crop yields that are projected for the future based on weather forecast data [16, 17]. According to a study on the effects of harsh weather on various parts of Europe, rainfall and air temperature thresholds are agricultural production's most accurate weather indicators [18]. The two climatic parameters that are most frequently employed in studies are precipitation and air temperature. Various machine learning models (MLMs) have been employed to predict crop yields by incorporating additional parameters such as solar radiation, air humidity, soil moisture, and wind speed. These MLMs include the Artificial Neural Network (ANN), Semiparametric Neural Network [19], Convolutional Neural Network (CNN) [20], Lasso, Kernel Ridge, Enet [21], Naïve Bayes, K-Nearest Neighbor [22], Recurrent Neural Network (RNN), Long short-term memory (LSTM) [23], and Random Forest [24]. These models utilize a range of techniques to analyze the given parameters and generate yield predictions for agricultural purposes. Table 1 shows the publication related to this problem.

3 System Model

The topography of many agricultural areas is complex and varied. Different soil qualities are favorable for other crops, and they frequently alter significantly in an area of one or several square kilometers. It is crucial but challenging from an agricultural management perspective to choose the right crop for each piece of land when planning crop cultivation in such a complicated agricultural area in order to maximize the predicted total profit. For farmers and governments to plan the different operations related to agriculture, crop output forecast is crucial. Farmers

Table 1 Review analysis

S.no	Reference	Crop	Method
1	Charoen-Ung and Mittrapiyanuruk [25]	Sugarcane	Random forest
2	Rao and Manasa [26]	Common	Neural network
3	Ranjan and Parida [27]	Paddy	Linear regression
4	Filippi et al. [28]	Common	Random forest
5	Xu et al. [29]	Wheat	Random forest, support vector machine
6	Shidnal S and Mrityunjaya [30]	Paddy	Two-layered CNN
7	Khaki L and Wang L [31]	Barley, wheat, bitter guard, apple, orange	Regression
8	Schwalbert R A [32]	Tomato, soyabean, corn	LSTM

rely on the conventional methods of crop management and crop production forecasting, which are based on conviction, actual visits, and reports. Deep learning has been used to estimate crop yield using a variety of methodologies, however accurate predictions have not been made. Even though the CNN models performed in a better way, an expert's deep knowledge is needed for designing the CNN model. In these situations, to avoid those problems, the metaheuristic-based optimizers are introduced for automatic hyperparameter and weight tuning process. These metaheuristic-based optimizers tune the hyperparameters without human intervention. This work presented a novel MLO, which is the upgraded variant of Lion optimization with the CNN model.

One of the mainframes in the field of AI is deep learning. Deep learning architectures can be broadly divided into generative (Deep Belief Network), discriminative (DNN, CNN, and RNN), and hybrid (a blend of discriminative and generative) models. Choosing the best hyperparameters is crucial when learning deep learning techniques like CNN architecture. In Fig. 1, the hyperparameters are listed.

The quantity of training samples utilized in a single iteration is referred to as "batch size" in machine learning and deep learning models. The ideal batch size is unknown; nonetheless, batch size primarily influences training efficiency and accuracy. The word "iteration" refers to the process of repeatedly running a batch size of data sample images through the whole model. Epoch is a technique used in the learning process to independently treat each image in the databases once going both forward and backward. One of the alterable hyperparameters in the training phase for efficient weight updating in the architecture model is learning rate, which has a minuscule positive range value. Typically, the learning rate falls between 0.001 and 0.01. This setting regulates how the weight update is modified based on the loss gradient.

Typically, the CNN model is trained with this hyperparameter through a trial-and-error process. The training model needs in-depth information and human

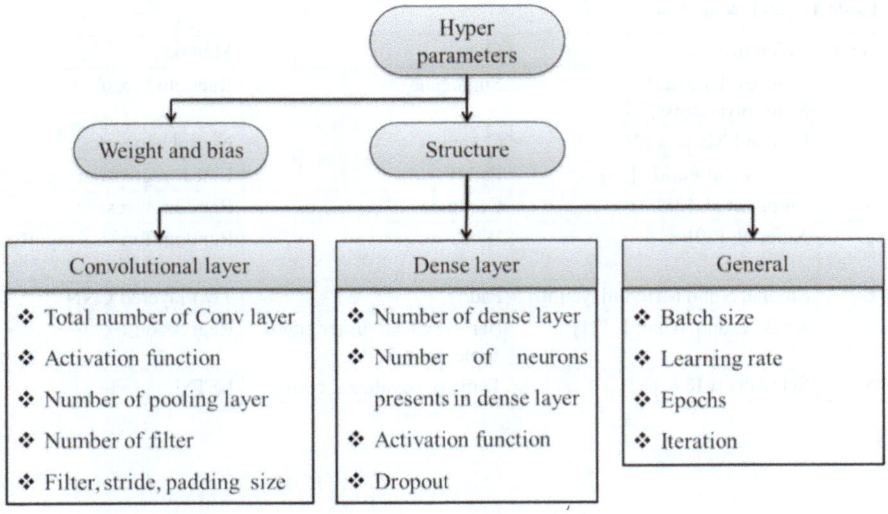

Fig. 1 Hyperparameters of CNN

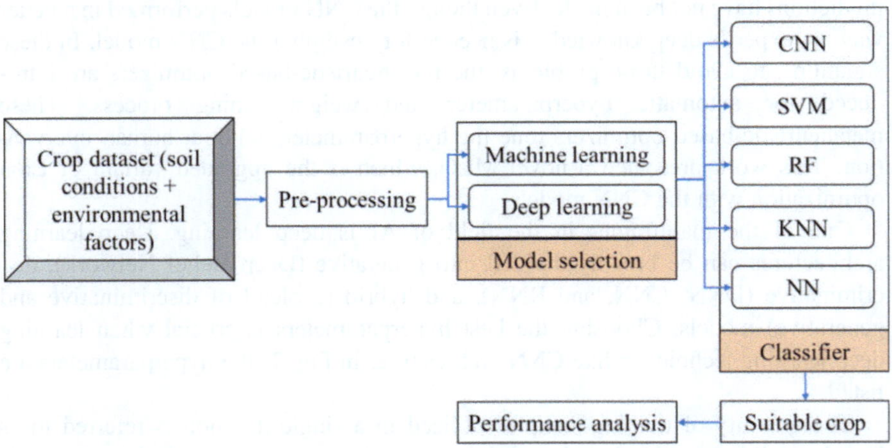

Fig. 2 Traditional model for crop prediction

professionals to tweak the hyperparameter. Given that deep learning uses so many hyperparameters, it is difficult to physically determine their optimal and optimum value. Finding the right hyperparameters is a complex problem; however, many researchers have handled these difficult optimization tactics for particular applications using metaheuristic-based algorithms. So many studies have been conducted on the computerization and updating of hyperparameters based on optimization. One of the top techniques for deep learning model hyperparameter adjustment is metaheuristic-based optimization. Figure 2 depicts the overall architecture for crop prediction.

4 Crop Prediction Using Metaheuristic Optimization

Since the concentration of researchers turns into the side of a metaheuristic-based optimization algorithm, the OCR research field also starts to utilize these types of optimizations. The planned system also uses this metaheuristic-based optimization algorithm to attain the finest solution with a high precision rate. There are lots of metaheuristic-based optimizations, from which the proposed chooses the MLO. Generally, many researchers are choosing the metaheuristic optimizer for any one of tuning such as the structure of the learning model or weight updation. But the proposed model is going to concentrate on both the forms, which means, the proposed system is going to use MLO for the structure and weight tuning of the CNN model. The MLO is going to concentrate on the number of dense (FC) layers, the number of hidden neurons present in the dense layers, and their weight. We did not include other types of hyperparameter tuning because of the limitation of available computing power. Figure 3 explains the CNN architecture with an MLO.

Many biological and naturally inspired optimizers have been developed recently for a variety of tasks. The suggested system addresses the problem of crop prediction using a unique, nature-inspired optimization method named the Lion Optimization method (LOA) [33]. The intended optimization method is based on the lion's social structure. Wang and Rajakumar introduce lion optimization in two distinct techniques that are motivated by the traits of a lion's survival ability. Following optimization technique, the crop prediction. The lion's social behavior of survival, which makes it the strongest member of the cat species, served as the basis for the LOA [34]. Finding the ideal answer in the search space in algorithmic form is made easier by understanding the perception underlying the behavior of the lion. A lion's pride behavior, or the process of updating the pride and living as a group, constitutes its social behavior. The term "pride" refers to a lion's social living structure in groups [35]. A pride typically consists of 1–3 lion pairings where the resident females give birth to young. "Territory" refers to the region that male and female lions with their

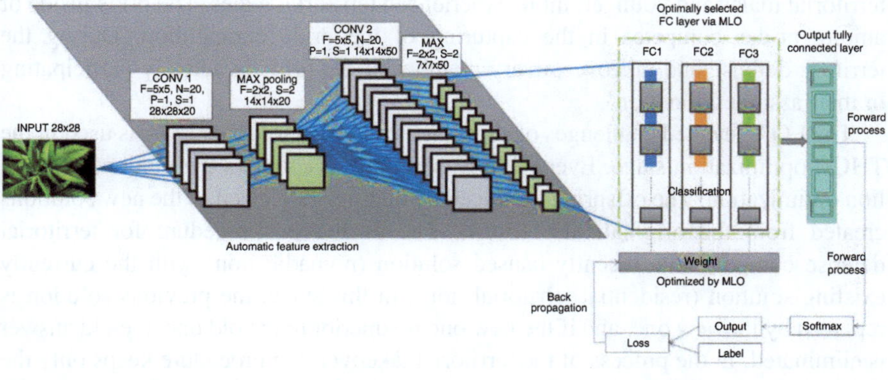

Fig. 3 CNN with modified lion optimizer

young share. The lions living in the area are known as residential lions [36]. The dominant lion of a region, also known as a territorial lion, governs the area by engaging in combat with other animals that come after it, including migratory lions. The territorial lions should defend the pride from the nomadic lions during this time because it takes the progeny of territorial lions 2–3 years to mature.

Lions that are nomads are born in different territories and abandon them once they reach adulthood. The nomadic lion roams erratically, either alone or in a pack. By engaging in combat with and driving out the domestic lions, these migratory lions may attempt to annex another pride's territory. As a form of territorial defense, the nomadic lion attacked the pride of another region. The residential and nomadic lions fight to control the pride during the territory defense; the residential lions attempt to maintain their proud position; while the nomadic lion attempts to seize the residential lion's pride. If the coalition of residential lions beats the nomadic lion, the territorial pride regains control of the residential lions; the nomadic lion resumes its roaming lifestyle and looks for a new area. Perhaps, if the nomadic lion defeats the domestic lion, he will take control of the pride by eliminating the domestic lion and its offspring. The nomadic lion evicts the residential lions from the land it has conquered and takes up residence there. The local lion changes into a new nomad lion and goes in search of another pride. The new resident lion bears the offspring of the new pride after becoming friends with them.

Takeover of territory is another intriguing social behavior seen by lion prides. "Territory take-over" refers to the conflict between the grown cubs of the domestic lions over territories in an effort to seize control of the pride. When the youngster reaches adulthood, the more powerful cub eliminates or drives the less aggressive neighborhood lions from the area. The newly grown and strongest cub establishes the new pride, which is prepared to breed, while the other lagging lions migrate as nomadic lions. Thus, lions can live either a nomad or a resident lifestyle; through the process of territorial defense, nomads can transform into residents, and vice versa. In light of these social behaviors, such as territory defense and takeover, the LOA looks for the best solution. Territorial defense entails fierce competition between male residents and male nomads, and territorial takeover is contested between veteran territorial males and younger, more experienced territorial males. The pride inside or among prides competes in the capturing of the pride competition. During the territory defense and takeover, every member of the pride is actively participating in their assigned function.

The LOA selected challenges of continuous optimization. The LOA is used in the THCR optimization stage. Every lion is taken into account as the input solution in lion optimization. The offspring produced by mating are viewed as the new solutions created from the original ones (lions). The evaluation procedure for territorial defense compares the recently caused solution (nomadic lion) with the currently existing solution (residential territorial lion). In this stage, the previous solution is replaced by the new one, and if the new one is superior to the old one, the old answer is eliminated. In the process of the territorial takeover, the procedure keeps only the finest derived solutions (strongest lion in territorial), which are competent as a solution to a certain extent, and the existing solutions disappear (weaken lions in

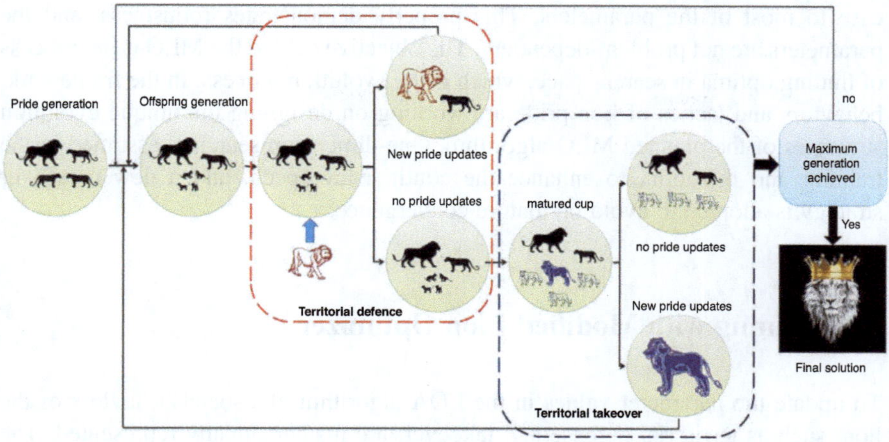

Fig. 4 Lion optimization algorithm

territorial) as happens in the pride. Many of the suggested nature-inspired and biologically inspired optimization techniques there follow the territorial takeover process for the purpose of the selection operation. The selection process is held to select the best solution, which saves it and removes the unsuitable solutions. But the idea of territorial defense is novel in the area of optimization. In real-time, from an extended terse evaluation, the territorial defense process only keeps the lion as the strongest animal. Similarly, this equivalent process of territorial defense only creates the LOA better than other optimization algorithms. The lion algorithm is displayed in Fig. 4.

The objective of this lion's social behavior in the actual world is to capture the position of the pride, but during the algorithmic phase, the objective is to capture the best possible solution. The process of territorial defense and takeover, which will replace the most inappropriate solution with the new and best solution, is the foundation for finding the new and best answer. Similar to mating, prey capture, territorial invasion, and migration, these subprocesses are crucial to uncovering a fresh approach. Each lion in the area is significant in the concept of lion pride behavior. Each lion is taken into account as a single solution in the algorithm. For the same reason that only the strongest member (solution) choose the best solution, only the strongest lion resides in the pride. The young pups and the weakest lions from the pride will be killed or thrown away if the new lion (nomadic or domestic) overwhelms them. The best solution will defeat the worst at the algorithmic phase. Here, the MLO is mostly presented through the pride's updated strategies and the deadly lion-on-lion battle. The most important aspect of the lion algorithm is the survival of the strongest (fittest) lion. Fewest fittest lions are often present in the territory to protect the pride. The MLO is created based on this best model. The MLO is working symbolically to the strategy of optimum searching space for solving continuous optimization problems. The MLO property mentions that which is not a sensitive case to most of the parameters. This property demonstrates robustness, and the parameters are not problem-dependent. The objective behind the MLO is the process of finding optima in search space, which is the evolution process. In the framework, behaviors and tactics of lion pride are working on designing the unique evolution strategies of the planned MLO algorithm. One-dimension search is assumed in the training and mutation, to enhance the contingency speed, and a new crossover strategy is adopted to avoid premature convergence.

5 Training with Modified Lion Optimizer

To update the parameter values in the LOA algorithm, the social behaviors of the lion, such as territorial defense and takeover, are mathematically represented. The MLO often comprises some of the significant building components, including (1) pride generation, (2) fitness and fertility appraisal, (3) offspring generation, (4) territory defense, and (5) taking over a territory. In the MLO, a pack of lions, which are randomly generated solutions, mold the original population. Only a certain percentage of the lions are designated as nomads; the remainder are classified as residential lions. A search space or pool of solutions is thought to exist in the production of pride. According to the algorithm, the solution (territorial lion) must be strong and fit enough to defeat a random solution (nomadic lion); the weak solution (weak lions, possibly including cubs), on the other hand, either vanishes from the solution pool or is driven out of it while processing (territorial defense or territorial takeover).

In recent times, modern metaheuristic optimizers are using different adaptation techniques to improve the convergence rate and speed. From that, this proposed work planned to use the self-adaptation technique with planned lion optimizer. Usually, the self-adaptation technique improves the convergence performance. The task of the Tamil handwritten character identification introduces the adaptive evaluations in MLO, which is noted as a self-adaptive lion algorithm (SALA). The pride generation method used for an novel mechanism for weighing individual leaders at each interaction in the lion optimizer, to improve the guidance method of best individuals and to obtain a more efficient exploitation phase.

5.1 Pride Generation

Pride is a solution pool (search space) that is constantly changing. Consider an optional (target) solution as an objective function, as shown in Eqs. 1, 2, and 3, where n denotes the *i*th solution variable in an n-dimensional solution search space. The x_i^{min}, x_i^{max} variables represent the solution's minimum and maximum bounds. X marks the solution vector.

$$X^{\text{target}} = \underset{x_i \in (x_i^{\min}, x_i^{\max})}{\arg\min} \; f(x_1, x_2, \ldots x_n); n \geq 1 \tag{1}$$

$$X^{\text{target}} = X : f(X) < f\left(X' | X' \neq X; x_i' \in \left(x_i^{\min}, x_i^{\max}\right.\right. \tag{2}$$

$$X = [x_1, x_2, \ldots x_n] \tag{3}$$

Make a random search area and fill it with answers. Each solution is represented by a "lion" in the approach. The "X" means the lion. In a search space, choose one piece of the nomad line and the residential lion at random. The domestic lion is separated once more into male and lioness (female lion). The algorithm starts the pride with a territorial lion known as X^B. The female lion is referred to as X^G, and the nomad lion is X^N. The nomad lion is not a member of the generated pride. The vector representations of male, female, and nomad lions are denoted as X_p^B, X_p^G, X_p^N. Here p = 1, 2,…P, in this case. The "P" is the length of the solution, which means the number of lions present in the pride. The value "P" is conveyed in Eq. 4.

$$P = \begin{cases} m; m > 1 & \text{(general case)} \\ n; n \text{ otherwise} & \text{(special case)} \end{cases} \tag{4}$$

where m and n are the integers used to determine the length. The length value is generated as one and zero when the m value is "one," as shown in Eqs. 5 and 6. In these equations, $g(x_p)$ is expressed mathematically in Eq. 7.

$$g(x_p) \in \left(x_p^{\min}, x_p^{\max}\right) \tag{5}$$

$$n\%2 = 0 \tag{6}$$

$$g(x_p) = \sum_{p=2}^{P} 2^{\left(\frac{P}{2} - p\right)} x_p \tag{7}$$

The cost function (fitness rate) of each solution (lion) is computed by the evaluation of the cost function. The fitness value is expressed in Eq. 8. The fitness value of lion, lioness, and nomad is mentioned as (X^B), f(X^G), and f(X^N). Depending on the fitness rate of the solution, the male and female lion values get saturated. Hunting, moving toward a safe place, and roaming are basic processes held in the generated pride.

$$f(\text{lion}) = f(X) = f(x_1, x_2, \ldots x_n) \tag{8}$$

5.2 Fertility Evaluation

During the serial process of LOA, each X^B and X^G begins to age or becomes sterile on occasion. This rate of maturation only makes the lion (solution) a laggard (weakest solution) in survival struggles or territorial takeover (most robust solution). If X^B and X^G become saturated by their fitness value, they will have reached the global or local optima and cannot lead us to better solutions. The fertility evaluation procedure can assist the key in avoiding local optimal value solutions. In this, the male lion X^B is discovered to be a laggard, and its laggardness rate is denoted as a_r. When $f(X^B) > f^r$, the laggardness rate La_r increases by a factor of "one." f^r indicates the reference fitness in this case. If the La_r exceeds its maximum boundary, the territorial defense will be activated. The sterility rate, St_r, confirms the fertility of the female lion X^G. After the crossover process, the sterility rate rises by one. If the sterility rate exceeds the tolerance sterility rate $\left(\mathrm{St}_r > \mathrm{St}_r^{\max}\right)$, the lioness X^B updates by X_p^{G+1}, as shown in Eq. 9. When the lioness X^B updated by X_p^{G+1}, the pride begins the mating process. The lioness X^B update is continuous till g_c reaches the $g_c^{\max}$. If there is no lioness update, the X^B is still fertile enough to produce better cubs. The mathematical equation for lioness update in cth is given in Eq. 10, where c is the random generator among the $[1, P]$, the lioness update function "∇" is given in Eq. 11; r_1 and r_2 are random integers among $[0, 1]$.

$$X_p^{G+1} = \begin{cases} X_c^{G+1}; \text{if } p = c \\ X_p^G; \text{otherwise} \end{cases} \tag{9}$$

$$X_c^{G+1} = \min\left[x_c^{\max}, \max\left(x_c^{\min}, \nabla_c\right)\right] \tag{10}$$

$$\nabla_c = \left[X_c^G + (0.1 r_2 - 0.05)\left(X_c^B - r_1 X_c^G\right)\right], \text{where} \begin{cases} r_1 = \left(\dfrac{\text{ma}}{\text{maxAge}}\right) \times \text{mutation rate} \\ r_2 = \text{rand} \times \text{mutation rate} \end{cases} \tag{11}$$

5.3 Mating Process

The processes involved in mating are divided into two categories in the MLO algorithm: primary steps and supplementary steps. The two main processes in mating are the crossover process and mutation. As an additional technique, gender clustering is used. The male and female lions give birth to their offspring (lion cubs) through crossover and mutation. These offspring are considered as a newly derived solution from the previous solution element, such as X^B and X^G. The MLO considers the maximum natural littering rate (maximum of four cubs in a single female lion pregnancy). So the mutation process generates four cubs, which is mentioned as X^{cubs}; the crossover process generates the four cubs, which is mentioned as X^{new}. Totally eight cubs (eight new solutions) are generated newly in the pride (search space). The gender clustering is performed to group the newly born cubs into male cubs X^{B-cubs} and female cubs X^{G-cubs}. The grouped X^{B-cubs} and X^{G-cubs} are entering into the cub growth function to attain self-update. The cub growth function acts as a local search function with the mutation rate of G_r. After muted, the X^{B-cubs} and X^{G-cubs} are exchanged by the cub growth function rate in the factor of age A_{cub}. The value of the cub growth function rate A_{cub} is incremented by one for every growth updations of the cubs toward maturity. Initially, the age rate A_{cub} is selected as zero.

5.4 Territorial Defense with Proposed Update Strategy

Territorial defense is one of the key functions in the LOA-based optimization algorithm. It allows the solution to analyze in a broader method in a search space. Generally, the territorial defense includes the generating nomad coalition, survival war between the territorial male lion X^B and nomad lion X^N, and the updates of pride and nomad coalition. The normal lion optimization is based on the X_p^B, X_p^G, X_p^N. If we give the same weight to all types of the lions, and then the actual behavior of hunting will collapse. But at the same time according to the incident, the weight will depend on enemy prey availability.

The targeted effort is taking place in the generation of nomad lions. Based on the self-adaptation technique, a new variable, k, which is a weight variable and designated as a generation variable, is generated in order to share the same relevance for all types of lions. Each iteration's k value is unique to the search agent. A consistent assessment is not achievable, however, because some lion species have a better reputation than others on occasion. Then the fitness value of the nomad lion X^N is evaluated as $f(X^N)$, which is compared to the fitness value $f(X^B)$ of the territorial lion X^B. If the value of $f(X^N)$ is lesser than $f(X^B)$, then the initialized k value is one; otherwise it is zero. If the initialized k value is one, then the current search is using the Eq. 10 for fertility update, otherwise follow the Eqs. 12 and 13 for update function. Here q is X^B and X^Gma is the prime number. The X^N generation when k is one is given in Eq. 14; if k value is zero, then the X^N generation is based on the Eq. 15.

$$X_q^{G+1} = \mathrm{kX}_q^{G+1} + 0.1 \cdot \mathrm{rand}(0, 1), q \tag{12}$$

$$\mathrm{kX}_q^{G+1} = (1 - c).\mathrm{kX}_q^G + c.\mathrm{mean}\left(\mathrm{GoodX}_q\right) \tag{13}$$

$$k = \mathrm{mod}\left(x^{\max} * \mathrm{rand} + x^{\min}, \mathrm{ma}\right) \tag{14}$$

$$k = x^{\min} + \left(x^{\max} - x^{\min}\right) \times \mathrm{rand}(1, \text{size of the solution}) \tag{15}$$

The territorial defense instructs the MLO algorithm to avoid local optima; analogous to identifying diverse solutions with the same fitness rate value, the coalition can allow for a better choice of alternatives than previously created solutions. The nomad coalition is maintained together by the capacity rate X^{e-N}. The nomad alliance employs the winner-take-all strategy. The X^N is chosen using Eqs. 16, 17, and 18. When X^B is defeated (defeated) by X^N, the pride value is updated. In this case, updating X^B is substituted by updating X^N. Only one nomad is chosen at a time by the updating strategy.

$$f\left(X^{e\_N}\right) < f\left(X^{B}\right) \tag{16}$$

$$f\left(X^{e\_N}\right) < f\left(X^{B\_\mathrm{cub}}\right. \tag{17}$$

$$f\left(X^{e\_N}\right) < f\left(X^{G\_\mathrm{cub}}\right) \tag{18}$$

5.5 *Territorial Takeover*

The territorial takeover occurs when the male cubs $X^{B-\mathrm{cubs}}$ and female cubs $X^{G-\mathrm{cubs}}$ mature faster than the X^B and X^G, indicating that the age factor of cubs has exceeded the boundaries ($A_{\mathrm{cub}} > 3$). If the cubs matured, they may be considered lions as well. As a result, the old lions X^B and X^G must demonstrate their might to the developed cubs in order for the pride to survive. The territorial takeover is permitted if the following conditions are met, as given in Eqs. 19 and 20. The pride has been updated based on the success of adult cubs and old pride lions. If the cub achieves success, the pride is replaced by X^{cub}. The replaced cub is now ready to generate its offspring. If the X^{cub} is defeated by the territorial lion, then the pride will restart the cycle.

$$f\left(X^{B\_\mathrm{cub}}\right) > f\left(X^{B}\right); \text{when}\, f\left(X^{B\_\mathrm{cub}}\right) > f\left(X^{\mathrm{cubs}}\right) \tag{19}$$

$$f\left(X^{G\_\mathrm{cub}}\right) > f\left(X^{G}\right); \text{when}\, f\left(X^{G\_\mathrm{cub}}\right) > f\left(X^{\mathrm{cubs}}\right) \tag{20}$$

The above process is repeated till the maximum number of generations are obtained. The termination criteria of the MLO were held while at least one of the following conditions was met. The conditions are mentioned in the Eqs. 21 and 22, where e^T function is representing the value of the threshold (target error). The maximum count of function generation is mentioned by the notation of N_o. The workflow of LO is displayed in Fig. 5.

$$N_o > N_o N_o{}^{\max} \tag{21}$$

$$\left[f\left(X^{B}\right) - f\left(X^{\mathrm{optional}}\right)\right] \leq e^{T} \tag{22}$$

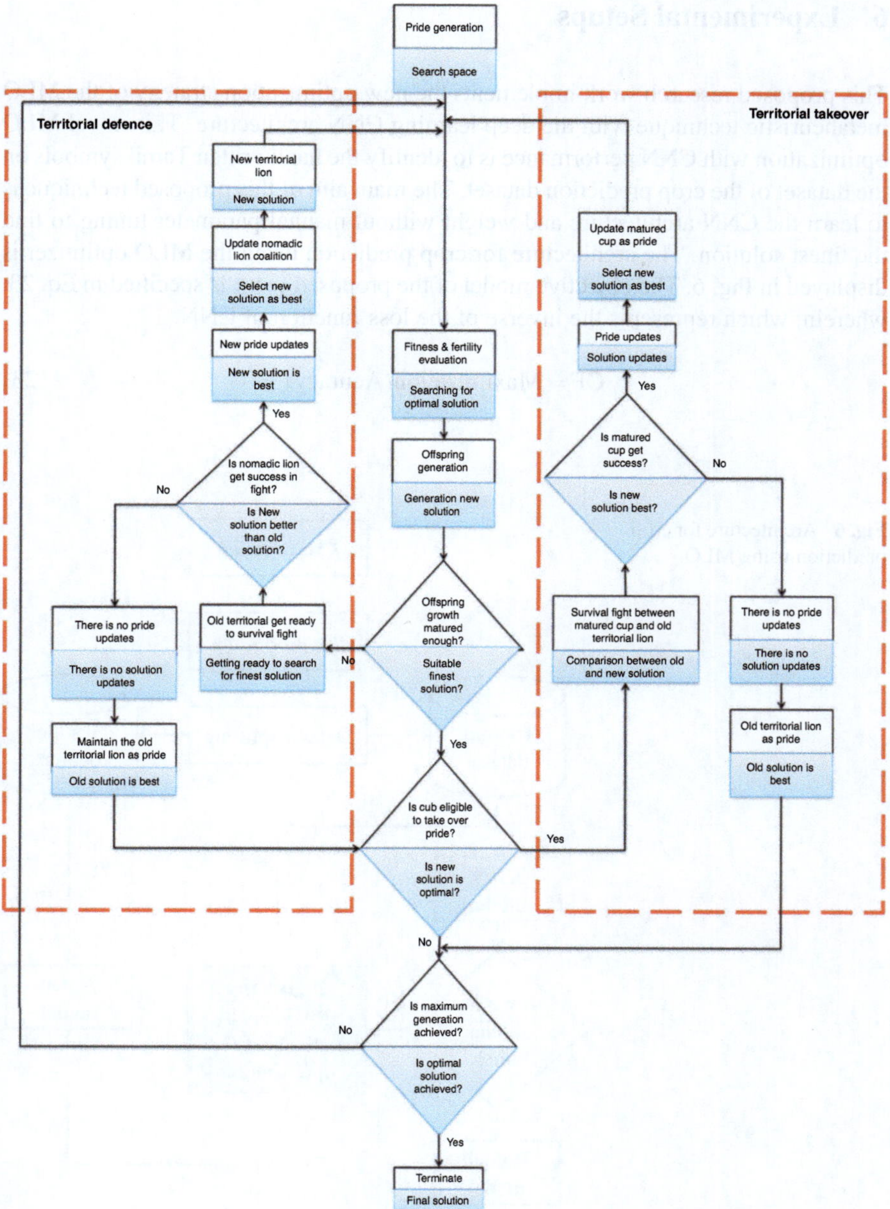

Fig. 5 Lion optimization work flow

6 Experimental Setups

This proposed research work implements the new optimization strategy of the MLO metaheuristic technique with the deep learning CNN architecture. This novel MLO optimization with CNN performance is to identify the handwritten Tamil symbols on the dataset of the crop prediction dataset. The main aim of this proposed technique is to learn the CNN architecture and weight without manual parameter tuning to find the finest solution. The architecture for crop prediction using the MLO optimizer is displayed in Fig. 6. The objective model of the proposed work is specified in Eq. 23, wherein, which represents the inverse of the loss function of CNN.

$$\mathrm{OF} = \mathrm{Maximization}(\mathrm{Acuracy}) \tag{23}$$

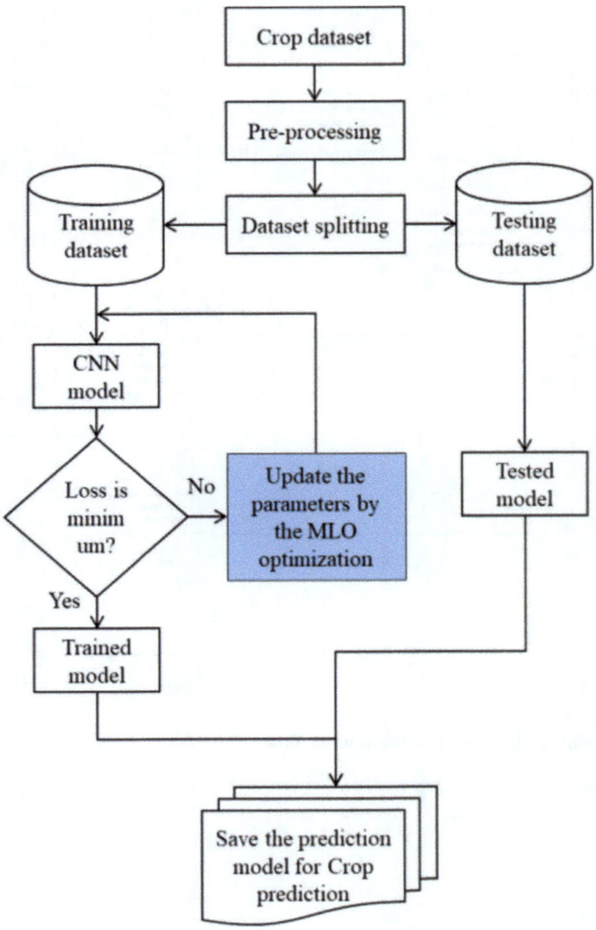

Fig. 6 Architecture for crop prediction using MLO

Before training the model, it is necessary to determine the tuned architecture of CNN with MLO. The proposed work chooses the number of dense layers, the number of neurons associated with that densely connected layer, and weight as the hyperparameter. The range of selected hyperparameter is given to determine the fittest solution. MLO involves a dense layer range as (FC1, FC2, FC3, FC4), the number of neurons present in a dense layer as (1–400), and weight as (−0.04 to +0.04). To formulate the hyperparameters into a suitable form of problem representation, the hyperparameter is represented in its binary form. For this purpose, a direct-encoding representation of the hyperparameters is performed where the value of a hyperparameter is represented in its binary form.

The accuracy value, which determines the number of correctly recognized characters, is the fitness function employed in this study. In order to produce the next successful generation of lions, the existing population of lions is operated upon based on this fitness function. The model first builds the structure using the MLO technique. That indicates that it determines the total number of dense layers and the number of neurons contained inside each individual dense layer. In order to achieve the best accuracy, the suggested algorithm selects the CNN model with two thick layers. The architecture of CNN with MLO is seen in Fig. 7. Following the creation of the structure, the weight is also optimized using the MLO method. Here, standard back propagation algorithm is used during the backward process.

The weights are often adjusted in accordance with the performance function's gradient's negative slope in the standard backpropagation algorithm. Because the optimization approach uses global searching, calculating gradients is substantially more difficult. Therefore, the optimization algorithm and the backpropagation technique are coupled to get the gradient value. The chosen basic CNN model is given the dataset. The cross-entropy loss, which is utilized as the fitness benchmark, is discovered after the first iteration. The weights are based on the suggested optimization approach in order to decrease the loss function update. The planned model

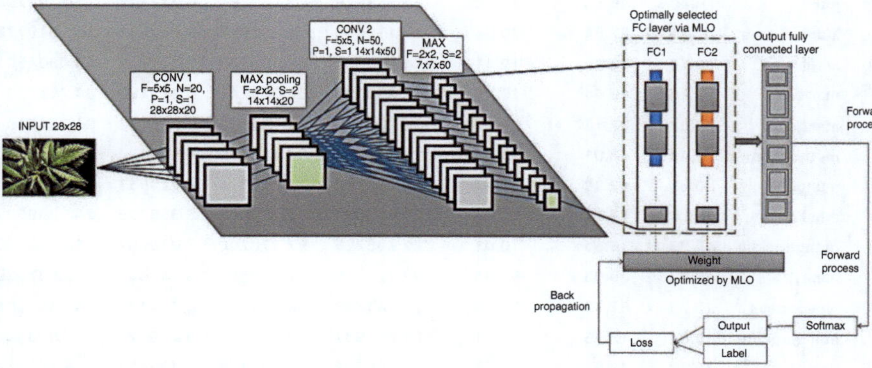

Fig. 7 Optimized CNN with MLO

takes into account a batch size of 32, 30 epochs, and 500 iterations. Finally, the proposed CNN+MLO model is compared with some other models such as CNN with Adam and CNN with LO.

To feed the model, a sizable generic crop dataset with agricultural parameters is used. As the feature dataset, another dataset is used. The datasets are gathered from the kaggle.com website. The crop dataset is 7841 kb in size. This dataset's prediction parameters include area, temperature, rainfall, pH level, and relative humidity. This dataset includes a variety of crops, including wheat, rice, maize, millet, pea, pigeon pea, sugarcane, green gram, and others. For a single crop, a variety of values are accessible for each and every forecast parameter. For instance, if the crop is wheat, the prediction parameters can be set to any value from the range of values present in the dataset. Figure 8. Crop Dataset. In Fig. 2, a small part of the crop dataset is displayed. In this, crop pH value of soil, the temperature, the rainfall, and relative humidity. The grouped data are displayed in Fig. 8.

The dataset was split into train and test sets before the models were trained, with 85% of the data used for training and 15% for testing the model. We experimented with a variety of ratios, increasing the training portion while decreasing the test component, starting with equal portions (50% for both). Finally, we discovered that the ratios of 0.85 and 0.15 for the training and testing parts, respectively, produced

	Label	Nitrogen	Phosphorus	Potassium	Temperature	Humidity	pH	Rainfall
0	apple	20.80	134.22	199.89	22.630942	92.333383	5.929663	112.654779
1	banana	100.23	82.01	50.05	27.376798	80.358123	5.983893	104.626980
2	blackgram	40.02	67.47	19.24	29.973340	65.118426	7.133952	67.884151
3	chickpea	40.09	67.79	79.92	18.872847	16.860439	7.336957	80.058977
4	coconut	21.98	16.93	30.59	27.409892	94.844272	5.976562	175.686646
5	coffee	101.20	28.74	29.94	25.540477	58.869846	6.790308	158.066295
6	cotton	117.77	46.24	19.56	23.988958	79.843474	6.912675	80.398043
7	grapes	23.18	132.53	200.11	23.849575	81.875228	6.025937	69.611829
8	jute	78.40	46.86	39.99	24.958376	79.639864	6.732778	174.792798
9	kidneybeans	20.75	67.54	20.05	20.115085	21.605357	5.749411	105.919778
10	lentil	18.77	68.36	19.41	24.509052	64.804785	6.927932	45.680454
11	maize	77.76	48.44	19.79	22.389204	65.092249	6.245190	84.766988
12	mango	20.07	27.18	29.92	31.208770	50.156573	5.766373	94.704515
13	mothbeans	21.44	48.01	20.23	28.194920	53.160418	6.831174	51.198487
14	mungbean	20.99	47.28	19.87	28.525775	85.499975	6.723957	48.403601
15	muskmelon	100.32	17.72	50.08	28.663066	92.342802	6.358805	24.689952
16	orange	19.58	16.55	10.01	22.765725	92.170209	7.016957	110.474969
17	papaya	49.88	59.05	50.04	33.723859	92.403388	6.741442	142.627839
18	pigeonpeas	20.73	67.73	20.29	27.741762	48.061633	5.794175	149.457564
19	pomegranate	18.87	18.75	40.21	21.837842	90.125504	6.429172	107.528442
20	rice	79.89	47.58	39.87	23.689332	82.272822	6.425471	236.181114
21	watermelon	99.42	17.00	50.22	25.591767	85.160375	6.495778	50.786219

Fig. 8 Crop prediction dataset

the best outcomes. Similar to this, other parameters have been adjusted using Python, which gave the best results for the parameters required to train the models and produce accurate prediction results. The general steps for crop prediction are: First, load the crop dataset that contains various parameters. Next load the necessary packages and libraries. Data preprocessing is carried out in the next process. The data are split into a training set and a testing set in order to prepare the dataset. Following this, a model is built using deep learning (CNN) and nature-inspired optimization algorithm (MLO) techniques, which in turn forecasts the best crop to produce. The test set verifies the model's functionality. The model throws an error, "value mismatch or wrong prediction," if any erroneous values are provided as input.

7 Model Evaluation and Result Analysis

The dataset of gathered crops is loaded to begin the implementation of the study task. The relevant libraries and packages were then imported, and the process of doing data preprocessing continued. The data are divided into test data and training data. The optimum crop to plant on a particular plot of land is then determined by using the necessary AI algorithms in a model that has been built. The machine learning and deep learning algorithms require two sets of data, known as the Trained set and Test set, in order to create a prediction model. The survey data that has been gathered from previous occurrences makes up the trained data, whereas the data from the most recent poll is test data. After the code has been executed, a graphic representation of the response is shown for each of the original parameter values. These variables include precipitation, humidity, and PH. On the basis of the data for these parameters, the yield is anticipated.

The graphs drawn below are displaying the yield with respect to the features taken, i.e., rain fall, humidity, temperature, phosphorus, nitrogen. The distribution chart of different parameter condition is shown in Fig. 9. These graphs confirm that there are outliers present in the data. Also it helps us understand the overall distribution of the dataset. The data points in a scatter plot are plotted on a particular place that coordinate values, i.e., the x-value and the y-value. It means a data point is an intersection of x-coordinate and y-coordinate.

In Fig. 9, the yield is determined with respect to nitrogen, phosphorus, potassium, humidity, and temperature. In all these cases, the values of features are taken from the feature dataset and hence, the yield is calculated. There are certain parameters on which the dataset is taken into analysis. These parameters are temperature, rainfall, location of soil, relative humidity, and area. These are the features with the help of which the dataset is analyzed and the better yield is predicted. There are certain crops that are taken for yielding on a particular piece of land. These crops include wheat, rice, maize, pea, pigeon pea, green gram, potato, sugarcane, soyabean, etc. Comparison of Mean Attributes of various classes is given in Fig. 10.

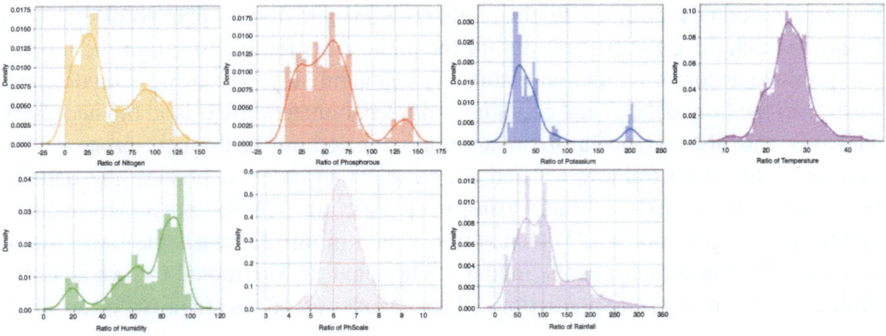

Fig. 9 Distribution chart for crop prediction dataset

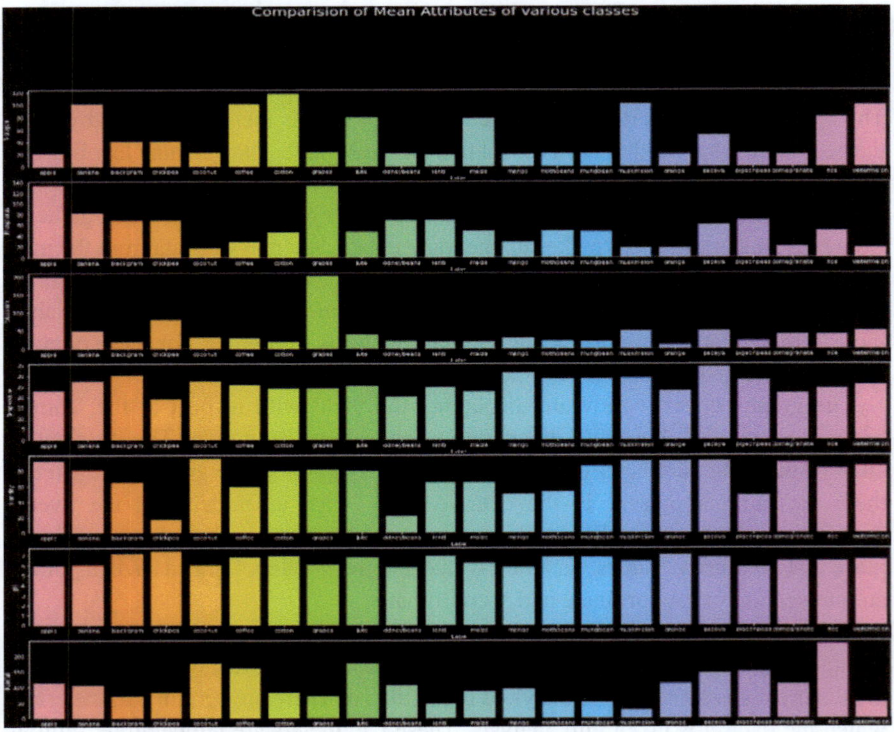

Fig. 10 Comparison of mean attributes

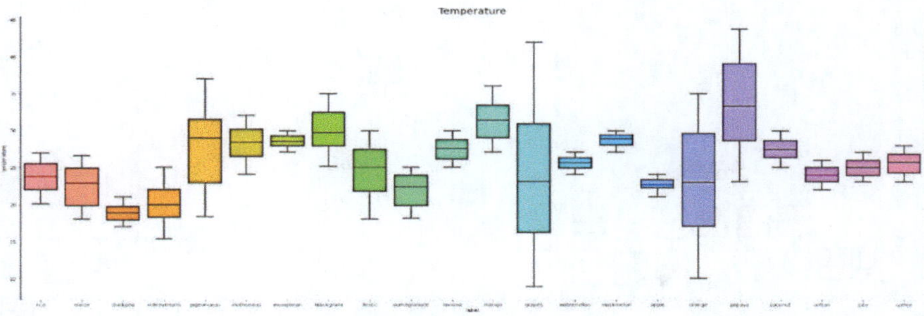

Fig. 11 Temperature variation for crop

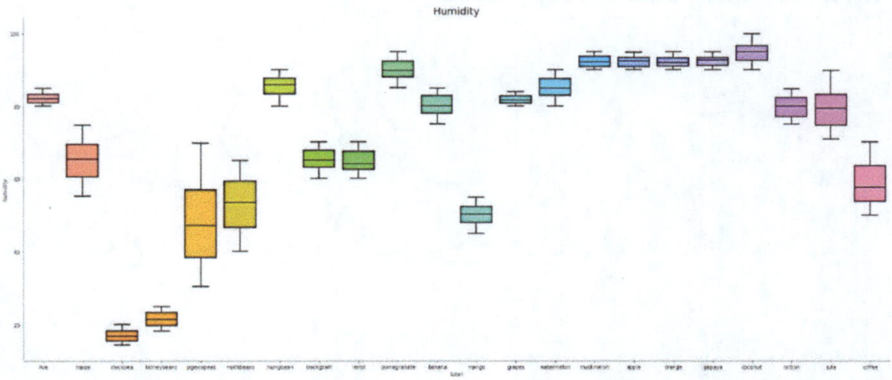

Fig. 12 Humidity variation for crop

The parameter relationship between with respect to temperature, humidity, phosphorus is displayed in Figs. 11, 12, and 13. The overall parameter relationship is displayed in Fig. 14.

The trained model produced 99% accuracy rate for crop prediction dataset. The result based on this algorithm is shown in Table 2.

From this result, we can conclude that the following observations are mentioned.

- Cotton requires most nitrogen.
- Apple requires most phosphorus.
- Grapes require most potassium.
- Papaya requires a hot climate.
- Coconut requires a humid climate.
- Chickpea requires high pH in soil.
- Rice requires huge amount of rainfall.

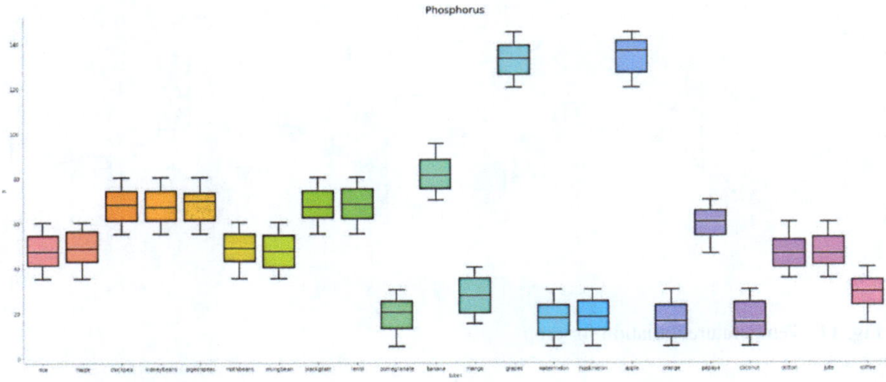

Fig. 13 Phosphorus variation for crop prediction

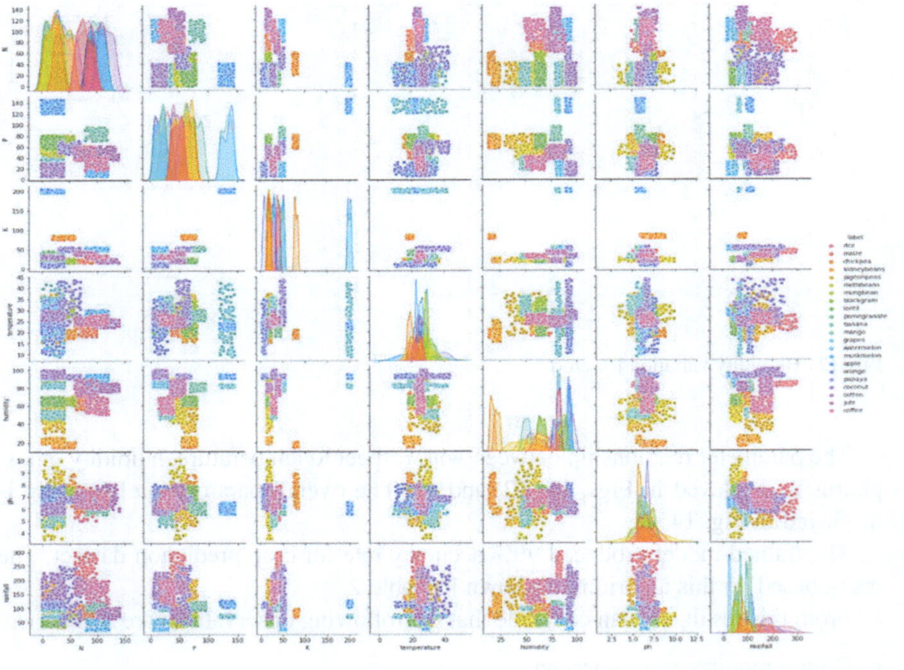

Fig. 14 Overall parameter variation versus crop prediction dataset

Table 2 Results based on algorithm

Parameter	Result analysis
Top 5 least nitrogen requiring crops:	Lentil—18.77 Pomegranate—18.87 Orange—19.58 Mango—20.07 Pigeon peas—20.73
Top 5 least humidity requiring crops:	Chickpea—16.8604394237 Kidney beans—21.6053567295 Pigeon peas—48.0616330847 Mango—50.1565726953 Moth beans—53.16041802790001
Top 5 least rainfall requiring crops:	Muskmelon—24.689952066 Lentil—45.680454204 Mung bean—48.403600902899996 Watermelon—50.7862189449 Moth beans—51.198487045700006
Top 5 least phosphorus requiring crops:	Orange—16.55 Coconut—16.93 Watermelon—17.0 Muskmelon—17.72 Pomegranate—18.75

Fig. 15 Accuracy graph for proposed work

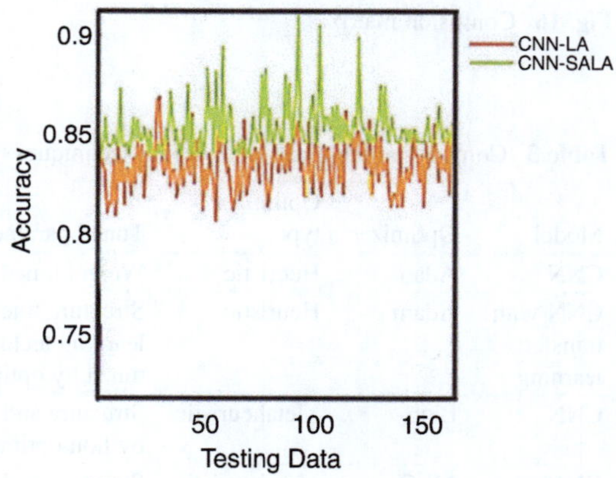

The accuracy of the proposed work is shown in Fig. 15. The accuracy is calculated for CNN with lion optimizer and CNN with modified lion optimizer. The confusion matrix is displayed in Fig. 16. From result analysis, the proposed work generates the accuracy rate with 96.5%. The comparison table is shown in Table 3.

From Figs. 17 and 18, we can conclude that the proposed system can produce the best performance in terms of accuracy than the existing models, even though the processing time is high. The processing time may increase than the simple CNN model due to model complexity, which means the high depth model such as the Hybrid CNN model, due to its global search, is the best fitness solution.

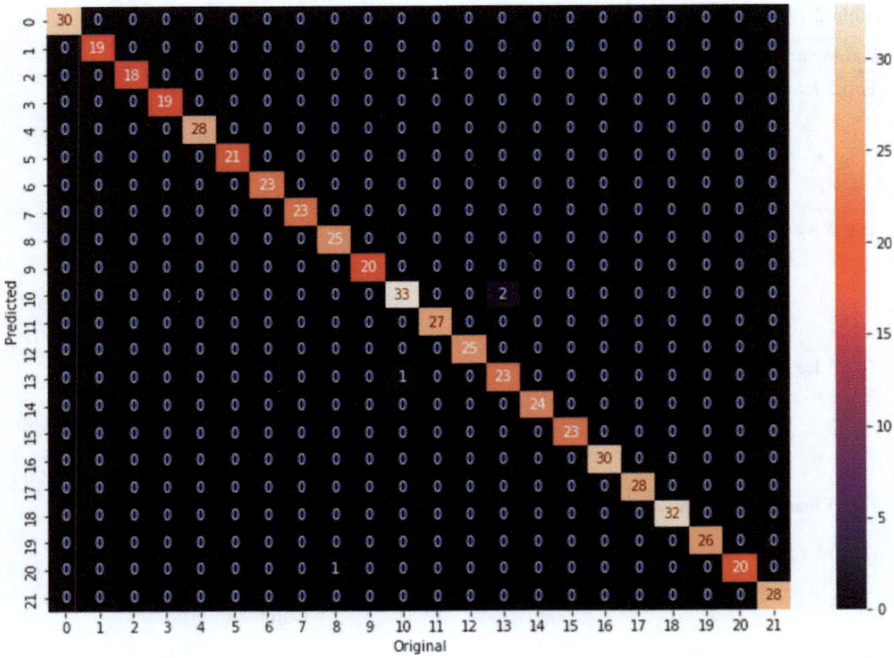

Fig. 16 Confusion matrix

Table 3 Comparison table between various techniques

Model	Optimizer	Optimizer type	Tuned parameter	Accuracy	Time / epoch
CNN	Adam	Heuristic	Weight tuned by optimizer	0.7546	2618 s
CNN with transfer learning	Adam	Heuristic	Structure fine-tuned by transfer learning technique; weight tuned by optimizer.	0.82338	2185 s
CNN	Lion	Metaheuristic	Structure and weight both tuned by lion optimizer	0.83597	3068 s
CNN	MLO	Metaheuristic	Structure and weight both tuned by SALA optimizer	0.85718	3219 s
Hybrid CNN	MLO	Metaheuristic	Structure and weight both tuned by sea lion optimizer	0.96547	2626 s

8 Conclusion

The proposed model utilizes AI algorithms to address the challenges faced by farmers, such as crop losses resulting from limited knowledge about cultivation in diverse soil and weather conditions. Deep learning models are employed to construct this model. By analyzing various prediction parameters, the model determines the

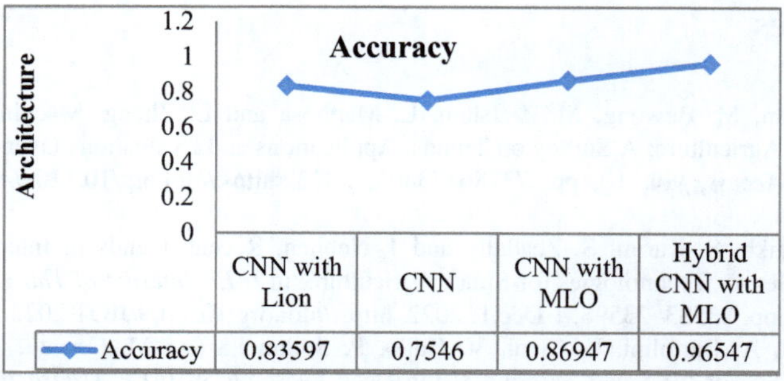

Fig. 17 Performance comparison of hybrid CNN with MLO in terms of accuracy

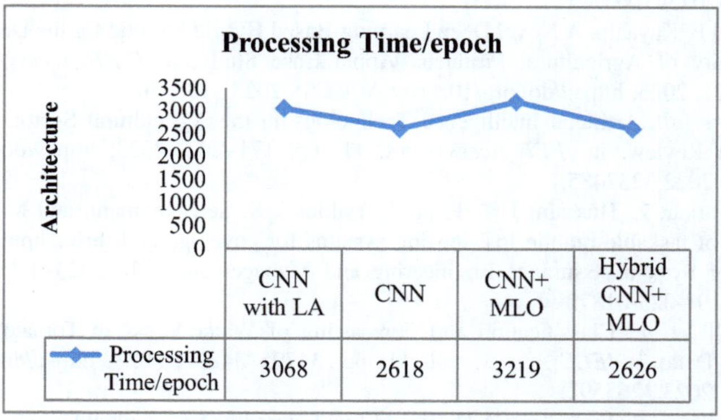

Fig. 18 Performance comparison of hybrid CNN with MLO in terms of processing time

most suitable and cost-effective crops to be cultivated on specific lands. Extensive research indicates that no existing work applies these techniques to crop prediction, demonstrating the enhanced accuracy of this research compared to alternative methods. The accuracy rate of the model stands at 96.5%. Furthermore, the model exhibits significant potential for future expansion and integration into a versatile and comprehensive application. This would enable farmers to access precise information on optimal crop yields directly on their mobile devices. Consequently, even when farmers are away from their fields, they can effectively manage their operations and mitigate potential losses. The advancements made in the agricultural industry through this model will greatly benefit farmers in optimizing crop production.

References

1. S. Condran, M. Bewong, M. Z. Islam, L. Maphosa and L. Zheng, Machine Learning in Precision Agriculture: A Survey on Trends, Applications and Evaluations Over Two Decades, in *IEEE Access*, vol. 10, pp. 73786–73803, 2022, https://doi.org/10.1109/ACCESS.2022.3188649.
2. F. K. Shaikh, S. Karim, S. Zeadally and J. Nebhen, Recent Trends in Internet-of-Things-Enabled Sensor Technologies for Smart Agriculture, in *IEEE Internet of Things Journal*, vol. 9, no. 23, pp. 23583–23598, 1 Dec.1, 2022, https://doi.org/10.1109/JIOT.2022.3210154.
3. G. Patrizi, A. Bartolini, L. Ciani, V. Gallo, P. Sommella and M. Carratù, A Virtual Soil Moisture Sensor for Smart Farming Using Deep Learning, in *IEEE Transactions on Instrumentation and Measurement*, vol. 71, pp. 1–11, 2022, Art no. 2515411, https://doi.org/10.1109/TIM.2022.3196446.
4. Ali-Khusein and Urquhart, Present and Future Applications of Robotics and Automations in Agriculture, Journal of Robotics Spectrum, vol.1, pp. 047–055, 2023. https://doi.org/10.53759/9852/JRS202301005.
5. F. Bal and F. Kayaalp, A Novel Deep Learning-Based Hybrid Method for the Determination of Productivity of Agricultural Products: Apple Case Study, in *IEEE Access*, vol. 11, pp. 7808–7821, 2023, https://doi.org/10.1109/ACCESS.2023.3238570.
6. E. Elbasi *et al.*, Artificial Intelligence Technology in the Agricultural Sector: A Systematic Literature Review, in *IEEE Access*, vol. 11, pp. 171–202, 2023, https://doi.org/10.1109/ACCESS.2022.3232485.
7. A. A. Agarkar, Z. Hussain, J. T. Raja, A. Haldorai, S. Selvakanmani, and M. Thangamani, Analysis of taskable mobile IoT sensing systems for coverage and throughput, International Journal of System Assurance Engineering and Management, Feb. 2023, https://doi.org/10.1007/s13198-023-01872-w.
8. M. Bettelli *et al.*, Classification and Forecasting of Water Stress in Tomato Plants Using Bioristor Data, in *IEEE Access*, vol. 11, pp. 34795–34807, 2023, https://doi.org/10.1109/ACCESS.2023.3265597.
9. E. Moupojou *et al.*, FieldPlant: A Dataset of Field Plant Images for Plant Disease Detection and Classification With Deep Learning, in *IEEE Access*, vol. 11, pp. 35398–35410, 2023, https://doi.org/10.1109/ACCESS.2023.3263042
10. K. Roy *et al.*, Detection of Tomato Leaf Diseases for Agro-Based Industries Using Novel PCA DeepNet, in *IEEE Access*, vol. 11, pp. 14983–15001, 2023, https://doi.org/10.1109/ACCESS.2023.3244499.
11. S. Das, A. Biswas, V. C and P. Sinha, Deep Learning Analysis of Rice Blast Disease Using Remote Sensing Images, in *IEEE Geoscience and Remote Sensing Letters*, vol. 20, pp. 1–5, 2023, Art no. 2500905, https://doi.org/10.1109/LGRS.2023.3244324.
12. Laura Schaefer, An Emerging Era of Artificial Intelligence Research in Agriculture, Journal of Robotics Spectrum, vol.1, pp. 036–046, 2023. https://doi.org/10.53759/9852/JRS202301004.
13. A. Haldorai, A Review on Artificial Intelligence in Internet of Things and Cyber Physical Systems, Journal of Computing and Natural Science, pp. 12–23, Jan. 2023, https://doi.org/10.53759/181x/jcns202303002.
14. C.-C. Lin, D.-J. Deng, J.-R. Kang, and W.-Y. Liu, A Dynamical Simplified Swarm Optimization Algorithm for the Multiobjective Annual Crop Planning Problem Conserving Groundwater for Sustainability, IEEE Transactions on Industrial Informatics, vol. 17, no. 6, pp. 4401–4410, Jun. 2021, https://doi.org/10.1109/tii.2020.3029258.
15. M. M. E. Alemany, A. Esteso, Á. Ortiz, and M. del Pino, Centralized and distributed optimization models for the multi-farmer crop planning problem under uncertainty: Application to a fresh tomato Argentinean supply chain case study, Computers; Industrial Engineering, vol. 153, p. 107048, Mar. 2021, https://doi.org/10.1016/j.cie.2020.107048.

16. S. Javadinejad, S. Eslamian, and K. O. A. Askari, The analysis of the most important climatic parameters affecting performance of crop variability in a changing climate, International Journal of Hydrology Science and Technology, vol. 11, no. 1, p. 1, 2021, https://doi.org/10.1504/ijhst.2021.112651.
17. D. Chakraborty et al., Usability of the Weather Forecast for Tackling Climatic Variability and Its Effect on Maize Crop Yield in Northeastern Hill Region of India, Agronomy, vol. 12, no. 10, p. 2529, Oct. 2022, https://doi.org/10.3390/agronomy12102529.
18. D. Beillouin, B. Schauberger, A. Bastos, P. Ciais, and D. Makowski, Impact of extreme weather conditions on European crop production in 2018, Philosophical Transactions of the Royal Society B: Biological Sciences, vol. 375, no. 1810, p. 20190510, Sep. 2020, https://doi.org/10.1098/rstb.2019.0510.
19. A. Crane-Droesch, Machine learning methods for crop yield prediction and climate change impact assessment in agriculture, Environmental Research Letters, vol. 13, no. 11, p. 114003, Oct. 2018, https://doi.org/10.1088/1748-9326/aae159.
20. Y. Kang, M. Ozdogan, X. Zhu, Z. Ye, C. Hain, and M. Anderson, Comparative assessment of environmental variables and machine learning algorithms for maize yield prediction in the US Midwest, Environmental Research Letters, vol. 15, no. 6, p. 064005, May 2020, https://doi.org/10.1088/1748-9326/ab7df9.
21. J. Sun, L. Di, Z. Sun, Y. Shen, and Z. Lai, County-Level Soybean Yield Prediction Using Deep CNN-LSTM Model, Sensors, vol. 19, no. 20, p. 4363, Oct. 2019, https://doi.org/10.3390/s19204363.
22. P. S. Nishant, P. Sai Venkat, B. L. Avinash, and B. Jabber, Crop Yield Prediction based on Indian Agriculture using Machine Learning, 2020 International Conference for Emerging Technology (INCET), Jun. 2020, https://doi.org/10.1109/incet49848.2020.9154036.
23. D. J. Reddy and M. R. Kumar, Crop Yield Prediction using Machine Learning Algorithm, 2021 5th International Conference on Intelligent Computing and Control Systems (ICICCS), May 2021, https://doi.org/10.1109/iciccs51141.2021.9432236
24. X. Wang, J. Huang, Q. Feng, and D. Yin, Winter Wheat Yield Prediction at County Level and Uncertainty Analysis in Main Wheat-Producing Regions of China with Deep Learning Approaches, Remote Sensing, vol. 12, no. 11, p. 1744, May 2020, https://doi.org/10.3390/rs12111744
25. P. Charoen-Ung and P. Mittrapiyanuruk, Sugarcane Yield Grade Prediction using Random Forest and Gradient Boosting Tree Techniques, 2018 15th International Joint Conference on Computer Science and Software Engineering (JCSSE), Jul. 2018, https://doi.org/10.1109/jcsse.2018.8457391.
26. S. S.Panchal and L. Sharma, Soil Classification and Suitable Crop Yield Prediction Using Support Vector Machine, International Journal Of Engineering And Computer Science, Jun. 2016, https://doi.org/10.18535/ijecs/v5i6.16.
27. A. K. Ranjan and B. R. Parida, Paddy acreage mapping and yield prediction using sentinel-based optical and SAR data in Sahibganj district, Jharkhand (India), Spatial Information Research, vol. 27, no. 4, pp. 399–410, Jan. 2019, https://doi.org/10.1007/s41324-019-00246-4.
28. P. Filippi et al., An approach to forecast grain crop yield using multi-layered, multi-farm data sets and machine learning, Precision Agriculture, vol. 20, no. 5, pp. 1015–1029, Jan. 2019, https://doi.org/10.1007/s11119-018-09628-4.
29. X. Xu et al., Design of an integrated climatic assessment indicator (ICAI) for wheat production: A case study in Jiangsu Province, China, Ecological Indicators, vol. 101, pp. 943–953, Jun. 2019, https://doi.org/10.1016/j.ecolind.2019.01.059.
30. S. Shidnal, M. V. Latte, and A. Kapoor, Crop yield prediction: two-tiered machine learning model approach, International Journal of Information Technology, vol. 13, no. 5, pp. 1983–1991, Nov. 2019, https://doi.org/10.1007/s41870-019-00375-x.
31. S. Khaki and L. Wang, Crop Yield Prediction Using Deep Neural Networks, Frontiers in Plant Science, vol. 10, May 2019, https://doi.org/10.3389/fpls.2019.00621.

32. R. A. Schwalbert, T. Amado, G. Corassa, L. P. Pott, P. V. V. Prasad, and I. A. Ciampitti, Satellite-based soybean yield forecast: Integrating machine learning and weather data for improving crop yield prediction in southern Brazil, Agricultural and Forest Meteorology, vol. 284, p. 107886, Apr. 2020, https://doi.org/10.1016/j.agrformet.2019.107886.
33. S. V. Shinde, R. Shastri, A. K. Dwivedi, A. Haldorai, V. Sahni, and B. Adusumalli, Multi sensor data and temporal image fusion cross validation technique for Agri yield monitoring system, Sep. 2021, https://doi.org/10.21203/rs.3.rs-943821/v1.
34. R. Boothalingam, Optimization using lion algorithm: a biological inspiration from lion's social behavior, Evolutionary Intelligence, vol. 11, no. 1–2, pp. 31–52, Sep. 2018, https://doi.org/10.1007/s12065-018-0168-y.
35. B. R. Rajakumar, Lion algorithm for standard and large scale bilinear system identification: A global optimization based on Lion's social behavior, 2014 IEEE Congress on Evolutionary Computation (CEC), Beijing, China, 2014, pp. 2116–2123, https://doi.org/10.1109/CEC.2014.6900561.
36. F. A. Saif, R. Latip, Z. M. Hanapi and K. Shafinah, Multi-Objective Grey Wolf Optimizer Algorithm for Task Scheduling in Cloud-Fog Computing, in *IEEE Access*, vol. 11, pp. 20635–20646, 2023, https://doi.org/10.1109/ACCESS.2023.3241240.

Part II
Artificial Intelligence for Smart Healthcare and Monitoring

An Investigation on Different Approaches for Medical Imaging

1 Introduction

Science and technology have advanced humanity significantly due to the third industrial revolution's enormous achievement [1]. Public awareness of the need for healthcare in this situation has increased [2]. The daily tasks of medical staffs are under more stress and strain as a result of the expanding healthcare market volume [3]. The idea of the Internet of Medical Things (IoMT) was derived from the need for the healthcare business to use intelligent and autonomous technology to improve the situation as it currently stands [4]. The use of Artificial Intelligence (AI) within the field of medical imaging is not limited to image interpretation and processing, it is one of the most intriguing fields of health innovation [5]. Indeed, AI has a variety of potential applications, including assisted reporting, planning, monitoring, storage of information, data mining, and many others. AI is anticipated to have a significant impact on the daily lives of the radiologist due to its numerous uses. AI utilizes different methods to evaluate a particular dataset to extract essential features or highlight from big data, rendering it challenging to identify the best organization of significant features and eliminating repeated ones. It is critical to consider these features, or precision measures will be inaccurate. Therefore, selecting just a few attributes from a variety of features will enhance the effectiveness of the model. Thus, eliminating redundant and disturbing features will minimize the dimensionality of data and speed up the trained model in a manner reminiscent of boosting [6].

A rapidly developing area of research in the field of contemporary medical images is the intelligent segmentation recognition as well as identification of digital pathological images as a result of the expansion of healthcare and medical facilities, which has led to an increase in digital utilization of pathological samples using pathological full-section imaging equipment at major hospitals. By deftly integrating worldwide business data, particularly varied complex medical images, the IoMT is

© European Alliance for Innovation 2024
A. Haldorai et al., *Artificial Intelligence for Sustainable Development*,
EAI/Springer Innovations in Communication and Computing,
https://doi.org/10.1007/978-3-031-53972-5_3

able to provide an intelligent application infrastructure service for health care. Medical images are frequently utilized during diagnosis of diseases as a type of significant presenting format of pathological visualization. Automated IoMT's medical image recognition technology developed into a significant area of smart medicine in recent years, along with the rapid growth of image processing technologies [7]. It appears hopeful that by fully utilizing machine intelligence, strong scheduling IoMT platforms and efficient recognition algorithms would reduce the effort of medical professionals.

Any area of the body could be impacted by the growth of malignant cells. As the malignant growth develops throughout the body, normal cells are squeezed out, making it challenging for the body to function normally [8]. Cancer is a group of diseases rather than a single illness. There are numerous varieties of malignant growths. Malignant development is not localized to one region; it can develop in blood cells as well as any internal organ. There is a definite distinction between them whenever it comes to the growth and spread for unwanted growths. Breast cancerous growth starts with cells that are developing abnormally. These cells frequently form knots or x-beams, which are hallmarks of tumors. If the cells within tumors develop into novel, highly potent compounds or potentially spread across the body, tumors may be dangerous (diseases). Breast cancer can spread to different areas of the breast. The milk ducts are where the majority of cancers of the breast start before moving to the areola. Some of them have their roots in the milk-producing organs. There are numerous ways that breast cancer can develop, some of which are more frequent than others. There are a few types of breast cancer that develop in particular tissues.

The computer-aided design framework can classify colon polyps and malignant breast & lung growth. However, these methods can still be used if the master human audience is not present. Using a collection of patient drawings, computer-assisted design models were used to depict radiologist places that looked to be abnormalities. As CAD can operate significantly depending on the preferences, we should keep this in mind and make any required adjustments to obtain the most accurate results [9]. Despite the technical advancements that define the modern period, some medical jobs, like picture annotation, are still carried out by hand, frequently through laborious and operator-dependent processes. For instance, a wholly manual process is typically used to delineate the segmentation of Gross Tumor Volume (GTV) for radiation treatment. This manual procedure is very time-consuming because dozens of slices need to be shaped. Additionally, operator dependence is crucial for the reproducibility of results. The fact that these manual processes heavily rely on the expertise and knowledge of the clinician indicates that a considerable variability between operators might negatively impact the results of the segmentation along with quantification.

The characteristics in medical images are particularly fine-grained in comparison to typical image recognition tasks [10]. When comparing medical photographs with and without pathogenic information, there may not always be a noticeable difference. Only sound professional judgment can make crucial distinctions, necessitating precisely robust recognition techniques. Deep machine learning (DML), which has strong feature abstraction and expression capabilities, is frequently employed in a variety of medical image processing and recognition tasks. By using multilayer feature mapping, the foundation of DML models is able to create the goal works together with large degrees of freedom and acquire adaptivity to various settings. For the most effective implementation of the comparatively challenging recognition tasks, high-level extraction of features may additionally convey the latent aspects of images to the fullest extent. Contrarily, because their parameters can communicate precise meanings, conventional machine learning (CML) frameworks such as support vector machines and logistic regression have superior explainability. This makes the entire process of learning and training simpler to comprehend and master. In actuality, CML models served as a foundation for the advancement and success of DML frameworks. In order to sufficiently mine the benefits of the two modes, fresh medical image recognition models can be developed using the hybrid intelligence that combines DML and CML. In this investigation, four well-known unsupervised segmentation techniques—k-means, Fuzzy C-Means (FCM), Split-and-Merge with Region Growing (SMRG), and spatial FCM (sFCM)—were used to delineate breast lesions in DCE-MRI series. It is easy to see from experimental results that fuzzy clustering algorithms greatly outperform crisp k-means, and direct region discovery methods (such as split-and-merge and region expanding). The equal result is produced by FCM and sFCM; however, sFCM performed better due to the incorporation of spatial information. The optimum computational aspects technique for the issue at hand was therefore one that explicitly managed segmentation uncertainty through several degrees of class memberships and spatial information. This article presents an evolutionary approach that utilizes image processing and deep machine learning to accurately detect and classify breast cancer. This model employs image augmentation, machine learning, image preprocessing, and image segmentation techniques to classify and diagnose breast cancer.

The main objectives of this paper are as following:

1. This study examines the significance of deep learning and intelligent segmentation system based on recognition for medical images.
2. In this article, data processing, fuzzy c-means, and methodology regarding the investigation are given in this article.

The following is how paper is organized. The Sect. 2 consists of related work. Section 3 discusses about Deep Learning for Medical Imaging and intelligent segmentation recognition system. Section 4 gives methodologies and result for diagnosis of breast cancer. Section 5 gives the discussion and conclusion of the research paper.

2 Related Works

An analysis of the development of technologically assisted identification of breast cancer from the start of the research is conducted through a systematic review [11]. The review was vital for different literature works within the field and may be used as a reference for various technical databases. However, the article's focus was strictly on scientific and academic publications, and it did not take into account any commercial factors. The survey's findings give a summary of the present condition of computer-assisted diagnosis models in regard to the visual modes and machine learning classifiers that are being used [12]. The purpose of this article is to demonstrate the advantages of unsupervised recognition of patterns methods, which does not require learning and is more easily incorporated into daily medical practice, particularly from a clinical acceptability standpoint. Because of this, supervised methods like CNNs were not covered in this study. We choose the compare-enhancing mass segmentation using DCE-MRI as a pertinent case study. It is important to note that in [13], we were able to achieve outcomes that were on par with or better than those produced by deep learning methods using traditional unsupervised procedures that were automatic and semiautomatic.

In the majority of pathological studies, the pathologist is focused on identifying particular anatomical regions. The authors in [14] created a double CGAN network using a dictionary-based learning framework to recognize tumor locations in cancer patients' pathological photos. The scholars in [15] demonstrated a method that could possibly identify the region to be addressed based on a convolution neural network in order to accomplish automatic segmentation within a gland area. The distinction between various gland adhesion scenarios and the network structure's design must continue to be adjusted, though. In [16], a new computer model was developed and assessed to predict the risk of breast cancer in the near future. The model was based on a quantitative examination of changes in bilateral mammographic picture elements within a dataset of complete-field digitized mammography images that showed unfavorable findings. The historical dataset included 335 digital mammograms of women, collected from four consecutive time periods. Three support vector machines with replicated risks have been constructed and validated using leave-one-instance-out-based cross-validation. The study presents a risk model for breast cancer based on differentiating mammographic features. It also reveals an increasing trend of risk for breast cancer diagnosed by mammography over time.

The technique proposed by [17] for cancer of breast detection employs enhanced convolutional neural network for the aim of categorizing breast cancer. A convolutional neural network enhances the categorization of breast cancer lesions, hence assisting medical professionals in diagnosing the condition more effectively. The CCNI-BCC method is applicable for classifying medical imaging into categories of benign, malignant, or healthy individuals. Timely identification and categorization of breast cancer, as elucidated by [18], might assist patients in obtaining appropriate medical attention. The authors have introduced a novel deep learning framework that utilizes transfer learning to accurately identify and classify various forms of breast cancer based on breast cytology images. Transfer learning is the process of using knowledge gained from one specific task to solve another task in the future, which may be related or similar, as opposed to current learning methods. The diagnosis of breast cancer has been the subject of related research [19–22]. A network model has been created by the authors to categorize medically related data. Additionally, security procedures are being developed to protect wireless sensor data related to medicine.

3 Overview of Different Techniques in Medical Imaging

3.1 Deep Learning for Medical Imaging

In the last 20 years, deep learning models have significantly improved the accuracy and efficiency of types and numbers of issues that neural networks could address [23]. Artificial intelligence (AI) is the capacity of a computer to acquire knowledge of techniques to think and complete tasks like writing and reading, having conversations, resolving problems, and decision-making processes. Figure 1 illustrates the vast variety of applications for AI in the breast cancer field, including tumor screening, staging, diagnosis, medication, follow-up, and drug discovery.

The machine learning (ML) branch of AI is primarily used to develop a mathematical framework for analysis and predictions of unidentified information and to extract characteristics data from the training set. Although DL does not depend on

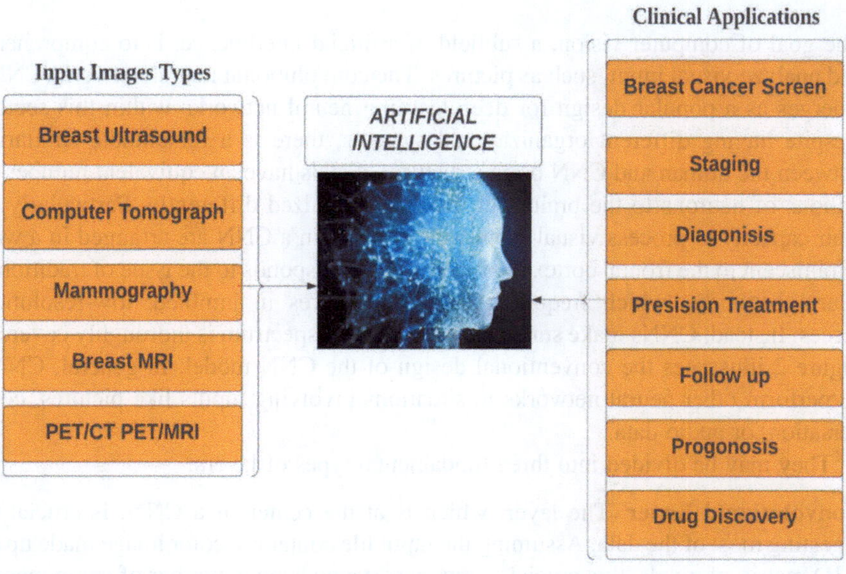

Fig. 1 AI application in breast cancer field

manually created characteristics and instead learns more like a human, it still requires on input data or datasets to develop decision-making abilities. DL relies on several, intricately coupled computing devices (gadgets for computing) that resemble the neurons in the brain of a human being and have been influenced by the biological nervous system. These units are made up of layers, and the artificial neural networks are created by fully connecting each layer. Deep learning is not a single approach but rather a group of calculations and geographical frameworks that can be used to solve a variety of problems [24]. Although connectionist structures have been around for more than 70 years, contemporary architectures and GPUs have propelled them to the forefront of artificial intelligence. While no acknowledged industry-accepted standard of excellence for image segmentation techniques, it has been identified that the automatic image segmentations methods often comprise edge, thresholding, region, clustering-based methods or similar segmentation methods. CNN is the most advanced DL algorithms, in which featuring of the computational pipeline includes convolutional, nonlinear in-nature pooling, and fully linked layers. DL algorithms typically refer to different methods but most common one is neural network methods. AI used in breast cancer screening can lessen or perhaps eliminate some obvious lesions being missed or misdiagnosed. Increasing monitoring for those at higher risk is the current emphasis of breast cancer prevention and management initiatives, and early detection is a crucial part of breast cancer management strategies [25].

3.2 Convolutional Neural Network

The goal of computer vision, a subfield of artificial intelligence, is to comprehend and analyze visual input, such as pictures. The convolutional neural network (CNN) emerges as a popular design for deep learning neural networks within this sector. Despite having different organizational systems, there is a remarkable similarity between the human and CNN brain structures. CNNs have an equivalent number of billions of neurons to the brain, but they are organized differently. Particularly in their capacity to process visual inputs, the neurons in a CNN are arranged in a way reminiscent to the frontal cortex. This architecture responds to the issue of traditional neural networks, which frequently analyze pictures in jumbled, low-resolution pieces. Instead, CNNs make sure that the full visual spectrum is thoroughly covered. Figure 2 illustrates the conventional design of the CNN model. In general, CNNs outperform other neural networks in situations involving inputs like pictures, conversation, or audio data.

They may be divided into three fundamental types of layers:

Convolutional Layer The layer, which is at the center of a CNN, is crucial to digesting most of the data. Assuming the input file contains a color image made up of a 3D matrix of pixels, this crucial component depends on a number of components, including the input data, a selection procedure, and a feature map. As a result, the

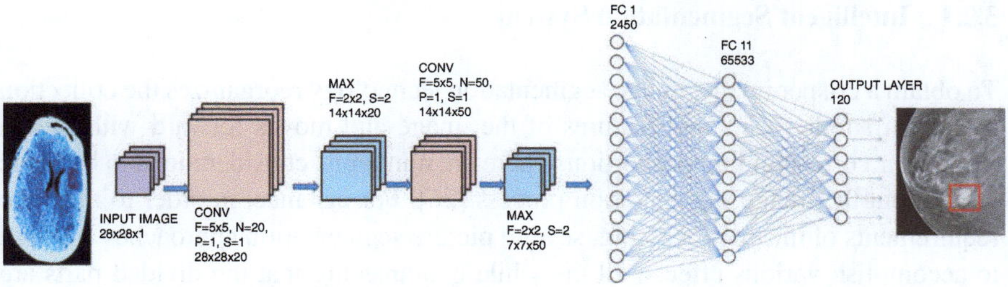

Fig. 2 Traditional model of CNN

input has three dimensions, which are analogous to the RGB elements in an image: height, width, and depth. The receptive fields of an image are rigorously examined by a specialized detector, known as a kernel or filter, to identify certain characteristics. Frequently, a subsequent convolutional layer comes before the convolutional layer, which is where the majority of calculations are done. Subsequently, a kernel or filter located beneath this layer scrutinizes the receptive fields of the picture to ascertain the presence of the desired feature before to commencing convolution.

Pooling Layer The pooling layer utilizes a pixel kernel or filter on the input image, similar to how the convolutional layer works. The pooling layer loses some information even though it has fewer input parameters than the convolutional layer. However, by simplifying its processes and improving overall efficiency in a positive way, this layer optimizes CNN.

Following are two of the primary pooling types:

1. Max Pooling: The filter scans the input and selects the pixel with the greatest value; this information is then sent to the array that results. Due to its added benefits, this technique, also known as max pooling, is used increasingly frequently.
2. Average Pooling: The filter calculates the standard deviation of the values inside the chosen receiving area as it moves through the input and sends this result to the final array.

Fully Connected Layer The CNN's FC (fully connected) layer is in charge of classifying pictures using the attributes deduced from earlier layers. In the context of the FC layer, "fully connected" refers to the arrangement where every input and node from the preceding layer is linked to each activation unit and node in the subsequent layer. Complete network connectivity, on the other hand, would result in unnecessary complexity, higher computing costs, more information losses, and subpar output quality. Therefore, for economy and improved overall performance, CNNs eliminate full linkages between all layers.

3.2.1 Intelligent Segmentation System

To obtain a crisper image, image segmentation technology reorganizes the collection of certain similar traits and features of the image and moves forward with this in mind their corresponding proportion. There are numerous considerations to be made throughout the image segmentation process [26]. For instance, in order to meet the requirements of image processing, several picture segmentation approaches are used to accomplish various effects, all the while guaranteeing that the divided parts are appropriately connected and that the segmented areas are unique. Images that are coherently segmented and have visually consistent information both before and after segmentation remain unchanged. For static photos, you may use the static image segmentation approach; for dynamic images, you can use the dynamic image segmentation technique. To deal with various image types, different industries have different needs for image segmentation technology. For example, the medical field needs to segment images of the human body, the transportation industry needs to process important intersections using image segmentation, and the military needs to segment images of military bases. Data manipulation users are primarily active in the data subsystem because they annotate experimental data and save it in the database. Figure 3 depicts the system's general architecture.

Three of the system's subsystems are shown in the above diagram. The component that injects different files like annotation files or original images, and preprocesses data is known as the data subsystem. Model subsystem is a component of model training or model parameter creation. The image recognition subsystem performs slice recognition. In a Python environment, data images and annotations files are kept. The Python Open Slide functions library, which includes a number of techniques including processing and interpreting medical images, is the foundation upon which data preprocessing is built. The model subsystem's parameters are

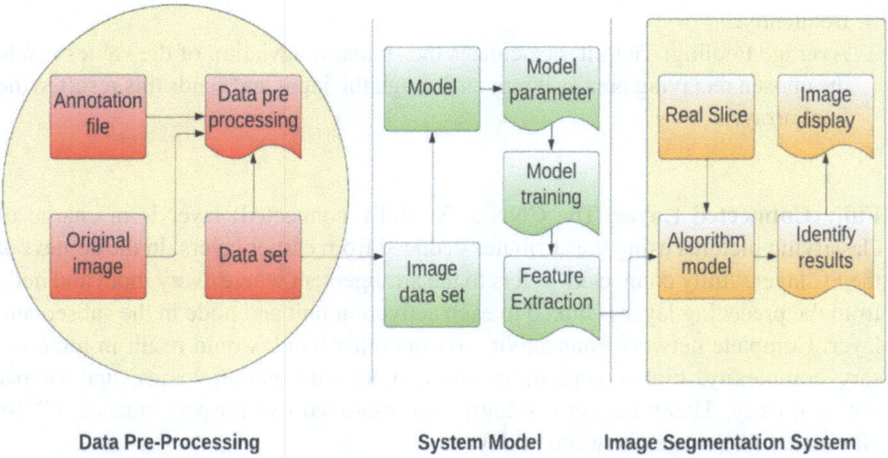

Fig. 3 General architecture of image segmentation

mostly used by the image recognizing subsystem to evaluate actual slices and present the outcomes, which are shown on the front page. The data user, who must gather breast cancer slices, is the system's main function of the data subsystem. An RGB image in the WSI format that may be saved in a database on a computer is created from the original slice through processing.

3.2.2 Data Processing

Within the computer vision field, noise reduction is typically accomplished through filtering, in which helpful frequencies are retrieved or inappropriate frequencies have been eliminated using a filter throughout the digital signal processing process. Convolutional operation is used in image processing to achieve image filtering. The low echo and distinct characteristics of the tissues and lesions that need to be separated from surrounding tissues in the picture are visible. Traditional isotropic filters, on the other hand, impair the performance of the image's edges while also destroying the textural elements of the image and erasing some details. Two conditions must be met for image filtering to be effective: first, the image's edge contour and other details must remain; second, the image's clarity and contrast must be improved [27].

After filtering an input image *P* by an example image, *I*, where *I* and *P*, are the algorithm's inputs, the output image for image *P* is obtained. As depicted here, guided filtering describes a linear filtering procedure. The filtered result for pixels at position *I* is of weighted average:

$$q_i = \sum_j wij(\mathrm{I})p_j \tag{1}$$

where, correspondingly, i and j stand for pixel subscripts. A filter kernel called W_{ij} only has connections to the guiding image I. The coefficients *a* and *b* are deterministic constants that only determine the behavior of a deterministic window with a radius of *r*. The loss function determined by the window *wk* to discover linear correlation is:

$$E(a,b) = \sum_{i=wk} \left[(aI_i - b_k + P)^2 - \varepsilon a^2 \right] \tag{2}$$

where wk denotes the adoption of regularization compensation when a value is excessively high. Among these, a^2 shows that the suppressing value is too high and can change the edge detection efficiency and fuzzy level. If there is no edge in the guide picture I, the result is mean fuzzy. If there are edges in the guide picture I, those edges are preserved in the output. A linear regression is performed for formula 2 and solved as follows:

$$a = \frac{(I/|w|)\sum_{i\in w_k}(I_i P_i + \mu p)}{\delta^2 + \varepsilon} \tag{3}$$

$$b = p + a\mu \tag{4}$$

$$P = \frac{1}{|w|}\sum_{i\in w_k}(p_i + \mu) \tag{5}$$

The segmentation process includes dividing an image into uniform and significant subregions. According to a formal perspective, segmenting an image I entails recognizing of a finite number of regions R1, R2..., RN according to Eq. 6:

$$\bigcup_{i=1}^{N} R_i = I' \tag{5}$$

With some constraints:

$$R_i \cap R_j = \varnothing, for\ i \neq j, \tag{6}$$

$$P(R_j) = \text{True, for } i = 1, 2, \ldots, N \tag{7}$$

$$P(R_i \cup R_j) = \text{False, for } i \neq j \tag{8}$$

P is a logical predicate that is appropriate for directing the segmentation process. The combined result of all the segmentation process-generated subregions is shown in Eq. 5. The point of intersection of two distinct subregions is the empty set, as shown by Eq. 6, indicating that the segmented subregions do not overlap. Equation 7 states that the conclusion of the logical predicated P on all of the pixel values remaining within a similar subregion is TRUE; alternatively put, the predicate P states that all of the image pixels pertaining to the solitary part may have similar properties. The P on union results between two separate subparts is therefore FALSE, according to Eq. 8.

3.2.3 Fuzzy C-Means

Clustering techniques can exploit the spectral or/and spatial pixel features in digital photography. The unsupervised clustering method known as K-means aims to partition the N input observation set into the k cluster [27]. Consider X = {x1, x2..., xn} be a collection of vector observation where xi = 1, 2,..., N and xi = Rn. If segmentation relies exclusively on grayscale intensity, the N-dimensional observations xi in Rn degrade to the xi scalar value within R, which represents the i-th pixel intensity. During the process of picture segmentation, the vector x is divided into individual components, each representing a specific numerical property of a pixel. By utilizing the cost function depicted in Eq. 9, the k-means algorithm converts the task of classification into an optimization issue.

Unsupervised clustering methods like the FCM algorithm look for the best way to divide a set of input data. In consideration to the distance metric between feature vectors, the principle driving the FCM method of classification is that of minimizing the intraclusters variability as well as maximizing the interclusters variance. The Fast Clustering (FCM) technique seeks to partition a data set of N objects, X = (x1, x2,..., xN), into the C clusters in the most optimal manner. The FCM algorithm enables each item to be assigned to various clusters with various membership levels, in contrast to the k-means technique that allocates every point exclusively to a single cluster. Since the FCM-described optimization problem lacks a closed form solution, it is necessary to iteratively find the optimum value of the *Jm (U, V; X)* function. In particular, Eq. 9 dictates that the centroid values for the prototype set V (t) and matrix U (t) are altered throughout each iteration t.

$$v^i t = \frac{\sum_{j=1}^{N} \left(u_{ij}^{(t)}\right)^m x_j}{\sum_{j=1}^{N} \left(u_{ij}^{(t)}\right)^m} \tag{9}$$

Each object xk is allotted to the nearest cluster at each iteration by being compared to the components of the centroid vector. The procedure concludes when either the highest number of iterations Tmax are achieved or the convergence condition is met. The convergence criterion is defined as the matrix norm difference from $U(t + 1)$ to $U(t)$ being lower than a given value, or the smallest advancements in the target component J in two subsequent iterations. After reaching convergence, every pixel is de-fuzzified and allocated to the clusters with a higher degree of membership, results to binary categorization.

4 Methodology and Result

The experiment was conducted using the training and testing method for the recognition system as well as intelligent segmentation of breast cancer. The line charts were used to demonstrate the rate of segmentation accuracy of the predictive model and the variance in the segmentation Dice coefficient. Additionally, a smoothing process was performed to enhance the visualization. It is important to note that each step's optimization for the particular case study received special attention, essentially on three distinct levels. About 50 patients without breast cancer made up the clinical DCE-MRI dataset used for this study's analysis, which resulted in the processing of 599 slices. Table 1 summarizes the key information about MRI

Table 1 Features of DCE-MRI dataset

Characteristics	Value
MRI sequence	DCE-MRI
Scanner manufactures	GE (General electric)
Scanner systems	Signa HDxt
Time repetition [ms]	37.720
Slice thickness [mm]	2
Time echo [ms]	17.6400
Slice pixel spacing [mm]	0.6875
Interslice spacing [mm]	1
Matrix size [pixels]	512 × 512

acquisition settings. For the subsequent processing steps, there are four steps: preprocessing, segmentation, optimization of the examined segmentation techniques, and postprocessing, which involves defining the proper region the characteristics (based on associated elements) to enhance the segmentation outcomes.

The following phrase sets up the segmentation process as a whole:

- Dataset loading: Ground truth and all DCE-MRI pixel images associated with the specified patients are loaded.
- ROI selection: In the central image, a bound box encompassing the lesion is traditionally extracted. As a result, the technique only analyzes the pixels inside the box, speeding up calculation and preventing categorization errors.
- Preprocessing is essential to lowering noise and supplying images with comparable properties for subsequent processing processes.
- All the MR stack pictures of each patient are segmented using several approaches such as k-means, SMRG, FCM, and sFCM, after performing lesion segmentation.
- Afterwards, postprocessing is used to enhance the segmentation and precisely identify the Region of Interest (ROI) from each binary mask obtained in the previous step.
- Performance evaluation: Following the storage of the final masks, performance metrics such as area-based metrics and distance-oriented measurements are calculated and compared to the ground truth.

We conducted many experiments on the CCKS2020 dataset to validate the effectiveness of the pseudo-data-generation technique proposed in this research. Initially, the program generated 2000 pseudo-labeled data. Subsequently, the algorithm used these data to train the neural network model, which was then applied to the CCKS2020 dataset. Different configurations of training data were utilized, following the methodology outlined in this paper [25]. The training data used in CCKS2020 is 1000 records, and the correct quantity of data with erroneous labels is also supplied. Furthermore, this work runs ten iteration rounds of the previously indicated experimental approach. The average F1 value derived from these iterations is used as the final F1 value, taking into account the unpredictability of the power source pseudo-labeled data. The approach used in this study attains an F1 score of 74.68 by including 1000 instances of pseudo-labeled data into the training process, surpassing the F1 score (73.52) achieved by the method utilized in the CCKS2020 medical event extraction and assessment challenge. Since just a certain number of the entire image is processed, cutting out regions with pixels that resemble the lesion can speed up processing time and prevent the algorithm from performing as intended, which would invalidate the results. The initial ROI comprising the lesion can be manually selected. Following preprocessing, segmentation entails dividing an image stored electronically into various sets of pixels in accordance with the particular clinical objective. One of the most difficult image processing problems nowadays is segmenting digital images; in fact, segmenting medical images is frequently still done manually using laborious, operator-dependent processes.

They are used to look at the several breast cancer mammography databases. The two databases that are utilized the most commonly are DDSM and MIAS. The MIAS dataset is used in this investigation. The MIAS database has 322 mammography images depicting both the left and right breasts. About 207 pictures from 161 patients were found to be normal, 64 to be benign, and 51 to be cancerous. In MIAS database images, the pectoral muscles, different noises, and background details are frequently visible. The background of this image has a high level of noise. In order to get a more exact and accurate assessment and interpretation of breast pictures, it is crucial to exclude any extraneous variables [28]. This study assesses the effectiveness of many algorithms by analyzing three key parameters: accuracy, sensitivity, and specificity.

$$\text{Accuracy} = \frac{(\text{TP} + \text{TN})}{(\text{TP} + \text{TN} + \text{FP} + \text{FN})}$$

$$\text{Sensitivity} = \frac{\text{TP}}{(\text{TP} + \text{FN})}$$

$$\text{Specificity} = \frac{\text{TN}}{(\text{TN} + \text{FP})}$$

$$\text{Precision} = \frac{\text{TP}}{(\text{TP} + \text{FP})}$$

$$\text{Recall} = \frac{\text{TP}}{(\text{TP} + \text{FN})}$$

TP, TN, FP, and FN stand for true positive, true negative, false positive, and false negative, respectively. Spatial region-based metrics were used to make a comparison of the areas that were semiautomatically segmented (RA) and manually segmented (RT) by determining the percentage of region that overlaps between the two masks created by image *I* segmentation.

$$R_{\text{TP}} = R_A \cap R_T$$

$$R_{\text{FP}} = R_A - R_T$$

Table 2 False-positive and -negative ratio

Method	FP ratio	FN Ratio
K-Mean	0.25 ± 0.20	0.25 ± 0.09
FCM	0.16 ± 0.13	0.23 ± 0.12
sFCM	0.14 ± 0.12	0.22 ± 0.09

$$R_{\mathrm{FN}} = R_T - R_A$$

$$R_{\mathrm{TN}} = I - R_A - R_B$$

The false-negative ratio (FNR) and false-positive ratio (FPR), as described in Eqs. 10 and 11 respectively, indicate the presence of incorrect positive and negative results in relation to the reference area.

$$\mathrm{FPR} = \frac{R_{\mathrm{FP}}}{R_{\mathrm{FP}} + R_{\mathrm{TN}}} \tag{10}$$

$$\mathrm{FNR} = \frac{R_{\mathrm{FN}}}{R_{\mathrm{FN}} + R_{\mathrm{TP}}} \tag{11}$$

Equations 12 and 13 precisely define sensitivity and specificity as the proportion of properly detecting the foreground (positive pixel), and background (negative pixel), respectively, by segmentation algorithm in reference to ground truth.

$$\mathrm{Sensitivity} = \frac{R_{\mathrm{TP}}}{R_{\mathrm{TP}} + R_{\mathrm{FN}}} \tag{12}$$

$$\mathrm{Specificity} = \frac{R_{\mathrm{TN}}}{R_{\mathrm{TN}} + R_{\mathrm{FP}}} \tag{13}$$

It is evident that clustering algorithms, particularly sFCM, improve lesion identification compared to k-means. However, k-means has a higher likelihood of missing lesion segment (signified by FN), particularly when the lesion is not similar, and integrates regions exterior to the lesion (signified by FP) [29]. Table 2 displays the rates of false positives and false negatives achieved by the unsupervised segmentation methods.

In the pathological collection of cancer, three evaluation indices are compared in the following figure. Among them, the network's segmentation performance improves with increasing specificity and sensitivity. In comparison to the conventional machine learning method for segmenting breast tumors, the recognition and segmentation approach using digital image processing has undergone significant advances since its inception. Because there are frequently far more instances of noncancer than there are of cancer, cancer detection is a well-known illustration of an unbalanced classification problem.

The mammography dataset, Fig. 4 shows detection of cancer candidate which involves identifying breast cancer from radiographic scans, specifically through the detection of clusters of the microcalcifications that show brilliant on a mammogram,

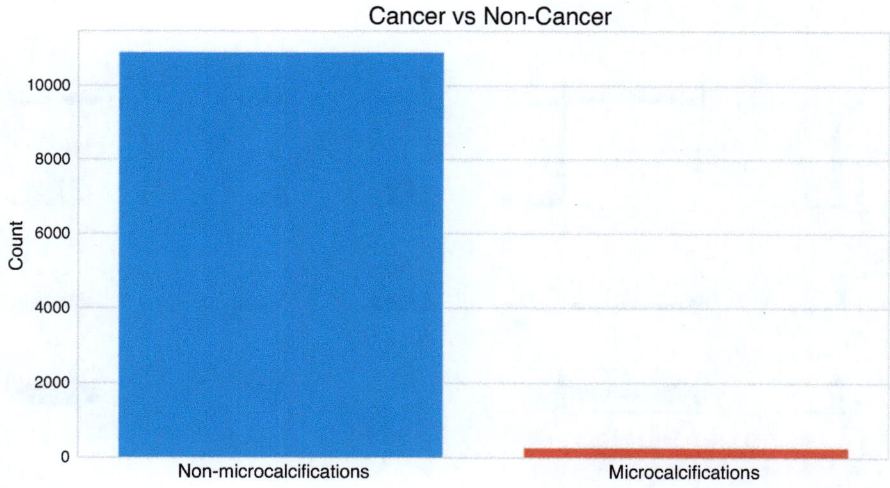

Fig. 4 Cancer versus noncancer

is a typical unbalanced classification dataset. By scanning the photos, segmenting them into groups of candidate objects, and then using algorithms for computer vision to describe each candidate object, this dataset was created. Modern deep learning methods are at the forefront of technology, allowing us to accurately and effectively segment lesions in datasets with a large number of patients [30]. Image segmentation deep learning systems often need supervised approaches, including extensive computations periods and big training data [31]. The data must accurately reflect all possible circumstances that the deep neural network may meet, which are not typically present in SMEs and are not therapeutically practical. Additionally, it is worth mentioning that scholars in [32] indicated that semiautomatic systems using conventional unsupervised approaches yielded results that are superior to or comparable to those achieved by deep CNNs, such as U-Net, and SegNet.

In Fig. 5, we observe that the distributions for many variables do differ indicating that it will be possible to make some logical distinctions between cases of cancer and cases of no cancer and depicts the phases of a few benign and malignant tumors. The dataset includes individuals diagnosed with breast cancer at different stages, hence allowing us to encompass a diverse array of clinical settings. This dataset contains scans with varying degrees of challenges associated with segmentation, with some exhibiting minimal contrast and significant dissimilarities. Included are also lesions with a necrotic core, masses composed of irregular shapes, or nonhomogeneous enhancement regions.

Remarkably, the IoU scores for the Validation Data surpass those of the Training Data, indicating that the model has the potential for far higher performance. The loss of the training set converges more rapidly and effectively throughout the training process, and its segmentation accuracy yields superior results compared to the test set, within the specified range, as seen in Fig. 6. There were no significant

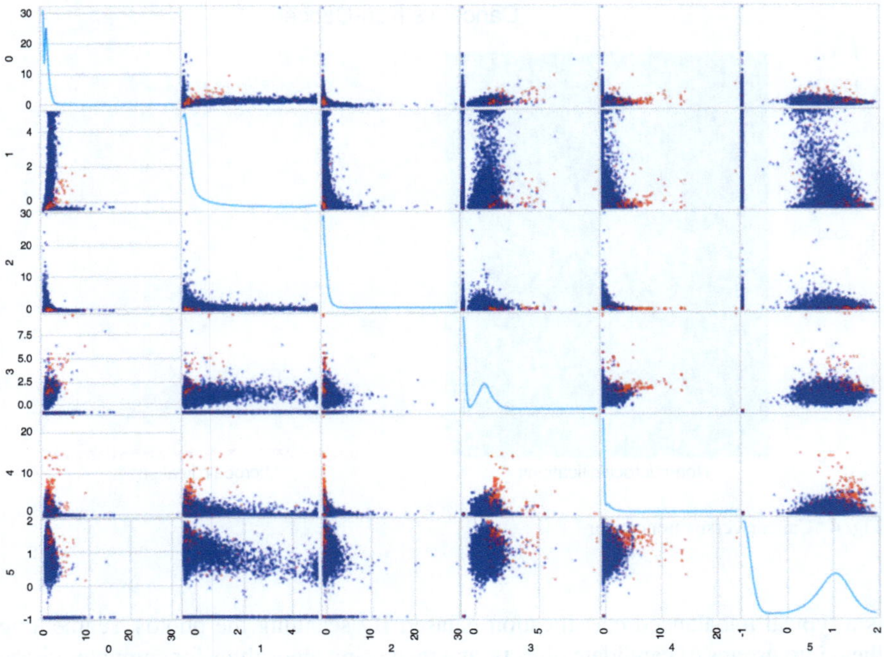

Fig. 5 Dataset distribution

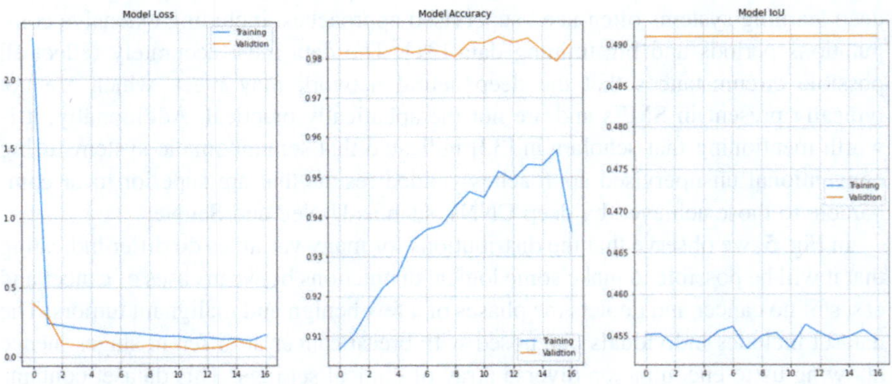

Fig. 6 Accuracy and loss for the proposed work

fluctuations in the data throughout the trial; they changed rather steadily and smoothly. After almost 2500 iterations, the experiment finally attained convergence, and the final experimental findings met expectations and performed well on the test set.

5 Discussion

An approach that shows promise for the diagnosis of breast cancer is AI-assisted imaging as well as FCM. A few intriguing possible applications for AI include: (1) AI-based diagnosis refers to the use of minimally invasive or noninvasive procedures to assess practical efficacy. (2)The process entails discovering distinctive molecular biological characteristics in primary or metastatic/recurrent sites. (3) Artificial intelligence facilitates the comprehension of tumor heterogeneity. (4) It facilitates the identification of treatment efficacy or tumor advancement, specifically in the context of immunotherapy. Nevertheless, the current methods are insufficient and inadequate in enabling the utilization of AI-assisted imaging diagnosis for conducting autonomous imaging or clinical diagnosis. This encompasses various areas, including imaging detection techniques, lesion segmentation, and analysis of lesion picture characteristics. Contemporary deep learning methods are at the forefront, allowing us to accurately and effectively separate lesions in datasets including hundreds of patients. Image segmentation deep learning algorithms are often supervised techniques that need extensive computational time and abundant training data. These data must be almost impossible to get as they are not consistently available in smaller hospitals, must include all possible situations in which the deep neural network may be used, and must accurately reflect all potential scenarios.

Currently, there is no universally acknowledged operating standard for the procedure, which encompasses imaging detection, tumor segmentation, and image feature extraction, to assure consistent results. In order to manage various sorts of pictures with varying characteristics, as well as the distinct variances among patients, it is necessary to use secondary, innovative algorithms that provide enhanced specificity. Moreover, while AI diagnosis based on imaging has the capability to decrease the false-positive rate (FPR) for breast cancer, its diagnostic precision relies on the image quality of detection methods such as mammography and ultrasound. Furthermore, the algorithm needs additional improvement.

6 Conclusion

Breast cancer affects about one out of every eight women worldwide, making it the most lethal form of cancer. Currently, there are no effective medications available for the prevention or treatment of breast cancer due to the unknown etiology of the illness. Early diagnosis and treatment of breast cancer are very efficacious approaches for diagnosing and treating the illness, with the potential to enhance the chances of complete recovery. The empirical findings affirm the suitability of using unsupervised pattern recognition techniques for the segmentation of medical images.

References

1. Lakhani P, Prater AB, Hutson RK et al (2018) Machine learning in radiology: applications beyond image interpretation. J Am Coll Radiol 15:350–359
2. Russell S, Bohannon J (2015) Artificial intelligence. Fears of an AI pioneer.Science 349:252
3. Q. Zhang *et al.*, "Graph neural networks-driven traffic forecasting for connected internet of vehicles," *IEEE Trans. Netw. Sci. Eng.*, to be published, https://doi.org/10.1109/TNSE.2021.3126830.
4. K. Yu, L. Tan, X. Shang, J. Huang, G. Srivastava, and P. Chatterjee, "Efficient and privacy-preserving medical research support platform against COVID-19: A blockchain-based approach," *IEEE Consum. Electron. Mag.*, vol. 10, no. 2, pp. 111–120, Mar. 2021.
5. H. Li, K. Yu, B. Liu, C. Feng, Z. Qin, and G. Srivastava, "An efficient ciphertext-policy weighted attribute-based encryption for the Internet of Health Things," *IEEE J. Biomed. Health Informat.*, to be published, https://doi.org/10.1109/JBHI.2021.3075995.
6. Kalaiselvi Balaraman and S.P.Angelin Claret, "Hybrid Resnet and Bidirectional LSTM-Based Deep Learning Model for Cardiovascular Disease Detection Using PPG Signals, Journal of Machine and Computing, vol.3, no.3, pp. 351-359, July 2023. https://doi.org/10.53759/7669/jmc202303030.
7. K. Yu, L. Tan, L. Lin, X. Cheng, Z. Yi, and T. Sato, "Deep-learningempowered breast cancer auxiliary diagnosis for 5GB remote E-health," *IEEE Wireless Commun.*, vol. 28, no. 3, pp. 54–61, Jun. 2021.
8. P. Gope, Y. Gheraibia, S. Kabir, and B. Sikdar, "A secure IoT-based modern healthcare system with fault-tolerant decision making process," *IEEE J. Biomed. Health Informat.*, vol. 25, no. 3, pp. 862–873,Mar. 2021
9. S. Shanthi, S. Saradha, J. A. Smitha, N. Prasath, and H. Anandakumar, "An efficient automatic brain tumor classification using optimized hybrid deep neural network," International Journal of Intelligent Networks, vol. 3, pp. 188–196, 2022, https://doi.org/10.1016/j.ijin.2022.11.003.
10. P. Kaur, G. Singh, and P. Kaur, "Intellectual detection and validation of automated mammogram breast cancer images by multi-class SVM using deep learning classification," *Informatics in Medicine Unlocked*, vol. 16, Article ID 100151, 2019.
11. J. Wang, C. Xia, A. Sharma, G. S. Gaba, and M. Shabaz, "Chest CT findings and differential diagnosis of mycoplasma pneumoniae pneumonia and mycoplasma pneumoniae combined with streptococcal pneumonia in children," *Journal of Healthcare Engineering*, vol. 2021, pp. 1–10, Article ID 8085530, 2021.
12. S. S. Mohant y and P. K. Mohant y, "Obesity as Potential Breast Cancer Risk Factor for Postmenopausal Women." *Genes & Diseases*, vol. 8, 2019.
13. H. Wang, B. Zheng, S. W. Yoon, and H. S. Ko, "A support vector machine-based ensemble algorithm for breast cancer diagnosis," *European Journal of Operational Research*, vol. 267, no. 2, pp. 687–699, 2018.
14. M. Tan, B. Zheng, J. K. Leader, and D. Gur, "Association between changes in mammographic image features and risk for near-term breast cancer development," *IEEE Transactions on Medical Imaging*, vol. 35, no. 7, pp. 1719–1728, 2016.
15. A. M. Hemeida, S. Alkhalaf, A. Mady, E. A. Mahmoud, M. E. Hussein, and A. M. B. Eldin, "Implementation of natureinspired optimization algorithms in some data mining tasks," *Ain Shams Engineering Journal*, vol. 11, no. (2), 2019.
16. G. Durgadevi and H. Shekhar, "An intelligent classification of breast cancer images," Indian Journal of Science and Technology, vol. 9, no. 28, p. 382, 2016.
17. G.-R. Sinha, "CAD based medical image processing: emphasis to breast cancer detection," i-Manager's Journal on Software Engineering, vol. 12, no. 2, p. 15, 2017.
18. A. Suresh, R. Udendhran, and M. Balamurgan, "A novel internet of things framework integrated with real time monitoring for intelligent healthcare environment," Journal of MedicalSystems, vol. 43, no. 6, pp. 1–10, 2019.

19. F. F. Ting, Y. J. Tan, and K. S. Sim, "Convolutional neural network improvement for breast cancer classification," *ExpertSystems with Applications*, vol. 120, pp. 103–115, 2019.
20. S. Chaudhury, N. Shelke, K. Sau, B. Prasanalakshmi, and M. Shabaz, "A novel approach to classifying breast cancer histopathology biopsy images using bilateral knowledge distillation and label smoothing regularization," in *Computational* and Mathematical Methods in Medicine, D. Koundal, Ed., vol. 2021, Hindawi Limited, Article ID 4019358, 11 pages, Hindawi Limited, 2021.
21. W. Wang, C. Qiu, Z. Yin et al., "Blockchain and PUF-based Lightweight Authentication Protocol for Wireless Medical Sensor Networks," *IEEE Internet Of @ings Journal*, vol. 1-1, 2021.
22. R. 'ippa, S. Bhattacharya, P. K. R. Maddikunta, and S. Hakak, "Antlion re-sampling based deep neural network model for classification of imbalanced multimodal stroke dataset," *Multimed Tools Appl*, Springer, Berlin, Germany,2020.
23. Zheng D, He X, Jing J. Overview of Artificial Intelligence in Breast Cancer Medical Imaging. *Journal of Clinical Medicine*. 2023; 12(2):419. https://doi.org/10.3390/jcm12020419.
24. X. Tang, Y. An, and C. Li, "Intelligent Segmentation and Recognition Method of Breast Cancer Based on Digital Image Processing Technology," Wireless Communications and Mobile Computing, vol. 2021, pp. 1–9, Dec. 2021, https://doi.org/10.1155/2021/2256316.
25. June Huh Eddie, "A Review of Medical Data Sources, and Advanced Data Analytics in the Medical Sector", Journal of Biomedical and Sustainable Healthcare Applications, vol.3, no.2, pp. 106-117, July 2023. https://doi.org/10.53759/0088/JBSHA202303011.
26. R. Arulmurugan and H. Anandakumar, "Region-based seed point cell segmentation and detection for biomedical image analysis," International Journal of Biomedical Engineering and Technology, vol. 27, no. 4, p. 273, 2018, https://doi.org/10.1504/ijbet.2018.094296.
27. Kristina Olson, "A Comprehensive Review on Healthcare Data Analytics", Journal of Biomedical and Sustainable Healthcare Applications, vol.3, no.2, pp. 095-105, July 2023. https://doi.org/10.53759/0088/JBSHA202303010.
28. V. D. P. Jasti et al., "Computational Technique Based on Machine Learning and Image Processing for Medical Image Analysis of Breast Cancer Diagnosis," Security and Communication Networks, vol. 2022, pp. 1–7, Mar. 2022, https://doi.org/10.1155/2022/1918379.
29. C. Militello et al., "On Unsupervised Methods for Medical Image Segmentation: Investigating Classic Approaches in Breast Cancer DCE-MRI," Applied Sciences, vol. 12, no. 1, p. 162, Dec. 2021, https://doi.org/10.3390/app12010162.
30. G. Jimenez and D. Racoceanu, "Deep learning for semantic segmentation vs. classification in computational pathology: application to mitosis analysis in breast cancer grading," Frontiers in Bioengineering and Biotechnology, vol. 7, p. 145, 2019.
31. S. Shakya, "Analysis of artificial intelligence based image classification techniques," Journal of Innovative Image Processing (JIIP), vol. 2, no. 1, pp. 44–54, 2020.
32. H. Zerouaoui and A. Idri, "Reviewing machine learning and image processing based decision-making systems for breast cancer imaging," Journal of Medical Systems, vol. 45, no. 1,pp. 1–20, 2021.

AI-Based Smart Decision System for Early and Accurate Brain Tumor Prediction

1 Introduction

A brain tumor is a potentially fatal condition caused by the uncontrolled spread of malignant cells. Because of the diversity and size of the lesion, diagnosing this condition is difficult [1]. Tumors have traditionally been discovered utilizing time-consuming invasive approaches that may result in human errors. The possibility that the patient would not survive could be compromised by such an event. The billions of cells that make up the human brain act as the body's central nervous system. The establishment of unwanted cell clusters in the brain leads to the aberrant proliferation that results in brain tumors. This abnormal cell may disrupt the normal functioning of the brain and the body's usual healthy processes. This brain tumor cell has the potential to attack neighboring brain cells. The brain tumor could disrupt the humanoid body's central nervous system, affecting the nervous system as a whole. Because this disrupts the body's functioning, it has recently become a life-threatening disease in children and adults [2].

Brain abnormalities are typically detected manually by medical experts using MRI imaging. MRI is a medical imaging analysis method normally used to spot and trace tumors in the brain. Because of weariness and an oversupply of MRI slices, the large-scale physical assessment process frequently leads to misunderstanding. Furthermore, it is unique, resulting in intra/inter reader heterogeneity. To address these concerns, there is a need to develop an innovative approach for a detection system that can diagnose various brain disorders. This approach will not only facilitate rapid and reliable analysis but also assist clinical professionals in making informed decisions. In recent years, deep learning techniques have played a significant role in creating and organizing tools with remarkable success. The diagnosis of brain tumors poses several challenges due to factors such as the intricate synaptic structure, varying sizes and shapes, and the overall complexity of the brain.

© European Alliance for Innovation 2024

A. Haldorai et al., *Artificial Intelligence for Sustainable Development*,
EAI/Springer Innovations in Communication and Computing,
https://doi.org/10.1007/978-3-031-53972-5_4

Brain tumors are classified based on how quickly they develop and grow back following treatment. Brain tumors are often categorized into two classes. They are as follows: (1) Grade-1 (low impact) and (2) Grade-2 (significant impact). Grade-1 and Grade-2 tumors are classified as benign or malignant. Benign is a noncancer form that does not progress, grows slowly, and cannot spread; thus, this type of brain tumor is considered less aggressive. As a result, benign tumors are treatable [3]. Malignant tumors, on the other hand, include cancer cells, overgrow, have the potential to spread throughout the body, and are very aggressive forms. Pituitary tumor is another prevalent kind of brain tumor. The importance of timely detection, accurate treatment planning, and correct treatment in increasing the patient's life span cannot be overstated.

The early detection and timely assessment are important in the effective treatment of any disease. The identification of disease is a vital phase in the healthy life of human beings. Advanced finding entails analysis, measurement, and decision-making. Supercomputers play an important part in medical-based decision assistance. Brain cancer is particularly tough among the lethal diseases because it is frequently not identified until it is too late for prognosis. Because of its usefulness and safety, MRI is being used highly specifically [4]. The automated categorization of brain tumors is essential for formulating efficient therapeutic approaches and streamlining the tumor assessment procedure. The interpretation of brain MRI images and tumor identification has become increasingly dependent on deep learning algorithms. Their innate ability to learn from experience is what has made them so popular; it allows for constant performance enhancement over time. Deep learning has proven to be more reliable and better to classic machine learning techniques in a number of applications, most notably medical image segmentation [5, 6]. These algorithms solve the problem of inaccurate brain tumor predictions by outperforming human accuracy. This paper presents a thorough examination of a brand-new deep learning system created especially for the identification of brain tumors.

Artificial neural networks (ANNs) are the main component of the machine learning area known as deep learning. These techniques are inspired by the structure and operation of the human brain. These algorithms essentially aim to mimic the functioning of human brains. Similar to how the nervous system is organized, deep learning algorithms convey information from one neuron to the next. A typical deep learning model has at least three layers and operates in layers. Data are transferred from one layer to the next by way of each accepting it [7].

In order to detect tumors, this article also considers transfer learning techniques and deep learning algorithms. The success of deep learning techniques in various small-data scenarios depends heavily on transfer learning. In research, deep learning is frequently employed. Millions of labeled data points may not be available in many real-world contexts, making model training difficult [8, 9]. Massive amounts of data are needed for deep learning techniques in order to adjust a neural network's millions of parameters [10]. This means that you'll need a lot of (expensive) labeled data, especially for supervised learning. This process requires access to an existing network's internal workings and involves fine-tuning a model that has already been trained. Users first add newly found classes to the already-existing network.

New tasks with the ability to categorize data more precisely can be carried out when network changes are finished. This method has the benefit of requiring minutes or hours less computing time than alternatives due to the large reduction in data requirements. The following are the research's main contributions:

- For the purpose of classifying and predicting brain tumors using Magnetic Resonance Imaging (MRI), a hybrid deep learning (DL) model with a quicker R-CNN basis is suggested.
- In order to efficiently filter noise and extract important characteristics, the model makes use of transfer learning techniques and a CNN architecture.
- Large amounts of high-dimensional data can be effectively managed by this hybrid deep learning approach, which improves model performance.
- The suggested model is a potentially useful diagnostic tool that can help radiologists and other specialists make decisions by identifying brain tumors in MRI brain pictures.

An extensive overview of brain tumor consequences is given in Sect. 1. In Sect. 4, the results are discussed in detail, while in Sects. 2 and 3, the suggested work and experimental setup are further explained. Lastly, the results are discussed in Sect. 4. The Sect. 5 gives the conclusion.

2 Literature Review

An overview and conclusions of some recent studies are offered below. Sasikala et al. [11] selected features for effective feature dimension reduction using genetic machine learning methods. The artificial neural network is given these chosen features. However, the results used only four of the 29 elements and had a maximum accuracy of 97%. Another study in [12] offered a modified version of the AlexNet CNN frame for detecting noninvasive glioma brain lumps. The classification's efficiency can be tested using whole-brain MRI scans. As a result, the provided strategy produces modest performance with a 90% accuracy. Muhammad et al. [13] presented a new model for the data augmentation approach integrated with the CNN framework for improved accuracy. This novel technique is used to classify multigrade MRI brain pictures. Furthermore, for the robust classification, the presented study used pretrained GG-19 CNN architecture with a transfer learning idea. After augmentation, the proposed system attained 90.34% accuracy. As a result, the augmentation procedure improves accuracy even further.

A brain convolution-based method using Glioma Brain Tumor Segmentation Networks in Magnetic Resonance Imaging was presented by Fatemeh Derikvand [14]. This approach uses a hybrid framework made up of several different architectures of convolutional neural networks. These networks estimate each pixel's label by utilizing both local and global brain tissue information, which improves segmentation results. Conversely, Tonmoy Hossain's [15] technique starts with the input picture's skull stripped, which eliminates the skull region from the Magnetic

Resonance Imaging image. The paper only requires the brain section for segmentation rather than the entire skull. Thus we perform morphological procedures on our pictures. The image is now subjected to feature extraction, including texture-based (Dissimilarity, Homogeneity, Energy, Correlation) and statistical-based (Mean, Centroid, Standard Deviation, Entropy) characteristics. The retrieved features are now fed into a five-layer Convolution Neural Network model, a seven-stage model that includes a hidden layer and produces good results.

Amin et al. [16] recommended a novel approach for detecting and classifying MR brains. To remove noise, the Gaussian filter was active first. Then features of brain image were recovered using embedded, contrast, cyclic, block presence, and other techniques for segmentation step and cross-validation skill for classification process. The author Chen et al. [17] proposed a mixture of fuzzy clustering with MRF. This hybrid technique is the subdivision of subsidiary information and works well. To separate enhanced signal zones in fluid-attenuated inversion retrieval, the MRI images, the authors Korfiatis et al. [18] devised a deep CNN technique based on autoencoders. They used the BRATS Benchmark dataset to train automatic neural encoders to section the brain tumor image, and a dataset containing three retail specialists was used to assess the accuracy.

Deep Nayaka et al. [19] discussed deep neural network framework to identify several classes of brain abnormalities. The cornerstones of their proposed SRVFL-AE are the autoencoders RVFL. To increase learning generalizability and speed in comparison to deep learning techniques based on the autoencoder, RVFL was chosen in place of a crucial module of the SRVFL-AE. They also added a Rectified Linear Unit (ReLU) activation algorithm to the deep NN, which would improve the input features' concealment and speed. It used two common MD-1 and 2 MRI data datasets to evaluate their method's efficacy. A deep learning model was proposed by Amin et al. [20] to categorize input samples as tumor (unhealthy) or nontumor (healthy). After that, the slices were combined using a median filter. They enhanced the output slices' quality by highlighting and smoothing the edges of the sample input. It divided the component into two layers of the stacked sparse autoencoder (SSAE) techniques that were finely calibrated. After conducting numerous trials, they chose the model's hyperparameters. They employed 200 concealed components in the primary sheet and 400 hidden elements in another layer. They experimented with the model using a Softmax function to forecast from tumor, no tumor-containing images.

Mishra et al. [21] planned an effective strategy for classifying MRI brain images. To extract features and choose coefficients for categorization using SVM classifiers, dual-tree M-band wavelet transform (WT) transform was utilized. Through Deep WT, SWT, they separated the features of the ordinary and pathological MRI brain images. In order to classify brain tumors, Raja et al. [22] used a Bayesian fuzzy clustering-based subdivision technique with a hybrid deep autoencoder. Alzheimer's disease neuroimaging (ADNI) study was put up by Osama et al. [23]. The analysis made use of 138 people from an fMRI data set.

Novel subdivision research for brain tissues utilizing MR images was proposed by Arunkumar et al. [24]. An algorithm that uses ANN to pinpoint an ROI precisely serves as the foundation for a completely robotic MRI-based brain tumor location

Table 1 Brain tumor detection review

S. no	Paper	Dataset	Transfer learning	Method
1.	Rehman et al. [29]	FLAIR scans MRI	No	RF, SVM, AdaBoost1, and RUSBoost
2.	Kumar et al. [30]	MICCAI and BRATS	No	K-nearest neighbor
3.	Alrashedy et al. [31]	Kaggle	Yes	ResNet152V2 and MobileNetV2
4.	Amin et al. [32]	2020-BRATS, local images, and Kaggle	Yes	Quantum variational classifiers (QVR)
5.	Zeineldin et al. [33]	BraTS 2019	No	NeuroXAI
6.	Qureshi et [34]	T1-weighted MRI	No	Ultra-light learning architecture
7.	Alsaif et al. [35]	Kaggle	Yes	VGG, ResNet, and AlexNet
8.	Han, C et al. [36]	BRATS	No	GAN
9.	Ghassemi et al. [37]	NHGH MRI	Yes	Pretrained GAN
10.	Mondal, M et al. [38]	BRATS	Yes	VGG_19

detection and classification method. A differential feature map (DFM) block was suggested by Huang et al. [25] to use MR images to detect brain tumors. A Differential Characteristic (NN) is created by combining DFM blocks with SE (stress and excitation) blocks. The regularity axis of the brain's MRI picture was first roughly paralleled to the vertical axis using an automatic image correcting technique. DFNN was also developed to separate brain MRI pictures into "normal" and "abnormal" categories. Their testing findings demonstrate that the suggested system's average accuracy on the two databases can achieve 99.2% and 98%. To find unsupervised medical anomalies (MADGAN), Han [26] suggested an unsupervised medical anomaly network model. In the study conducted by Alanazi et al. [27], a novel transfer deep learning model was employed to enable early diagnosis of brain tumors (Table 1).

An unknown brain MRI dataset was used to apply the transfer learning method, with impressive results. The proposed model displayed an astonishing 96.89% accuracy. Similar to this, Senan et al. [28] used a transfer learning-based strategy on an unreported brain MRI dataset and saw encouraging results. The proposed model had a 96.89% accuracy rate, which is quite high. The merging of deep learning and machine learning techniques for the analysis of brain tumors was closely examined in this work. To discover and classify brain tumors, the authors used Support Vector Machine (SVM) approaches, especially in combination with the AlexNet and ResNet-18 designs. They applied the average filter technique to improve the MRI images of brain tumors, and deep convolutional neural network (CNN) layers were used to extract important characteristics. SVM and Softmax were then used to classify these retrieved characteristics.

3 Proposed Work

Deep Learning has significantly improved in recent years. Deep learning techniques assist in the automated detection and classification of brain tumors; they can lighten the burden on radiologists who must interpret many pictures. This can reduce diagnostic errors and subsequently increase the overall accuracy rate. Deep learning methods in the field of radiology that had previously achieved the best performance rate for some disorders have been made public by recent studies. In this proposed work, faster R-CNN model with transfer learning technique is considered for brain tumor detection.

Faster R-CNN, which debuted in 2015, is the most popular modern iteration of the R-CNN (Regions with CNN) family. The advancement of the R-CNN often involved decreased test time, increased performance, and increased computing efficiency (integrating the many training steps). These neural network architectures characteristically contain the subsequent components: (1) The Region Proposal algorithm will make bounding box, which will mark the location of the substances from the given image. (2) The nest stage is feature generation, here the features are collected and selected from the objects with the selected CNN model. (3) In the object detection pipeline, the classification layer plays a crucial role by categorizing the object and determining its class. Additionally, the regression layer enhances the accuracy of the bounding box coordinates for the object. Previously, the region proposal method was a separate component in the Fast R-CNN architecture. To address this issue, a unique neural network called the Region Proposal Network (RPN) was presented in the Faster R-CNN. Furthermore, the region proposal step can exchange layers with later detection stages because of this connection. "Faster R-CNN"describes a detection pipeline structure that combines Fast R-CNN as the object detection network with the Region Proposal Network (RPN) for region proposal generation. The RPN architecture is shown in Fig. 1, which demonstrates the faster and more effective object detection capabilities that arise from this integration.

The input image undergoes initial processing within the convolutional neural network's backbone to initiate the Region Proposal Network (RPN). Initially, the input image is resized such that its smaller side measures 600 pixels, and its longer side does not exceed 1000 pixels. Based on the stride of the backbone architecture, the resulting output features (denoted as H × W) are typically much more compact compared to the original example. For both potential backbone networks utilized, the network stride is set to 16. This means that two points in the input image, spaced 16 pixels apart, correspond to two consecutive pixels in the output feature map of the backbone. The primary objective of the network is to identify object presence in the input image at specific locations on the feature map and estimate their dimensions. To accomplish this, a group of "Anchors" is positioned on the input sample at every position of the output features generated by the backbone system. These anchors act as representations of potential objects, encompassing a variety of sizes and aspect ratios that could potentially exist within the image.

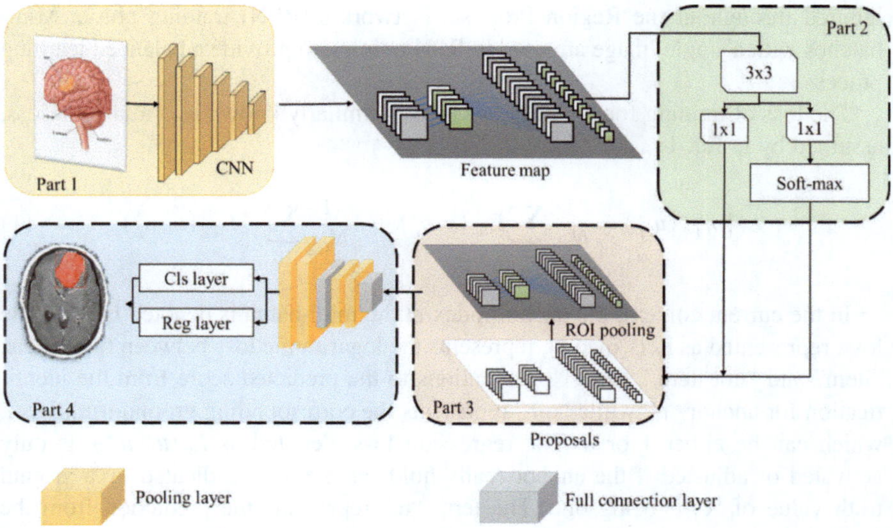

Fig. 1 RPN architecture

The network's job is to determine whether the k matching anchors spanning the input image actually cover objects as it moves through each pixel in the output feature map. It then refines these anchors' coordinates to produce bounding boxes known as "object proposals" or regions of interest. The resulting feature map is first subjected to a (3 × 3) Convolution Operation (CO) in order to produce 512-dimensional feature maps for each location. Then, two more parallel layers are added: a 1 × 1 CO for bounding box regression and a 1 × 1 CO for target identification. The target identification output has a size of (H, W, 18), where H and W stand for the feature map's height and breadth, respectively. The target identification output consists of 18 units. The four regression coefficients for each of the nine anchors connected to each point on the backbone feature map are computed using this output. The regression branch receives as input a feature map with dimensions of H × Width that was generated by the backbone architecture. The 36-unit regression branch produces an output with dimensions (H, W, 36).

About 40 by 60 make up the final feature map that is constructed, which equals 20,000 anchors (40 × 60 × 9). To avoid their contribution to the loss function, anchors that extend past the border are removed during training. As a result, each image retains about 6000 anchors for additional processing. Whether an anchor is regarded as a "positive" sample depends on two factors. First, an anchor is classified as positive if it shows the highest Intersection over Union (IoU) when paired with a ground truth box. Second, an anchor is likewise classified as positive if its IoU with any ground truth box is more than 0.7 [39]. It is feasible for the same ground truth box to yield positive labels for more than one anchor. On the other hand, an anchor is categorized as a "negative" sample if its Intersection over Union (IoU) with all ground truth boxes is less than 0.3. Anchors not classified as positive or negative are ignored throughout the Region Proposal Network's (RPN) training phase. Mini-batches with a single image are used in RPN training to provide a balanced learning process.

The loss of training for the RPN is which is similarly known as a multitask loss, assumed by in Eq. 1.

$$L(\{o_i\},\{u_i\}) = \frac{1}{N_{\text{cls}}}\sum_i L_{\text{cls}}(o, o_i^*) + \lambda\frac{1}{N_{\text{reg}}}\sum_i o_i^* L_{\text{reg}}(u_i, u_i^*) \tag{1}$$

In the current context, the anchor index at the mini-batch is denoted by "I." The loss, represented as Lcls(o_i, o_i*), represents the logarithmic loss between two labels: "item" and "not item." Here, "o_i" mentions to the predicted score from the identification for anchor "*i*," while "o_i*" represents the corresponding ground truth label, which can be either 1 or 0. The regression loss, denoted as $L_{re}(u_i, u_i*)$, is only activated or adjusted if the anchor really holds an element, indicated by a ground truth value of "o_i*" being one. The term "u_i" represents the prediction from the regression layer and consists of four variables: $[u_e, u_f, u_g, t]$. The regression target, denoted as "u_i*," is calculated according to Eqs. 2–5.

$$u_e^* = (e^* - e_a)/w_a \tag{2}$$

$$u_f^* = (f^* - f)/h_a \tag{3}$$

$$u_g^* = log(g^*/g_a) \tag{4}$$

$$u_h^* = log(h^*/h_a) \tag{5}$$

The height (h) and width (w) of a box, as well as the (e, f) coordinates of the box's center, are represented by the variables e, f, g, and h in the given context. The coordinates of the ground truth bounding box and the anchor box are represented by ea. and e*, respectively. It is crucial to remember that every anchor box has unique regressors with separate weights. Consequently, the regression loss for anchor "i" is supplied into the proper regressor if a sample is positive. The related anchor box (which is expected to be positive) can be used with the learnt regression output (ti) during testing. Parameters "*e*," "*f*," "*g*," and "*h*" for the estimated can then be calculated in reverse using Eqs. (6–9).

$$t_e = (e - e_a)/g_a \tag{6}$$

$$t_f = (f - yf_a)/h_a \tag{7}$$

$$t_g = \log(g/g_a) \tag{8}$$

$$t_h = \log(h/h_a) \tag{9}$$

Regression coefficient results are sent to anchors for accurate bounding box (BB) values in order to achieve exact localization. The cls values determine where each box should be placed. A nonmaximum suppression (NMS) with a threshold range of

0.7 is then implemented. Beginning with the highest score, any bounding box (BB) with an Intersection over Union (IoU) greater than 0.7 with extra BBs is eliminated. As a result, only the BB with the highest score is kept for a group of overlapping boxes. Over 2000 ideas are generated on every image, and cross-border BBs are retained and repositioned near the edge of the image. Based on these object ideas, all 2000 proposals from the Region Proposal Network (RPN) help train and understand the Fast R-CNN detection pipeline. Only the top N suggestions were tested.

The Region Proposal Network (RPN) in the Faster R-CNN architecture is a technique for making suggestions for regions, while the Fast R-CNN is the detection network. In addition to a convolutional neural network (CNN) backbone, the Fast R-CNN detector has two parallel branches for bounding box regression and classification, a fully connected layer, and a Region of Interest (ROI) pooling layer. The self-governing and automatic training is received in the RPN. Similarly, the network of Fast R-CNN is also independent training model. The initial/ starting layer weights are recommended from the Imagenet classification model, the following layer weights are fine-tuned based on the identification and detection requirement. Generally, weights of RPN are fixed, so the Faster R-CNN will do the learning process based on the suggestion of RPN. The weights of RPN are fine-tuned for the proposal task, at the initial stage the weights are initialized by Faster R-CNN [40]. In the current situation, the weights of RPN are adjusted in the common layers. The weights are adjusted between RPN and detector. The overall Faster R-CNN is presented in Fig. 2 (Table 2).

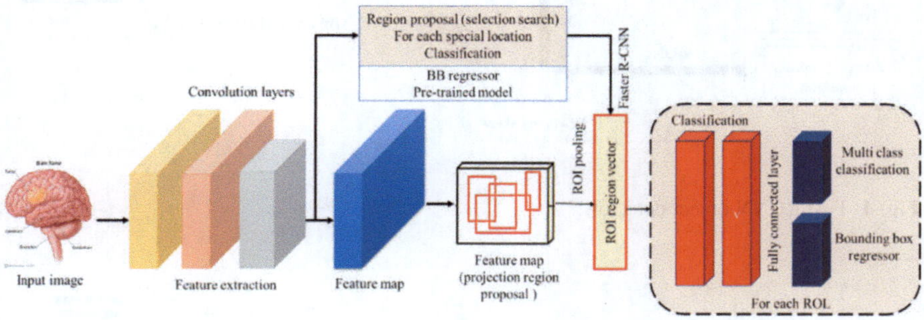

Fig. 2 Faster R-CNN

Table 2 Comparison of R-CNN models

S. No	R-CNN detector	Description
1.	Train RCNN object detector (Fig. 3)	Training and detection is slow process. Permits only the custom region proposal
2.	Train fast RCNN object detector (Fig. 4)	Permits custom region proposal
3.	Train faster RCNN object detector (Fig. 5)	Optimal run-time performance Will not support the custom region proposal

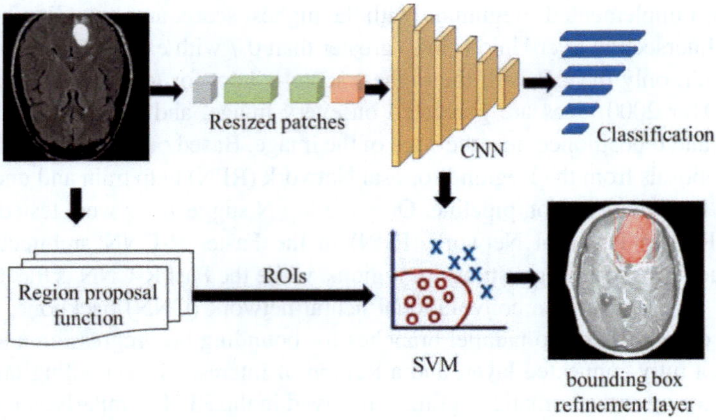

Fig. 3 RCNN object detector

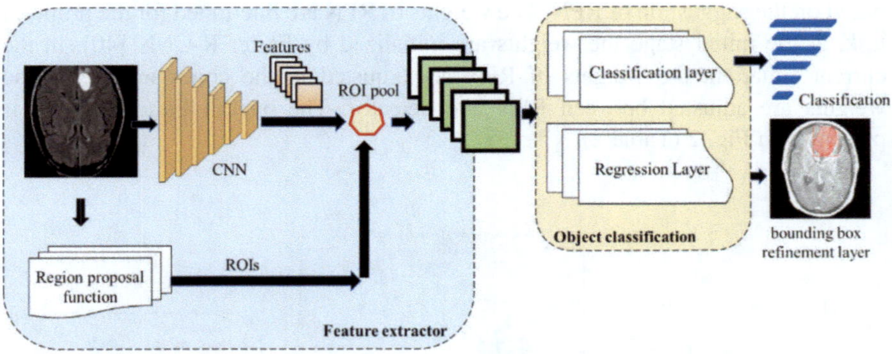

Fig. 4 Fast RCNN object detector

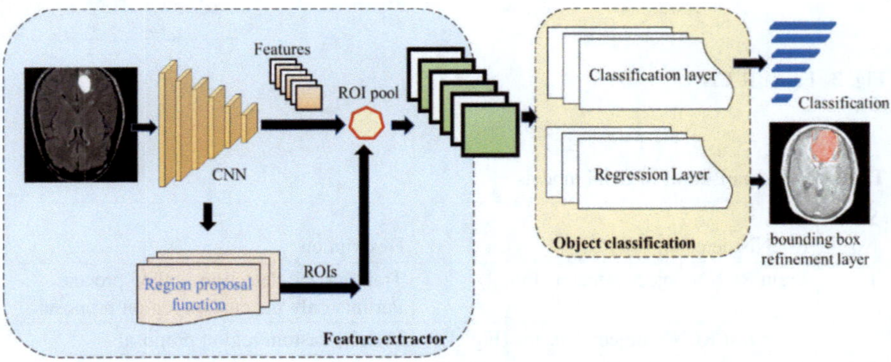

Fig. 5 Faster RCNN object detector

3.1 Transfer Learning with Faster R-CNN

One of the deep learning techniques that is probably most frequently utilized in business is object detection. Although this is the case, creating an object identification model from scratch is not precisely a simple operation. Transfer learning is frequently used in place of traditional instruction to save time and complexity. In this paper, we will examine how to train a pretrained torch-vision faster R-CNN model to become an expert at identifying brain tumor via transfer learning. This approach allows a machine to leverage knowledge gained from prior work to enhance predictions for a new task. In transfer learning, the focus is on utilizing the information learned from previous training to improve performance on a different but related task. In the field of computer vision, the initial layers of neural networks are typically specialized in detecting edges, while the middle layers are responsible for recognizing shapes. The final layers are designed to capture task-specific features. Transfer learning primarily focuses on reusing and retraining the initial and middle layers, while the later layers are often fine-tuned or adjusted to suit the new task. By utilizing pretrained models and leveraging the knowledge encoded in the lower layers, transfer learning can significantly accelerate training and improve performance on new tasks.

It is considered the tagged database from the task that assisted as its training ground. The important benefits of the transfer learning process is less training time, improvement in selected network performance. It is well suitable at satiation like less data samples; sometimes the database contains less amount of training data. At that time, the model can consider transfer learning technique. It will avoid scratch training. This transfer learning will present successful machine/deep learning models with minimum training image samples; transfer learning is widely utilized in NLP projects, primarily due to the requirement for large amounts of data to achieve improved performance. Furthermore, it significantly reduces training time, as developing a deep neural network from scratch for complex tasks can take days or even weeks. Figure 6 illustrates the implementation of transfer learning in brain tumor detection using the Faster R-CNN approach.

4 Experimental Setup

Training process, testing process, and validation procedure are the three critical stages for any classifier. The training/testing activities are frequently conducted till the outcomes of the training presentation are acceptable. This study considers the deep learning model because it has been demonstrated to be an effective machine-learning technique for resolving challenging issues. Noise removal, cropping, and scaling are all part of the preprocessing stage. A CNN-based classifier was employed for the classification portion.

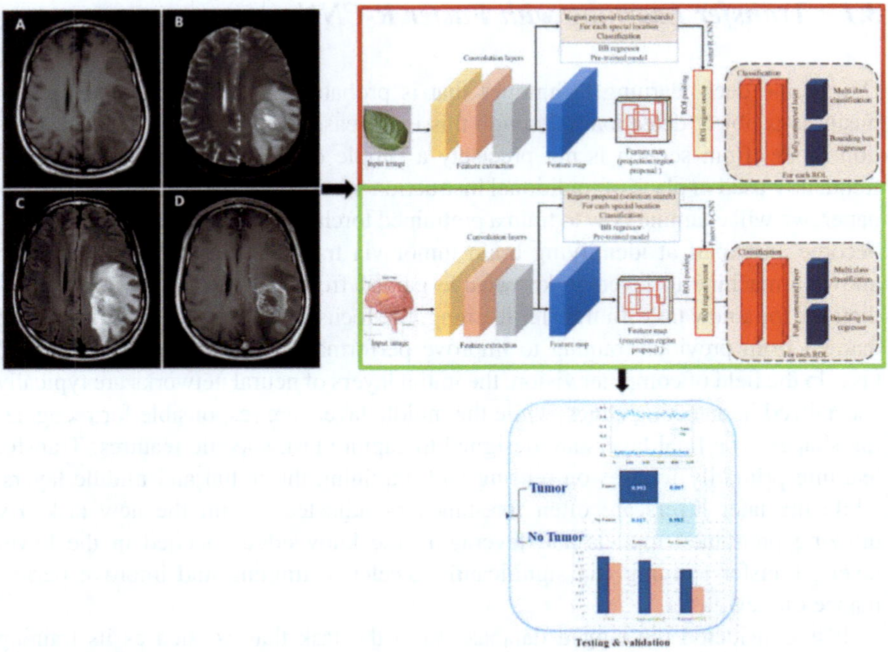

Fig. 6 Faster R-CNN with transfer learning

Fig. 7 Experimental setup

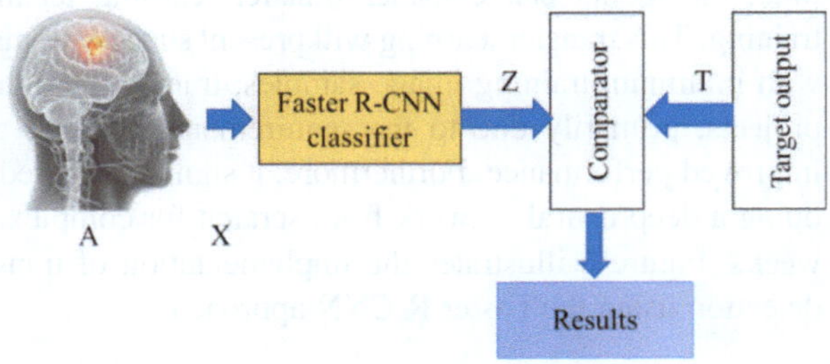

The Faster R-CNN architecture is considered in this study to categorize the type of tumor. Figure 7 depicts the experimental configuration. The input images must be the exact size because the proposed strategy is based on Faster R-CNN. As a result, the input brain tumor MRI dataset images are scaled. Because using input photos of the same size makes for incredibly fine-grained features. As a result, every image in the input brain tumor MRI dataset is shrunk. In order to enhance the caliber of the preprocessing stage, the experiments also contain a data augmentation procedure. Selected models are used to train the photographs. The Adam optimizer is used to optimize the parameters. The categorical cross-entropy is used to determine the error function. Since the suggested system uses the transfer learning technique, reuse the lightweight, freeze the base layer, and only retrain the top few layers. Only a few

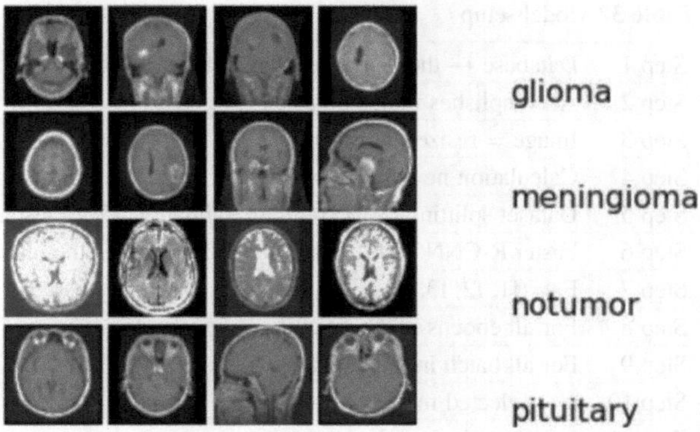

Fig. 8 Samples from the dataset

dense layers of the train remained. A Softmax activation function is used in the classification to generate the probability value. A model file is used to hold the finished trained model.

Dataset Collection The four classes in this study—no tumor, meningioma, glioma, and pituitary—are classified using a deep learning-based classifier. On the membrane, a Meningioma tumor could be seen. This type of malignancy also impacts the brain and spinal cord. The spinal cord and glial tissues are where glioma type breeds. The pituitary type gives birth in the vicinity of the pituitary gland. The dataset was gathered using resources that were open to the public. The four classes—no tumor, meningioma, glioma, and pituitary—are described on the Brain Tumor Classification (MRI) | Kaggle page. The dataset includes MRI images from the aforementioned four classes. The Fig. 8 shows a sample image from the dataset.

Preprocessing Preprocessing is done before the MRI pictures are entered into the suggested structure. Initially, during the image preprocessing stage, batches of tensor images with real-time data augmentation are produced using the ImageDataGenerator method from the Keras package. This guarantees that a variety of photos are shown to the algorithm. Furthermore, the ImageDataGenerator technique in the Keras library's preprocess_input option is set up to display an MRI image in the format that the model requires. The ImageDataGenerator method is part of the Keras package and performs many necessary preprocessing operations, including cropping, rotating, determining the brightness range, flipping, and determining the four rectangular boundary coordinates of the cropped image.

Data Modeling The research divided the dataset into a training set (80%) and a test set (20%) after standardizing the full set. The ImageDataGenerator is then used to divide the image data into training and test sets. The binary cross-entropy loss function and the Adam optimizer are employed in the creation of the Faster R-CNN model. The Adam optimizer, which combines features from RMSProp and AdaGrad, performs well when dealing with noisy sparse gradients. To reduce total loss, the binary cross-entropy loss function is used. The model learns complex patterns and reaches a peak accuracy of 93% after undergoing 150 training epochs using the Atlas basic model. For an exhaustive synopsis of the model setup, please consult the algorithm described in Table 3.

Table 3 Model setup

Step 1	Database ← the database is taken from Kaggle
Step 2	Accomplishes image preprocessing steps
Step 3	Image = resized sample
Step 4	Calculation new image = image (extleft, extright, cropping)
Step 5	Dataset splitting (80% training/ validation +20% testing
Step 6	Faster R-CNN ← É classifying and extracting the feature through deep learning model
Step 7	F = (f1, f2, f3, ………..) automatic feature map extraction
Step 8	For all epochs in the quantity of epochs do
Step 9	For all batch in the batch-size do
Step 10	x = selected model (F)
Step 11	Loss = calculate the loss value
Step 12	Optimization of the function, applied for validation/ training
Step 13	Calculate the validation metrics
Step 14	End
Step 15	End model
Step 16	Return results

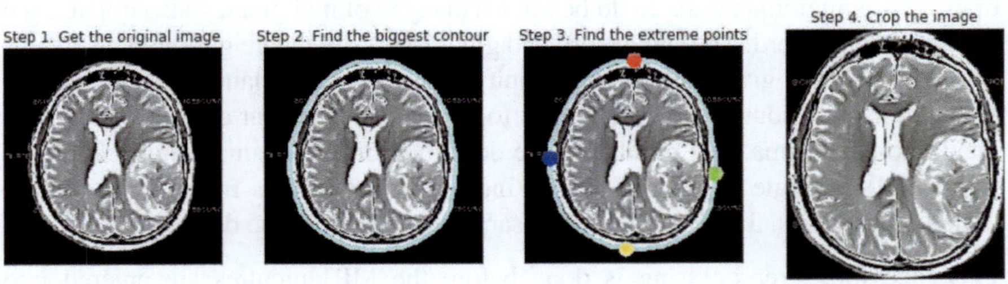

Fig. 9 Preprocessing image

5 Result and Discussion

The brain tumor database is used to train the chosen Faster R-CNN model. According to the specifications, the photographs are preprocessed and resized. Figure 9 displays the preprocessed image. Here, following contour and extreme point discovery, the photos are cropped.

The image augmentation phase, which includes artificially growing or enlarging the dataset, is applied to the preprocessed images. This method is very helpful when working with datasets that contain a small number of data samples. There is a chance of overfitting when the model is trained on a smaller sample, which can have detrimental effects. To counteract this, new data examples are produced using image augmentation techniques including zooming, rotation, cropping, and others. When these parameters are incorporated into the training of a deep learning model,

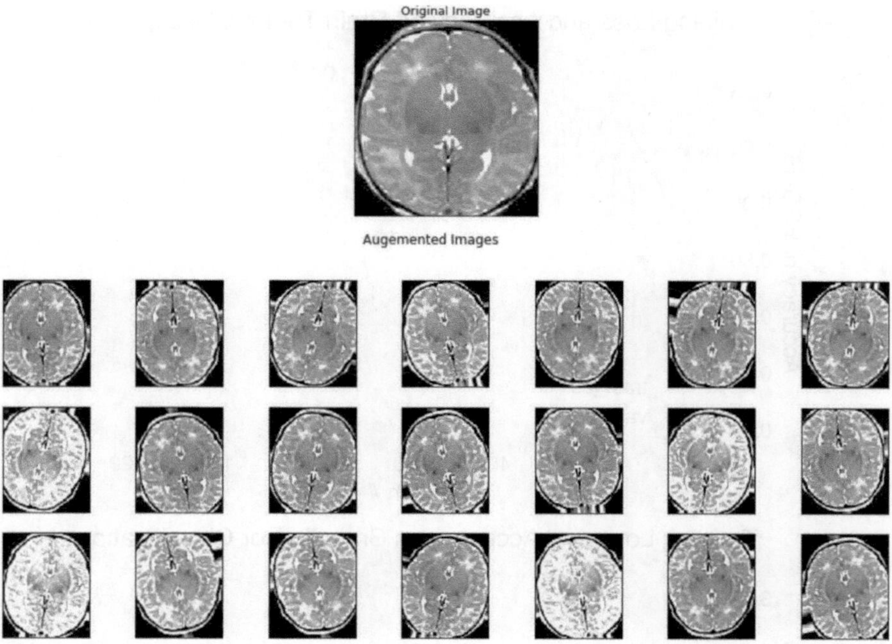

Fig. 10 Augmented image

images with these augmented characteristics are produced. In general, the use of image augmentation leads to an approximately three- to fourfold increase in the sample size of the existing data. The augmented image is illustrated in Fig. 10.

An algorithm for deep learning is trained using a training model, which is a dataset. It contains groups of related input data; the output also may vary based on these input database. To evaluate the planned outcome with sample outcome, the proposed model is trained on these selected databases. The iteration process is known as model fitting. The training and loss values of Faster R-CNN are displayed in Fig. 11.

The trained model is stored as model file. Based on the stored file, the test image is verified. First the model is classified. In the proposed model, four types of classes are there. If the test image is classified as tumor, then the location of tumor is detected. The test image classification is shown in Fig. 12. Figure 13 presents the hidden details about the images; which means that feature map. The location of brain tumor is detected in Fig. 14.

A confusion matrix is used in the assessment of the suggested Faster R-CNN model. The calculation of several performance indicators, such as accuracy rate, loss, sensitivity rate, precision value, and specificity range, is made easier by this matrix. Taken together, these measures assess how well brain tumors are classified using the generated confusion matrix. True positive (TP), false negative (FN), false positive

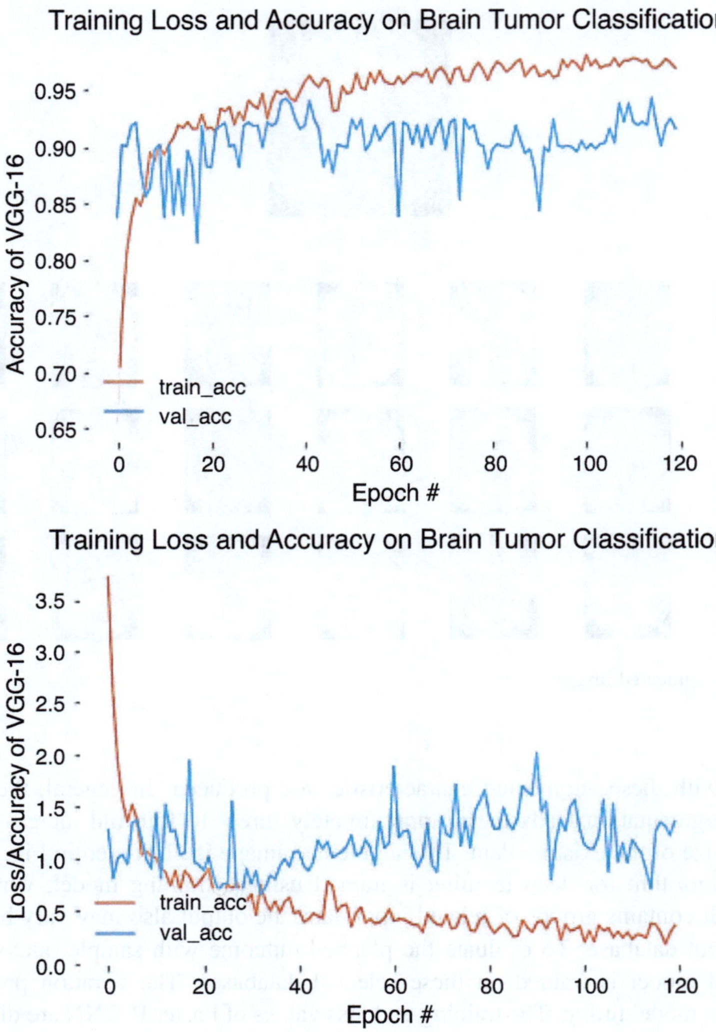

Fig. 11 Faster R-CNN model training

(FP), and true negative (TN) are the four main statistical indices used in the evaluation. The study takes into account and makes use of four significant statistical indices in order to evaluate the effectiveness of the proposed categorization system: true positive (TP), false negative (FN), false positive (FP), and true negative (TN). Tables 4 and 5 provide more information about the evaluation in relation to the other models, emphasizing criteria such as recall, F1-score, false-positive rate (FPR), false-negative rate (FNR), sensitivity, specificity, accuracy, and precision. Figures 15 and 16 offer a graphic depiction of the comparison.

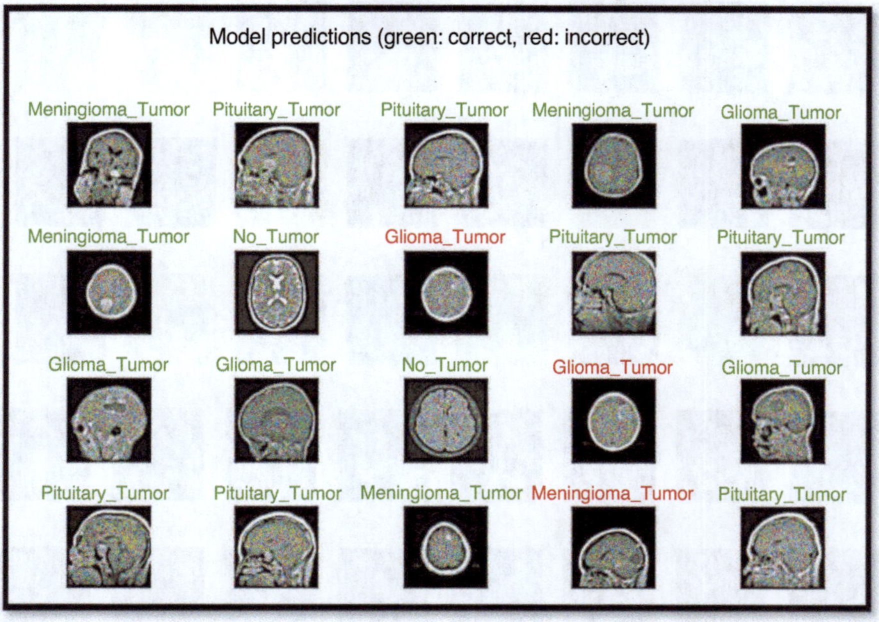

Fig. 12 Brain tumor classification

6 Conclusion

Usually, the medical-based image analysis is difficult task, since the identification of image is complex. In recent years, magnetic resonance imaging (MRI), which has uses such as the identification of brain tumors, has developed into an efficient tool for clinical research. The proposed method for locating brain tumors is quite successful. One of the main goals of the proposed research is to spot and trace the tumors automatically, which will help the neurosurgeons/ experts to identify speedily. When applied to these MRI pictures, deep learning algorithms aid in the tumor detection process. We have used the MRI image dataset to implement the Faster R-CNN approach. According to investigational conclusions, the projected model performs more accurately than other CNN. The system will undoubtedly be beneficial in the healthcare industry. It is possible to determine whether or not there is a tumor using the suggested procedure. However, the system can be improved more in the future to make therapy recommendations properly. Comparing the experimental results to state-of-the-art technologies, accuracy is 96%. In future reduction of time and system complexity, memory requirement will reduce. This same approach will be applied for eyes, liver, breast cancer identification. Similarly the research will try to increase accuracy with minimum loss.

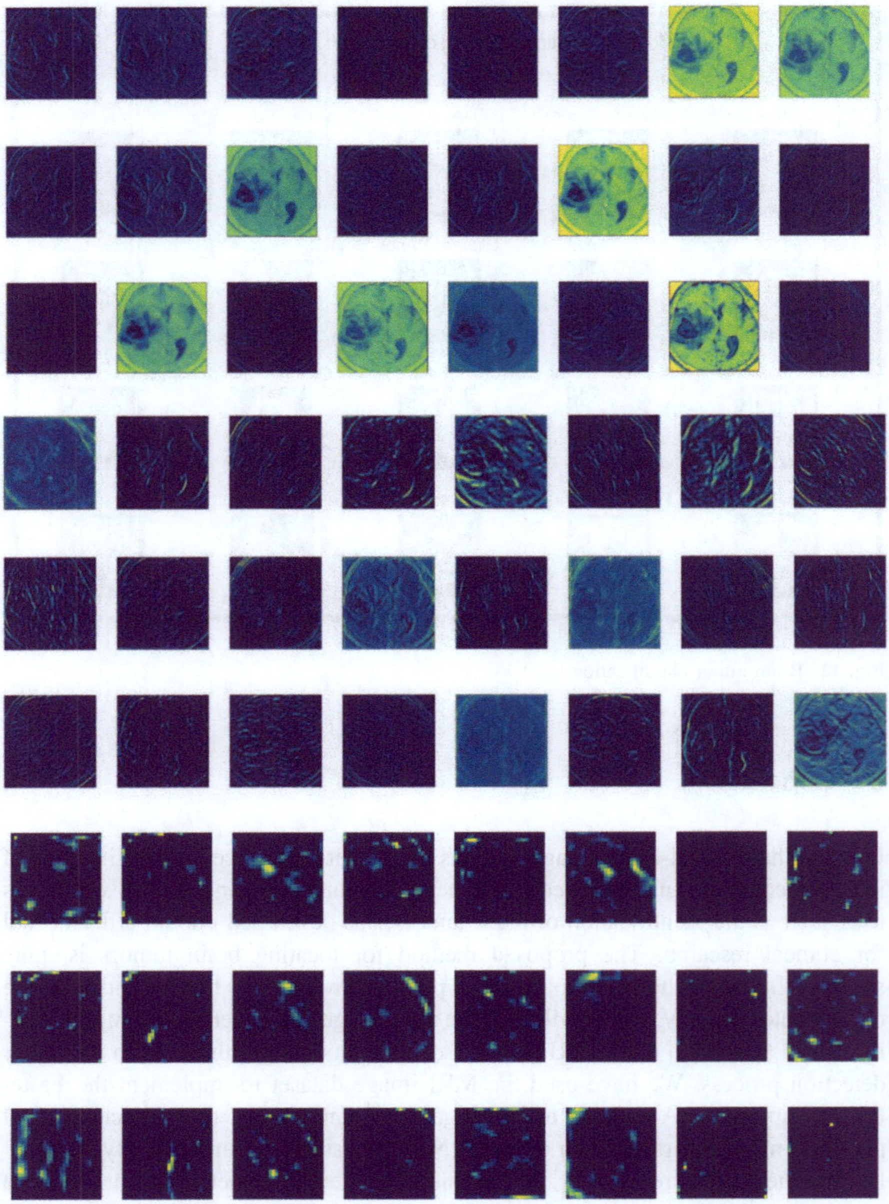

Fig. 13 Feature map visualization

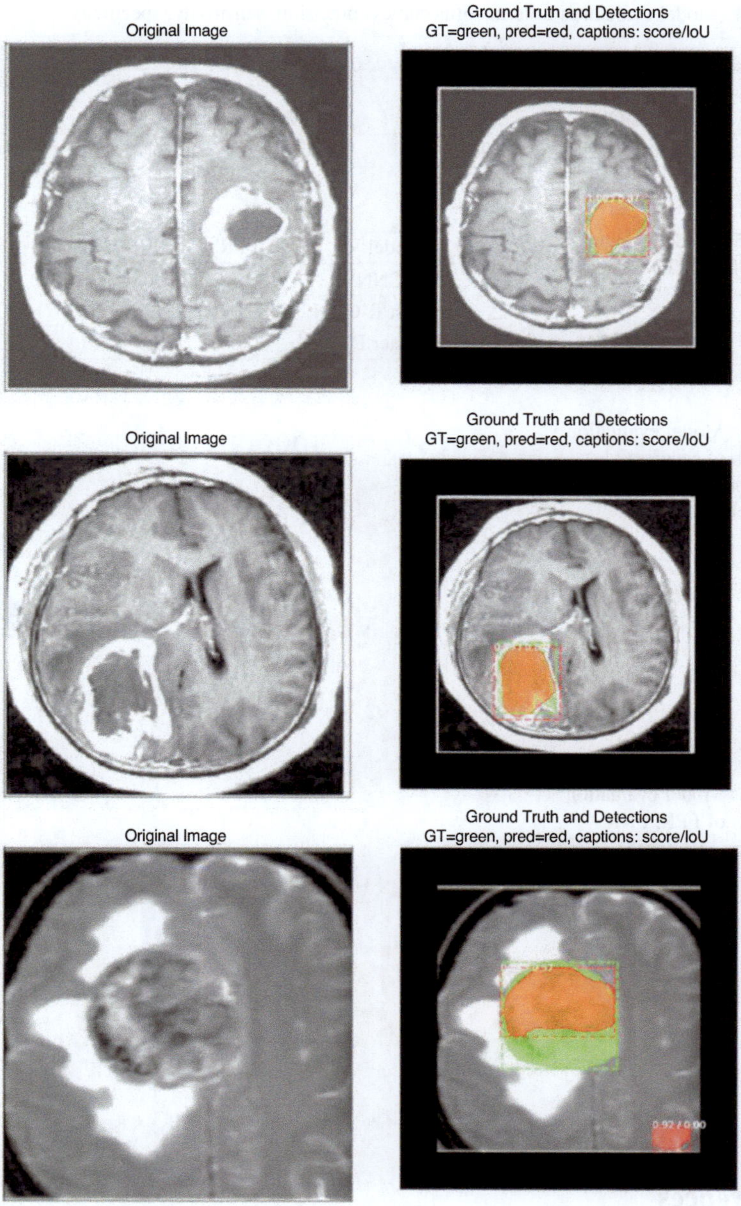

Fig. 14 Brain tumor location detection

Table 4 Model evaluation in terms of accuracy, precision, sensitivity, specificity

S.no	Model	Accuracy	Sensitivity	Specificity	Precision
1	R-CNN	89%	85.85%	88.48%	89.66%
2	Fast R-CNN	91.7%	90.7%	92.31%	92.12%
3	Faster R-CNN	96%	95.6%	96.9%	96.9%

Table 5 Model evaluation in terms of FPR, FNR, recall, and F1-score

S.no	Model	FPR	FNR	Recall	F1-score
1	R-CNN	0.16	0.06	0.86	0.81
2	Fast R-CNN	0.07	0.12	0.91	0.90
3	Faster R-CNN	0.06	0.03	1	0.97

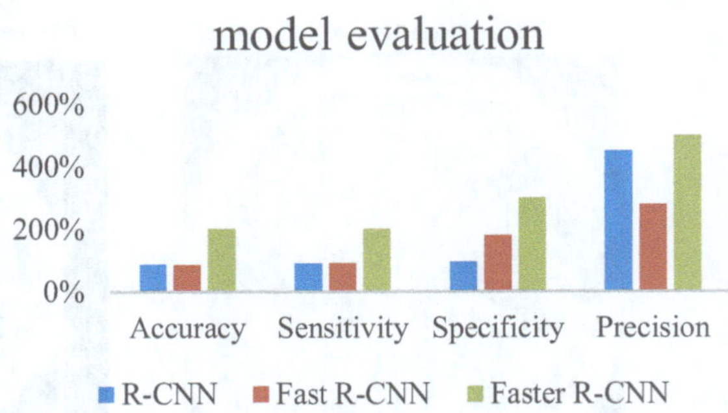

Fig. 15 Model evaluation in terms of accuracy, precision, sensitivity, specificity

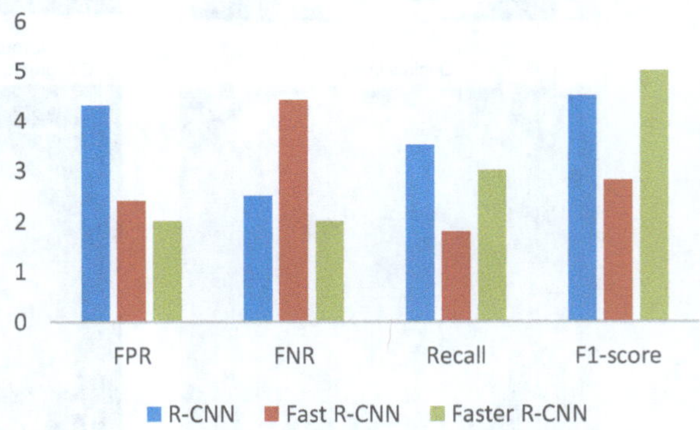

Fig. 16 Model evaluation in terms of FPR, FNR, recall, and F1-score

References

1. T. A. Soomro et al., Image Segmentation for MR Brain Tumor Detection Using Machine Learning: A Review, in IEEE Reviews in Biomedical Engineering, vol. 16, pp. 70–90, 2023, doi: https://doi.org/10.1109/RBME.2022.3185292.
2. S. Solanki, U. P. Singh, S. S. Chouhan and S. Jain, Brain Tumor Detection and Classification Using Intelligence Techniques: An Overview, in IEEE Access, vol. 11, pp. 12870–12886, 2023, doi: https://doi.org/10.1109/ACCESS.2023.3242666.

3. W. Yu, H. Kang, G. Sun, S. Liang and J. Li, Bio-Inspired Feature Selection in Brain Disease Detection via an Improved Sparrow Search Algorithm, in IEEE Transactions on Instrumentation and Measurement, vol. 72, pp. 1–15, 2023, Art no. 2500515, https://doi.org/10.1109/TIM.2022.3228003.
4. Anbumani K, Murali Dhar M S, Subramanian P, Jasmine J, Mahaveerakannan R and John Justin Thangaraj S, Analysis of Missing Health Care Data by Effective Adaptive DASO Based Naive Bayesian Model, Journal of Machine and Computing, vol.3, no.4, pp. 582–590, October 2023. https://doi.org/10.53759/7669/jmc202303049.
5. D. Hemanand, N. P. G. Bhavani, S. Ayub, M. W. Ahmad, S. Narayanan, and A. Haldorai, Multilayer vectorization to develop a deeper image feature learning model, Automatika, vol. 64, no. 2, pp. 355–364, Dec. 2022, https://doi.org/10.1080/00051144.2022.2157946.
6. Ignacio Villegas Vergara and Liza Chung Lee, A Schematic Review of Knowledge Reasoning Approaches Based on the Knowledge Graph, Journal of Enterprise and Business Intelligence, vol.3, no.3, pp. 179–189, July 2023. https://doi.org/10.53759/5181/JEBI202303018.
7. Mia Ella Grace, Cloud Computing for Decision Making within the Clinical Sector, Journal of Biomedical and Sustainable Healthcare Applications, vol.3, no.2, pp. 087–094, July 2023. https://doi.org/10.53759/0088/JBSHA202303009.
8. K. S. Umadevi, K. S. Thakare, S. Patil, R. Raut, A. K. Dwivedi, and A. Haldorai, Dynamic hidden feature space detection of noisy image set by weight binarization, Signal, Image and Video Processing, vol. 17, no. 3, pp. 761–768, Aug. 2022, https://doi.org/10.1007/s11760-022-02284-2.
9. M. Rizwan, A. Shabbir, A. R. Javed, M. Shabbir, T. Baker and D. Al-Jumeily Obe, Brain Tumor and Glioma Grade Classification Using Gaussian Convolutional Neural Network, in IEEE Access, vol. 10, pp. 29731–29740, 2022, doi: https://doi.org/10.1109/ACCESS.2022.3153108.
10. 10. A. Anaya-Isaza and L. Mera-Jiménez, Data Augmentation and Transfer Learning for Brain Tumor Detection in Magnetic Resonance Imaging, in IEEE Access, vol. 10, pp. 23217–23233, 2022, https://doi.org/10.1109/ACCESS.2022.3154061.
11. M. Sasikala and N. Kumaravel, A wavelet-based optimal texture feature set for classification of brain tumours, Journal of Medical Engineering & Technology, vol. 32, no. 3, pp. 198–205, Jan. 2008, https://doi.org/10.1080/03091900701455524.
12. S. Khawaldeh, U. Pervaiz, A. Rafiq, and R. Alkhawaldeh, Noninvasive Grading of Glioma Tumor Using Magnetic Resonance Imaging with Convolutional Neural Networks, Applied Sciences, vol. 8, no. 1, p. 27, Dec. 2017, https://doi.org/10.3390/app8010027.
13. M. Sajjad, S. Khan, K. Muhammad, W. Wu, A. Ullah, and S. W. Baik, Multi-grade brain tumor classification using deep CNN with extensive data augmentation, Journal of Computational Science, vol. 30, pp. 174–182, Jan. 2019, https://doi.org/10.1016/j.jocs.2018.12.003.
14. Fatemeh Derikvand, Hassan Khotanlou, Brain Tumor Segmentation in Magnetic resonance imaging Images Using a Hybrid Deep Network Based on Patch and Pixel, 2020 International Conference on Machine Vision and Image Processing (MVIP), ISBN: 978-1-7281-6832-6.
15. T. Hossain, F. S. Shishir, M. Ashraf, M. A. Al Nasim, and F. Muhammad Shah, Brain Tumor Detection Using Convolutional Neural Network, 2019 1st International Conference on Advances in Science, Engineering and Robotics Technology (ICASERT), May 2019, https://doi.org/10.1109/icasert.2019.8934561.
16. J. Amin, M. Sharif, M. Yasmin, and S. L. Fernandes, A distinctive approach in brain tumor detection and classification using MRI, Pattern Recognition Letters, vol. 139, pp. 118–127, Nov. 2020, https://doi.org/10.1016/j.patrec.2017.10.036.
17. M. Chen, Q. Yan, and M. Qin, A segmentation of brain MRI images utilizing intensity and contextual information by Markov random field, Computer Assisted Surgery, vol. 22, no. sup1, pp. 200–211, Oct. 2017, https://doi.org/10.1080/24699322.2017.1389398.
18. P. Korfiatis, T. L. Kline, and B. J. Erickson, Automated Segmentation of Hyperintense Regions in FLAIR MRI Using Deep Learning, Tomography, vol. 2, no. 4, pp. 334–340, Dec. 2016, https://doi.org/10.18383/j.tom.2016.00166.

19. D. R. Nayak, R. Dash, B. Majhi, R. B. Pachori, and Y. Zhang, A deep stacked random vector functional link network autoencoder for diagnosis of brain abnormalities and breast cancer, Biomedical Signal Processing and Control, vol. 58, p. 101860, Apr. 2020, https://doi.org/10.1016/j.bspc.2020.101860.
20. J. Amin et al., Brain Tumor Detection by Using Stacked Autoencoders in Deep Learning, Journal of Medical Systems, vol. 44, no. 2, Dec. 2019, https://doi.org/10.1007/s10916-019-1483-2.
21. S. K. Mishra and V. H. Deepthi, RETRACTED ARTICLE: Brain image classification by the combination of different wavelet transforms and support vector machine classification, Journal of Ambient Intelligence and Humanized Computing, vol. 12, no. 6, pp. 6741–6749, Jul. 2020, https://doi.org/10.1007/s12652-020-02299-y.
22. P. M. Siva Raja and A. V. rani, Brain tumor classification using a hybrid deep autoencoder with Bayesian fuzzy clustering-based segmentation approach, Biocybernetics and Biomedical Engineering, vol. 40, no. 1, pp. 440–453, Jan. 2020, https://doi.org/10.1016/j.bbe.2020.01.006.
23. M. Odusami, R. Maskeliūnas, R. Damaševičius, and T. Krilavičius, Analysis of Features of Alzheimer's Disease: Detection of Early Stage from Functional Brain Changes in Magnetic Resonance Images Using a Finetuned ResNet18 Network, Diagnostics, vol. 11, no. 6, p. 1071, Jun. 2021, https://doi.org/10.3390/diagnostics11061071.
24. N. Arunkumar, M. A. Mohammed, S. A. Mostafa, D. A. Ibrahim, J. J. P. C. Rodrigues, and V. H. C. Albuquerque, Fully automatic model-based segmentation and classification approach for MRI brain tumor using artificial neural networks, Concurrency and Computation: Practice and Experience, vol. 32, no. 1, Oct. 2018, https://doi.org/10.1002/cpe.4962.
25. Z. Huang et al., A computer-aided diagnosis system for brain magnetic resonance imaging images using a novel differential feature neural network, Computers in Biology and Medicine, vol. 121, p. 103818, Jun. 2020, https://doi.org/10.1016/j.compbiomed.2020.103818.
26. C. Han et al., MADGAN: unsupervised medical anomaly detection GAN using multiple adjacent brain MRI slice reconstruction, BMC Bioinformatics, vol. 22, no. S2, Apr. 2021, https://doi.org/10.1186/s12859-020-03936-1.
27. M. F. Alanazi et al., Brain Tumor/Mass Classification Framework Using Magnetic-Resonance-Imaging-Based Isolated and Developed Transfer Deep-Learning Model, Sensors, vol. 22, no. 1, p. 372, Jan. 2022, https://doi.org/10.3390/s22010372.
28. E. M. Senan, M. E. Jadhav, T. H. Rassem, A. S. Aljaloud, B. A. Mohammed, and Z. G. Al-Mekhlafi, Early Diagnosis of Brain Tumour MRI Images Using Hybrid Techniques between Deep and Machine Learning, Computational and Mathematical Methods in Medicine, vol. 2022, pp. 1–17, May 2022, https://doi.org/10.1155/2022/8330833.
29. Z. U. Rehman, M. S. Zia, G. R. Bojja, M. Yaqub, F. Jinchao, and K. Arshid, Texture based localization of a brain tumor from MR-images by using a machine learning approach, Medical Hypotheses, vol. 141, p. 109705, Aug. 2020, https://doi.org/10.1016/j.mehy.2020.109705.
30. D. M. Kumar, D. Satyanarayana, and M. N. G. Prasad, MRI brain tumor detection using optimal possibilistic fuzzy C-means clustering algorithm and adaptive k-nearest neighbor classifier, Journal of Ambient Intelligence and Humanized Computing, vol. 12, no. 2, pp. 2867–2880, Sep. 2020, https://doi.org/10.1007/s12652-020-02444-7.
31. H. H. N. Alrashedy, A. F. Almansour, D. M. Ibrahim, and M. A. A. Hammoudeh, BrainGAN: Brain MRI Image Generation and Classification Framework Using GAN Architectures and CNN Models, Sensors, vol. 22, no. 11, p. 4297, Jun. 2022, https://doi.org/10.3390/s22114297.
32. J. Amin, M. A. Anjum, M. Sharif, S. Jabeen, S. Kadry, and P. Moreno Ger, A New Model for Brain Tumor Detection Using Ensemble Transfer Learning and Quantum Variational Classifier, Computational Intelligence and Neuroscience, vol. 2022, pp. 1–13, Apr. 2022, https://doi.org/10.1155/2022/3236305.
33. R. A. Zeineldin et al., Explainability of deep neural networks for MRI analysis of brain tumors, International Journal of Computer Assisted Radiology and Surgery, vol. 17, no. 9, pp. 1673–1683, Apr. 2022, https://doi.org/10.1007/s11548-022-02619-x.

34. S. A. Qureshi et al., Intelligent Ultra-Light Deep Learning Model for Multi-Class Brain Tumor Detection, Applied Sciences, vol. 12, no. 8, p. 3715, Apr. 2022, https://doi.org/10.3390/app12083715.
35. H. Alsaif et al., A Novel Data Augmentation-Based Brain Tumor Detection Using Convolutional Neural Network, Applied Sciences, vol. 12, no. 8, p. 3773, Apr. 2022, https://doi.org/10.3390/app12083773.
36. C. Han et al., Infinite Brain MR Images: PGGAN-Based Data Augmentation for Tumor Detection, Smart Innovation, Systems and Technologies, pp. 291–303, Sep. 2019, https://doi.org/10.1007/978-981-13-8950-4_27.
37. N. Ghassemi, A. Shoeibi, and M. Rouhani, Deep neural network with generative adversarial networks pre-training for brain tumor classification based on MR images, Biomedical Signal Processing and Control, vol. 57, p. 101678, Mar. 2020, https://doi.org/10.1016/j.bspc.2019.101678.
38. M. Mondal, F. Faruk, N. Raihan, and P. Ahammed, Deep Transfer Learning Based Multi-Class Brain Tumors Classification Using MRI Images, 2021 3rd International Conference on Electrical & Electronic Engineering (ICEEE), Dec. 2021, https://doi.org/10.1109/iceee54059.2021.9719003.
39. Y. Li, S. Zhang and W. -Q. Wang, A Lightweight Faster R-CNN for Ship Detection in SAR Images, in IEEE Geoscience and Remote Sensing Letters, vol. 19, pp. 1–5, 2022, Art no. 4006105, https://doi.org/10.1109/LGRS.2020.3038901.
40. R. Gonzales-Martínez, J. Machacuay, P. Rotta and C. Chinguel, Hyperparameters Tuning of Faster R-CNN Deep Learning Transfer for Persistent Object Detection in Radar Images, in IEEE Latin America Transactions, vol. 20, no. 4, pp. 677–685, April 2022, https://doi.org/10.1109/TLA.2022.9675474.

Detecting Alzheimer's Disease Using Deep Learning Framework for Medial IoT Application

1 Introduction

Over the last few years, the research and development of wearable technologies have tremendously expanded. Wearable devices define a device which humans can wear in the form of wristwatches, chest straps, glasses, rings, and prosthetic sockets. Wearable technology belongs to the Internet of Medical Things (IoMT). The wearable devices have the property of implantable, ambient, and stationary machines that can be readily available at hospitals. That wearable device can communicate remotely through an internet connection, as shown in Fig. 1. The wearable devices consist of a sensor that can detect human signals such as temperature, optical, and biometric signals [1]. Sensors in IoMT devices and human interaction between these sensors are considered massive data sources. From these significant sources of data, features can be extracted for machine learning algorithms to detect and learn new patterns. Integrating IoMT and machine learning will benefit healthcare and elderly care applications such as fall detection, health state assessment, stress detection, fitness tracking, and disease diagnosis. Study from human signals by using machine learning algorithms and stored the same in wearable devices for the effective analysis of a patient's body is an active research area since the last decades [2]. Even though there may be many wearable devices in the market, there is always an expectation to have wearable technology with machine learning techniques in the market. For example, an Apple watch sends notifications about irregular heart rhythms. That product was approved by U.S Food and Drug Administration (FDA) with lots of warnings and precautions [3]. At present, this product is no longer available in the market. Hence, practical and reliable use of machine learning techniques-based wearable devices still faces many challenges.

In this work, the accurate detection and diagnosis of neurodegenerative disease based on deep learning has been performed. The disease called neurodegenerative disorder is commonly found among older adults. This disorder occurs when

© European Alliance for Innovation 2024

A. Haldorai et al., *Artificial Intelligence for Sustainable Development*,
EAI/Springer Innovations in Communication and Computing,
https://doi.org/10.1007/978-3-031-53972-5_5

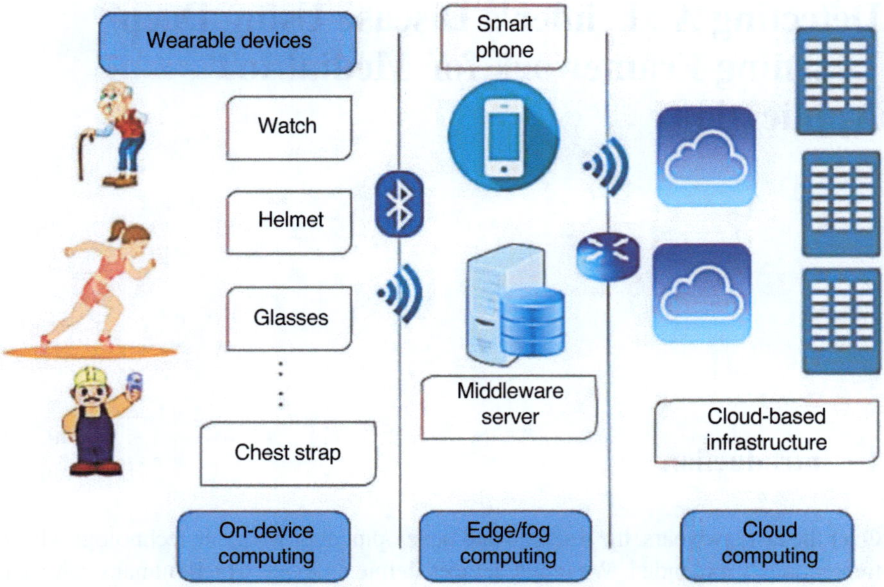

Fig. 1 Various types of wearable devices

progressively deterioration of the central nervous system. Although the clinical procedure is there for diagnosing the disorder, as the patients needed help remembering the symptoms, it would be hard to analyze at practice. Therefore, there is an unmet need for more thorough and ongoing symptom monitoring in NDs. New technical solutions are needed to address the issue for computerized diagnosis and persistent illness progression tracking [4]. Wearable technology has become a breakthrough method of healthcare as a result. Wearable technology provides a more personalized method for diagnosing and disease management. There are several uses for this wearable technology for neurodegenerative disorders, including better diagnosis, affordable, noninvasive diagnostic tools, disease progression monitoring, and continuous illness management. Furthermore, the development of wearable technology recently makes it possible to collect high-dimensional data from numerous fields while performing daily tasks [5]. Hence, wearable devices can support deep learning algorithms for analyzing large quantities of data and accurately detecting disease manifestations.

Accurately diagnosing any disease depends on continuous monitoring, objective measurement, and remote monitoring. All the objectives mentioned above can be satisfied using wearable technology with a deep learning algorithm. Hence, the proposed wearable devices will be able to predict the disease earlier and more accurate detection, thereby enhancing the outcome of the product. The wearable technology may also monitor several other indicators, including brain activity, blood pressure, mobility, and sleep patterns. Additionally, the technology aids in diagnosing NDs and offers insights into cognitive function. Furthermore, the data collected from wearable devices is used to implement machine learning algorithms to study the patterns and develop predictive models. These might assist professionals in providing care and therapy [6]. In conclusion, wearable technology has excellent potential for NDs since it allows for high-dimensional data DL analysis while offering continuous and objective monitoring. Therefore, wearable technology is projected to become more crucial in diagnosing and managing NDs as it develops.

We worked specifically on Alzheimer's disorders and computed the results of our suggested strategy. It is a dementia-causing factor that includes memory loss, loss of spatial orientation, loss of timing, and behavioral problems. Acute stages of Alzheimer's disease have retrograde amnesia and modest cognitive impairment. Alzheimer disorders are brought on by distinctive aggregates called neurofibrillary tangles that are seen in patient's brains [7, 8]. The entanglements mentioned above may impact how quickly brain networks deteriorate, resulting in the gradual loss of physical functions. Thus, Alzheimer disease leads to mental and physical disability [9]. Therefore, it affects families and caregivers.

Current developments in healthcare have improved life expectancy, increasing the aged population. Hence, the probability of susceptibility to Alzheimer disease also has increased. Therefore, computerized techniques are based on various technologies such as neural networks, optimization, and machine learning to solve medical field issues. Some researchers have begun to initiate sensors using machine learning algorithms [10]. Field experts can manually extract and label features in the existing machine learning algorithms. Most notably, in the computer vision domain, an advanced machine learning algorithm called deep learning can perform more than a machine learning algorithm to detect inclusive structures in complex and high-dimensional data. The main advantage of using deep understanding is that it can learn high-dimensional properties from their own brain imaging data. Thus, deep learning eliminates the need for domain expertise. Furthermore, deep learning outperforms machine learning in handling heavy data volume with precision care, while machine learning requires a specific processing step.

The main contribution of this paper is listed below:

- The paper presented an Internet Of Things-based layered architecture. Particularly this paper considers Triboelectric Nanogenerator (TENG) Sensor.
- The Inception V3-based AD detection technique is introduced.
- The abnormality of the AD patient is consciously tracked with IoT mechanism.

The Sect. 1 presents the introduction about needs of IoT and details about AD. The existing works are discussed in Sect. 2. The model architecture is clearly evaporated in Sect. 3. This section presents the architecture details and TENG sensor. Similarly, the Sects. 4 and 5 explain about the materials and methods, and results respectively. The conclusion is given in Sect. 6.

2 Literature Review

Alzheimer means mental loss. Memory-related issues are main symptoms of this disease. The many researchers introduced many techniques for AD prediction. Some of the techniques are discussed in this section. S. N. Siri et al. [11] introduced a IoT-based AD prediction device. The proposed work is combined with mobile application for speedy tracking. This is one of the successful solutions in terms of cost, portable, and efficiency. S. Kamakshi et al. [12] suggested early prediction and analysis of the various brain diseases such as Alzheimer and Mild Cognitive Impairment. Initially the authors consider SVM for training, then they compare this SVM-based model with CNN-based trained model. It analyzes the models for various age groups.

N. V. K, R. Ch et al. [13] planned a smart wearable device for Alzheimer patients, women, and children. They introduced the device with location tracker. GPS was considered for location tracking in this work. Twilio API was used for message sharing. This is a type of smart watch, which is low cost and portable. A. Yalamanchili et al. [14] came up with new idea for Alzheimer disease prediction and tracking. The RCNN-PSO technique was utilized by the author for better performance. Always, IoT is the best option for elder people, patients, women, and child tracking. V. Patil and B. Chikkoppa [15] used this IoT-based tracking for AD patients. The tracking system is combined with CNN model. Pavithra T et al. [16] designed a tracking system for elderly people. This tracking system measures the health parameters like heart rate, temperature, and pressure. Another one AD tracking system is introduced by the R. Chokri et al. [17]. The psychological status is predicted by this proposed system. Also this model predicts the stages of AD. The Mini Mental State Examination of AD patients was tested by machine learning-based models in the paper B. V. Chowdary et al. [18]. S. Patil et al. [19] track the AD patients by the smart specs. The smart specs was designed by the fusion of IoT and Artificial intelligence technology. B. Singh et al. [20] used a novel fusion model with Deep Learning, Blockchain, and IoT for AD patient tracking. The model used the deep learning model for training purpose. The conditions, health parameters, and stages are tracked successfully with this model.

In our proposed model, we are using TENG sensor. Recently, the requirement for a self-driven, reliable, and less costly health management system has become more inevitable than before. Hence, the nanogenerator paves a new path toward biomedical applications through its nanostructured and eco-friendly materials. The nanogenerator works with the concept of harvesting biomechanical energy. In this overview, many TENG applications, including wearable, small-scale, and straightforward fixes that use human motion as their power source, have been compiled. The triboelectric nanogenerator was devised in 2012 [21]. Since then, TENG got attracted researchers for numerous research ranges. A TENG device transforms mechanical energy into electrical energy using a combination of the triboelectric effect and electrostatic induction. Moreover, TENG is a self-powered healthcare device that obtains the required power from various human body and organ motions

[22–24]. The primary characteristics of TENG are induced to use in healthcare applications. They are (1) For the sensors or equipment used for external health monitoring, TENG is used as an energy collecting source [25]. (2) It can function as an autonomous sensor that does not require any external power sources [26] (3) and autonomous-powered electrical simulators to diagnose a human condition [27]. Initially, the TENGs are used for filtering purposes.

Next, the TENG technology [28] will be used for water disinfectors. Water is a vital nutrient for human survival because it makes up 70% of the human body [29]. According to 2017 WHO reports, 2.2 billion people could not get clean water for drinking [31]. In addition, many water-borne diseases such as cholera, typhoid, and many others will affect people and children under age 14 due to unclean water [30]. For that, a researcher called Ding et al. developed an affordable and one-step disinfection model using TENG. The water disinfector has been used in a hand-powered Trio Pump system composed of three parts. A disc TENG, a coaxial mechanical structure that houses the water pump, and a tubular coaxial-electrode copper ionization cell are all present. With D-TENG as the power supply, the coaxial-electrode ionization cell functions as a disinfection device. This manually controlled equipment is incredibly inexpensive and has a two-year lifespan. As a result, this water disinfection device is best suited for rural areas without access to electricity and safe drinking water.

Then, subsequently, the TENGs are used as a healthcare monitoring sensor for early diagnosis of a medical condition. Early diagnosis is always beneficial for treating disorders before they cause irreparable harm. TENG has developed several noninvasive and wearable techniques to study the elements in real time. Applying TENG in the medical field included heart, breathing, and bodily movements. A common reason for old injuries that result in impairment and a poor quality of life is falling or slipping. So, with that in mind, a self-powered and reasonably priced fall detection system is designed using triboelectric technology. Jeon et al. [32] devised a working model consisting of electrically connected 15-cell arrays of TENGs. When a cell is pressed, electric impulses are produced and categorized as falls or not falls by a processor. The positioning of the body fall has been cleverly covered by a 15-cell linked TENG array. Tests on the fall detection system's ability to distinguish between falls and routine activities produced a high accuracy of $>95\%$. To facilitate aging management centers, such technology may be immediately used in smart homes and hospitals.

Because of our changing lifestyle, the sleeping quality could be much better. Poor sleeping habits cause depression, obesity, cardiovascular, pulmonary, and diabetes disorders. A self-powered bedsheet based on a TENG array was disclosed by Lin et al. [33] for monitoring sleep. Three layers comprise the sleeping sheet: two Ag-coated conductive layers sandwiched between a thin layer of PET with waves. While the conductive fibers have positive triboelectric characteristics, the PET layer has negative ones. The smart bedsheet generates electrical signals when pressed and released by outside pressure, enabling it to detect sleeping patterns (0.77 V ⟦Pa⟧^(−1)). The smart bedsheet will be able to state the body position, postures, and pressure distribution and also show the sleep patterns. And it provides a warning when emergency situations, like falling out of bed, occur. This robust, machine-washable, simple, and tightly integrated innovative textile-based sleep monitoring system is practically appropriate for mass manufacturing. Numerous research areas, including motion tracking, wireless remote healthcare systems, and tactile sensing, are expected to be impacted.

3 Model Design

The proposed model includes five layers for AD tracking. They are user layer, IoT/sensor layer, Fog layer, core network layer, and cloud layer. The user layer includes the users with Alzheimer patients. The next IoT/ sensor layer consists of sensors to collect the information generated by the Alzheimer patients.

The different types of sensors are available in the market such as emotion, voice, and acceleration sensors. In this work, TENG-based sensors are considered for data collection. The next layer is Fog. In this, the collected information is analyzed by the various nodes and servers. The next layer transmits the data from the Fog layer to the cloud layer. Finally the information is stored in cloud layer. The model design is displayed in Fig. 2.

3.1 Triboelectric Nanogenerator (TENG) Sensor

As technology advances are being developed, the Internet of Things (IoT) field can integrate sensors beyond 200 billion by 2025. Thus, the demand for sustainable energy is also highly desired for many medical applications, because charge storage devices do not need to be recharged or replaced with sustainable energy sources. Due to its many qualities, the triboelectric nanogenerator has become more popular than other energy harvesters. TENGs have several outstanding attributes, including

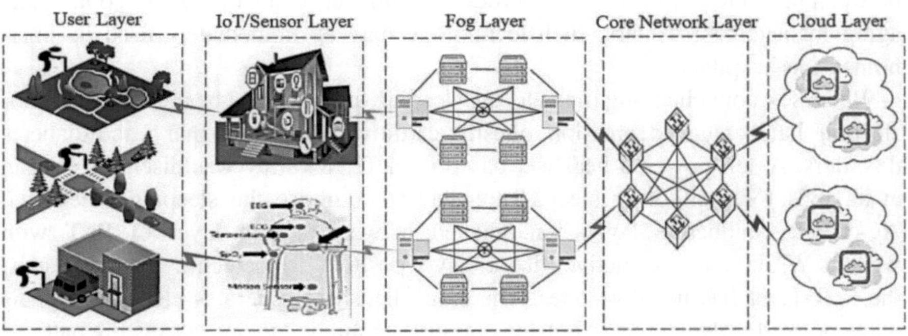

Fig. 2 Model design for AD tracking

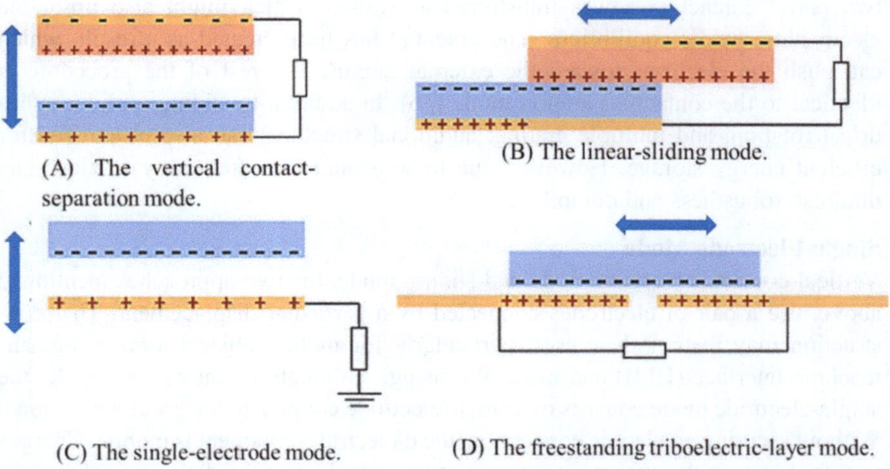

Fig. 3 Different modes of operation principles of TEN. (**a**) The vertical contact-separation mode. (**b**) The linear-sliding mode. (**c**) The single-electrode mode. (**d**) The freestanding triboelectric-layer mode

greater instantaneous output power, the need for various readily available materials, environmental friendliness, the lowest construction method, and many customized working models for target applications. TENGs are powered by discarded mechanical energy from the surrounding environment. Electrical energy can be produced from mechanical energy [34]. Additionally, TENGs come in two different configurations: single-electrode and double-electrode structures. These two configurations give rise to three operating modes: contact separation, sliding, and freestanding. The Triboelectric Nanogenerator works in different modes, which is displayed in Fig. 3.

Contact-Separation Mode: TENG's most fundamental operating mode is the vertical contact-separation mode, as shown in Fig. 2a. It induced opposing charges on the surfaces of two materials with electron affinity when they come into contact. Suppose both the materials got separated at a given space utilizing applied external force, an equal amount of potential will be developed at the interfacial region. Their electrodes receive the created potential, which drives the electrons to the external circuit to maintain the electrostatic field's equilibrium. As soon as the two materials came into touch, the triboelectric potential disappeared. As a result, the electrons generated fed back. The alternating output can be found under the periodically applied force if the external load is applied. This model is relatively straightforward to build and has a straightforward structure [35]. Hence simple to manufacture, but the need for a volume to account for the fluctuating gap always creates a packaging challenge.

Sliding Mode

The vertical contact-separation connection between the two material sections in this mode can be converted to lateral relative sliding, as depicted in Fig. 2b. Thus, the two parts' contact area was transferred as sliding, which might also upset the electrostatic field's equilibrium. The potential has been created as a result, which can push the electrons toward the external circuit. The rest of the procedure is identical to the contact-separation mode [36]. In addition to sliding motion, cylindrical rotation, and multiple grating, additional structures can also be utilized for efficient energy storage. However, due to wear and tear, frequency friction may diminish robustness and durability.

Single-Electrode Mode

Vertical contact-separation mode and sliding mode, the two approaches mentioned above, use a pair of electrodes connected by a particular displacement. That construction may restrict their uses, particularly for mobile objects used in human-machine interface (HMI) and medical sensing. Although, as shown in Fig. 2c, the single-electrode mode consists of a single electrode coupled to the ground via a load. Without securing an electric connection, the dielectric component is mobile. Charges transfer between the stationary electrode and ground when the dielectric component periodically moves vertically or laterally toward and away from the electrode. Although this mode has greater flexibility, the output performance could be better [37].

Freestanding Triboelectric Layer Mode

The two electrodes are stable in this mode, and a ground-free connection has been established. Hence, the Freestanding mode is depicted in Fig. 2d. Using a load, the two symmetric electrodes are connected. An asymmetric charge is produced through induction when the dielectric component moves between two electrodes. Through an external load, this leads to a charge distribution between two electrodes. Based on its characteristics, this mode was created as a grating structure and is always used for sliding motions. These four modes are sensibly selected and incorporated into sophisticated designs for excellent performance in particular applications. Triboelectrical Nanogenerator sensors have so far been employed to improve human health. The use of TENG as a minimally invasive device for various medical diseases is now established. Among the advancements are heart disorders that can be treated with electrical stimulation, neural tissue engineering, nerve simulators, hearing loss repair, and skin wound healing. Electrical stimulation therapy is typically referred to as a treatment strategy that can treat any medical condition by converting the organs' biomechanical energy into electrical energy. The necessary electric fields for cellular and tissue activity are listed in Table 1.

Triboelectrical nanogenerators have been used as piezoelectric materials in the medical field, as has been explained up until now. However, human-machine interfaces (HMIs), crucial for effective interaction between humans and the digital world in the 5G and Internet of Things (IoT) age, have experienced a boom in recent years. Today, HMI can be created in many ways, including as a glove, pair of glasses, touchpad, exoskeleton, electronic skin, and more. Nearly all HMIs require hands to operate; for instance, a computer keyboard takes ten fingers to tap, and a mouse requires hands to click or move the mouse pointer, because hands have the

Table 1 Summary of electric field and cellular characteristics

Cells and tissues	Required electrical fields	References
Movement of nerve cells	10 Mv/mm	[25]
Movement of embryonic cells	153 Mv/mm	[26]
Movement of neural crest cell	160 Mv/mm	[27]
Passage of human keratinocytes	10–100 Mv/mm	[28]
Wound healing	45–180 Mv/mm	[29]
Cultivation of human bone marrow mesenchymal stem cells	100–600 Mv/mm	[30]
Proliferation of osteoblastic cells	20 mv/cm	[31]
Proliferation, migration, and differentiation of muscle precursor cells	Several to tens of Na/cm^3	[32]
Cardiac adipose tissue-derived progenitor cells	7 Mv/mm	[23]
Muscle stimulation	Ma-level current	[24]

highest proportion of nerve regions compared to other body parts. Because hands and fingers have large sensory and motor nerve areas concealed [38], the bending and stretching of the fingers are flexible. The quiet complex postures have been made by the degree of flexibility of the fingers. The design of HMI in the form of a glove is considered a standard way of the wearable device. Initially, the HMI glove device follows the mature sensing techniques, which consist of accelerometers and gyroscopes. However, this method exhibits more power consumption due to the rigid sensing elements and complex data format. To overcome that, resistive optical fiber-based sensor has been used. But it has its issues such as temperature effect, limited sensing range. The subsequent version of low-power data gloves can therefore be designed using the upcoming triboelectric sensing technology, which has the advantages of a wide range of material selection, a straightforward design, and self-powered properties. The work on [39] describes how the researcher created the most straightforward glove-based HMI using the TENG textile sensor. Here, there were two different sensor sets in use. The index finger may be precisely identified from any angle using these two sensor setups.

Additionally, this glove-based HMI may mimic the functionality of a mouse for web browsing and writing the alphabet, offering a more straightforward, power-friendly interaction manner in daily life. However, the humidity of the environment can occasionally cause the HMI gloves to perspire, which could impact the triboelectric output. Therefore, the TENG-based glove coated with thermoplastic elastomer described in [40] was developed to address the issue. This particular brand of gloves features anti-sweat properties. Machine learning technology allowed for the minimal sweat effect realization of the complicated gesture detection function, which was effectively applied to real-time VR/AR control applications such as baseball pitching, shooting games, and floral design. The TENG-based HMI gloves discussed above included arch-shaped strain sensors with a significant air gap between two contact layers. The result will be a smaller sensing area. TENG strain sensors with unique yarn structures were developed by Zhou et al. [41] to address this issue. The sensing efficiency for this technology will be between 20 and 90 percent.

Due to their high output, lightweight, and increased flexibility properties, alternative structural designs of HMIs were available in addition to glove-based HMIs. Besides eye motion, additional facial or body motions are acceptable possibilities for translation into control signals as HMI communication aids. The rapid development of innovative electronic devices makes it possible to create a smart home environment with intelligent monitoring and communication systems for monitoring healthcare, detecting falls, recognizing body motions, automating tasks, and ensuring security. Humidity/temperature sensors and camera-based image recognition have been used in intelligent home monitoring. However, this current methodology has rigidity, large size, high-power consumption, and video privacy issues. Therefore, alternative HMI technology with maximum suitability, low price, and autonomous-powered sensing ability is needed for the innovative home environment. Nowadays, the world has become more industrialized. Hence, people are under much pressure due to the unbalanced nature between work and life. It may lead to sleeping disorders, which further enhances other health-related issues such as weight loss, cardiopathy and diabetes. To ensure long-term sleep, many electronic devices have been developed. That is exhibited low sensitivity, high fabrication cost, and nonwashability, which are restricted for practical use. Therefore, TENG-based textile technology provides less cost, high sensing capability smart bed sheet for monitoring sleeping disorders.

For complex data analysis, an ordinary sensor only works efficiently regarding functionalities. But with the help of machine learning, the advancement in feature extraction opens new research direction for realizing diversified applications such as gesture/pose estimation, voice recognition, and object recognition for advanced human-machine interactions. Currently, several AI-enabled TENG-based HMIs have been created employing SVM, NN, and other techniques. The intricate characteristics of the triboelectric waveform, such as contact sequence and impact vibration, are buried in these AI-based HMIs. This technique uses an intelligent sensory system to offer precise recognition. In contrast to traditional recognition techniques, which rely on many resistive/capacitive sensor nodes for a high-precision recognition rate. The effectiveness of cybersecurity has been enhanced using keystroke dynamic-enabled authentication systems nowadays. In [24], a TENG-based keystroke dynamic monitoring method has been developed. The TENG-based keystroke system consists of 16 silicon keys. The keys are responsible for making contact-separation mode to convert motion signals into electrical signals. With a high accuracy rate of 98.7%, this method demonstrates the viability of a keystroke dynamics-enabled authentication system for actual use. Gait analysis-based detection is also a promising biometric authentication technology. The results from gait analysis with the help of Artificial Intelligent determine the secure level of personal information such as identity and health status. Zhang et al. [7] devised deep learning-based TENG socks for gait analysis. Five participants with different weights could be identified with a high recognition accuracy of 96% using an improved four-layer one-dimensional (1D) convolutional neural network (CNN) model. The accuracy could still be higher than 93.5%.

Sensory units are essential for efficient Human-Machine Interface (HMI) modules and require specialized feedback modules. The feedback module provides the necessary stimulations to aid in manipulator status cognition. A large portion of feedback research is also concentrated on biomimetic stimulators because most of the feedback system depends on actual human sensation. In particular, kinesthetic feedback may portray the temporal mobility of various bodily parts, and dermatological feedback, delivers tactile and thermal stimulations through different skin terminals. TENG-based sensing technology has been thoroughly investigated for low-power HMIs in this time due to the advantages of a variety of structural alternatives and a wide range of material options. In this review, we thoroughly highlight the key technologies and development of TENG-based HMIs with respect to a variety of application scenarios, such as wearable HMIs, HMIs for robotic-related applications, and HMIs for smart home applications. In order to enable multimodal sensor fusion with AI analysis for multipurpose applications or more dependable performance in a variety of environmental scenarios, TENG sensors can also be employed with other sensing technologies. This would undoubtedly illuminate the peaceful cohabitation of humans and machines in the future of the IoT era, as well as the efficient and immersive interactions in a range of settings, with the seamless deployment of low-power/self-sustainable TENG-based HMIs.

4 Methods and Materials

In the existing works, low-order neurodynamic only computed the mono-band frequency of resting-state functional magnetic imaging (rs-fMRI). But there is a necessity of taking care of high-order neurodynamic. If we consider high-order neurodynamic imaging, the deep learning will outperform. In our experimental approach, we introduce an automated system for diagnosing Alzheimer's disease (AD) by constructing a sophisticated neurodynamic functional network that utilizes multiple levels of resting-state functional magnetic resonance imaging (rs-fMRI) data. The general architecture of AD tracking is shown in Fig. 4.

4.1 ADNI Dataset

The ADNI dataset, developed by the National Institute of Aging (NIA), is a comprehensive collection of multimodal neuroimages obtained from individuals with Alzheimer's disease. This dataset consists of three classes of biological markers: AD (Alzheimer's disease), MCI (mild cognitive impairment), and NC (normal control). For our study, we selected a total of 153 baseline subjects, and their demographic details are provided in Table 2.

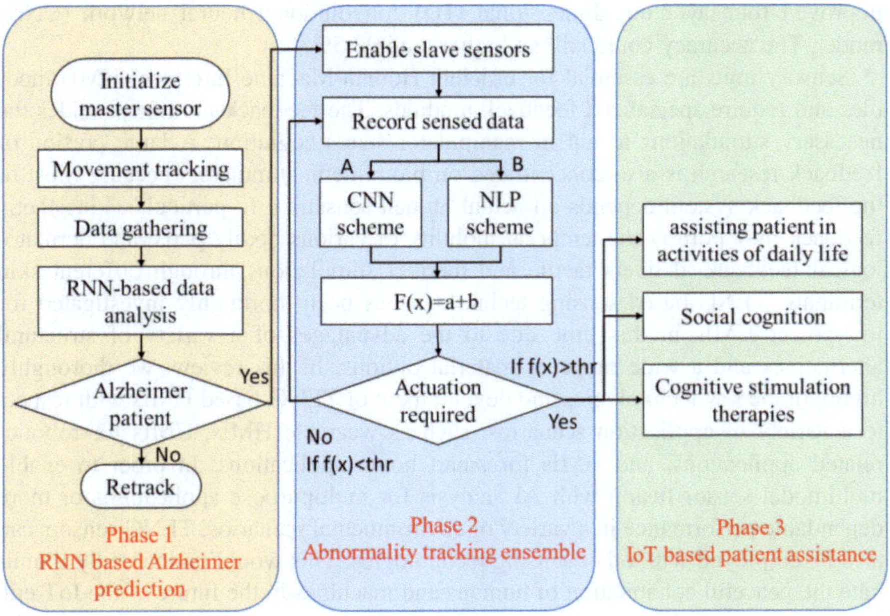

Fig. 4 General architecture of AD tracking

Table 2 The demographic details of the rs-fMRI

	AD (n = 51)		MCI (n = 51)		NC (n = 51)	
Number of samples	Average	Standard deviation (SD)	Average	Standard deviation (SD)	Average	Standard deviation (SD)
Age	72.5	7.2	71.4	6.0	72.4	4.2
Education	13.2	3.2	15.8	2.2	14.2	3.1
MMSE	22.1	2.1	26.3	1.5	27.3	1.1
CDR	0.6	0.3	0.5	0.0	0.0	0.0

The ADNI1 dataset focuses on consistent longitudinal structural imaging using 1.5T scanners. This MRI protocol employs T1 and dual-echo T2-weighted sequences. However, approximately one-fourth of the ADNI1 dataset images were acquired using T3 scanners while following the same protocol. On the other hand, the ADNI-GO and ADNI2 datasets were obtained using 3T scanners with the same T1-weighted parameter as ADNI1. Instead of dual-echo T2-weighted images from ADNI1, the 2D PLAIR technique was used to incorporate T2-weighted images across all sites. Both datasets contain fully sampled and accelerated T1-weighted images.

4.2 Preprocessing

Preprocessing is the process of preparing the images for further image processing steps. SPM 12 is a preprocessing algorithm used to preprocess the ADNI dataset. Following the preprocessing stage, the image was divided into gray, white, and cerebrospinal fluid planes. Ten amounts of subjects were initially eliminated in order to preserve the dynamic equilibrium in each subject. The slice-time correction was also applied to the arrangement and resampling of all the slices. That arrangement and resampling procedure will provide uniformity in a particular period. For resampling purposes, consider the middle slice as a reference slice. After that, a realignment approach based on the reference slice will be used. According to the MRI dataset, the individual averaged functional slice can be registered using a landmark-based registration technique. The method of segmentation will then be used to separate the various sections of the brain. We care for the White Matter (WM), Gray Matter (GM), and cerebrospinal fluid (CSF) here for Alzheimer's disease. Moreover, each r MRI dataset image slice was resized and normalized related to Montreal Neurological Institute (MNI). And also, the resampling would occur with $3 \times 3 \times 3\ \text{mm}^3$ setting.

The processed image was not in the form for clarity. Thus, we need to smooth the processed image. In this work, Gaussian kernel will be using for smoothing process. Finally, low-frequency images are segregated according to their ranges. They are: slow4 is the range between 0.024 and 0.04 Hz and slow5 is the range between 0.01 and 0.025 Hz. And full band is the range between 0.01 and 0.06 Hz. The intermediary relationship between the down-order and up-order brain networks for the improvement of accuracy is shown by Zhang et al. [25]. Even yet, the Zhang model was unable to explain how the brain changes on a dynamic basis. This work develops a high-order neurodynamic functional network by employing an automated AD detection system to suggest a unique strategy. The higher order is achieved utilizing rs-fMRI time-series data at several frequency levels (slow4, slow5, and full-band). The conversion of time series into visuals is another transition. The benefit of this transformation is that, given any two time-series points, it is simple to obtain the data. Convolutional neural networks can be used to classify these image time series [26].

The ability to discern between AD and MCI from the average level of health will be much improved by higher-order functional brain connections across a variety of frequency bands. As a result, a new area of research for AD diagnosis is opened up by merging higher-order dynamic and various frequency band brain networks. In order to effectively improve the performance of the suggested model, we have adopted a collaborative process in our work. A variety of models will be combined through collaboration into a single rigid model. The suggested deep neural network model's whole design flow is depicted in Fig. 5. To process the input scanned images, the SPM 12 software [34], the required toolboxes for data processing assistant for rsMRI (DPARSF), and resting-state fMRI data analysis toolkit (REST) are required. Initially ten volumes of images were removed for maintaining

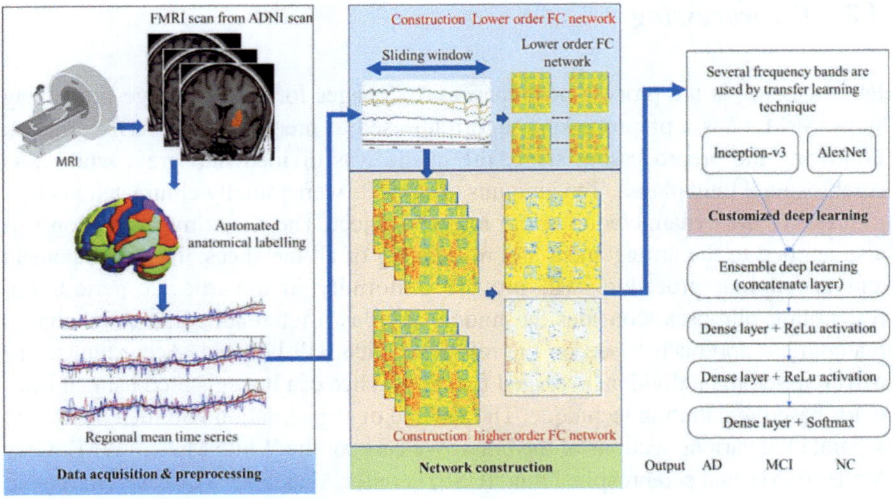

Fig. 5 Proposed model using deep neural network

dynamic equilibrium. Then, the normalization has been performed on images from 0 to 1. In our proposed work, Inception V3 architecture has been used for detecting abnormalities in the brain scan with more accuracy. Also, this architecture yields very less computational efforts. The significant goal of this Inception network is to select specific layer at each level. This Inception network's main objective is to choose certain layers at each level. The Inception V3 employs max pooling action along with a single filter size of the MRI image (1 × 1).

In addition to quick and effective identification utilizing rs-fMRI images, the suggested neurodynamic functional network offered improved accuracy, sensitivity, and specificity. Recall, specificity, and total accuracy have all been used to assess the performance of the ADNI dataset:

- The disease can be identified by classifying the AD images collected from medical expert during scanning.
- The computer-based diagnosis depends on rs-fMRI dataset.
- The biomarkers are used to study the AD disease's severity level.

4.3 Inception V3 Model

Noninvasive techniques and T1-weighted MR imaging can both be used to detect Alzheimer's disease. The proposed Inception V3 can effectively identity the abnormalities among brain images and detect them. Finding the aberrant brain development or block can produce accurate results with minimal computing effort. This Inception V3 framework's primary task is to choose a certain layer at each level.

Moreover, the Inception V3 uses a single filter for processing an MRI image of size (a1 × a1). Also, the framework will perform max-pooling operation for the effective result. The Inception V3 functions the four pipes at a same time. The proposed novel Inception V3 have (1 × 1 × 1) filters in the initial stage for reducing the dimensions. The proposed Inception V3 architecture consists of "n" parameters as shown in Fig. 4. The size of the traditional network model is (b3 × b3), which can be divided into (a1 × b3) and (b3 × a1) convolutions. For example, convolution in the form of (b3 × b3) or (c5 × c5) is similar to the convolution in the form of (a1 × b3) or (a1 × c5). The advantage here is, this transformation requires less computation than (b3 × b3) dimension. In addition to that the suggested framework consists of only two convolutions (a1 × a1 × a1); the third layer of the proposed architecture contains only the pooling layer and also the fourth layer has only (a1 × a1 × a1) filter of convolution.

The customized model still produces some faulty or erroneous result. Hence, in this work pretrained models have been used. The pretrained Inception model frequently generates useful features. The way the problem is formed differs between a customized and a pretrained model. Inception in particular was built for multiclass classification. The system is able to pick up a lot of useless information when doing multiclass categorization. The inner layers of the Inception V3 architecture can be adjusted to fix any problems that may arise. The intranetwork layers of AD versus MCI, MCI versus NL, and NL versus AD must therefore be changed by the system. The proposed Inception V3 framework is designed with Adam optimizer for diminishing the parameters for the fine-tuning and high processing speed. The architecture of Inception V3 Convolutional Neural Network has been presented in Fig. 6.

In the traditional training process, the learning rate is constant. Migration learning has not used this method, despite recent suggestions that the learning rate be changed gradually. Frequent learning rate modifications and a gradually declining learning weight not only speed up network convergence but also address the problem of an

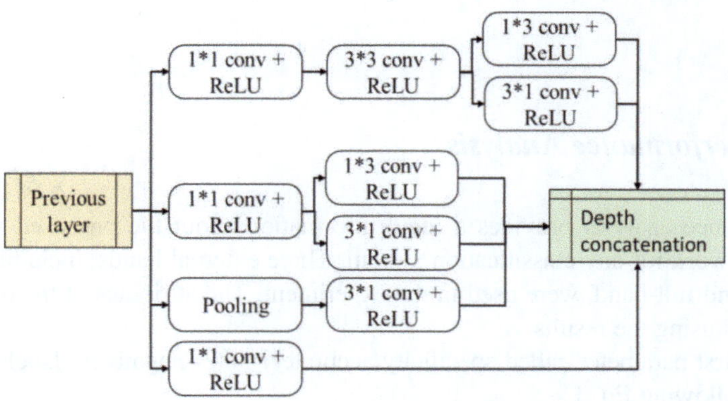

Fig. 6 The proposed Inception V3 framework

Table 3 Training parameters for Inception V3

Hyperparameter detail	Value
Batch size	32
Epochs	100
Learning rate	1×10^{-5}
Optimizer	Adam
β_1	0.9
β_2	0.999
ε	1×10^{-7}

unstable loss value that hinders convergence. Better weight parameters are discovered during top-level network training. To guarantee the model's better recognition accuracy and increased convergence rate, a series of experiments are conducted. In Table 3, the hyperparameter value is displayed.

5 Experimental Setup and Result Analysis

The entire design has been worked on and analyzed by using the software python for comparative analysis. Also, the proposed work requires the typical Windows 10 operating system with 8 GB RAM memory. This experiment uses the ADNI dataset for its evaluation. In Sect. 2, the preprocessing steps for ADNI images were described. For all of the processing, the same set of ADNI photos will be used as the input. An objective estimated image will be produced from this analysis for performance improvement. One hundred fifty-three groupings of brain pictures are selected and used for additional processing in consolidation. To calculate the accuracy model, tenfold cross-validation was performed. A total number of cross-validation sets have been produced from the entire dataset. All those average values can be used to evaluate the precision's result. For various low frequencies, the deep learning models' input array varies. Figure 7 provides the entire execution process model.

5.1 Performance Analysis

Performance analysis provides a lucid information about the proposed work and existing work for any classification system. Three external bands, including slow4, slow5, and full-band, were used in the experiment. The attributes of the dataset are extracted using the results.

The first parameter called specificity accurately notes negatively labeled classes by the following Eq. 1

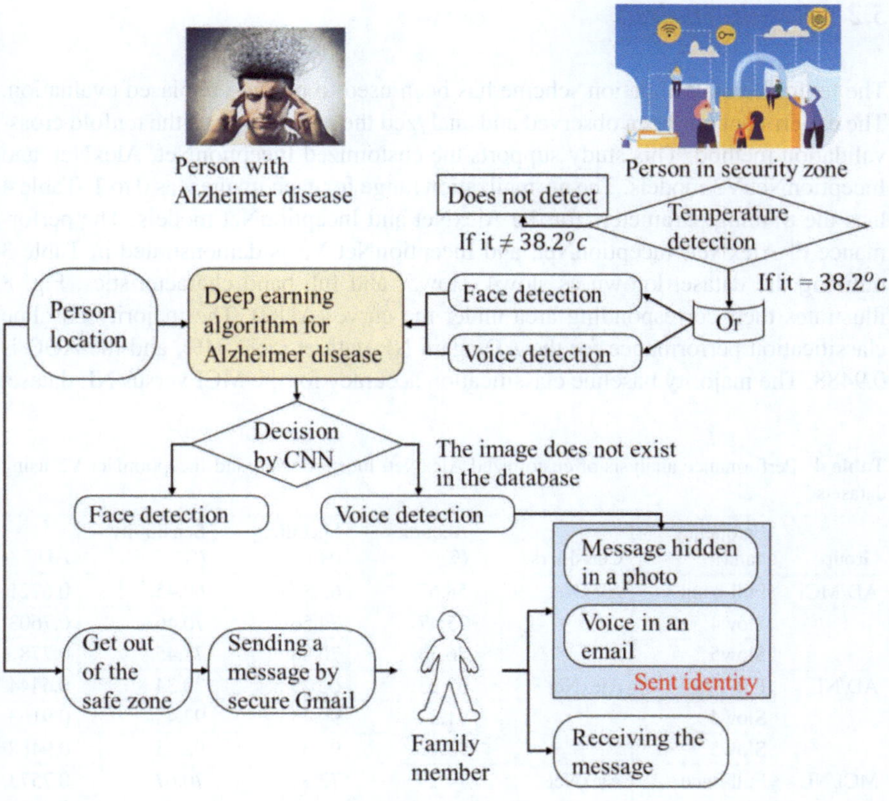

Fig. 7 The workflow of the overall execution steps

$$S = \frac{\text{TrueNegative}}{\text{TrueNegative} + \text{FalsePositive}} \tag{1}$$

The Recall or Sensitivity picked all the positive labeled classes by the following relation (2)

$$R = \frac{\text{TruePositive}}{\text{TruePositive} + \text{FalseNegative}} \tag{2}$$

The final parameter called Accuracy is defined as the overall accuracy of true positive and negative out of the total number of observations that are termed as Eq. 3

$$\text{Accuracy} = \frac{\text{TP} + \text{TN}}{\text{TP} + \text{FP} + \text{FN} + \text{TN}} \tag{3}$$

5.2 Result Analysis

The tenfold cross-validation scheme has been used to provide unbiased evaluation. The experiment has been observed and analyzed the mean result of the tenfold cross-validation method. This study supports the customized InceptionNet, AlexNet, and InceptionNet V3 models. The normalization range for each image was 0 to 1. Table 4 lists the training parameters for the AlexNet and InceptionNet models. The performance of AlexNet, InceptionNet, and InceptionNet V3 is demonstrated in Table 3 utilizing the dataset known as slow4, slow5, and full-band characteristics. Fig. 8 illustrates their corresponding area under the curve (AUC). The majority baseline classification performance for the AD again NL dataset is 93.34%, and the AUC is 0.9488. The majority baseline classification accuracy for the MCI versus NL dataset

Table 4 Performance analysis of customized AlexNet, InceptionNet, and InceptionNet V3 using datasets

Group	Frequency band	Classifiers	Accuracy (%)	Specificity (%)	Sensitivity (%)	AUC
AD/MCI	Full-band	AlexNet	58.67	65.37	68.45	0.6221
	Slow4		75.47	74.56	70.46	0.7603
	Slow5		76.34	76.88	77.45	0.778
AD/NL	Full-band	AlexNet	63.23	60.23	70.34	0.6444
	Slow4		91.34	90.45	92.42	0.9163
	Slow5		93.43	92.34	92.45	0.9488
MCI/NL	Full-band	AlexNet	74.29	72.34	70.67	0.7573
	Slow4		75.23	77.45	77.95	0.7675
	Slow5		76.56	77.56	78.34	0.7754
AD/MCI	Full-band	InceptionNet	78.89	77.37	75.56	0.7967
	Slow4		79.79	74.37	75.56	0.8021
	Slow5		80.45	82.45	81.45	0.8196
AD/NL	Full-band	InceptionNet	65.9	62.38	72.56	0.6878
	Slow4		92.38	92.36	91.46	0.9390
	Slow5		94.47	93.87	93.56	0.9518
MCI/NL	Full-band	InceptionNet	62.67	64.38	62.47	0.6320
	Slow4		75.59	78.59	72.57	0.7668
	Slow5		79.67	78.56	76.34	0.7989
AD/MCI	Full-band	InceptionNet V3	76.67	77.37	75.88	0.7782
	Slow4		78.58	75.31	75.96	0.7959
	Slow5		82.67	81.15	80.16	0.8445
AD/NL	Full-band	InceptionNet V3	64.67	61.2	73.67	0.6540
	Slow4		94.56	92.12	94.96	0.9663
	Slow5		96.61	94.34	94.96	0.9663
MCI/NL	Full-band	InceptionNet V3	66.99	67.65	62.4	0.6758
	Slow4		74.19	77.12	71.25	0.7514
	Slow5		81.87	79.86	75.47	0.8221

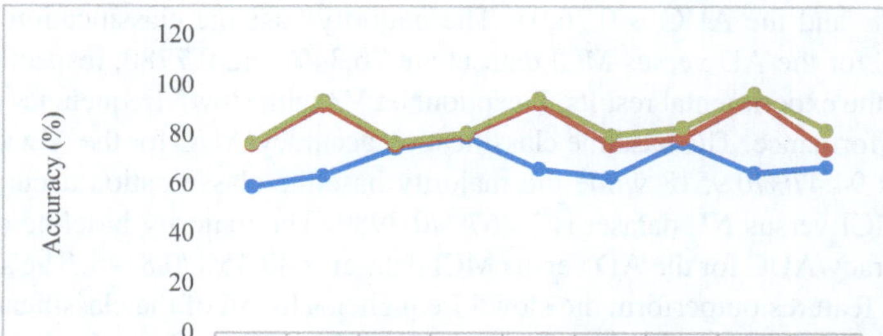

	AD/MCI	AD/NL	MCI/NL	AD/MCI	AD/NL	MCI/NL	AD/MCI	AD/NL	MCI/NL
Full-band	58.67	63.23	74.29	78.89	65.9	62.67	76.67	64.67	66.99
Slow4	75.47	91.34	75.23	79.79	92.38	75.59	78.58	94.56	74.19
Slow5	76.34	93.43	76.56	80.45	94.47	79.67	82.67	96.61	81.87

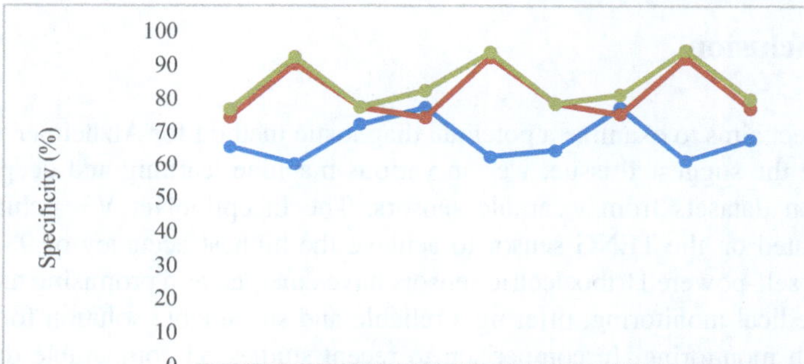

	AD/MCI	AD/NL	MCI/NL	AD/MCI	AD/NL	MCI/NL	AD/MCI	AD/NL	MCI/NL
Full-band	65.37	60.23	72.34	77.37	62.38	64.38	77.37	61.2	67.65
Slow4	74.56	90.45	77.45	74.37	92.36	78.59	75.31	92.12	77.12
Slow5	76.88	92.34	77.56	82.45	93.87	78.56	81.15	94.34	79.86

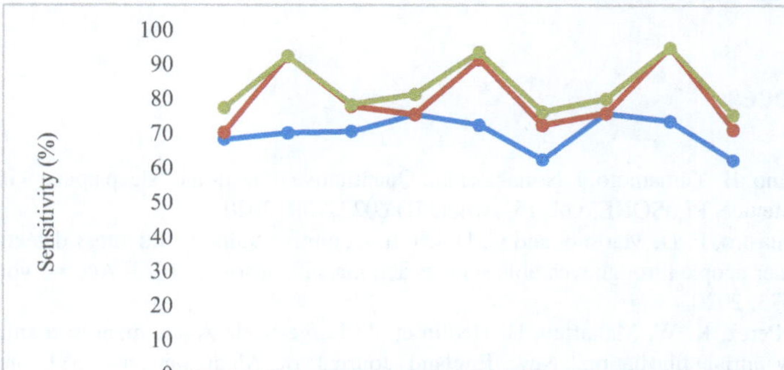

	AD/MCI	AD/NL	MCI/NL	AD/MCI	AD/NL	MCI/NL	AD/MCI	AD/NL	MCI/NL
Full-band	68.45	70.34	70.67	75.56	72.56	62.47	75.88	73.67	62.4
Slow4	70.46	92.42	77.95	75.56	91.46	72.57	75.96	94.96	71.25
Slow5	77.45	92.45	78.34	81.45	93.56	76.34	80.16	94.96	75.47

Fig. 8 **(a–c)** shows the ROC curve obtained from the AD/NL, MCI/NL, and AD/MCI

is 76.56%, and the AUC is 0.76.03. The majority baseline classification accuracy and AUC for the AD versus MCI dataset are 76.34% and 0.7780, respectively.

From the experimental results, InceptionNet V3 with slow5 frequencies offers the better performance. The baseline classification accuracy/AUC for the AD versus NL dataset is 94.47%/0.9518 while the majority baseline classification accuracy/AUC for the MCI versus NL dataset is 76.67%/0.7989. The majority baseline classification accuracy/AUC for the AD versus MCI dataset is 80.45%/0.8145. The slow4 and full-band features outperform the slow5 frequencies for all of the classification tasks when compared to the other two features. The D2 model performs better than other traditional models, it is also mentioned. It demonstrates the variety of outcomes generated by the AlexNet and InceptionNet models.

6 Conclusion

This project aims to examine a potential diagnostic method for Alzheimer's disease. To tackle the suggested issue, we ran various machine learning and deep learning models on datasets from wearable sensors. The InceptionNet V3 architecture is implemented on the TENG sensor to achieve the highest accuracy of 98.1%. The wearable self-powered triboelectric sensors have emerged as a promising technology for biomedical monitoring, offering a reliable and sustainable solution for continuous health monitoring. In comparison to recent studies on comparable data, these results are excellent. We have estimated the learning rate and performed some hypertuning on the parameters to match the final performance of the models. Data from healthy subjects were gathered for the experiment. Testing this model in real-world scenarios, though, might be difficult.

References

1. J. Hayano, H. Yamamoto, I. Nonaka et al., Quantitative detection of sleep apnea with wearable watch device, PLoSONE, vol. 15, Article ID e0237279, 2020.
2. F. Delmastro, F. D. Martino, and C. Dolciotti, Cognitive training and stress detection in MCI frail older people through wearable sensors and machine learning, IEEE Access, vol. 8, Article ID 65573, 2020.
3. M. V. Perez, K. W. Mahaffey, H. Hedlin et al., Large-Scale Assessment of a smartwatch to identify atrial fibrillation, New England Journal of Medicine, vol. 381, no. 20, pp. 1909–1917, 2019.
4. Li, T.; Li, J.; Liu, J.; Huang, M.; Chen, Y.-W.; Bhatti, U.A. Robust watermarking algorithm for medical images based on log-polar transform. EURASIP J. Wirel. Commun. Netw. 2022, 2022, 1–11.
5. Mao, S.; Zhang, C.; Gao, N.;Wang, Y.; Yang, Y.; Guo, X.; Ma, T. A study of feature extraction for Alzheimer's disease based on resting-state fMRI. In Proceedings of the 2017 39th Annual International Conference of the IEEE Engineering in Medicine and Biology Society (EMBC), Jeju, Republic of Korea, 11–15 July 2017; pp. 517–520.

6. K. Ashok, M. Ashraf, J. Thimmia Raja, M. Z. Hussain, D. K. Singh, and A. Haldorai, Collaborative analysis of audio-visual speech synthesis with sensor measurements for regulating human–robot interaction, International Journal of System Assurance Engineering and Management, Aug. 2022, https://doi.org/10.1007/s13198-022-01709-y.
7. S. Shanthi, S. Saradha, J. A. Smitha, N. Prasath, and H. Anandakumar, An efficient automatic brain tumor classification using optimized hybrid deep neural network, International Journal of Intelligent Networks, vol. 3, pp. 188–196, 2022, https://doi.org/10.1016/j.ijin.2022.11.003.
8. Odusami, M.; Maskeli ¯ unas, R.; Damaševi˘cius, R.; Krilavi˘cius, T. Analysis of Features of Alzheimer's Disease: Detection of Early
9. Stage from Functional Brain Changes in Magnetic Resonance Images Using a Finetuned ResNet18 Network. Diagnostics 2021, 11, 1071.
10. S. S. C, B. L. R and D. S, Design and Analysis of CNN based Residue Number System for Performance Enhancement, 2023 Third International Conference on Artificial Intelligence and Smart Energy (ICAIS), Coimbatore, India, 2023, pp. 1182-1187, doi: 10.1109/ICAIS56108.2023.10073805.
11. S. N. Siri, H. B. Divyashree and S. P. Mala, The Memorable Assistant: An IoT-Based Smart Wearable Alzheimer's Assisting Device, 2021 IEEE International Conference on Computation System and Information Technology for Sustainable Solutions (CSITSS), Bangalore, India, 2021, pp. 1-6, https://doi.org/10.1109/CSITSS54238.2021.9682788.
12. S. Kamakshi, P. Penchalaiah and P. Bhasha, Classification of Brain Disease & MRI-Based Age Estimation Using Deep Learning Algorithms, 2023 International Conference on Intelligent Systems for Communication, IoT and Security (ICISCoIS), Coimbatore, India, 2023, pp. 345-348, https://doi.org/10.1109/ICISCoIS56541.2023.10100478.
13. V. Ranganayaki, P. S. Ramesh, K. Rammohan, A. Deepak, Machine Learning Approaches on Pedestrian Detection in an autonomous vehicle, 2023 Fifth International Conference on Electrical, Computer and Communication Technologies (ICECCT), Erode, India, 2023, pp. 1-8, https://doi.org/10.1109/ICECCT56650.2023.10179836
14. A. Yalamanchili, D. V. Sekhar, G. V. Kumar and T. U. Rani, Region-based Convolutional Neural Networks with IoT-based Alzheimer's disease detection and classifications, 2022 IEEE North Karnataka Subsection Flagship International Conference (NKCon), Vijaypur, India, 2022, pp. 1-5, https://doi.org/10.1109/NKCon56289.2022.10126627.
15. S. S. C, B. V and B. L. R, High-Performance Computing Based on Residue Number System: A Review, 2023 9th International Conference on Advanced Computing and Communication Systems (ICACCS), Coimbatore, India, 2023, pp. 639-647, https://doi.org/10.1109/ICACCS57279.2023.10112959.
16. P. T, R. M, V. K. K, S. T, M. B. E and K. Devanathan, Geriatric Assistive Device for Remote Health Monitoring using IoT based Caplet System, 2022 IEEE Delhi Section Conference (DELCON), New Delhi, India, 2022, pp. 1-5, https://doi.org/10.1109/DELCON54057.2022.9752851.
17. R. Chokri, W. Hanini, W. B. Daoud, S. A. Chelloug and A. M. Makhlouf, Secure IoT Assistant-Based System for Alzheimer's Disease, in IEEE Access, vol. 10, pp. 44305-44314, 2022, https://doi.org/10.1109/ACCESS.2022.3168311.
18. Parthasaradi, V., Kavitha, M., Sridevi, A. et al. Novel rare-earth Eu and La co-doped ZnO nanoparticles synthesized via co-precipitation method: optical, electrical, and magnetic properties. J Mater Sci: Mater Electron (2022). https://doi.org/10.1007/s10854-022-09272-9.
19. S. Patil et al., Medical Assistance for Alzheimer's Disease Using Smart Specs, 2022 4th International Conference on Circuits, Control, Communication and Computing (I4C), Bangalore, India, 2022, pp. 363-366, https://doi.org/10.1109/I4C57141.2022.10057684.
20. B. Singh, M. Tatiya, A. Shrivastava, D. Verma, A. Pratap Srivastava and A. Rana, Detection of Alzheimer's Disease Using Deep Learning, Blockchain, and IoT Cognitive Data, 2022 2nd International Conference on Technological Advancements in Computational Sciences (ICTACS), Tashkent, Uzbekistan, 2022, pp. 863-869, https://doi.org/10.1109/ICTACS56270.2022.9988058.

21. Lin Wang, Self-Powered Electrostatic Adsorption Face Mask Based on a Triboelectric Nanogenerator, ACS Applied Materials & Interfaces 2018 10 (8), 7126-7133 https://doi.org/10.1021/acsami.7b18732
22. Barnali Ghatak et al, Design of a self-powered triboelectric face mask,Nano Energy,Volume 79,2021. https://doi.org/10.1016/j.nanoen.2020.105387
23. Dale A Schoeller, Changes in total body water with age, The American Journal of Clinical Nutrition, Volume 50, Issue 5, 1989, Pages 1176-1181. https://doi.org/10.1093/ajcn/50.5.1176.
24. Yuan, M.; Li, C.; Liu, H.; Xu, Q.; Xie, Y. A 3D-Printed Acoustic Triboelectric Nanogenerator for Quarter-Wavelength Acoustic Energy Harvesting and Self-Powered Edge Sensing. Nano Energy 2021, 85, 105962.
25. Chen, F.; Wu, Y.; Ding, Z.; Xia, X.; Li, S.; Zheng, H.; Diao, C.; Yue, G.; Zi, Y. A Novel Triboelectric Nanogenerator Based on Electrospun Polyvinylidene Fluoride Nanofibers for Effective Acoustic Energy Harvesting and Self-Powered Multifunctional Sensing. Nano Energy 2019, 56, 241–251.
26. Kang, S.; Cho, S.; Shanker, R.; Lee, H.; Park, J.; Um, D.-S.; Lee, Y.; Ko, H. Transparent and Conductive Nanomembranes with Orthogonal Silver Nanowire Arrays for Skin-Attachable Loudspeakers and Microphones. Sci. Adv. 2018, 4, eaas8772.
27. Chan, M.; Campo, E.; Estève, D.; Fourniols, J.-Y. Smart Homes—Current Features and Future Perspectives. Maturitas 2009, 64, 90–97.
28. M. T, S. Upadhyay, R. Beaulah Jeyavathana and A. Gopatoti, Big Data Analytics with Deep Learning based Intracranial Haemorrhage Diagnosis and Classification Model," 2022 International Conference on Augmented Intelligence and Sustainable Systems (ICAISS), Trichy, India, 2022, pp. 877-883, https://doi.org/10.1109/ICAISS55157.2022.10010826
29. Yan, Z.; Wang, L.; Xia, Y.; Qiu, R.; Liu, W.; Wu, M.; Zhu, Y.; Zhu, S.; Jia, C.; Zhu, M.; et al. Flexible High-Resolution Triboelectric Sensor Array Based on Patterned Laser-Induced Graphene for Self-Powered Real-Time Tactile Sensing. Adv. Funct. Mater. 2021, 31, 2100709.
30. Seung-Bae Jeon, Young-Hoon Nho, Sang-Jae Park, Weon-Guk Kim, Il-Woong Tcho, Daewon Kim, Dong-Soo Kwon, Yang-Kyu Choi, Self-powered fall detection system using pressure sensing triboelectric nanogenerators, Nano Energy, Volume 41, 2017, Pages 139-147, https://doi.org/10.1016/j.nanoen.2017.09.028.
31. Lin, Z., Wu, Z., Zhang, B., Wang, C., Guo, H., Liu, G., Chen, C., Chen, Y., Yang, J., & Wang, Z. L. (2019). A Triboelectric Nanogenerator-Based Smart Insole for Multifunctional Gait Monitoring. Advanced Materials Technologies, 4(2), 1800360. https://doi.org/10.1002/admt.201800360
32. Liangxu Xu, Zheng Zhang, Fangfang Gao, Xuan Zhao, Xiaochen Xun, Zhuo Kang, Qingliang Liao, Yue Zhang, Self-powered ultrasensitive pulse sensors for noninvasive multi-indicators cardiovascular monitoring, Nano Energy, Volume 81, 2021, https://doi.org/10.1016/j.nanoen.2020.105614.
33. Indira Rustempasic, Mehmet Can, Diagnosis of Parkinson's Disease using Fuzzy C-Means Clustering and Pattern Recognition, Southeast Europe Journal of Soft Computing, Vol 42, 2013.
34. Siva Kumar, S. Raja, N. Pritha, Havaldar Raviraj, An adaptive transformer model for anomaly detection in wireless sensor networks in real-time, Measurement: Sensors, Volume 25, 2023,100625. https://doi.org/10.1016/j.measen.2022.100625
35. Chang, Q., He, Y., Liu, Y., Zhong, W., Wang, Q., Lu, F., and Xing, M. (2020). Protein gel phase transition: toward superiorly transparent and hysteresis-free wearable electronics. Adv. Funct. Mater. 30, 1910080.
36. Dong, B., Yang, Y., Shi, Q., Xu, S., Sun, Z., Zhu, S., Zhang, Z., Kwong, D.L., Zhou, G., Ang, K.W., et al. (2020). Wearable triboelectric-human-machine interface (THMI) using robust nanophotonic readout. ACS Nano 14, 8915–8930.
37. Chen, X.P., Xie, X.K., Liu, Y.N., Zhao, C., Wen, M., and Wen, Z. (2020). Advances in healthcare electronics enabled by triboelectric nanogenerators. Adv. Funct. Mater. 30, 2004673.

38. Suzuki, K.; Yataka, K.; Okumiya, Y.; Sakakibara, S.; Sako, K.; Mimura, H.; Inoue, Y. Rapid-Response, Widely Stretchable Sensor of Aligned MWCNT/Elastomer Composites for Human Motion Detection. ACS Sensors 2016, 1, 817–825.
39. Wei, P.; Yang, X.; Cao, Z.; Guo, X.; Jiang, H.; Chen, Y.; Morikado, M.; Qiu, X.; Yu, D. Flexible and Stretchable Electronic Skin with High Durability and Shock Resistance via Embedded 3D Printing Technology for Human Activity Monitoring and Personal Healthcare. Adv. Mater. Technol. 2019, 4, 1900315.
40. Kim, S.; Oh, J.; Jeong, D.; Bae, J. Direct Wiring of Eutectic Gallium–Indium to a Metal Electrode for Soft Sensor Systems. ACS Appl. Mater. Interfaces 2019, 11, 20557–20565.
41. Shi, Q.; Dong, B.; He, T.; Sun, Z.; Zhu, J.; Zhang, Z.; Lee, C. Progress in Wearable Electronics/Photonics—Moving toward the Era of Artificial Intelligence and Internet of Things. InfoMat 2020, 2, 1131–1162.

Risk Prediction of Maternal Health by Model Analysis Using Artificial Intelligence

1 Introduction

According to estimates from the International Diabetes Federation (IDF), 463 million individuals between the ages of 20 and 79 have diabetes, and 374 million people have impaired glucose tolerance. This will double to 592 million by 2035. Among pregnant women, gestational diabetes mellitus is prevalent. There are numerous conventional techniques for diagnosing diabetes that rely on laboratory and physical testing. GDM stands for gestational diabetes mellitus, which is diabetes connected to pregnancy. The higher level of blood sugar that is noticed during the initial stage of pregnancy helps in identifying GDM [1, 2]. The hormonal changes during pregnancy increase insulin resistance in the mother's body to accommodate the increased energy needs of the growing fetus. GDM can impose high risks to both fetal growth and maternal health, and it requires careful management and monitoring.

Type 2 diabetes can develop from gestational diabetes, particularly if there is a family history of the disease and high blood glucose levels. The majority of instances of diabetes are type 2, which is the most common type worldwide. It appears when the body either stops producing enough insulin to keep blood sugar levels within the normal range or becomes resistant to the effects of insulin. Type 2 diabetes is frequently associated with lifestyle variables such genetic predisposition, bad food, obesity, and sedentary behaviors. It can be managed with dietary changes, oral medicines, and, in certain cases, insulin therapy.

© European Alliance for Innovation 2024

A. Haldorai et al., *Artificial Intelligence for Sustainable Development*,
EAI/Springer Innovations in Communication and Computing,
https://doi.org/10.1007/978-3-031-53972-5_6

1.1 Challenge of AI in Healthcare

AI-driven technology and related applications have the potential to completely transform healthcare in the modern digital age by providing timely reminders and tailored interventions based on a person's vital signs. With the use of numerous digital gadgets, these interventions are easily incorporated into our everyday routines. This technological breakthrough has the power to fundamentally alter healthcare organizations by changing their service delivery models, patient engagement tactics, operational paradigms, and optimization initiatives [4].

The revolutionary influence of AI on the healthcare environment goes beyond improving overall patient outcomes. AI is able to proactively identify dangers, offer suitable therapies, and even provide real-time feedback by continuously monitoring vital signs and other pertinent health data. In addition to enabling people to take control of their health, this degree of individualized attention and care guarantees the effective use of healthcare resources. AI innovations strive to emulate human cognitive processes, and computers that employ AI algorithms to comprehend and work with massive information have come a long way. In the age of big data, one benefit of using AI in healthcare is that intricate linkages and patterns may be found in the data algorithmically, saving money by reducing the need for additional healthcare personnel to handle data analysis [6, 7].

Figure 1 illustrates how AI has had a major impact on healthcare which is elaborated below.

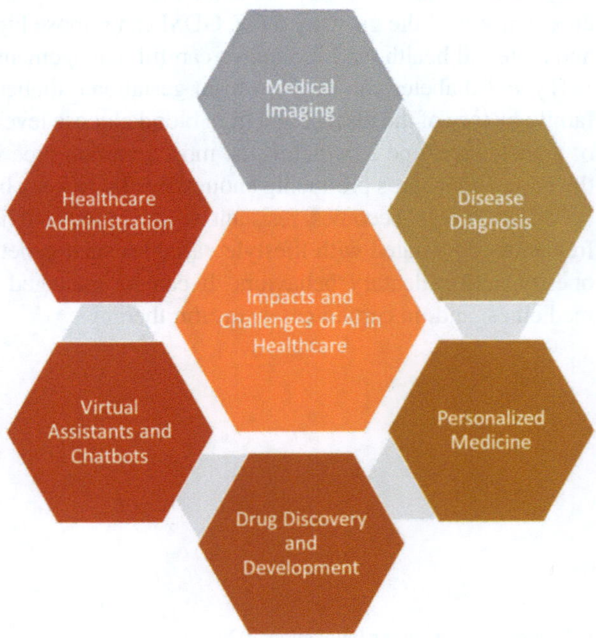

Fig. 1 Overview of AI challenges in healthcare

- Medical Imaging: The analysis of medical images such as CT, MRI, and X-rays has demonstrated AI's significant promise. Early disease identification and diagnosis can be aided by deep learning algorithms' ability to recognize shapes and anomalies in photos. However, challenges include the need for large annotated datasets, potential biases in training data, and ensuring the reliability and interpretability of AI-generated results.
- Disease Diagnosis: Large volumes of patient data, like as genetic information, medical records, and symptom data, can be analyzed by AI algorithms to help diagnose diseases. AI can provide more accurate and timely diagnoses, leading to improved treatment outcomes. The incorporation of AI systems into current healthcare workflows, protecting data security and privacy, and handling regulatory issues remain obstacles, though.
- Customized Medical Care: AI can help tailor treatment plans according to a patient's genetic composition, medical history, and lifestyle preferences. This opens the door to more focused treatments, better patient outcomes, and lower medical expenses. The difficulty of incorporating AI into clinical decision-making procedures, moral issues with patient data use, and the requirement for strict regulation and validation of AI-driven personalized medicine solutions are some of the difficulties.
- Drug Research: As AI is capable of analyzing large amounts of biomedical data and forecast the potential efficacy of treatment options, AI is a critical tool in the drug discovery process. It can therefore require a lot less time and money to introduce new drugs to the market. Notwithstanding, there exist certain obstacles such as insufficient data of superior quality, certain AI algorithms being opaque, and the requirement for cooperation between academics specializing in particular domains and AI specialists.
- Robotic chatbots: AI can provide 24/7 support using chatbots and virtual assistants, respond to patient inquiries, and help with appointment scheduling and triage. They can improve access to healthcare services and reduce the burden on healthcare professionals. Challenges include the need for accurate and reliable algorithms, ensuring patient privacy and confidentiality, and addressing concerns related to the loss of human interaction in healthcare delivery.
- Healthcare Administration: AI can streamline administrative tasks such as medical coding, billing, and resource allocation, leading to increased efficiency and cost savings. It can automate repetitive tasks so that medical staff can concentrate more on patient care. The possibility of biases in algorithmic decision-making, maintaining transparency in AI-driven administrative systems, and resolving labor concerns about job displacement are among the challenges.
- Ethical and Legal Challenges: Using AI in healthcare presents a number of ethical and legal issues, such as algorithmic bias and fairness, data security and privacy protection, transparency and comprehensibility of AI systems, liability for AI-driven decisions, and the possibility of aggravating already-existing healthcare disparities. It is crucial to guarantee the moral and conscientious advancement and application of AI in the healthcare sector.

AI has the potential to revolutionize healthcare, but in order to maximize its benefits and reduce any risks, it must carefully address a number of issues. These include issues with data quality, interpretability, integration with current systems, regulatory frameworks, and ethical considerations. To overcome these obstacles and fully utilize AI in healthcare, cooperation between policymakers, researchers, healthcare practitioners, and other stakeholders is crucial [8].

The remainder is made up of the following: Sect. 2 presents machine learning and its importance for GDM; Sect. 3 presents the system design of the suggested work. Various performance metrics and the findings from Sect. 5 are described in Sect. 4. Lastly, a conclusion wraps up this debate in Sect. 6.

2 Machine Learning for GDM

The use of machine learning (ML) techniques [3] in clinical prediction models has increased dramatically due to the increasing availability of data and the potential for improved predictive performance. Machine learning techniques have developed into highly favored tools for managing and preventing diseases as outlined in Fig. 2. These techniques have been used by an increasing number of studies to identify GDM risk variables and develop early prediction models for the illness [9].

- Predicting Maternal Risk: Machine learning algorithms can examine a range of risk indicators, such as maternal age, BMI, medical history, and glucose levels, to predict the chance of GDM development. Medical practitioners can identify high-risk patients who may need more frequent monitoring or early intervention by using these models [5].
- Mother Blood glucose Monitoring: It is recommended that individuals diagnosed with gestational diabetes maintain normoglycemia by participating in dietary counselling, diligently monitoring their blood glucose levels, and engaging in regular physical activity. For individuals who require insulin therapy, the ideal frequency of blood glucose testing is yet unknown. However, it is standard practice to perform four daily glucose checks. To rule out fasting hyperglycemia,

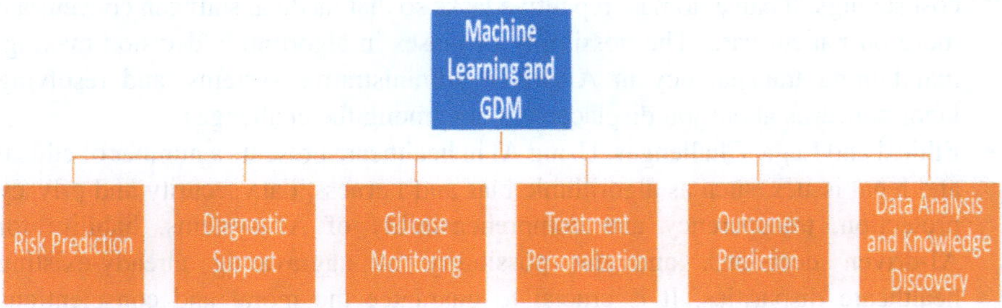

Fig. 2 Machine learning for GDM

this protocol usually starts with a fasting morning glucose level measurement. One to two hours following meals, follow-up examinations aid in ensuring efficient glucose management [12].

- Personalize Treatment: For women with GDM, machine learning can help create individualized treatment programs. Models that analyze patient data, such as glucose levels, food details, and response to medication, might suggest tailored interventions, including meal planning, insulin dosage, or lifestyle changes [13].
- Predict Outcomes: To anticipate possible negative consequences related to GDM, machine learning models can be trained on sizable datasets that contain maternal features, medical history, and fetal monitoring data. Preterm birth, preeclampsia, and neonatal problems are among the difficulties that these models can detect in women who are more vulnerable [14].
- Data Mining and Discovery: Large-scale GDM datasets can yield insightful information by revealing new risk factors, associations, or subgroups within the GDM population through the application of machine learning algorithms. This can enhance GDM management techniques and help researchers and medical professionals comprehend the underlying mechanisms [15].

The development and validation of machine learning models must adhere to strict scientific protocols, which include suitable data pretreatment, model selection, and evaluation strategies [11]. The interpretation of results and clinical decision-making should involve the collaboration of medical professionals, taking into account the context and individual patient characteristics [10]. Machine learning is a tool that can augment healthcare providers' expertise but should not replace clinical judgment and human decision-making.

3 System Design

Risk prediction using machine learning during gestational diabetes mellitus involves training AI models on relevant input to recognize individuals at a higher risk of developing GDM [14]. By combining insights from many machine learning techniques, the project aims to develop a diagnostic system for expectant mothers that provides more accurate diabetes detection and can forecast the probability of the condition developing into type 2 diabetes. Figure 3 depicts the proposed overall system model using machine learning for GDM and each of the process is outlined below.

3.1 Dataset Collection

The initial source of the dataset was the Pima Indians Diabetes Database, which can be accessed through Kaggle [18]. It is made up of one target variable and multiple medical analyst variables. Determining the patient's diabetes status is the dataset's

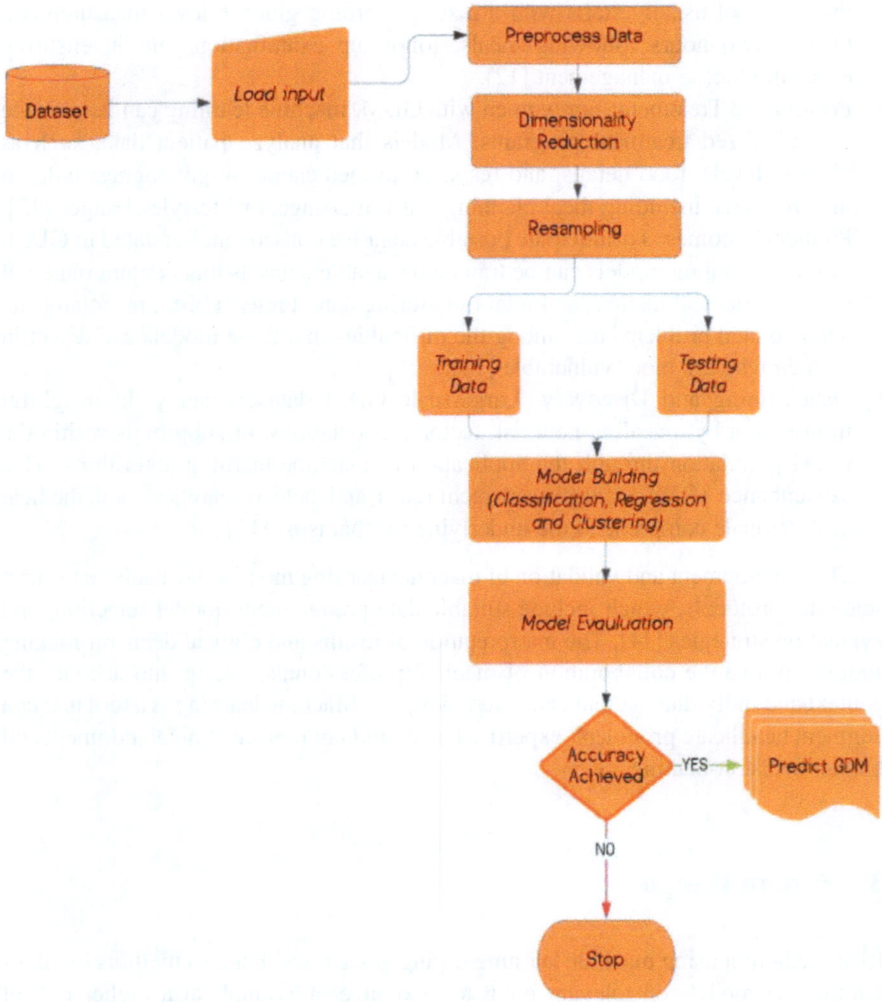

Fig. 3 Overview of system model

goal The procedure comprises data gathering and analysis to reveal hidden patterns and trends that support outcome evaluation and prediction. The dataset, as shown in Fig. 4, consists of nine columns that contain all the features and 156 rows that contain all the data that are currently accessible [16]. The original dataset is used to create a new dataset that includes a novel attribute known as "type 2." There are several independent variables in this dataset, and just one dependent variable—designated as "type 2." Here, "type 2" is the feature that is being studied; a value of 0 means that it will not develop into type 2 diabetes, but a value of 1 means that it may develop into type 2 diabetes. Crucially, the dataset has no null or missing values [17].

#	Column	Non-Null Count	Dtype
0	Pregnancies	156 non-null	int64
1	Glucose	156 non-null	int64
2	BloodPressure	156 non-null	int64
3	SkinThickness	156 non-null	int64
4	Insulin	156 non-null	int64
5	BMI	156 non-null	float64
6	DiabetesPedigreeFunction	156 non-null	float64
7	Age	156 non-null	int64
8	Outcome	156 non-null	int64
9	Type2	156 non-null	int64

Fig. 4 Dataset description

3.2 *Data Preprocessing*

In order to produce more exact and accurate results, this phase of the model manages inconsistent data. For example, we deleted the feature since the Id in this dataset is inconsistent. There are no missing values in this dataset. Because certain attributes cannot have zero values, we imputed missing values for a few chosen attributes, including age, skin thickness, blood pressure, glucose level, and BMI. We then used StandardScaler to scale the data. No feature selection was done because the features were crucial for prediction and there were less of them.

3.2.1 Missing Value Identification

Finding missing values is a crucial step in preparing and analyzing data. If missing values are not appropriately handled, the findings may be biased or erroneous. The missing values in the datasets were acquired using the Panda library and SK-learn, and those missing values were then filled with the matching mean value [19].

3.2.2 Feature Selection

A frequently used method for determining which qualities or features are most pertinent is the Pearson's correlation method. The correlation coefficient, which assesses the connection between input and output quality, is computed. The range of the coefficient value is from −1 to 1. A value over 0 or below −1 indicates a significant correlation, whereas a value of zero indicates no correlation at all.

3.2.3 Scaling and Normalization

The algorithm's processing was sped up by normalizing the data within the 0–1 range in order to do the feature scaling. Data are adjusted by the process of scaling such that it falls into a predetermined range, like 0–100 or 0–1. Scaling the data is crucial when using techniques like support vector machines (SVM) and k-nearest neighbors (KNN), which rely on the measurement of distances between data points. This guarantees that different numerical feature values of "1" are given equal weight in these methods.

3.3 Splitting of Data

The dataset is split up into distinct subgroups for training and testing. This technique is frequently used in data analysis and machine learning to assess and enhance the performance of models. The train-test split, which divides the dataset into two parts—one for training the model and the other for evaluating its performance—is the most common division technique. This method uses 30% of the total dataset for testing and the remaining 70% for training the model.

3.4 Model Algorithm Analysis

In this section, we outline diverse machine learning models that were used for evaluating the proposed system.

3.4.1 Logistic Regression

When dependent variables can be classified, machine learning techniques such as logistic regression are employed. Based on the provided features, the logistic regression produced results. Here, the output is categorized using a sigmoidal function.

3.4.2 K-Nearest Neighbors

By leveraging "feature similarity," the K-nearest neighbor algorithm predicts the values of new datapoints. This suggests that the new datapoint will be assigned a value based on how closely it resembles the points in the training set. Predicting an instance (x) requires sifting through the whole training set, identifying the K instances that are nearest to it, and then summarizing the output variable for each of those K instances.

3.4.3 SVM

For classification, supervised learning algorithms like SVM are employed. For SVM to correctly categorize the data, we must determine which hyper plane to use. The right values for the parameters must be entered here. In order to determine the proper hyper plane, we must determine the proper margin, which requires us to select the radial basis function (RBF) kernel values and a gamma value of 0. The hyper plane will fail to be classified if we choose it with a little margin.

3.4.4 Naive Bayes

Naive Bayes classifiers are a type of classification algorithms that rely on Bayes' Theorem. It is not a single algorithm; rather, it is a family of algorithms bound by the notion that every pair of features to be classified stands alone.

3.4.5 Decision Tree

A nonparametric classifier in supervised learning is the decision tree. This approach represents all the information as a tree, with internal nodes representing the attributes and leaves representing the class labels. The nodes were divided using the Gini Index.

3.4.6 Random Forest

An ensemble learning technique for categorization is called random forest. The quantity of tree structures in the data is used to estimate the accuracy of this technique, which uses trees. In this case, the leaves correlate with the class labels, and the characteristics correspond with the internal tree nodes. The Gini index is utilized to divide the nodes in this instance, and there are 100 trees in the forest.

3.4.7 Boosting Classifier

The Gradient Boosted decision tree is implemented by XGBoost. In several Kaggle competitions, XGBoost models predominantly win. An algorithm is developed that creates decision trees sequentially. In terms of XGBoost, weights are crucial. Following the assignment of weights to each independent variable, the decision tree is used to forecast the outcome. The second decision tree receives the variables that the first decision tree mispredicted, and their weight is increased. Next, these separate predictors/classifiers combine to produce a robust and accurate model. It is applicable to user-defined prediction problems, regression, classification, and ranking.

4 Performance Metrics

4.1 Confusion Matrix

A confusion matrix is a table that shows the quantity of true-positive (TP), true-negative (TN), false-positive (FP), and false-negative (FN) predictions. It functions as an overview of the effectiveness of a classification model. In terms of cases that are correctly and erroneously classified, it enables us to comprehend the model's performance. The matrix acquired is displayed in Fig. 5, and the accuracy values of the total positive computations of all nine characteristics for two classes are acquired across the diagonal.

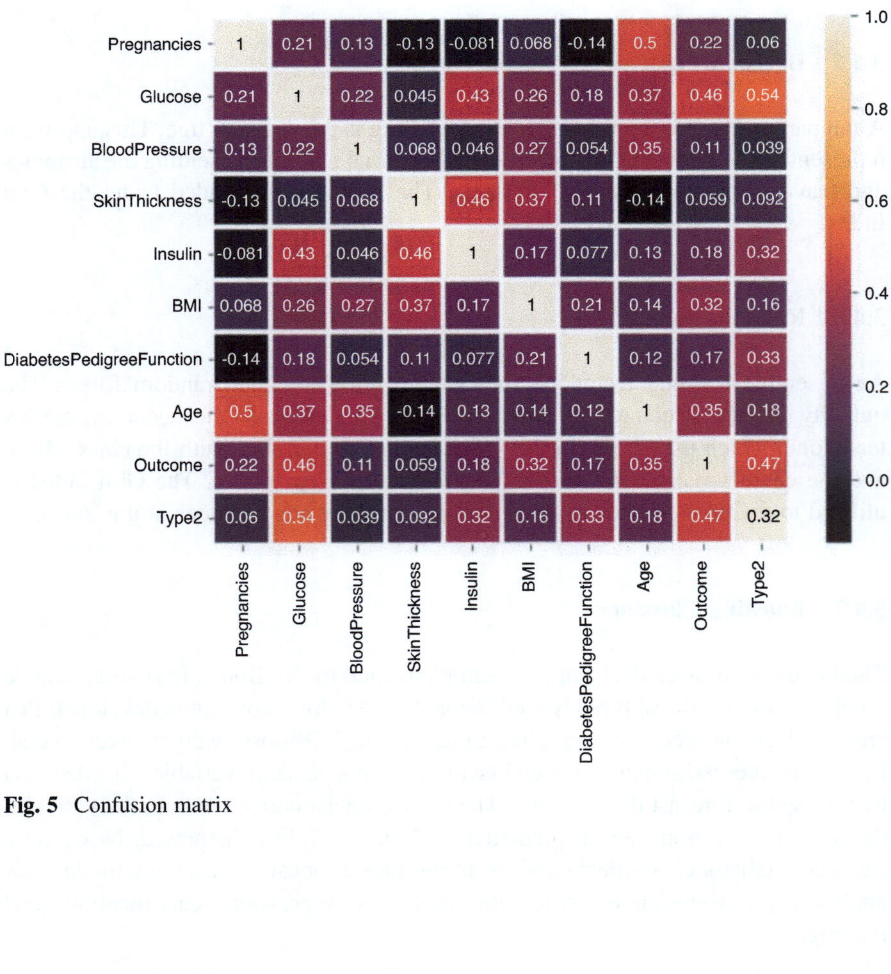

Fig. 5 Confusion matrix

4.2 Precision

Precision is a metric used to assess how well a model predicts good outcomes. True positives (TP) divided by the sum of false positives (FP) and true positives (TP) is the ratio. Precision measures the percentage of accurately predicted positive cases out of all cases expected to be positive.

$$\mathbf{Precision} = \mathbf{TP}/(\mathbf{TP} + \mathbf{FP})$$

4.3 Accuracy

The total correctness of a model's predictions is gauged by its accuracy. It is the proportion of accurate predictions (TP + TN) to all instances in the dataset. A model's accuracy offers a broad evaluation of its performance in every class.

$$\mathbf{Accuracy} = (\mathbf{TP} + \mathbf{TN})/(\mathbf{TP} + \mathbf{TN} + \mathbf{FP} + \mathbf{FN})$$

4.4 Recall

A model's recall serves as a gauge for how comprehensive its forecasts are. The ratio of genuine positives to all positives—true positives as well as false negatives—is used to compute it. A high recall indicates that the model is effective at identifying every positive case in a dataset, while a low recall indicates that the model is missing part of the positive examples in a dataset.

$$\mathbf{Recall} = \mathbf{TP}/(\mathbf{TP} + \mathbf{FN})$$

5 Results and Discussion

The testing data are used to evaluate the performance of the proposed study and determine the efficacy of our system utilizing a number of evaluation measures. Every state-of-the-art (SOTA) machine learning model [15] is compared with one another. Finally based on the experimental results obtained, the superior model is suggested for the proposed problem. Table 1 indicates the value of accuracy obtained for each model for the considered dataset. Out of the eight methods,

Table 1 Accuracy of all models

Model	Accuracy
Logistic regression	94.423%
SVM-linear	93.653%
Random forest	97.564%
XGB	96.025%
Decision tree	95.128%
SVM-RBF	94.487%
KNN	93.653%
Naïve Bayes	83.141%

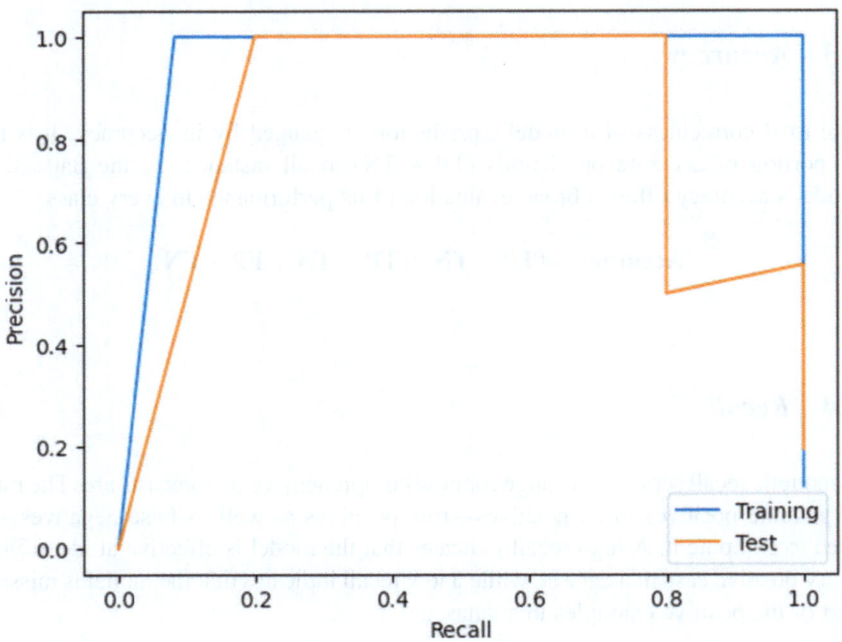

Fig. 6 Precision-recall curve

Random Forest appears to be performing the best with an accuracy of 97.56% and it can be used to predict whether a patient suffers from Gestational DM or not more accurately and efficiently compared to other SOTA models. Figure 6 shows the precision versus recall curve obtained for the train and test dataset.

Based on the results, it can be understood that the random forest method can be used to predict whether a patient suffers from Gestational DM or not, and we maximize recall while sacrificing accuracy and precision to minimize the number of false negatives, the probability threshold can be set to 0.2.

6 Conclusion

This study emphasizes the importance of AI to support healthcare analysis and how AI models can be efficiently used in prediction of a diabetic mellitus among maternal women using machine learning algorithms. In this chapter, the significance of machine learning algorithm for prediction of diabetics was presented by comparing several SOTA machine learning techniques. Based on the results, the random forest method is concluded to perform better with 97.56% accuracy which is higher compared to other models. Further, this model can support in effective diabetes disease prediction system among maternal women and also suggest the level of change from type 1 to type 2 diabetic based on their healthcare data.

References

1. C. Bayliss, "Machine learning based simulation optimisation for urban routing problems," Applied Soft Computing, vol. 105, p. 107269, Jul. 2021, https://doi.org/10.1016/j.asoc.2021.107269.
2. Kabir Ali, "A Review of Emotions, Behavior and Cognition", Journal of Biomedical and Sustainable Healthcare Applications, vol.3, no.2, pp. 165–176, July 2023. https://doi.org/10.53759/0088/JBSHA202303016.
3. E. Jyothi Kiranmayi, N.V. Rao and K.S. Nayanathara, "Hybrid Interval Type-2 Fuzzy AHP and COPRAS-G-based trusted neighbour node Discovery in Wireless Sensor Networks, Journal of Machine and Computing, vol.3, no.3, pp. 251–263, July 2023. https://doi.org/10.53759/7669/jmc202303023.
4. H. Li and J. Sansalone, "Implementing machine learning to optimize the cost-benefit of urban water clarifier geometrics," Water Research, vol. 220, p. 118685, Jul. 2022, https://doi.org/10.1016/j.watres.2022.118685.
5. H. Tanim, C. B. McRae, H. Tavakol-Davani, and E. Goharian, "Flood Detection in Urban Areas Using Satellite Imagery and Machine Learning," Water, vol. 14, no. 7, p. 1140, Apr. 2022, https://doi.org/10.3390/w14071140.
6. L. Wei, W. Tian, E. A. Silva, R. Choudhary, Q. Meng, and S. Yang, "Comparative Study on Machine Learning for Urban Building Energy Analysis," Procedia Engineering, vol. 121, pp. 285–292, 2015, https://doi.org/10.1016/j.proeng.2015.08.1070.
7. Anna Alsufieva and Daria Aleeva, "A Review of the Definition and Significance of the Brain Health", Journal of Biomedical and Sustainable Healthcare Applications, vol.3, no.2, pp. 129–140, July 2023. https://doi.org/10.53759/0088/JBSHA202303013.
8. T. Thamaraimanalan et al., "Machine Learning based Patient Mental Health Prediction using Spectral Clustering and RBFN Algorithms," 2022 8th International Conference on Advanced Computing and Communication Systems (ICACCS), Mar. 2022, https://doi.org/10.1109/icaccs54159.2022.9785142.
9. S. Satheesh Kumar and Manjula Sanjay Koti, "Deep Learning Based Hybrid Network Architecture to Diagnose IoT Sensor Signal in Healthcare System", Journal of Machine and Computing, pp. 103–114, April 2023. https://doi.org/10.53759/7669/jmc202303011.
10. John Huria Nderitu, "AMental State Adaptive Interfaces as a Remedy to the Issue of Long-term, Continuous Human Machine Interaction", Journal of Robotics Spectrum, vol.1, pp. 078–089, 2023. https://doi.org/10.53759/9852/JRS202301008.

11. S. Ayub, R. Boddu, H. Verma, S. Revathi B, B. K. Saraswat, and A. Haldorai, "Health Index Estimation of Wind Power Plant Using Neurofuzzy Modeling," Computational and Mathematical Methods in Medicine, vol. 2022, pp. 1–8, May 2022, https://doi.org/10.1155/2022/9535254.
12. K. Nandini and G. Fathima, "Urban Air Quality Analysis and Prediction Using Machine Learning," 2019 1st International Conference on Advanced Technologies in Intelligent Control, Environment, Computing and Communication Engineering (ICATIECE), Mar. 2019, https://doi.org/10.1109/icatiece45860.2019.9063845.
13. Z. Lv, J. Li, C. Dong, H. Li, and Z. Xu, "Deep learning in the COVID-19 epidemic: A deep model for urban traffic revitalization index," Data and Knowledge Engineering, vol. 135, p. 101912, Sep. 2021, https://doi.org/10.1016/j.datak.2021.101912.
14. M. S. Hossain and G. Muhammad, "Environment Classification for Urban Big Data Using Deep Learning," in IEEE Communications Magazine, vol. 56, no. 11, pp. 44–50, November 2018, https://doi.org/10.1109/MCOM.2018.1700577.
15. M. S. Hossain and G. Muhammad, "Environment Classification for Urban Big Data Using Deep Learning," in IEEE Communications Magazine, vol. 56, no. 11, pp. 44–50, November 2018, https://doi.org/10.1109/MCOM.2018.1700577.
16. C. Karyotis, T. Maniak, F. Doctor, R. Iqbal, V. Palade and R. Tang, "Deep Learning for Flood Forecasting and Monitoring in Urban Environments," 2019 18th IEEE International Conference On Machine Learning And Applications (ICMLA), Boca Raton, FL, USA, 2019, pp. 1392–1397, https://doi.org/10.1109/ICMLA.2019.00227.
17. G. Garcia-Zanabria et al., "Mirante: A visualization tool for analyzing urban crimes," 2020 33rd SIBGRAPI Conference on Graphics, Patterns and Images (SIBGRAPI), Porto de Galinhas, Brazil, 2020, pp. 148–155, https://doi.org/10.1109/SIBGRAPI51738.2020.00028.
18. J. Samuel et al., "UD-SV: Urban data services and visualization framework for sharing multidisciplinary research," Transactions in GIS, vol. 27, no. 3, pp. 841–858, Apr. 2023, https://doi.org/10.1111/tgis.13049.
19. Pérez-Messina, E. Graells-Garrido, M. J. Lobo, and C. Hurter, "Modalflow: Cross-Origin Flow Data Visualization for Urban Mobility," Algorithms, vol. 13, no. 11, p. 298, Nov. 2020, https://doi.org/10.3390/a13110298.

Hemorrhage Detection from Whole-Body CT Images Using Deep Learning

1 Introduction

There are numerous direct and indirect reasons for a brain hemorrhage, also known as an intracranial hemorrhage (ICH), which is a potentially fatal disorder. There are several causes of brain hemorrhages, including head injuries, high blood pressure, certain drugs, blood vessel anomalies, and bleeding diseases [1]. Depending on the location and extent of the bleeding, there could be a range of symptoms, including severe headaches, nausea, vomiting, numbness, muscle weakness, difficulty speaking, impaired vision, seizures, and loss of consciousness. Figure 1 shows several types of brain hemorrhages.

- Intracerebral hemorrhage: When blood arteries inside the brain burst, blood spills into the surrounding tissue.
- Subarachnoid hemorrhage: Blood leaking into the space between the brain and its coverings is the manifestation of this type of hemorrhage.
- Epidural hemorrhage: This type of hemorrhage occurs due to bleeding between the dura mater that is the outermost layer that surrounds the brain, and the skull.
- Subdural hemorrhage: This type of hemorrhage is caused due to bleeding that takes place between the human brain surface and dura mater.

Artificial Intelligence (AI) can now be widely used in hospital settings to quickly detect the presence of bleeding within the skull or brain. But a new AI tool is aiming to spot signs of internal bleeding much earlier, before a person with head trauma has even arrived at the hospital. The library of algorithms is growing as researchers in computer science and radiology work to convert artificial intelligence‘s promise into workable solutions for tackling problems [2]. There are several methods in which artificial intelligence can enhance the diagnosis of brain hemorrhage.

© European Alliance for Innovation 2024

A. Haldorai et al., *Artificial Intelligence for Sustainable Development*, EAI/Springer Innovations in Communication and Computing,
https://doi.org/10.1007/978-3-031-53972-5_7

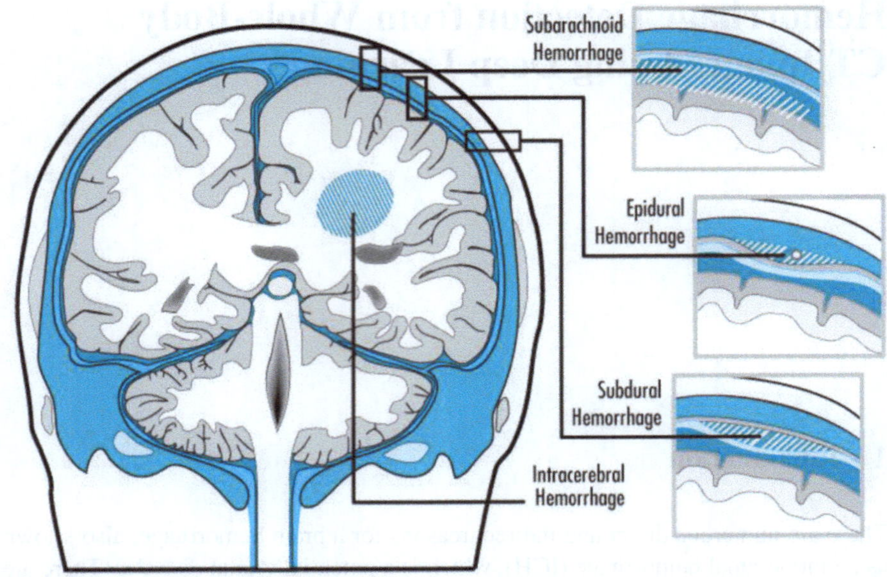

Fig. 1 Types of hemorrhage

- Medical imaging analysis: AI-based systems can be trained to analyze CT or MRI scans, as well as other types of medical imaging scans, in order to quickly and accurately identify signs of brain hemorrhage, such as abnormal brain bleeding.
- Pattern recognition: Large datasets containing cases of brain hemorrhages can be used to train AI models, enabling them to recognize patterns and characteristics that might point to a hemorrhage. By comparing new cases to the taught information, these systems can provide insights and indicate potential scenarios for further review by medical personnel.
- Decision Supporting: Artificial intelligence (AI) algorithms can provide suggestions or probabilities about the possibility of a brain hemorrhage based on patient data, including symptoms, medical history, and test findings. This helps physicians make well-informed judgments.
- Risk Examination: AI is useful in predicting the likelihood of brain hemorrhage in particular patient populations. AI models are capable of identifying risk variables and calculating individualized risk scores by analyzing large amounts of patient data, including demographics, medical histories, and lifestyle aspects. Early intervention and preventive efforts can be aided by this knowledge.

To build the predictive AI model, the data from the electronic health records of more than 3000 patients were collected as DICOM whole-body CT images. They trained the algorithms to detect links between certain data points and patients who were ultimately found to have intracranial hemorrhages. Research results indicate that an AI model can be used to immediately determine the best possible medical facility for a patient, possibly reducing further transfers and time-consuming evaluations.

1.1 Medical Imaging and DICOM

Using a variety of imaging methods and techniques to take pictures of the human body in order to help with diagnosis and treatment is known as medical imaging [3]. It also helps to keep an eye on continuing medical concerns, which makes it easier to create successful rehabilitation plans. Medical imaging comprises a wide range of techniques, each using a different technology to accomplish certain goals in the creation of images. Medical image analysis has become an increasingly important topic in healthcare, with a key role it plays in both scientific research and patient care.

In general, medical imaging refers to a variety of procedures and methods that are used to see the body's internal organs and tissues for therapeutic, health monitoring, diagnostic, and treatment reasons in relation to illnesses and injuries. On the other hand, some instruments make use of radio-guided surgery (RGS), such positron emission mammography (PEM). Other imaging modalities include planar x-ray equipment, which come in digital, analog, and portable versions, computed tomography (CT), ultrasonic imaging (US), and magnetic resonance imaging (MRI). A CT scanner was used as the imaging modality for this specific study. As shown in Fig. 2, the DICOM files produced by the CT scanner are saved with the ".dcm" extension and can be viewed and edited with any DICOM viewer [15].

Many healthcare facilities use the DICOM (Digital Imaging and Communications in Medicine) protocol, which makes it easier to transmit medical images and related data. Originally created in partnership with the American College of Radiology (ACR) and the National Association of Electrical Manufacturers (NEMA), DICOM is a registered trademark of the NEMA. With the common goal of standardizing the interchange of medical image data, users from various medical imaging professions collaborate to oversee DICOM's standards and practices through the DICOM Standards Committee.

Two core components of the DICOM definition are the formats used to store digital medical images and associated data, as well as the protocols that facilitate the integration of various communication systems that are necessary for the medical imaging workflow. In the field of medical imaging, DICOM has cemented its standing as the industry standard. Recently, a sizable percentage of digital medical imaging systems sold by large vendors—including servers, libraries, acquisition devices, workstations for diagnostics, medical printers, and more—comply with and incorporate different DICOM standard segments based on their unique functionalities. Furthermore, a wide range of medical facilities, including testing centers, research labs of all kinds, and public and private hospitals, have acknowledged and used DICOM.

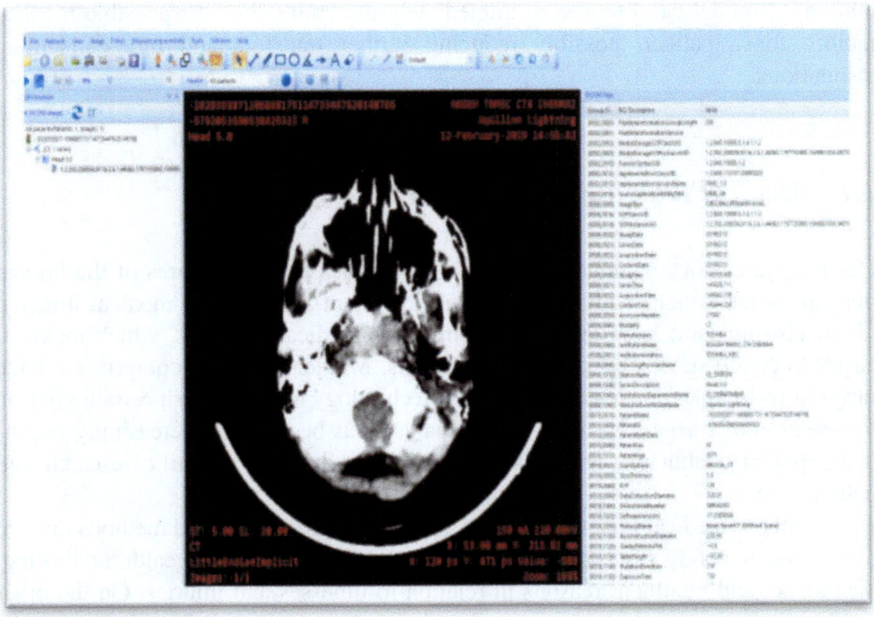

Fig. 2 DICOM visualizer

1.2 *Deep Learning for Medical Imaging*

Deep learning (DL) has been a potent and quickly developing topic in recent years. Remarkable generalization powers have been shown by deep learning-based networks, especially when it comes to complex medical problems like disease identification, medical image analysis, medical image categorization, and organ recognition. The focus on feature learning, which makes it possible for data representations to be automatically acquired, is a fundamental characteristic of deep learning techniques. The ability of deep learning-driven automated diagnosis systems to execute complicated cognitive tasks that previously required specialized knowledge and experience, as well as their faster inference, have attracted a lot of interest in recent years. Deep learning systems are trainable from start to finish because they are excellent at finding significant patterns and features in large datasets without the need for explicit instructions. When interpreting and carrying out complex cognitive activities, like interpreting CT scans, which have historically needed the knowledge of a radiologist, this machine learning skill proves essential [4].

Convolutional Neural Networks (CNNs) are the most efficient deep networks because of their structural paradigms that are inspired by biology, as seen in Fig. 3. In this work, we utilize a deep convolutional neural network to identify and categorize brain DICOM images with intracranial hemorrhage (ICH) [5]. CNN

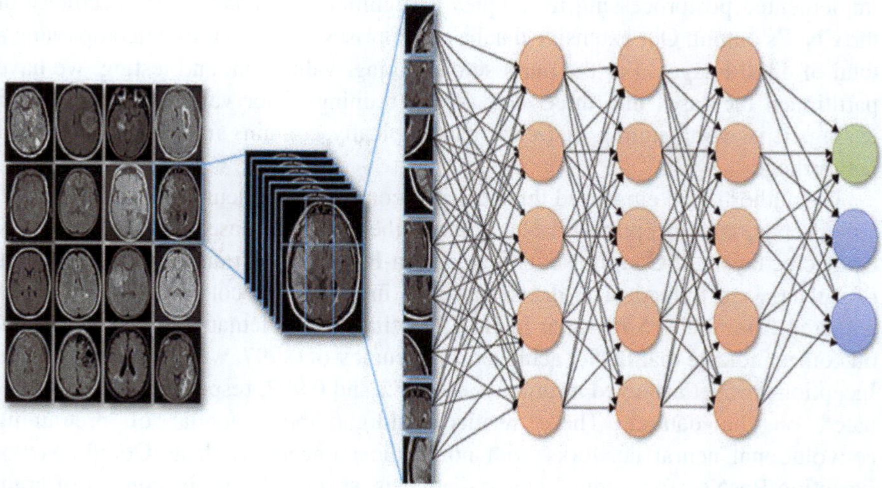

Fig. 3 Medical imaging using deep neural network

stands out as the most efficient deep network because of its architectural paradigms, which are naturally influenced by biology. CNNs are not like typical networks; instead, their architecture imitates the hierarchical feature detection of the human visual system. Through layers of convolutions, activations, and pooling, they are adept at catching complicated visual patterns that allow them to learn complex representations from input. It is important to remember, though, that CNN models usually use fully connected layers, which are similar to a traditional feedforward neural network that was trained using the backpropagation algorithm. CNNs have many benefits, but they could also have some of the disadvantages of less complex neural networks. The possibility of overfitting, heavy processing overhead, and the requirement for significant quantities of labeled data for efficient training are some of these possible drawbacks. CNNs are therefore susceptible to some of the difficulties that come with using neural network-based techniques, despite the fact that they are extremely effective tools for image-related tasks [16]. To maximize their potential and allay these worries, careful planning and training methods are required.

2 Literature Survey

In work [6], author utilized a sophisticated deep convolutional neural network that possesses the ability to learn features and perform classification simultaneously. This groundbreaking approach eliminates the need for multiple labor-intensive manual adjustments. We have further enhanced the network's performance by calculating the average output for rotated versions of the input image. Additionally, we have implemented postprocessing techniques to significantly enhance the specificity of the CNN's output. Our extensive database comprises 134 CT cases, encompassing a total of 4300 images. For the purpose of training, validation, and testing, we have partitioned the cases into three sets: 60 for training, 5 for validation, and 69 for testing. It is worth noting that each case typically contains multiple instances of hemorrhages.

The author in [7] employed three distinct convolutional neural networks: LeNet, GoogLeNet, and Inception-ResNet. During the training phase, only the last fully connected layers of GoogLeNet and Inception-ResNet were trained, while all layers of LeNet were trained. The dataset utilized in this study consisted of 100 cases gathered from the 115 Hospital in Ho Chi Minh City, Vietnam. The experimental outcomes indicate that LeNet achieved an accuracy of 0.997, while GoogLeNet and Inception-ResNet achieved accuracies of 0.982 and 0.992, respectively, when evaluated on the dataset. These results highlight the potential of pretraining convolutional neural networks with nonmedical images, such as GoogLeNet or Inception-ResNet, for medical image diagnosis, specifically in the context of brain hemorrhage diagnosis.

In [8], author has proposed a model that undergoes a transformation through the integration of a dilated convolutional neural network (CNN) with hypercolumn features. This innovative approach involves sampling a select number of pixels and concatenating the corresponding features from multiple layers. By allowing the freedom to sample individual pixels instead of entire image patches, the model focuses exclusively on the brain region while disregarding CT background padding. This strategy significantly enhances convergence time and accuracy as it enables the model to concentrate solely on healthy and defective brain tissues. To address the challenge of class imbalance, we employ a sampling technique that ensures an equal number of pixels are selected from each class.

The author in [9] created and evaluated a ResNet-50-based prediction model for bleeding type classification. With an average recall score of 76% for every class, our model's performance in correctly predicting numerous classes was impressive, coming in at 93.3%. We show that it is possible to get an average recall of 86% with a precision level of 70% by varying the prediction thresholds. This demonstrates how prediction performance in multiclass classification tasks can be balanced, maximizing sensitivity and precision to satisfy particular needs.

Deep convolutional neural networks and autoencoders were employed in the study [10] to accomplish this goal. The scientists have found that these models' performance varied when they were tested and trained on a dataset consisting of 2527 photos. Significantly, the three hidden layers of the stacked autoencoder they utilized in their study allowed it to beat the other networks, resulting in the lowest Mean Squared Error (MSE) and the maximum classification accuracy. The authors came to the conclusion that the CNN lower training volume might be the reason for the stacked autoencoder's better performance. CNNs usually need a large number of training instances in order to converge efficiently, therefore the study's relatively short dataset may have hurt the CNN's performance when compared to the stacked encoder.

In a different study, Mahajan and Mahajan [11] used the watershed method in conjunction with an artificial neural network (ANN) to classify different types of brain hemorrhage based on CT scan results. This allowed for a more thorough examination of brain hemorrhage. This study's authors used a feature extraction method before feeding photos into the neural classifier. The gray-level co-occurrence matrix (GLCM) was utilized by them to extract different attributes from the pictures. To identify the precise type of bleeding, these features were then classified using a traditional backpropagation neural network. The study found that the use of efficient image processing techniques, such as noise removal and sophisticated segmentation approaches, is necessary for the correct identification of different forms of bleeding.

3 System Design and Implementation

The primary aim of this project is to employ deep learning techniques for the efficient and automatic segregation of brain images from a vast archive of whole-body image data [12]. As the available DICOM images are unlabeled and manual labeling by trained radiologists is prohibitively expensive, the proposed approach leverages feature vectors encompassing all pixels of the images for an automated clustering process. The entire workflow, from dataset collection to the ultimate classification using a CNN model, is depicted in Fig. 4, with each step thoroughly explained below.

3.1 Anonymization

Patient demographic details and various other image research information are stored in the DICOM tags within an image header. To safeguard patient privacy when utilizing images for presentations, teaching materials, or publications, it is crucial to implement an "Anonymization" process in accordance with DICOM regulations. This process involves the removal or renaming of confidential patient information from the DICOM header when uploading a DICOM file for processing. By adhering to these standards, patient privacy is effectively protected during image usage.

3.2 Image Segmentation

Image segmentation refers to the process of dividing an image into multiple sub-regions. It holds a vital role in image processing, and when it comes to medical images, the segmentation task can be particularly challenging. Medical images often exhibit properties that make them difficult to segment, such as multiple tissues with similar intensities, overlapping and occluding organs, and significant physical

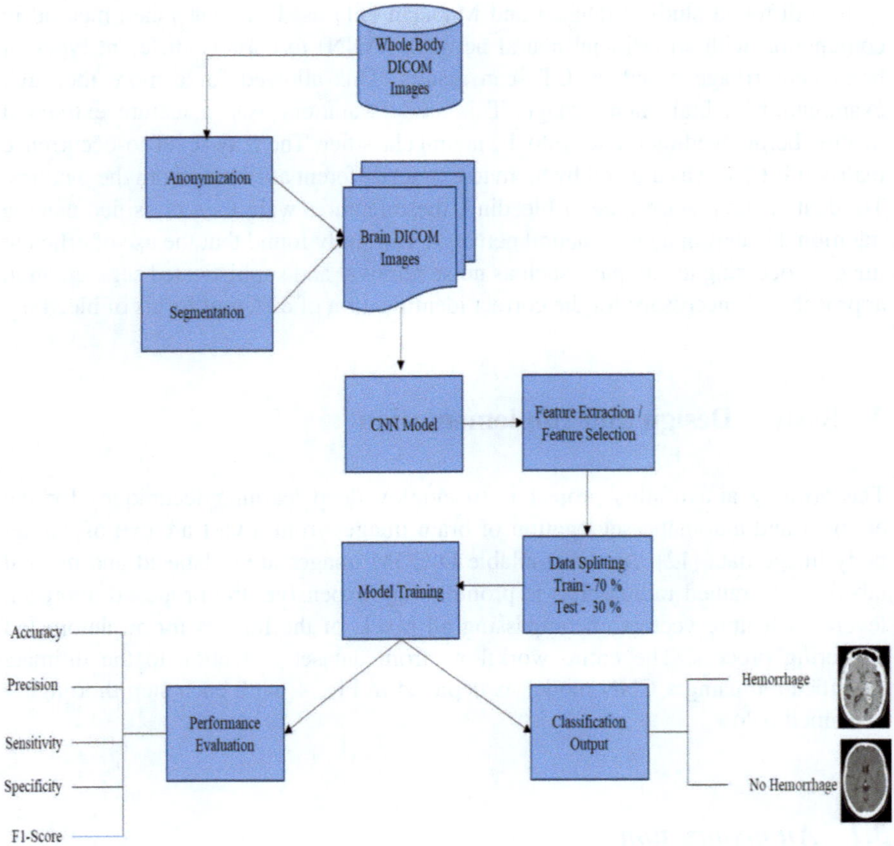

Fig. 4 Overall system design

variations among individuals. Over the past nearly 50 years, image segmentation has undergone continuous development and has now become a primary target for deep learning approaches in the field of medical imaging. In the present study, the focus is specifically on segmenting brain images. This segmentation is achieved by utilizing the attribute called "Protocol stack" from the DICOM tags, and the segmented brain images are collected and saved into a separate file for further analysis [13].

3.3 *CNN Architecture*

The identification of brain hemorrhages is among the many medical imaging tasks in which Convolutional Neural Networks (CNNs) have proven to be remarkably effective. The use of CNNs in this situation takes advantage of their innate capacity to independently extract hierarchical features from input images, which enables them to recognize complex patterns and critical attributes that are necessary for accurate

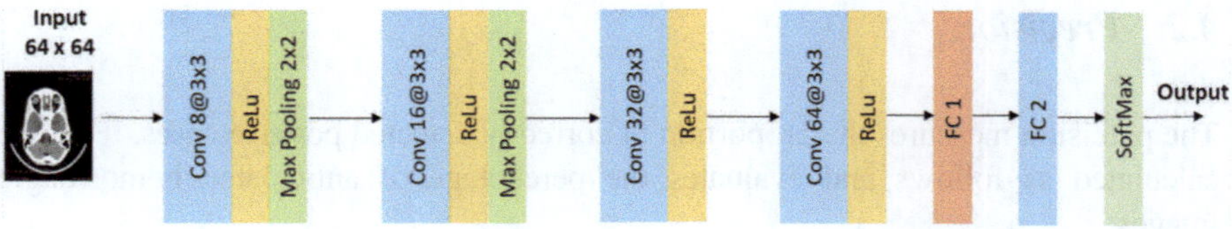

Fig. 5 Proposed CNN architecture

identification. The image's raw pixel data are converted into features using the CNN's tiered architecture. The architecture used in this study consists of multiple layers, as shown in Fig. 5, including the input layer, max-pooling layer, global average pooling layer, 2D convolutional layer, and dense layer. After extracting and honing features, each layer transfers them to the following layer. The network is then able to identify and classify pertinent data for precise brain hemorrhage detection by using these extracted features to be classified within the model.

The CNN utilizes [14] the large and diverse dataset of brain images with labeled hemorrhage annotations obtained from the previous step. This dataset will be used to train and validate the CNN model. Then it performs necessary preprocessing steps, such as resizing, normalization, and augmentation, to enhance the robustness and generalization of the CNN model. The preprocessed inputs are fed to the proposed CNN architecture which consists of four convolutional layers with ReLu activation function and two maxpool layers. Then it consists of two fully connected layers and Softmax layer at the output. The model is trained on the preprocessed dataset using an appropriate loss function and optimization algorithm called Adam optimizer. The performance of the trained CNN model is evaluated using final performance metrics.

4 Evaluation Metrics

A number of assessments were conducted to assess the accuracy of the CNN model such as precision, recall, and F1 score.

4.1 Classification Accuracy

The values of true positives (TP), false positives (FP), true negatives (TN), and false negatives (FN) are used to calculate accuracy, which evaluates a binary classification model's overall correctness. It can be calculated as follows and indicates the proportion of correctly identified hemorrhagic images:

$$\text{Accuracy} = (\text{TP} + \text{TN})/(\text{TP} + \text{TN} + \text{FP} + \text{FN}) \tag{1}$$

4.2 Precision

The precision measures the proportion of correctly detected positive cases. It can be calculated as follows and evaluates the percentage of anticipated hemorrhagic images:

$$\text{Precision} = \text{TP}/(\text{TP} + \text{FP}) \tag{2}$$

4.3 Sensitivity

Sensitivity measures the proportion of real positive cases detected correctly by the model out of all positive cases in the dataset. It is possible to calculate it as follows:

$$\text{Sensitivity} = \text{TP}/(\text{TP} + \text{FN}) \tag{3}$$

4.4 Specificity

Specificity is the fraction of true negative cases identified properly by the model out of all negative cases in the dataset. It is calculated as follows:

$$\text{Specificity} = \text{TN}/(\text{TN} + \text{FP}) \tag{4}$$

A high specificity value indicates that the model has a low rate of false positives.

4.5 F1-Score

The F1-score represents the harmonic mean of precision and recall, offering a more comprehensive evaluation of model performance. It measures the robustness of the model having a maximum score of 1, meaning perfect precision and recall, and having a minimum score of 0.

$$\text{F1-Score} = (2 \times \text{Precision} \times \text{Recall})/(\text{Precision} + \text{Recall}) \tag{5}$$

5 Results and Discussion

In order to identify brain hemorrhages in whole-body CT scans, we used Convolutional Neural Networks (CNN) in this study. With a predetermined set of hyperparameters—30 training epochs, 16 batch sizes, and an initial learning rate of 0.001—the CNN model we constructed was trained using the Adam optimizer. With an accuracy of 96% on the test dataset, the CNN model showed impressive results following this training program. This achievement highlights the model's ability to distinguish between photos of hemorrhagic and healthy brains with accuracy. The CNN has the potential to be a useful tool for radiologists and other healthcare professionals in helping them diagnose brain hemorrhages quickly and effectively, as evidenced by its high degree of accuracy in distinguishing between these serious medical diseases. The study's findings demonstrate the potential of CNNs to increase the precision and effectiveness of medical image processing, which will benefit patients and advance medical imaging diagnosis (Table 1).

The results obtained in this study reveal the robustness and accuracy of the CNN model with the Adam optimizer and the chosen hyperparameters for brain hemorrhage detection in medical images. The model's strong F1 score, sensitivity, specificity, accuracy, and precision show how well it can identify brain hemorrhages. When taken as a whole, these measures show how the model can help medical practitioners diagnose brain hemorrhages without producing false positives or false negatives. The model's performance was greatly enhanced by the selection of the Adam optimizer and suitable hy-perparameters. The model was able to effectively modify its weights during training thanks to the initial learning rate of 0.001, which enhanced convergence and increased generalization on the validation set. The excellent sensitivity and specificity of 100% indicate that the model has learned to effectively differentiate between hemorrhagic and nonhemorrhagic brain images. This is essential for making accurate diagnoses and ensuring timely and appropriate medical interventions for patients. The high precision of 95.7% indicates that the model is cautious in making positive predictions, minimizing the risk of misdiagnosing healthy patients. This precision is especially important in medical settings, where false positives can lead to unnecessary procedures and treatments.

Table 1 Outcome of proposed CNN model at each epoch

Epoch	Accuracy	Precision	Sensitivity	Specificity	F1-Score
05	86	76.8	100	90	96
10	94	90.6	91	93	95
15	95	94.3	90	91	90
20	**97**	**95.7**	**100**	**100**	**96.2**
25	78	69.4	100	80	92.3
30	96	92.3	100	90	94.5

Table 2 Comparative result analysis of proposed model with other state-of-the-art (SOTA) techniques

Deep Learning Algorithm	Accuracy	Precision	Sensitivity	Specificity	F1-Score
AlexNet	90.6	93.5	92.4	93.7	89.2
SVM	79.3	78.6	76.2	94.6	90.1
ResNet	93.5	95.3	94.3	95.7	92.3
Proposed CNN	**97**	**95.7**	**100**	**100**	**96.2**

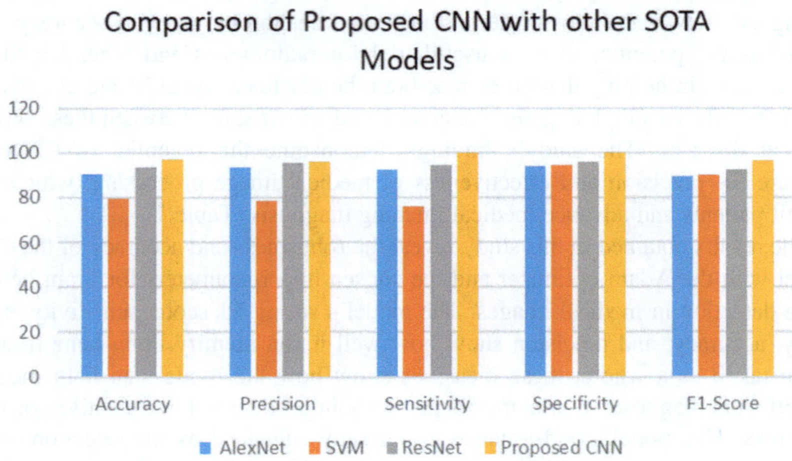

Fig. 6 Comparative result analysis of proposed model

The F1 score of 96.2% at epoch 20 demonstrates the model's balanced performance, taking into account both precision and sensitivity. This balanced evaluation is crucial in medical diagnosis, where both avoiding false positives and correctly identifying true positives are of paramount importance (Table 2 and Fig. 6).

6 Conclusion

The proposed CNN model demonstrated remarkable accuracy and efficiency in detecting brain hemorrhages from whole-body DICOM medical images. The CNN's ability to automatically learn and identify relevant patterns and features from brain images contributes significantly to its performance. While the results are promising, further research and refinement are necessary before such models can be effectively integrated into clinical practice for assisting healthcare professionals in diagnosing and treating patients with brain hemorrhages.

References

1. Phan, A. C., Vo, V. Q., & Phan, T. C. (2018, February). Automatic detection and classification of brain hemorrhages. In Asian Conference on Intelligent Information and Database Systems (pp. 417–427). Cham: Springer International Publishing.
2. Ch. B. Revathi, J. M. S. V. R. Kumar, and B. Sujatha, Intracranial hemorrhage detection in human brain using deep learning, AIP Conference Proceedings, 2023, https://doi.org/10.1063/5.0131258.
3. Murugan, S., Sumithra, M. G., & Murugappan, M. (2021). Efficient Clustering of Unlabeled Brain DICOM Images based on similarity. Journal of Physics: Conference Series, 1916(1), 012017. https://doi.org/10.1088/1742-6596/1916/1/012017
4. Elen, A. Diker, and A. Subasi, Brain hemorrhage detection using computed tomography images and deep learning, Applications of Artificial Intelligence in Medical Imaging, pp. 283–303, 2023, https://doi.org/10.1016/b978-0-443-18450-5.00014-1.
5. Toğaçar, M., Cömert, Z., Ergen, B., & Budak, Ü. (2019, November). Brain hemorrhage detection based on heat maps, autoencoder and CNN architecture. In 2019 1st International Informatics and Software Engineering Conference (UBMYK) (pp. 1–5). IEEE.
6. Majumdar, L. Brattain, B. Telfer, C. Farris and J. Scalera, Detecting Intracranial Hemorrhage with Deep Learning, 2018 40th Annual International Conference of the IEEE Engineering in Medicine and Biology Society (EMBC), Honolulu, HI, USA, 2018, pp. 583–587, https://doi.org/10.1109/EMBC.2018.8512336.
7. Mohamed Abed, A Comprehensive Examination of Human Brain Disorders, Journal of Biomedical and Sustainable Healthcare Applications, vol.3, no.2, pp. 141-152, July 2023. https://doi.org/10.53759/0088/JBSHA202303014.
8. J. Amin, M. Sharif, A. Haldorai, M. Yasmin, and R. S. Nayak, Brain tumor detection and classification using machine learning: a comprehensive survey, Complex & Intelligent Systems, vol. 8, no. 4, pp. 3161–3183, Nov. 2021, https://doi.org/10.1007/s40747-021-00563-y.
9. Francisco Pedro, Theoretical Analysis of the Brain and Artificial Intelligence, Journal of Robotics Spectrum, vol.1, pp. 024-035, 2023. https://doi.org/10.53759/9852/JRS202301003.
10. A. K. Gupta, T. Maity, A. H, and Y. K. Chauhan, An electromagnetic strategy to improve the performance of PV panel under partial shading, Computers Electrical Engineering, vol. 90, p. 106896, Mar. 2021, https://doi.org/10.1016/j.compeleceng.2020.106896.
11. Douglas Klutz, Neurofeedback for Cognitive Enhancement, Intervention and Brain Plasticity, Journal of Biomedical and Sustainable Healthcare Applications, vol.3, no.1, pp. 045-055, January 2023. https://doi.org/10.53759/0088/JBSHA202303005.
12. Al-Ayyoub, M., Alawad, D., Al-Darabsah, K., &Aljarrah, I. (2013). Automatic detection and classification of brain hemorrhages. WSEAS transactions on computers, 12(10), 395–405.
13. Jnawali, K., Arbabshirani, M. R., Rao, N., & Patel, A. A. (2018, February). Deep 3D convolution neural network for CT brain hemorrhage classification. In Medical Imaging 2018: Computer-Aided Diagnosis (Vol. 10575, pp. 307–313). SPIE.
14. Davis, V., &Devane, S. (2017, December). Diagnosis & classification of brain hemorrhage. In 2017 international conference on advances in computing, communication and control (ICAC3) (pp. 1–6). IEEE.
15. Gautam, A., & Raman, B. (2021). Towards effective classification of brain hemorrhagic and ischemic stroke using CNN. Biomedical Signal Processing and Control, 63, 102178.
16. Ozaltin, O., Coskun, O., Yeniay, O., & Subasi, A. (2023). Classification of brain hemorrhage computed tomography images using OzNet hybrid algorithm. International Journal of Imaging Systems and Technology, 33(1), 69–91.

Deep Learning for Mental Health Disorder Via Social Network Analysis

1 Introduction

Depression is a prevalent and significant medical mental disorder that has an important effect upon an individual's thoughts, emotions, and behavior. Numerous difficulties, physical as well as mental in character, may result, reduce your capacity to perform at place of employment and at home, and induce feelings of grief as well as a loss of enthusiasm in past hobbies. It is one of the most frequently disregarded causes of pain and fatalities worldwide, particularly among young adults. It is critical to comprehend and appreciate the grave consequences of this silent murderer [1]. According to the World Health Organization, there are currently 264 million infected people in the world, and that figure is rising daily. Depression happens every day, unlike regular mood fluctuations and fleeting emotional reactions [2, 3]. According to other research, 6.7% of people are suffering depression annually, or approximately one out every fifteen people. Furthermore, 16.6% of the population of people will develop depression during some time in their life. Though it may happen at any time, depression typically strikes around late adolescence and middle age. Women experience depression more frequently than males. One-third of women, according to some research, are going through a serious depressive episode right now. There is a higher degree of heredity (around 40%) when first-degree family members such as parents, kids, or brothers and sisters also experience depression. Many individuals today refrain from bringing up the topic of depressive disorders in conversation and avoid conceptualization of it as a sickness or a serious illness that requires treatment. People who have this ailment are afraid to even bring it forth because of this frame of mind. They are forced to suffer in silence and without any type of medical care as a consequence.

Finding depression symptoms among individuals often the most crucial step in the medical management of all illnesses, especially mental disorders like depression. The likelihood of reducing depression symptoms and underlying disorders can be

© European Alliance for Innovation 2024

A. Haldorai et al., *Artificial Intelligence for Sustainable Development*,
EAI/Springer Innovations in Communication and Computing,
https://doi.org/10.1007/978-3-031-53972-5_8

greatly increased by identifying depressive symptoms early on before they are assessed and treated. Additionally, it can lessen detrimental effects on social, economic, and academic life as well as happiness and good health. Potential diagnostic methods need the patient provide information to an emotional care specialist and require patients or caregivers to describe the symptoms. These methods are frequently expensive for the underprivileged populations of impoverished nations. This component of diagnostic approaches frequently results in patients' depression not being recognized and leaves them untreated [4]. The demand for a low- or no-cost diagnostic technique without the patient's physical participation is exceptional. Social media data have reportedly been used to identify, gauge, and monitor changes in the transmission of diseases in recent years, according to reports. Social media's increasing popularity has brought in, there is a significant possibility to enhance the information that is available to doctors and researchers in the field of mental health, making it more competent and equipped. Additionally, unpleasant emotions spread through social media have an adverse effect on people and raise the likelihood of depression and other mental illnesses. One of the biggest risk factors for suicide is mental illness, which affects over 80% of those who attempt suicide.

Social media has been a widely used tool for the sharing of thoughts and knowledge over the past 10 years. A tiny passage of text might reflect a person's mental state. Because of this, practitioners can infer a lot about a person's mental health and well-being from their Facebook, Twitter, or Instagram posts. Numerous books have been written about the physiology of social interactions in relation to social media, including breakups in relationships, quitting Nicotine and alcohol consumption, celebrating suicide, sexual assault, and mental diseases. The chance for success of employing social media as an indicator for depression diagnosis is made possible by the steadily rising use of digital media worldwide [5]. These social media platforms have developed into a forum for individuals to discuss their opinions, ongoing feelings, their emotions (how they feel) with their family members, friends, and close associates. Social media users' posting and sharing habits provide insight into their daily routines and mental health. An individual may experience depression at various levels or intensities. This intensity is typically minimal at the beginning, and the severity is determined by how the issue develops over time. According to the degree of depression, the affected person will often need treatment. Therefore, a domain expert may not always be able to prescribe the best course of treatment simply by identifying depression in a patient. Patients with early-stage depression and low intensities would benefit from taking precautions. Figure 1 shows the analysis of depression using social media. The technology will initially gather information from social media, such as tweets submitted by users of social networks. The algorithm will then process and analyze the textual data while interacting with a knowledge base that contains words for depressive sentiment. A friendly user interface will then warn the monitor (such as social workers or the parents of children) to any difficulties surrounding probable depression as a consequence of the analysis.

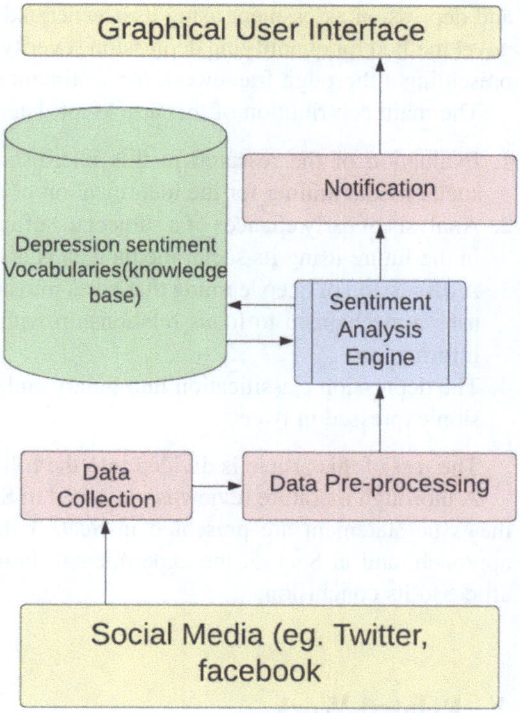

Fig. 1 Flow chart for depression detection via social media

This paper suggests an approach that applies deep learning techniques to analyze user-generated data from social media sites in order to objectively gauge the severity of depression. To be more precise, a modified shallow Long Short-Term Memory (LSTM) network is created to predict the degree of depression. Notably, this represents the first attempt to quantify depression in this way. The project involves gathering a dataset of 527 elements, each of which represents a different aspect of depressive illnesses as defined by the body of existing literature. Emotional, event-triggered, behavioral, user-level, and depression-related aspects are five categories into which these elements are divided. A modified shallow LSTM network is trained to predict depression severity using this combined dataset [6, 7]. The study also investigates several feature extraction methods, pretrained word embeddings, and deep learning algorithms to tackle the classification problem. Convolutional Neural Network (CNN) and LSTM layers are used in this framework, often referred to as the Social Sentiment Classification using Lexicon-based Method (SSCL), to improve the tasks of emotion detection and classification. In order to capture contextual information, the framework also makes use of self-attention mechanisms and Gated Recurrent Unit (GRU) models. GRU models were chosen because they outperform other recurrent neural network models like LSTM in terms of efficiency and efficacy. In addition to using GRU and self-attention techniques to extract contextual and implicit data, this system integrates CNN and LSTM layers to capture Twitter patterns and semantics. In conclusion, the study advances sentiment analysis and depression assessment using user-generated social media data by introducing a novel method for quantifying depression severity using deep learning techniques and presenting a thorough framework for sentiment classification.

The main contribution of this article for detection of Depression is as follows:

1. Evaluation of the research in this interdisciplinary field of mental health and social media mining for the identification of depression.
2. Analysis of early chances of a subject to suffer from depression and monitor them in the future using its social media data is discussed.
3. A discussion of deep learning that takes into account the weights of the words and uses a mechanism to focus relationship within the words as well as contextual information.
4. The depression classification into binary and ternary tasks is analyzed in depression expressed in tweets.

The rest of this article is divided into the following sections.

A thorough literature review is presented in Sect. 2, and then some key ideas and the issue statement are presented in Sect. 3. In Sect. 4, we show the suggested approach, and in Sect. 5, the experimental findings. In Sect. 6, we finally put this article to its conclusion.

2 Related Work

An interdisciplinary field that combines social media mining, mental health, and depression is the detection of depression using social data. Considering the fact that depression has become more prevalent in our society, early detection is crucial. It is a widespread mental condition that has serious consequences, including suicide, financial hardship, societal unrest, physical harm, and several grave health issues. In this section, we review the literature on depression identification and contemporary developments in the field, moving from early traditional methods to modern social media data exploitation.

2.1 Traditional Approaches on Depression Detection

In [8], depression screening tools consist of a series of 21 questions that primarily examine the respondents' psychological health. The CES-D Scale [9] uses a similar system of 20 questions to assess depressed symptoms. The development of these assessment tools is proof that research into depression and the difficulties it poses have a lengthy history that precedes the internet era [8, 9]. Long before the internet era, depression's substantial effects were recognized, prompting research initiative. The majority of these inquiries center on individuals' mental health, such as their guilt sentiments and sleeping habits. Either multiple choice questions with varying scores are used in the questionnaires, or users' input and the severity of their circumstances are also requested [11]. Based on the overall result, the severity of the depression is further diagnosed. Estimating a depression's strength is a long-standing research topic that helps in depression detection. The Beck Depression Inventory-II (BDI-II) [10] uses a classification system to evaluate depression based on questionnaire responses and is regarded as a key benchmark across a broad spectrum. This classification divides people into various groups: Scores between 0 and 13 indicate no or very little depression, scores between 14 and 19 indicate mild depression, scores between 20 and 28 indicate moderate depression, and scores between 29 and 63 indicate severe depression. However, the Diagnostic and Statistical Manual of Mental Disorders (DSM) was used as a substitute strategy in their analysis. The study [12] proposed nine different types of depressed signs as a standard criterion for depression diagnosis. Clinical psychologists typically use this technique to make a preliminary diagnosis of depression over time. In the earlier study, a different strategy known as the snowball sampling approach was utilized to compile a list of unusual words like transgender. The data were gathered using a number of Python tools. A user extracts a group of features from a given set of features in order to better understand the data that have been provided. The feature extraction was built on the findings of the data analysis. For each type of measure, a few features were retrieved, including "N Writings," "Avg Words," "Dev Words," "Min Words," "Max Words," and "Med Words" for Textual Spreading. The data were cleaned up by word encoding before being supplied to the word embedding models. To enable word embedding optimization, these models are built via unsupervised training [13]. The state-of-the-art early detection of depression using social media is proposed. To do this, innovative techniques that are explicit for immediate detection are taken into consideration. Additionally, text features are mined in a more precise manner than the current methodologies, which rely on TF-IDF vectorization [14].

2.2 Social Media and Depression Detection

Many academic investigations of depression analysis have made use of social media platform-derived datasets. Investigations in this area have looked into datasets obtained from popular websites including Twitter, Facebook, Reddit, and Reachout.com. The analysis of textual material that demonstrates signs of depression has been a prominent focus of these efforts. The research of [6] serves as an example of this type of study. The researchers focused their investigation on literary material that included both explicitly depicted symptoms of sadness and self-perceived indicators of the illness. Using a Long Short-Term Memory (LSTM) Recurrent Neural Network (RNN), Uddin et al. made it easier to classify and analyze these textual contents. The used model was good at distinguishing between postings that included descriptions of depression symptoms and those that did not. Finally, depressive postings were automatically predicted using the trained RNN. While providing a mean accuracy of 98%, these RNN and LSTM did not support specific depression detection [15]. Researchers have recently begun utilizing generated by users' data on these platforms for depression identification and mental health care since social media has come to be widely acknowledged as a medium for one's inner-self-expression. First, the study [16] examined the vocabulary used on Twitter to convey melancholy emotions. They conducted additional research on the depressive attitudes and behaviors of Twitter users. Face-to-face interviews were used in their study as a mode of inquiry with the goal of establishing a correlation between the interviews and data obtained from Twitter. The prediction of "major depressive disorders" was investigated separately [17], taking into account the integration of social media data. This information was used to analyze various user behavior patterns associated with depression and connect them to psychological studies.

In a recent endeavor [19], the investigation took the risk of employing depression analysis to learn more about how people use social media to express their emotional states. Another recent study [20] that concentrated on building a painstakingly labeled depression dataset collected from Twitter expanded this even further. Labels for cases of depression and nondepression were added to this dataset. The study identified six separate feature categories that were linked to depression, including both online behaviors displayed on social media sites and diagnostic criteria for clinical depression. To achieve their goals, they created a depressed lexicon system, an integrated learning tool that identifies Twitter users who are showing symptoms of depression. A further study [18] focused on the challenging process of moving depression labels from a source domain, Twitter, to a unique target domain, Weibo. Recent research in [20] evaluated the delay in diagnosis of depression made possible by social media. They used criteria including early risk detection errors (ERDE), a risk window-based method, and latency to assess a model's capacity to forecast depression based on Reddit data. The results of these research taken together highlight the potential of analyzing social media behavior patterns for depression. A cutting-edge project by [21] also suggested a deep learning-centric technique for the early identification of depression on social media sites. This method entailed comparing categorization techniques based on user-level linguistic metadata to the results of utilizing Convolutional Neural Networks (CNN) to evaluate various word embeddings.

2.3 Emotions and Mental Health Analytics

Depression and mental health are directly tied to emotions. It is possible to infer user emotion from their textual data thanks to social media trends. Users frequently publish a lot of messages and tweets that express their current emotional state. The data are transformed to a vector format and mined in order to decipher the syntactic and semantic content of a specific textual post. These representations can be made using a variety of techniques. The use of prepared dictionaries with useful labels for each keyword is common in keyword- or lexicon-based approaches [22, 23]. The labels either adhere to the category or dimensional model. Either manually or automatically, the dictionary is produced, which necessitates linguistics domain expertise. Sometimes unfavorable experiences in people's real lives have an impact on their mental health. Such an occurrence sets off mental health issues, which then precipitate depression. Personal event recognition within social media data has emerged as an interesting research area, related to the larger topic of event detection. The study [24] researched into compiling the life events encountered by users by taking inspirations from people's Twitter profiles. They also proposed an unsupervised framework that aims to build a vast archive of unique individual events. In a different project, the study [25] used a supervised classifier for a related objective. In this instance, the classifier was trained using manually defined textual features as inputs. This method made it possible to group user tweets into predetermined categories that matched different life events. Parallel to these initiatives, other literary works have focused on the effects of stress brought on by social media. They explore the complicated interplay between people and social media sites, including potential pressures brought on by online interactions and their effects on mental health. Word2Vec produces word representations primarily using one of the two model architectures: continuous bag-of-words (CBOW) or continuously skip-gram.

The model in the CBOW architecture predicts the current word based on the contextual words around it. Similar to the bag-of-words method, the prediction is typically unaffected by the order of contextual terms. The model performs the exact opposite of CBOW in the continuous skip-gram architecture. It forecasts the nearby contextual words based on the present word. The skip-gram architecture prioritizes close contextual terms over far-off ones. In terms of speed, CBOW beats skip-gram. When it comes to uncommon words, the skip-gram model outperforms CBOW. The RNN-based model was shown to be less accurate than the CNN model with Max that used optimized embedding [26]. Numerous models from the Scikit package were used by the researchers in this research section. These models were employed to determine whether the user required depression treatment. The following is the result of these models: The accuracy obtained during the boosting procedure was 75.13%, which was the highest. The bagging algorithm, on the other hand, had the lowest accuracy of all the algorithms used, at roughly 69%.

3 Detection and Monitoring of Early Depression

According to the American Psychiatric Association (APA) [27], depression is a significant condition that adversely affects our feelings and behavior. It is classified as a major depressive disorder. The main effects are a sense of sorrow and a lack of interest in anything. A person's regular capacity to operate at work and/or at home is likely to be hampered since it results in a variety of mental and physical issues. Some of the typical symptoms that depression patients typically encounter range from mild to severe intensity. The American Psychological Association (APA) identifies a number of depression's defining symptoms [27]. These include having a constant sense of sadness or melancholy; losing interest in or enjoyment from previously enjoyed activities; changes in appetite that result in unintended weight gain or loss, unrelated to dieting; sleep pattern disruptions that are accompanied by fatigue and decreased energy; sluggish physical movements and speech that are frequently noticed by others, along with a pervasive sense of isolation; and feelings of worthlessness or excessive guilt. The good news is that depression is a treatable disorder, despite the severity of these symptoms. It is highly recommended to seek a medical diagnosis of depression if someone persistently exhibits the aforementioned symptoms over a period of time longer than 2 weeks. This proactive strategy enables prompt diagnosis and appropriate therapy to lessen the negative effects of depression on a person's overall quality of life and well-being.

From social media, the system first gathers data (tweets). The machine then preprocessed and analyzed the data while interacting with a knowledge base that contained language for depressive sentiment. The analysis's findings, such as affirmative cases of potential depression, were then examined to gauge the severity of the condition [31]. Therapy recommendations or treatments were then assigned, along with aware warnings to the designated individual as a follow-up (e.g., human resources, social workers, or the parents of children), via the depression monitoring portal. The model was constructed over the course of three phases. Preprocessing, processing, and postprocessing are some of these stages. Features are transformed and extracted during the preprocessing stage, models are created and tuned based on the selected algorithm during the processing stage, and knowledge representation occurs during the postprocessing stage.

Figure 2 depicts a schematic representation of the depression monitoring paradigm. The Depression Sensitive Vocabulary (DSV), which was sourced from Myvocabulary.com and Thesaurus.com, was combined with the Twitter Application Programming Interface (API) during this phase of the research to collect data. The Depression Measurement Model (DMM) was subsequently developed and thoroughly evaluated using this collection of data as the core dataset.

The DMM's foundational knowledge base is safely kept on a MySQL server, which is necessary for its proper operation. The Depression Sensitive Vocabulary (DSV) and the related classification guidelines are both housed in this repository. PHP, a server-side programming language recognized for its adaptability and ability for such applications, powers the software used to conduct the analysis. The Valence Aware Dictionary for Sentiment Reasoning (VADER) model was used to assess the emotional nuanced inherent within the obtained data. This model accurately measures emotional intensity, which is important for understanding the levels of depression shown in the data. Depression is an emotional condition that is characterized by a low level of arousal and a negative valence. It entails the numbing of feelings, primarily those of sorrow, fear, rage, shame, and so forth. Sometimes the pain from these emotions spiraling back on themselves might be unbearable. According to this theory, relabeling the data based on depression intensity analysis and projecting user messages from their Twitter timelines into the Valance-Arousal space will aid in identifying the severity of their sadness.

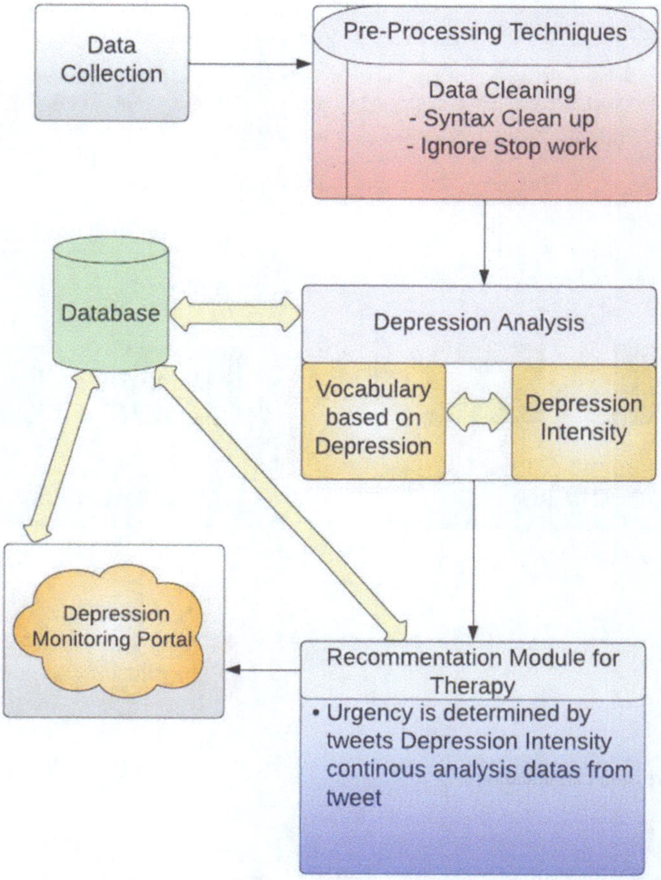

Fig. 2 Depression monitoring model

4 Mechanism Based on Deep Learning Hybrid Classification

The two data-related modules that we created were data preparation and data classification as shown in Fig. 3. The data were gathered, manually labeled as binary and ternary labels, then preprocessed in the first module to get the data ready for classification. We used a variety of feature extraction approaches on the prepared data in the second module, and we only used machine learning and deep learning classification algorithms on binary and ternary data. Finally, a hybrid paradigm for classifying depression is suggested as shown in Fig. 4.

The keyword "Depression," which played a crucial role in defining whether a certain tweet came into the category of being symptomatic of depression or not, served as the fundamental basis for categorizing tweets. A thorough manual

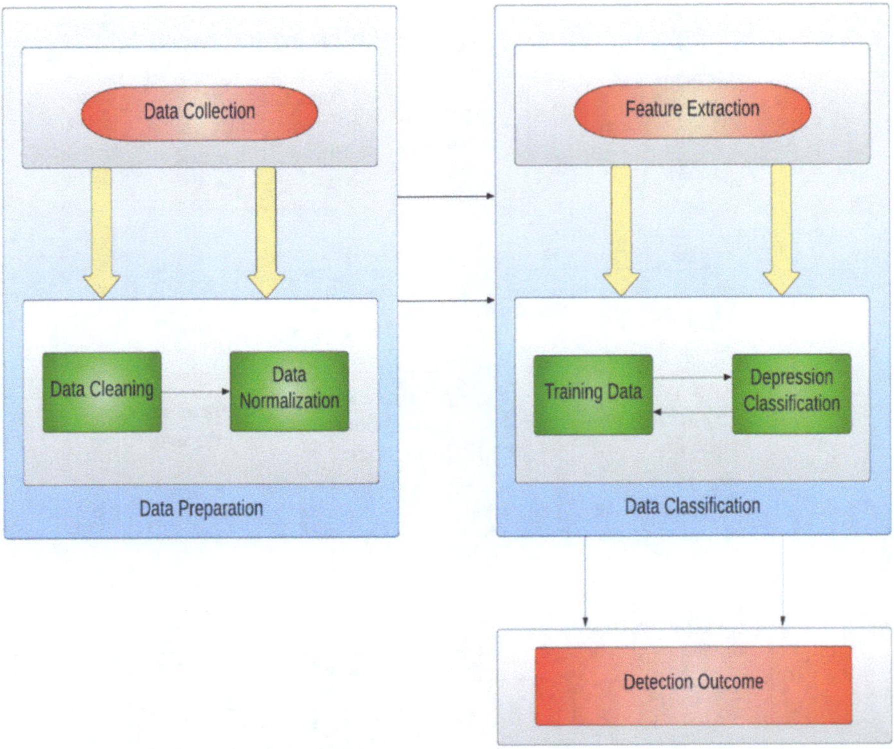

Fig. 3 Depression classification

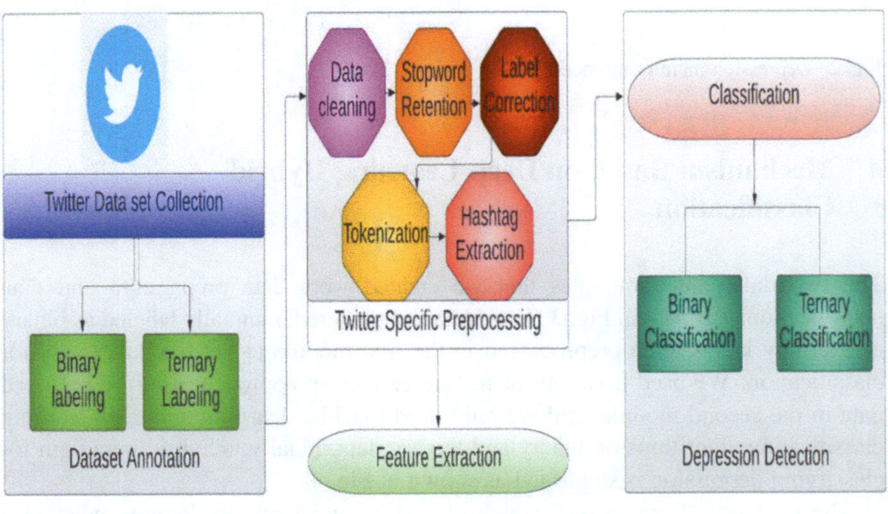

Fig. 4 Depression classification task and methodology

annotation procedure was started in order to fully capture the complex interplay of both implicit and explicit aspects within these two separate groups. This approach made it possible to fully comprehend the various ways that depression was being exhibited throughout the dataset. After this annotation procedure was finished, a custom preprocessing methodology was put into place that was made to prepare the data for analysis in the context of Twitter. As part of this thorough preprocessing procedure, repetitive comments, hyperlinks, hashtags, special characters, punctuation marks, unnecessary spaces, and text in other languages were all systematically removed. The task of binary and ternary categorizations was streamlined by utilizing the knowledge gained from the manual annotation process. This led to the creation of a clear set of labels that effectively divided tweets into two groups: those that reflected signs of depression and those that did not. This multistep procedure, which included both manual annotation and painstaking data preprocessing, provided a solid framework for later analysis and modeling, enabling a thorough investigation of content on depression inside the Twitter dataset.

The binary categorization separated tweets into those that suggested depression and those that did not. The ternary categorization further divided tweets into three categories: nondepressing tweets, first-person depression (tweets about one's own despair), and third-person depression (tweets about another person's misery). In addition, the text was lowercased while keeping the necessary stop words for recognizing first- and third-person references. Label correction made it possible to classify depressive remarks both explicitly and implicitly. Data were tokenized using a spacy tokenizer during preprocessing so that it could be examined further. We used a variety of feature extraction methods, such as TF-IDF, N-gram, and pretrained word embeddings. These methods were used on raw, binary, and ternary labeled datasets that had each been subjected to a different set of classification procedures [32]. Basic classification methods were initially used on the unlabeled raw data. Performance improved after switching to modified labeling, especially for real-time tweets. Deep learning algorithms were then investigated, incorporating N-gram theory and word embedding methods like Fast Text, Word2Vec, and GloVe. As a result, a unique hybrid model that included deep learning and an attention mechanism was created.

4.1 Dataset Preparation

The dataset [31] that was used in this study was obtained from Kaggle, an opensource platform. The dataset we selected consists of a file with 31,000 tweets with binary classifications for depressed and nondepressed participants. The labeling of the tweets was the issue that was seen in the raw dataset. The labeling process was set up so that tweets with keywords or terms associated with depression were considered depressed and given the label 1, while tweets without these terms or tweets that were off-topic were given the label 0 and were considered to be nondepressive tweets. As a result of the poor and imprecise labeling, the depression classification was incorrect. Additionally, the dataset did not take into account implicit remarks about depression, which occur when a person alludes to the illness but does not explicitly mention it in a tweet or statement. In summary, the depression-related phrases were used to identify the dataset. As an example, the label "1" was given to tweets that had clear terms like "depressed," "want to die," or "life is miserable." Notably, the tagging procedure ignored implicit comments and only took into account spoken displays of melancholy feelings. This annotation approach's main goal was to create a dataset that would help the model discriminate between True Positives (TPs) and True Negatives (TNs), which corresponds to the identification of both implicit and explicit findings. This strategy also attempted to reduce the number of False Positives (FPs) and False Negatives (FNs) in the model's classifications. The goal of this methodology is to build a solid foundation on which the model may learn. As a result, a more precise detection can be made.

Data sets [28] therefore contain errors, including false positives (FPs) and false negatives (FNs). False negatives, also referred to as false positives, are tweets tagged "0" but actually "1." False positives (FPs), false negatives (FNs), true positives (TPs), and true negatives (TNs) are all described in Table 1. Examples of FP and FN are also provided in Table 2. Since FP and FN have a negative impact on accuracy, they must be eradicated. As a result, labels have been created to greatly lessen the aforementioned issue.

In the initial dataset, there were tweets with the labels "1" and "0." False-positive and false-negative labels, however, presented difficulties for interpretation. An experienced mental health practitioner was consulted to provide a systematic approach for annotation correction in order to solve this problem. This procedure was rigorously carried out hand to assure accuracy. The original dataset included 31,000 tweets in total, of which 15,999 were classified as "1" and 15,000 as "0," as

Table 1 Elaboration of TP, FP, TN, and FN

TP—True positive	Positive examples of depression that have been adjudged to be positive
FP—False positive	Negative events that are classified as such and are not depressive
TN—True negative	Negative events that are deemed to be positive or depressed
FN—False negative	Instances of depression that are beneficial but are deemed negative

Table 2 Sample of label 1 and label 0

FP—Label 1, actually 0	For the first time in months, louis truly did manage to drag me out of my despair.
	Damn, have you ever improved to the point where you were able to overcome depression?
	I just cleaned my room, and it's already a shit good darn mess.
FN—Label 0, actually 1	Think your life is challenging because you have to conceal three chronic illnesses
	I've read a lot about the situation in Nigeria, but I haven't been able to make a difference because I don't have a job and my fears keep me up at night.
	My brain literally struggles to understand how life is real, like the fact that it exists, at times.

Table 3 Number of labels

Raw Tweets	No. of FPs: 930
	No. of FNs: 39
	No. of 1's: 15,999
	No of 0's: 15,000
Binary labeling	No. of 1's: 13,584
	No of 0's: 13,533
Ternary labeling	No. of 2's: 1742
	No of 1's: 11,843
	No. of 0's: 13,533

shown in Table 3. Duplicate comments were removed based on the depression detection domain to preserve fairness throughout the learning process and prevent distorted metrics. Additionally, tweets were cleaned up by deleting links, punctuation, excess spaces, non-English text, and special characters while enhancing them using hashtags to extract information. For consistency, the text was changed to lowercase. The term "pauses" was kept in order to improve the precision of learning models because it was crucial in capturing the emotional tone of both first-person ("I," "myself," "us") and third-person ("She," "himself") expressions. There were many often-used stop words omitted. Both machine learning (ML) and deep learning (DL) methods require tokenization, which was carried out. On the basis of white spaces, the Speedy Tokenizer divided the text. Additionally, it handled punctuation such as a single inverted comma and separated contractions like "shouldn't" into "should" and "n't." The Spacy tokenizer, which was created especially for deep learning and natural language processing, was used to tokenize our dataset [28]. The preprocessing stage included label rectification and/or data annotation; the statistics for label rectification is shown below. After label modification, Table 3 furnished the revised count of tweets for each class.

4.2 Analysis of Deep Learning Classification Models

The model's bottom section was where the classification process began. Bidirectional Long Short-Term Memory (BiLSTM) layer of output dimension received embedded characteristics as input. Here, BiLSTM was employed because, at any time step, it contained knowledge about the past and future, aiding the model in making more accurate predictions. Following that, hidden layer 1 received the output generated by the BiLSTM layer with an output size. With an output dimension, normalized metadata features were input into hidden layer 2. Deep learning classifiers and models have also been used to distinguish between depressive and nondepressive tweets in challenges involving depression detection. Category models were investigated using feedforward networks, convolutionary neural networks, long-term and short-term memory networks, recurrent gates, bidirectional LSTM (BiLSTM), and ensemble/hybrid models. A completely independent method with a

95.3 accuracy and 95 F1 scores is the convolutional neural network (CNN). The BiLSTM model performed well when combined with Word2Vec embeddings, attaining a noteworthy F1 score of 94.7 and an accuracy of 94.8. In addition, the CNN + LSTM architecture outperformed CNN + BiLSTM in the absence of embeddings and CNN + LSTM when Word2Vec embeddings were used. Fast Text embeddings combined with CNN + BiLSTM produced remarkable results, with an accuracy and F1-score of 96.6. When used in conjunction with CNN and applied to the ternary labeled data, the basic feedforward network demonstrated greater performance when compared to strategies like GRU and Word2Vec, displaying a strong accuracy and F1-score of 77. Similar results were produced by a straightforward CNN + LSTM model without embeddings. An F1-score of 80.1 and great accuracy were also achieved by Fast Text when paired with BiLSTM and CNN. A CNN + LSTM combination with Fast Text embeddings added produced an accuracy and F1-score of 79.4. With regard to comparison approaches, the addition of ternary labels, however, resulted in decreased accuracy and F1-score, which fell to 60.1. When GloVe embeddings were introduced, the results fell under the cutoff of 70. However, the model's performance greatly improved when it was supplemented with GRU and an attention layer, resulting in an accuracy and F1-score of 82.9.

The objective of an overview of learning models called word embeddings is to offer information on how words are represented or given meaning in our language. By mapping words onto a lower-dimensional semantic space, this is accomplished. These embeddings function fundamentally by maximizing an objective function that identifies connections between words that commonly occur together in particular contexts and those that are situated closely together in the semantic space. Some models in this category have an exceptional capacity to capture a broad range of lexical properties important to Natural Language Processing (NLP) applications. These include assessments of word similarity, word connections, and the overall prevalence of particular terms in texts or sentences. GloVe is unique since it was created by Stanford academics as an unsupervised learning method to produce meaningful vector representations of words. We included pretrained GloVe word embeddings in our analysis because they have already been trained on large corpora and are suited for a variety of NLP tasks. In our method, tweets were translated into a lower-dimensional vector space using a pretrained GloVe vector model. Our study collected textual properties from 2 billion tweets, or a total of 27 billion tokens and 1.2 million words, obtained from Twitter using the matrix's dimensions. It is important to note that the embedding matrix is made up of the word embeddings that were recovered using these models [29].

Self-attention appears as a potent instrument for collecting complex interactions between various sequence components, such as the syntactic differences between words. The proposed model can successfully stress contextual importance by fusing an attention mechanism with a GRU, acknowledging that each word inside a phrase contributes differently to the final identification. Using a self-attention layer with Softmax functions, which determine the proportionate importance of each word to the overall context of tweets, this is accomplished. A multiplication with multiple pertinent terms takes place in the process of understanding how these variables affect forecasts and in accordance with their relative weights. The Softmax function of the self-attention layer is crucial in accelerating the learning curve for semantic characteristics. In 32 learning units, or neurons, make up the initial FCL, which is paired with a dense layer that is triggered by the ReLU function. Eight neurons, all of which are turned on by the ReLU function, are present in the second FCL. In the final stage, an FCL with sigmoid activation function is included. This choice is appropriate because the function can scale output values between 0 and 1, which is exactly what categorization tasks call for. The sigmoid function is extremely useful for categorizing objects with both binary and multiple labels, which improves the model's versatility and applicability in a variety of classification settings.

5 Result and Discussion

The integration of attention processes is the main focus of this section's analysis of the results of the deep learning model. The discussion centers on a variety of performance indicators used to assess the model's efficacy and speedy detection of those displaying signs of depression. When using SVM and logistic regression for the binary classification problem, remarkable results were seen, especially when bi-grams were removed. The performance of the network saw very small gains after the addition of the swish activation mechanism [28]. Using feature subsets, we ran model training to further enlarge our inquiry. The emotion feature stood out among them for its higher performance in features-level prediction, which is supported by its natural association with depressive emotions. However, when working with user-level data, the situation changes due to its diversity, making it difficult to extract vital information on the severity of depression. To address this, we added user-level features to each feature set and ran parallel LSTM networks, using the user-level depression intensity as an extra element [30]. The importance of "first-person pronouns" and "TF-IDF" in accurately identifying cases of depression is stressed by prior studies. Experiments that carefully examined the relationship between various qualities and class labels lend validity to this assumption. To reach the best categorization outcomes, a wide range of feature combinations were investigated, with each iteration being specifically designed to achieve the highest level of accuracy in the identification process.

To ensure a thorough examination in both situations, the machine learning techniques used for binary-labeled data were expanded to data with ternary labels. The uni-bi-tri-gram method demonstrated extraordinary performance for Logistic Regression (LR) in the domain of ternary labels, producing a remarkable accuracy of 91.4% and an F1-score of 77.3. Simple feedforward networks, GRU, Word2Vec, and CNN-based architectures all have things in common with deep learning models, including simple feedforward networks. We support models that go beyond CNN and LSTM because of the need for increased precision in categorizing depression across implicit and explicit detection scenarios. To emphasize contextual clues within the text, this entails merging GRU and self-attention systems. The self-attention layer is essential for highlighting important and significant characteristics. This improves accuracy metrics and F1 scores while also making the depression classification task easier. Additionally, it decreases the frequency of false negatives while increasing the detection of actual positive occurrences. The self-awareness layer focuses on key characteristics at the same time that the GRU records complex long-distance interactions between words as they emerge. The role of the CNN in our proposed model is to stress semantics and important characteristics, departing from typical approaches that prioritize sequential and semantic information extraction. In addition, the LSTM acts as an effective method for sequence learning. GRU and a self-attention layer have been added in order to better detect implicit depression by revealing the contextual nuances that are hidden inside text.

Researchers conducted a thorough evaluation to determine how well user comments on Twitter could be identified by using the aforementioned categorization approaches. It used four different classifier algorithms to fully comprehend the varied weights of different feature categories. These techniques were purposefully created to take advantage of time patterns, language styles, emotional indicators, and a mix of all these characteristics. It is important to note that K-Nearest Neighbors (KNN) demonstrated impressive precision levels in this situation. The Decision Tree classifier, on the other hand, displayed remarkable skill in terms of recall and F-measure results. The Decision Tree classifier emerged as being the best at catching the pertinent signals, especially in the area of Facebook user comments indicating depressive attitudes. Additionally, the Decision Tree excels in terms of memory, precision, and F-measure for linguistic style. This encourages the use of a BERT (Bidirectional Encoder Representations from Transformers) model in an experiment to further improve accuracy. Both of the aforementioned problems are resolved by BERT, which has been demonstrated to perform not only more quickly. However, this experiment will only be used as a pilot study to confirm the results of the literature research. This experiment classified using a BERT-based model that Google had previously trained 27 times. The model was pretrained using the English Wikipedia and Book Corpus to assist it grasp English since it was created to enable translation.

The following procedures must be used to preprocess the data we extracted from the data collection. (1) Eliminate URLs. (2) Delete all emails. (3) Make all text lowercase. (4) Eliminate punctuation marks. Remove the stop word. Lemmatize the text. We started off by reading a piece of text which is shown in the Table 4. These classifiers have been run using for assessing matrices parameters (precision, recall, and F-measure). It was carried out using four different approaches. Depression instances that are true positive (TP) are expected to be true positive. True Negative (TN) cases of depression are those that are unfavorable and were expected to be unfavorable. Depression instances that are positive when they were expected to be negative are referred to as false negatives (FN). False-Positive (FP) depression instances are those that are real yet expected to be false positives. The following defines each evaluation metric. Precision is the ratio of actual positive instances to those that were predicted to be positive. The level of selected cases is what is correct.

Table 4 Text taken to lower case the words

['So, when you ask what the two illnesses are...they're similar in that they tend to have moodiness involved, impulsivity and self-damaging behaviors for compensatory measures. Otherwise, they're not really that similar, but they do seem to "like" one another and often show up in the same person...making their lives chaotic, difficult, dysfunctional, and intense. So, if your friend is concerned that they may have one (or both) of these illnesses, they should not be ashamed. It's definitely not their fault. They should, instead, get help from a professional...and INSIST that they are treated appropriately with the correct types of therapy and medication.']

$$\text{precision}(p) = \frac{\text{True Positive}}{\text{True positive} + \text{False Positive}}$$

Recall is calculated as the ratio of cases with true positive results to all cases. The selection criteria are the level of picked cases.

$$\text{Recall}(R) = \frac{\text{True Positive}}{\text{True positive} + \text{False Negative}}$$

The F-measure is the average of Precision and Recall. Both falsely positive and falsely negative results are recorded. Following is how F-measure is calculated:

$$F - \text{Measure} = 2\,\frac{\text{PR}}{P + R}$$

Lemmatized text has been removed from the text and the data, and both have been punctuation upper case letter free. By stacking a number of encoders on top of one another, BERT has a self-attention layer, which reduces the amount of time needed to train a model and enhances its performance at the expense of having a processing sequence length limit of 512 tokens. It is determined that trimming the sequence to the required length 29 is the best way to address this problem. However, this can

Table 5 Transmitted into lowercase words

```
[['so', 'when', 'you', 'ask', 'what',
'the', 'two', 'illnesses', 'are',
'theyre', 'similar', 'in', 'that',
'they', 'tend', 'to', 'have',
'moodiness', 'involved',
'impulsivity', 'and', 'self',
'damaging', 'behaviors', 'for',
'compensatory', 'measures',
'otherwise', 'theyre', 'not',
'really', 'that', 'similar', 'but',
'they', 'do', 'seem', 'to', 'like',
'one', 'another', 'and', 'often',
'show', 'up', 'in', 'the', 'same',
'person', 'making', 'their', 'lives',
'chaotic', 'difficult',
'dysfunctional', 'and', 'intense',
'so', 'if', 'your', 'friend', 'is',
'concerned', 'that', 'they', 'may',
'have', 'one', 'or', 'both', 'of',
'these', 'illnesses', 'they',
'should', 'not', 'be', 'ashamed',
'its', 'definitely', 'not', 'their',
'fault', 'they', 'should', 'instead',
'get', 'help', 'from', 'professional',
'and', 'insist', 'that', 'they',
'are', 'treated', 'appropriately',
'with', 'the', 'correct', 'types',
'of', 'therapy', 'and', 'medication']]
```

result in some crucial information being lost in the sequence. In order to reduce sequence length in this experiment, extractive summarization will be used for sequences larger than 512 tokens. Table 5 displays the raw data and the outcome of the summarization in addition to the data.

The model was adjusted using preprocessed data, containing 3412 data, of which 1884 data were tagged as depressing and 1528 other data as control, as was previously specified. The training is then finished using 8-batch sizes, 2×10^{-5} learning rates, and 1×10^{-8} epsilon, and the training test is evaluated using a split of 90% to 10%. About 10% of the training set is separated into a validation set in order to prevent the model from overfitting. As seen in Fig. 5, the model demonstrates a satisfactory outcome, and the configuration of summarization results in a training accuracy of 93% is crucial for performance optimization because it is thought to have an impact on one or more aspects of this model.

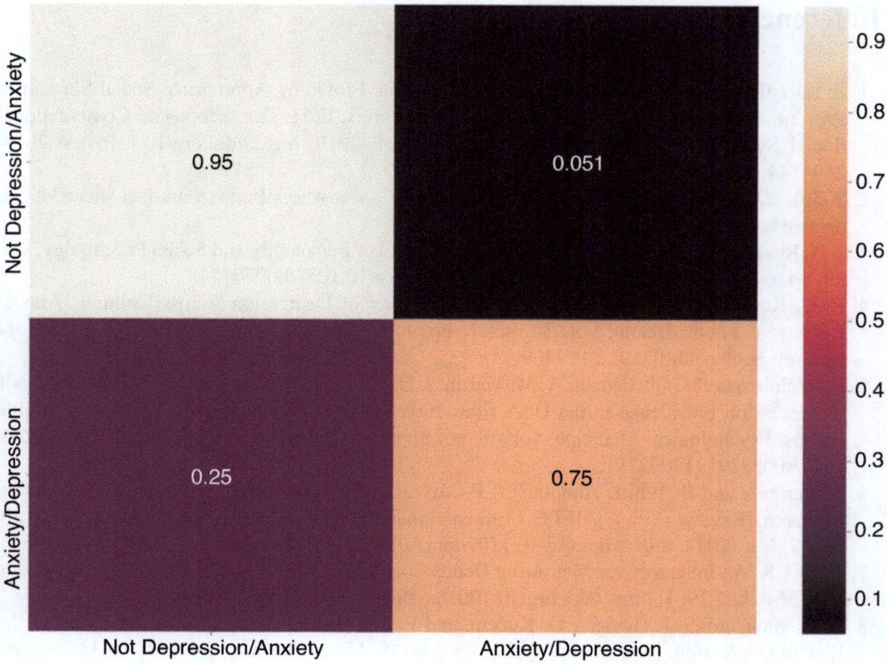

Fig. 5 Result of accuracy that shows decreased TPs

6 Conclusion

One of the most challenging problems to overcome is the detection of depression from text. However, we have explored a number of methods to identify depression in the text. Monitoring people's moods is possible via sentiment analysis. People that experience depression symptoms exhibit comparable behavior, which is reflected in the words and phrases shared on social media. As a result, this helpful information aids in identifying people who may be experiencing psychological disturbances like depression. Data annotation was done in two different ways: binary labeling and ternary labeling. To well to data triple classifications, tweets about other people's depression, and other depression data, tweets are categorized as depressed or not using binary annotations. Then the dataset and the related data preparation helped in the training process. Studies may also include a rating of the degree of depression to assess the severity of the condition.

References

1. F. Hao, G. Pang, Y. Wu, Z. Pi, L. Xia, and G. Min, Providing Appropriate Social Support to Prevention of Depression for Highly Anxious Sufferers, IEEE Transactions on Computational Social Systems, vol. 6, no. 5, pp. 879–887, Oct. 2019, https://doi.org/10.1109/tcss.2019.2894144.
2. 2020). *Depression.* [Online]. Available: https://www.who.int/newsroom/fact-sheets/detail/depression.
3. J. A. Russell, A circumplex model of affect., Journal of Personality and Social Psychology, vol. 39, no. 6, pp. 1161–1178, Dec. 1980, https://doi.org/10.1037/h0077714.
4. R. C. Kessler and E. J. Bromet, The Epidemiology of Depression Across Cultures, Annual Review of Public Health, vol. 34, no. 1, pp. 119–138, Mar. 2013, https://doi.org/10.1146/annurev-publhealth-031912-114409.
5. H. Weinberger, M. Gbedemah, A. M. Martinez, D. Nash, S. Galea, and R. D. Goodwin, Trends in depression prevalence in the USA from 2005 to 2015: widening disparities in vulnerable groups, Psychological Medicine, vol. 48, no. 8, pp. 1308–1315, Oct. 2017, https://doi.org/10.1017/s0033291717002781.
6. E. Cambria and B. White, Jumping NLP Curves: A Review of Natural Language Processing Research [Review Article], IEEE Computational Intelligence Magazine, vol. 9, no. 2, pp. 48–57, May 2014, https://doi.org/10.1109/mci.2014.2307227.
7. T. BECK, An Inventory for Measuring Depression, Archives of General Psychiatry, vol. 4, no. 6, p. 561, Jun. 1961, https://doi.org/10.1001/archpsyc.1961.01710120031004.
8. J. M. Havigerová, J. Haviger, D. Kučera, and P. Hoffmannová, Text-Based Detection of the Risk of Depression, Frontiers in Psychology, vol. 10, Mar. 2019, https://doi.org/10.3389/fpsyg.2019.00513.
9. L. S. Radloff, The CES-D Scale, Applied Psychological Measurement, vol. 1, no. 3, pp. 385–401, Jun. 1977, https://doi.org/10.1177/014662167700100306.
10. T. Beck, R. A. Steer, R. Ball, and W. F. Ranieri, Comparison of Beck Depression Inventories-IA and-II in Psychiatric Outpatients, Journal of Personality Assessment, vol. 67, no. 3, pp. 588–597, Dec. 1996, https://doi.org/10.1207/s15327752jpa6703_13.
11. Preface to DSM-5, Diagnostic and Statistical Manual of Mental Disorders, Mar. 2022, https://doi.org/10.1176/appi.books.9780890425787.x00b_preface_to_dsm-5.
12. Md. R. Islam, M. A. Kabir, A. Ahmed, A. R. M. Kamal, H. Wang, and A. Ulhaq, Depression detection from social network data using machine learning techniques, Health Information Science and Systems, vol. 6, no. 1, Aug. 2018, https://doi.org/10.1007/s13755-018-0046-0.
13. D. Hudson and A. Collins-Anderson, Understanding Perceptions of Depression and Depression Care across Culture and Context, International Journal of Environmental Research and Public Health, vol. 19, no. 18, p. 11720, Sep. 2022, https://doi.org/10.3390/ijerph191811720.
14. V. Leiva and A. Freire, Towards Suicide Prevention: Early Detection of Depression on Social Media, Lecture Notes in Computer Science, pp. 428–436, 2017, https://doi.org/10.1007/978-3-319-70284-1_34.
15. M. Z. Uddin, K. K. Dysthe, A. Følstad, and P. B. Brandtzaeg, Deep learning for prediction of depressive symptoms in a large textual dataset, Neural Computing and Applications, vol. 34, no. 1, pp. 721–744, Aug. 2021, https://doi.org/10.1007/s00521-021-06426-4.
16. Pabitha C, Kalpana V, Evangelin Sonia SV, Pushpalatha A, Mahendran G and Sivarajan S, Development and Implementation of an Intelligent Health Monitoring System using IoT and Advanced Machine Learning Techniques, Journal of Machine and Computing, vol.3, no.4, pp. 456–464, October 2023. https://doi.org/10.53759/7669/jmc202303037.
17. Li Hua Fang and Dong Yonggui, Definition, Challenges and Future Research for Internet of Things, Journal of Computing and Natural Science, vol.3, no.4, pp. 216–226, October 2023. https://doi.org/10.53759//181X/JCNS/202303020.

18. N. Gayathri, H. Anandakumar, R. Sathya, and S. Gowri, An Investigation on Spectrum Mobility Mechanisms in Cognitive Network Communication, 2023 9th International Conference on Advanced Computing and Communication Systems (ICACCS), Mar. 2023, https://doi.org/10.1109/icaccs57279.2023.10113068.
19. R. Xu and Q. Zhang, Understanding Online Health Groups for Depression: Social Network and Linguistic Perspectives, Journal of Medical Internet Research, vol. 18, no. 3, p. e63, Mar. 2016, https://doi.org/10.2196/jmir.5042.
20. F. Sadeque, D. Xu, and S. Bethard, Measuring the Latency of Depression Detection in Social Media, Proceedings of the Eleventh ACM International Conference on Web Search and Data Mining, Feb. 2018, https://doi.org/10.1145/3159652.3159725.
21. M. Trotzek, S. Koitka, and C. M. Friedrich, Utilizing Neural Networks and Linguistic Metadata for Early Detection of Depression Indications in Text Sequences, IEEE Transactions on Knowledge and Data Engineering, vol. 32, no. 3, pp. 588–601, Mar. 2020, https://doi.org/10.1109/tkde.2018.2885515.
22. Pak and P. Paroubek, Twitter for Sentiment Analysis: When Language Resources are Not Available, 2011 22nd International Workshop on Database and Expert Systems Applications, Aug. 2011, https://doi.org/10.1109/dexa.2011.86.
23. S. Ghosh, M. Chollet, E. Laksana, L.-P. Morency, and S. Scherer, "Affect-LM: A Neural Language Model for Customizable Affective Text Generation," Proceedings of the 55th Annual Meeting of the Association for Computational Linguistics (Volume 1: Long Papers), 2017, https://doi.org/10.18653/v1/p17-1059.
24. J. Li and C. Cardie, Timeline generation," Proceedings of the 23rd international conference on World wide web, Apr. 2014, https://doi.org/10.1145/2566486.2567969.
25. Li, A. Ritter, C. Cardie, and E. Hovy, Major Life Event Extraction from Twitter based on Congratulations/Condolences Speech Acts, Proceedings of the 2014 Conference on Empirical Methods in Natural Language Processing (EMNLP), 2014, https://doi.org/10.3115/v1/d14-1214.
26. Husseini Orabi, P. Buddhitha, M. Husseini Orabi, and D. Inkpen, "Deep Learning for Depression Detection of Twitter Users," Proceedings of the Fifth Workshop on Computational Linguistics and Clinical Psychology: From Keyboard to Clinic, 2018, https://doi.org/10.18653/v1/w18-0609.
27. *What is Depression?* Accessed: May 12, 2020. [Online]. Available:https://www.psychiatry.org/patients-families/depression/what-isdepression
28. Nadeem, M. Naveed, M. Islam Satti, H. Afzal, T. Ahmad, and K.-I. Kim, Depression Detection Based on Hybrid Deep Learning SSCL Framework Using Self-Attention Mechanism: An Application to Social Networking Data, Sensors, vol. 22, no. 24, p. 9775, Dec. 2022, https://doi.org/10.3390/s22249775.
29. Aijie Wang and Bin Liang, Analysis of Intelligent Decision Support Systems and a Multi Criteria Framework for Assessment, Journal of Enterprise and Business Intelligence, vol.3, no.4, pp. 224–235, October 2023. https://doi.org/10.53759/5181/JEBI202303022.
30. Allen Zhuo, Review of Computational Model from a Psychological and Neurophysiological Perspective, Journal of Biomedical and Sustainable Healthcare Applications, vol.3, no.1, pp. 001–012, January 2023. https://doi.org/10.53759/0088/JBSHA202303001.
31. K. S. Umadevi, K. S. Thakare, S. Patil, R. Raut, A. K. Dwivedi, and A. Haldorai, Dynamic hidden feature space detection of noisy image set by weight binarization, Signal, Image and Video Processing, vol. 17, no. 3, pp. 761–768, Aug. 2022, https://doi.org/10.1007/s11760-022-02284-2.
32. H. S. ALSAGRI and M. YKHLEF, Machine Learning-Based Approach for Depression Detection in Twitter Using Content and Activity Features, IEICE Transactions on Information and Systems, vol. E103.D, no. 8, pp. 1825–1832, Aug. 2020, https://doi.org/10.1587/transinf.2020edp7023.

Part III
AI for Urban Mobility and Smart City Applications

A Review on Smart Charging Approaches for Electric Vehicle

1 Introduction

Earth's temperature has risen above preindustrial levels as a result of global warming and carbon emissions. The major job in the near future is the mitigation of these problems. Road carbon dioxide emissions have grown as a result of the growing number of automobiles on the road. By June 2022, over 75% of carbon dioxide emissions will come through the transportation sector. The Fit for 55 packages were approved by the European Parliament (EP) in June 2022, making the CO_2 emissions objectives for light-duty vehicles and passenger cars respectively up to 2030., stricter by reducing them from −37.5% to −55% and from −31% to −50% in comparison to the levels in 1990. Paying attention to public roads and automobiles, which create approximately 70% of the transportation sector's CO_2 emissions [1], this measure will focus on reducing CO_2 emissions from road vehicles. Additionally [2], the plan seeks to attain carbon neutrality by 2050 and outlaw ICE-powered cars by 2035. The number of vehicles are rapidly increasing which causes drastic contamination of air in the highly populated area which was reported by the environment and climate change Canada in 2022 as shown in Fig. 1.

For the reduction of Carbon Dioxide (CO_2), Electric Vehicles (EVs) can play a vital role by promoting their advantages over other transportation, comfort, lower fuel cost, zero-emission, and cleaner atmosphere so that we can protect environmental sustainability. The benefit of EMS for MPS-EV as an onboard addition control software is that it has a low development cost and rapid development cycle compared to the technical complexity and high cost of developing hardware. The study of EMS has advanced significantly during the previous few decades. The EMS that has been created and put into operation may be referred to as governed by rules techniques, global optimization-based approaches, instant optimization-based strategies, or artificial intelligence-based/AI-based techniques. Rule-based EMS may be divided into three categories: fuzzy-based on-rules strategies [3], deterministic

© European Alliance for Innovation 2024

A. Haldorai et al., *Artificial Intelligence for Sustainable Development*,
EAI/Springer Innovations in Communication and Computing,
https://doi.org/10.1007/978-3-031-53972-5_9

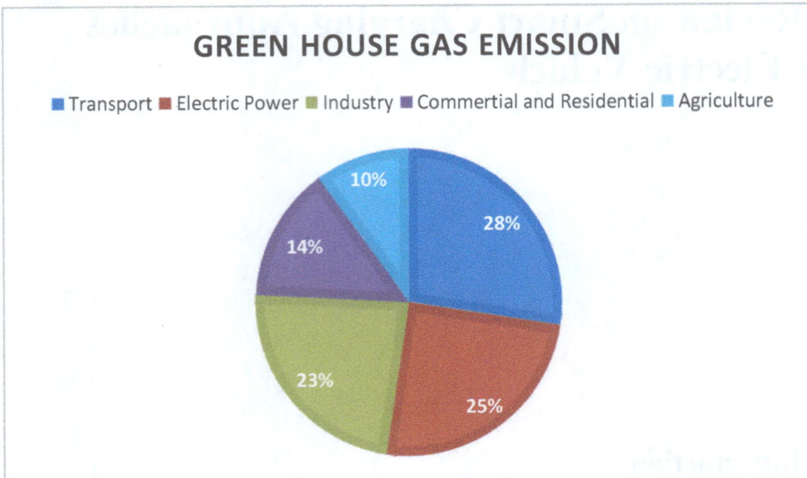

Fig. 1 Green house gas emission

theoretical rule-based strategies [4], and filter-based techniques [5]. Regarding the many benefits of electric vehicles, some of them include a decrease in greenhouse gas emissions, a reduction in the need for diesel or petroleum, a decrease in energy use at stationary locations, an increase in tank-to-wheel efficiency, a decrease in vehicle vibration and noise, the absence of gearboxes necessary for torque transformation, a simple mechanical structure that provides a higher power output over the full speed range, and others. The hypothetical situation claims that the rapid global supply of hundreds of thousands of electric cars has increased their practicality rate. About 6.8 million battery-powered cars were in operation worldwide in 2021, and about 3 million battery electric vehicles (BEVs) were produced for the first time.

One of the most prominent technologies for the achievement of the challenging targets powertrain electrification can be utilized, it can provide a better performance in lowering CO_2 emissions than traditional ICE power trains. For the purchase of new electrified cars benefited due to the growth of electrified power trains in the market which led to a 21% decrease in CO_2 emissions of new vehicles in Europe. Battery electric vehicles still have very strong limitations with short range, recharging time battery in long travel, high cost, and lacking of the infrastructure for charging [6]. As a stepping stone to net-zero carbon emission, the combination of electric and conventional power trains like hybrid electric vehicles with higher electrification with to adopting renewable sources of electricity production. The charging station (CSs) availability is a problem in the case of electric vehicles and should be addressed before deployment of the EVs on a large scale on the roads, this reason can lead to low preference for EVs in customers.

In order to make customers adopt EVs, a new approach that was proposed by researchers is energy transferring between electrical vehicles (Electric vehicles to Electric vehicles, a V2V approach) also known as the peer-to-peer charging approach (P2P), providing an EV fueling method that can be favorable for EV owners and the utility grid. As the required demand for Electrical vehicle charging, we concentrate on different analyses of different models and focus on different applications on models for predicting battery-operated vehicle charging demand by analyzing the distance traveled, time taken for charging, etc. The demand prediction charging can facilitate the company and the users to know the charging knowledge in terms of an extended period of time and the distance traveled. The only thing that could benefit from a comprehensive survey report addressing the present issues that EVs are facing, including driving range limitations, is to increase the acceptance of EVs among drivers, address data privacy, energy trading, not enough CSs, etc. The study of prior charging-demand-response strategies and an analysis of privacy-related characteristics are the main goals of this review.

The main objective of this chapter is as follows:

- An overview of energy management in electrical vehicles and the significance of data security within EV networks are been discussed.
- This paper gives a summary of EV charging planning and energy trading.
- Discussing federated learning and blockchain for data security of electrical vehicles.
- The article survey on machine learning approaches for electric vehicles and also some challenges of electric vehicles are mentioned.

The remainder of this chapter is divided into the following sections: A detailed literature review of energy management in the electrical vehicle is been explained and also existing strategies for optimally placing charging stations are covered in Sect. 2. Section 3 depending on EV categories, generally classifies various charging schedules and offers earlier research on increasing network security, including connecting EVs to other organizations for energy management, in order to better safeguard our system. Machine learning approaches for electric vehicles are given in Sect. 4. Section 5 classifies energy exchange and trade strategies between EVs and outlines the difficulties and potential directions for further study. Finally, this chapter ended in Sect. 6.

2 Literature Review

Jiarui Wang et al. in [7] elaborate that electric vehicles (EVs) are prominent to overcome the carbon dioxide emission, which they have proposed a load sensing method based on unsupervised learning for charging stations. The first step of the load sensing method is to establish the load sensing model properly with an unsupervised learning strategy and for the regulation capability, supervised learning with a neural network model is used. The effectiveness of the suggested method in assessing the load perception and controllability of charging stations is demonstrated by the verification of simulation data, and it offers a workable solution for the real-world application of auxiliary responses by electric vehicles.

Bassam Al-Hanahi et al. in [8] explain that the public's acceptance of EVs might rise as a result of well-positioned CSs' increasing station accessibility. It can help lessen the problem of power grid instability caused by the present, constantly rising demand for energy sources.

Tai-Yu Ma et al. [9] elaborate that though charging stations for electric vehicles are installed inside the transit system, the optimum CS positioning choices should take into account without going over the permissible limits of the power transmission network's features (without actually running over a specific voltage drop).

Yantao Huang et al. in [10] explain that a deep learning-based anomaly identification method is suggested. They have studied both supervised and unsupervised deep anomaly algorithms based on labels for anomalies. Deep anomaly detection is also used for charging data anomaly detection and is divided into deep anomaly detection for electric vehicles, deep anomaly detection for charging infrastructure, and deep anomaly detection based on the various charging data types.

Qi Feng et al. [11] elaborate that Artificial intelligence can be used for smart energy management, optimized energy storage operation, Data security, comprehension of AI technology's underlying concepts, cybersecurity, system retrofitting, and measuring the link between AI integration and economic gains are a few of the problems associated with adopting AI.

Amir Mosavi et al. in [12] elaborate that the application of AI models is applicable to all forms of renewable energy, including wind, solar, geothermal, hydro, ocean, bio, and hydrogen. Artificial neural networks (ANNs), support vector machines (SVM), wavelet neural networks (WNNs), decision trees, hybrid models, and ensemble models are a few examples of the several types of AI models. Hybrid ML models are frequently quicker, more precise, and simpler to use.

Pal Boza et al. in [13] explain that large amounts of data are needed to train these models; hence BD approaches are necessary. Blockchain technology is used to construct the energy management system for charging stations.

Sunil Kr. Jha et al. in [14] elaborate that this supports the system's demand side management. In the architecture, they are complemented by centralized and decentralized energy models. This aids in lowering peer-to-peer transfers of electricity from the power grid to electric automobiles.

C. P. Shirley et al. in [15] explain that the algorithm proposed is to schedule or provide electric vehicle charging on the basis of solar power to keep down the cost of charging.

E. Maximiliano Asensio et al. in [16] explain that autoregressive model for electrical vehicles is proposed on the basis of the power demand profile.

Yuanzheng Li et al. in [17] elaborate the RL that reinforcement learning has assisted with (DL) or deep learning framework of the probabilistic forecast for electric vehicle CS (charging stations) which incorporates the LSTM-AePPO forecast algorithm and the data transformer approach. There are some challenges that remain to get a higher prediction accuracy and the approach must also maintain the parameters that, like the cell state in LSTM, concurrently hold the data's and model's uncertainty.

Yu Liu et al. in [18] explain that the battery life may suffer if EV batteries are charged in an unauthorized way. Along with technological factors, arranging the decorously and quantity of battery charging at near accessible (CP) charging points might improve existing and potential consumers' quality of experience (QoE) with EVs. Numerous researchers are working to create techniques for estimating the driving range of EVs using information about the battery's state of charge (energy left inside the battery) and usage of energy from outside (external energy consumption) and considerations like air conditioning.

Can Berk Saner et al. in [19] explain the two main categories into which this research is divided, the Fact-based and paradigm-based forecasting or prediction methodologies.

Xingshuai Huang et al. [20] suggested an artificial intelligence predictive modeling framework for the series of hybrid electric vehicles energy management systems (EMS). For calculating the velocity profile in the future with energy consumption of the EV, considering route knowledge (information), the driver's characteristics, and the uncertainty of the traffic flow, a hybrid deterministic-stochastic approach has been developed.

H J Wang et al. in [21] suggested that to tackle the problem of Electric vehicle charging demand prediction employing a temporal encoder-decoder-LSTM (TLSTM- Ori-Time Features), concatenation has been developed. Also suggested a unique method for forecasting the remaining battery range of electric vehicles using two recurrent neural networks (RNNs).

3 Electrical Vehicle Charging Schedule and Data Security

3.1 Charging Schedule of Electric Vehicles

As a result, this predicted range number may fluctuate throughout the voyage. In contrast, model-based forecasts based on dynamic elements of the cars, such as route information, speed limits, driving habits, etc., apply computational methods to predict future energy use. The public will be able to adopt EVs in a sustainable way by providing cutting-edge solutions to the scheduling of charging amenities for EVs that are parked or shifting, as well as the best Charging Station recommendations with the shortest wait times.

3.1.1 Individually Used Electric Vehicle

A perfect suggestion model for a charging station needs information from individuals and also, the location of EV charging, detail about CS, and personal details of some users. In order to overcome this, many researchers have tried to discover a CS-based solution to the problem of location and charging. The attributes of several sides (EVs, CSs) can be combined to provide suggestions that are more precise. In applications where EV customers need notifications regarding the best time to charge, battery level, nearby charging stations, etc., a number of approaches have been proposed. They do, however, request personal data from consumers on both sides. Another study area focuses on prioritizing EV needs for charging scheduling and refilling. Several techniques and algorithms have been developed to enhance aspects, such as energy demand-response balance. The study in [22] investigated how long it takes for EVs that have been parked and placed in CSs to charge, and it suggested the best CSs with the shortest service wait times. A decision-making composite model was carried out by another source, by including evaluation theory for quantifiable aspects like charging length time, battery monitoring, etc.

3.1.2 Electric Bus Categories

In order for e-buses, a different type of EVs, to be broadly embraced by numerous governments, additional technology improvements are needed. These public cars could run into several common problems, such as extended charging times, unequal and inadequate geographical distribution of charging stations, extremely dynamic operation parameters, and so on [23]. Researchers and many research need to look into regional e-bus routes and stop networks to analyze operating and charging trends for the creation of real-time charging schedules in order to address the ongoing promotion of large-scale e-bus fleets. Fixed timetables are used to control the operation and charging schedules for e-buses, although sometimes these offline solutions don't work as well. The performance of e-buses will be impacted by dynamic elements, such as erratic congestion in traffic, changing weather/temperature, traffic-light situations, etc., making the formulation of the ideal approach difficult. Another difficult subject is the sizing of e-bus batteries since it affects the two ranges and the cost of operating a bus and is dependent on the type of transit service (duration and roads to use). For instance, according to the results of a case study in [24], the authorized electric city bus batteries are excessively huge given the local climate's short commutes and pleasant weather.

3.1.3 Categories of Electric Taxis

On the other hand, charging site modeling required the computation of additional variables, such as driving time, wait time, the ability to charge, and timely charging. E-taxi real-time charging site finder of the model was put out and is based on extensive data processing of GPS using real-time GPS directions and the electric taxi's history of recharging. Regarding trip durations, charging cycles, and station idleness, this reference sought to reduce the recharging initialization period. The requirements for electric taxis, which are growing in popularity, are different from those for personal EVs. Locating a fast-charging station at the closest location while maximizing profit is one important parameter for e-taxis, particularly during rush hours [25], a recommendation technique was put up to allocate e-taxis to the optimal charging place at the ideal moment. Variables for charging-time modeling include parameters such as electric-taxi (e-taxi) time unit revenue, charging capacity or volume, charging procedure duration, and duration-of-use power cost.

3.2 *Data Security of Electric Vehicles*

EVs are quickly establishing themselves as fixtures of daily life. The use of this renewable energy-powered mode of transportation is growing in popularity as a result of climate change and global warming. EVs have a variety of technological advancements that appeal to purchasers. Additionally, since the chargers used to refill these cars must connect to and interact with the car's computer to monitor charge level, voltage, and other parameters, they present difficulty. This might make EVs susceptible to malware or maliciously modified chargers that use the wrong voltage to harm vehicles. Although past evaluated research improved the recommender model's functionality, they neglected to take into account the data producers' desire to share their private information with other organizations due to some confidentiality issues.

3.2.1 Federated Learning

Federated learning or FL is known for its essential feature which is privacy, however, it provides meaningful privacy guarantees by using some techniques of privacy in FL. K-Anonymity is one of the security models which adds noise to hidden information (sensitive) from some others to produce them capable of reinstating the data. Figure 2 explains the basic architecture of federated learning. Additionally, FL or federated learning has employed (HL) or homomorphic encryption to protect user's private information by communicating training model parameters through an encryption method. The field of work known as Secure Multiparty Computation (SMC) also offers a framework for data security to guarantee accuracy.

According to the data-division qualities, the federated learning method is divided into three categories. When datasets have distinct samples but the same feature space, horizontal federated learning—also known as sample-based FL—is utilized. For instance, customers at two insurance company branches may differ (sample ID space), but the attributes of the firm are the same. The HFL methodology may be used to leverage the samples of data used in supervised ML algorithms to forecast driving behavior while maintaining the privacy and security of EV users' data whenever a single entity (EV) provides various data examples using the same attributes (driving time, GPS, acceleration). HFL model implementation does not require a complex algorithm because it is straightforward, however, it is a collaboration of multiple entities so HFL fails to function correctly from which similar samples of data are produced which have distinct features. However, Vertical Federated Learning or Situations where feature-based FL is appropriate when

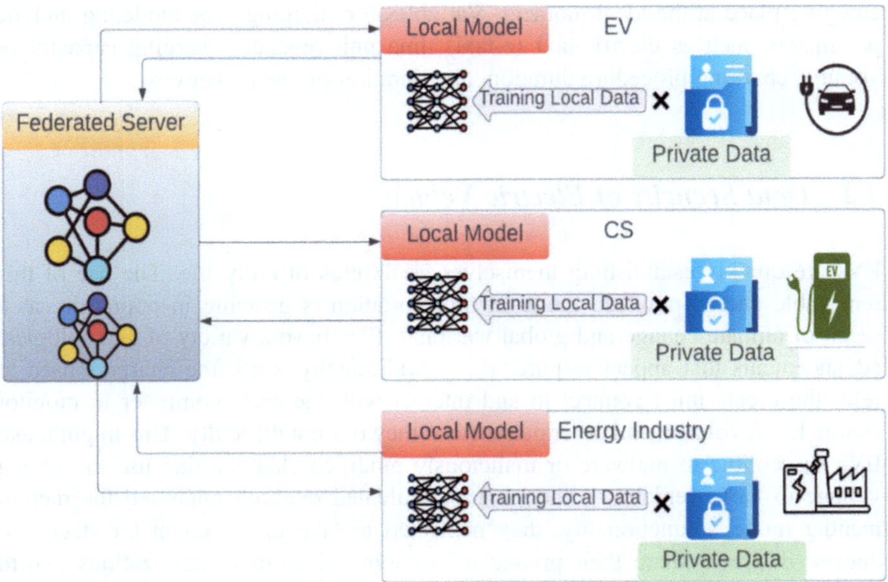

Fig. 2 Architecture of federated learning for secure data

comparable data samples have distinct feature scopes [26]. We need to create a model to cooperatively aggregate various properties for comparable samples in order to execute a calculation using the data from both sides. It appears that implementing VFL models is more difficult than implementing HFL since entity alignment between participants requires an additional step.

Therefore, to combine various entities in an ML model while taking user data safety into account, more complicated processes with increased computational complexity are added. Federated Transfer Learning may be used as an example with two distinct businesses that are spread out geographically and have minimal user group overlap. The transfer learning model served as the inspiration for FTL, which tries to offer ML methods when entities have inadequate data samples. For instance, a domain (electric bus) has some data accessible that may be employed in a forecasting or predicted model for a different EV domain with a sparse quantity of data. Federated Learning served as a foundational element for predicting energy demand within the EV network. This novel approach incorporates demand-based learning from charging station to facilitate energy management within the EV network. Here one CS will be in a central position that acts as a provider and collects all the data and information to perform the learning process. They built their model on FL; therefore, no confidential data were disclosed. The framework for learning was built using the CSS grouping method to improve model performance since it may increase prediction accuracy and reduce communication costs. A real-time FL was suggested by authors in [27] to anticipate the steering wheel angle of autonomous cars. In order to reduce communication costs and increase real-time streaming data flow, they included another sliding training window.

3.2.2 Blockchain for Security Threads

A sequential sequence of blocks are formed that contain the transactions because of the blockchain's structure, which uses a cryptographic signature mechanism to link every new block to the one before it. By taking transaction legitimacy into account, three basic consensus methods have been developed to promote agreement across completely decentralized nodes. Each newly inserted block is subjected to computationally difficult processes through Proof-of-Work (PoW). To add a new block to the blockchain, competing nodes compete to solve these challenging procedures, which constitute a cryptographic puzzle. An alternative technique called Proof-of-Stake (PoS) was introduced to overcome the issues with PoW's nonscalability & energy consumption. According to the Proof-of-Stake (PoS) consensus process, when validators secure their stake, they are chosen at random to set up an additional block in the blockchain network depending on the amount of staking made by the participating validators. Due to its better transaction throughput and reduced computational requirements, PoS is regarded as a cleaner and quicker protocol than PoW.

Within a blockchain network, representatives vote for their preferred validation servers to produce blocks that are new, using the Delegated Proof-of-Stake (DPoS) consensus algorithm. This protocol has a decreased probability of centralization since each representative has the ability to vote in proportion to their investment in the network. It is regarded as similar to the progressive PoS protocol. As a result, DPoS can handle more when compared to PoW and PoS, transactions are confirmed more quickly because fewer trustworthy nodes are needed to verify data in every newly created block of the network chain.

Subject to each application's requirements, other consensus techniques have been established. The five important metrics shown in Fig. 3 should be used to assess all of these treatments.

- *Security*: Security indicates with various attacks in protocol resistance, a Sybil and Denial of Service (DoS) attacks can crash the PoW-based model so this system should prevent the attacks using the feasible solution.
- *Throughput*: Indicates the number of transactions that are verified and released for every second (Tps) on the blockchain. For illustration, the throughput of Bitcoin is around 7 Tps.
- *Scalability*: In order to sustain the overall performance in terms of the transaction, node addition, and data. It denotes the consensus protocol's capacity.
- *Energy Consumption*: The main problem with blockchain technology's limited mainstream adoption is this statistic. As an illustration, the PoW consensus process consumes a lot of energy since choosing the best block miner needs a lot of computational power.

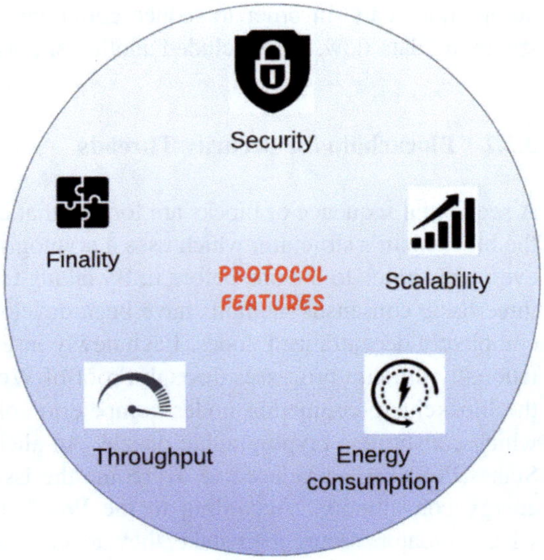

Fig. 3 Consensus protocol performance affecting properties

- *Finality*: Ensuring sure that blocks can't be intentionally reversed or modified once they have been added to the chain, ensures that the blockchain is deterministic.

The majority of earlier blockchain-based efforts attempted to reduce the vandalism of data by preserving local and global machine learning (ML) models throughout every relevant block [28]. This method is thought to work well as a solution for the safe exchange of information. It should be stressed, meanwhile, that participating nodes must acquire high-performance storage hardware. Additionally, the blockchain-based approach may have performance issues when additional users (EVs, CSs) join the network. This makes the concept unsuitable for real-time applications where ML results need to be generated quickly.

4 Machine Learning Approaches for Electric Vehicle

Depending on the charging technique used in the vehicle, some vehicle takes longer than half a day and some take at least 30 minutes. A completely charged EV, on the contrary hand, has a range for driving of around 180 to 663 km. A realistic charging point placement algorithm might become unexpected by taking into account this large range of driving. Therefore, instead of depending on a costly change in infrastructure, researchers are able to concentrate on different fields of investigation to offer an alternative for Vehicle charging (EV) while parked or on an excursion. This will help to minimize the inadequacy in the coverage of infrastructure for charging.

Table 1 Charging levels of EVs

Levels	Charging rate	Categories of EV charging
Level 1	Slow charging rate	120 V/used in residential
Level 2	Faster compared to level 1	240 V/both private and public
Level 3	Fast charging rate	480 V/ used in public

This part provides an overview of the background information regarding EV charging, machine learning methods, and algorithm assessment. The typical datasets that may be used to forecast charging behavior are also briefly described.

4.1 *Electric Vehicle Charging*

Flexible charging schedule approaches enable QoS and EV electricity requirements while reducing peak-period power consumption. In order to effectively lower electrical charging costs, reduce reach their highest power grid flows, and improve QoS, the idea of building multilevel hierarchical optimization frameworks manufactured up of distributed agents will achieve optimal time management through collaboratively computing a set of optimization problems [29]. Research in areas such as energy-efficient scheduling of routes, intelligent charge management, consumption of energy prediction, and suggested charging places will result in an enhanced EV customer experience, a greater range for driving, and a dependable energy supply in this situation.

As per the Electric Power Research Institute, Electric Vehicle (EV) charging is divided into three levels which are briefly described in Table 1.

With a normal working voltage of 120 V/15 A, level 1 charging offers the slowest charging rate. In this instance, plug and cord sets are used to deliver power from the charging equipment to the EV. On the other side, level 2 charging, which has been used in both private and public facilities, needs 240 VAC. Compared to Level 1, it offers quicker charging, however installing specialized charging equipment is necessary. Level 3, often known as "fast charging," makes use of 480 VAC and is frequently used in public and commercial settings with the aim of offering ICE cars a "grab and go" service comparable to petrol stations. Level three gives a charging in 30 minutes that makes customers not wait for a long time. The charging of EVs is divided into residential and nonresidential. Both levels 1 and 2 can be used in residential which is more predictable than the other. Nonresidential charging is unpredictable traffic at a public charging station close to a mall will vary depending on several variables, such as the day of the week, the weather, and the deals the mall is offering. Consequently, it may be more crucial to give more accurate scheduling, it is necessary to comprehend the dynamic recharging characteristics of nonresidential electric vehicle charging stations.

4.2 Machine Learning Approaches

Computer systems' capacity to learn from their experience without the need for explicit programming is machine learning (ML). The algorithms practice on their own by using the dataset and started recognizing the patterns and trends in that dataset. These models offer predictive analytics because, after a successful learning process, they can make precise predictions. Both these supervised and unsupervised machine learning (ML) techniques are usually separated into two groups. Depending on the type of response variable, additionally referred to as the variable that has been predicted, further Classifying data can be performed. If the variable that responds is continuous, the problem is commonly referred to as a regression problem. On the other hand, if the solution variable is categorical, the problem is referred to as a classification problem. Supervised and nonsupervised-based electric vehicle charging techniques are been discussed further in this part [30].

4.2.1 Supervised Learning

In supervised learning, machine learning (ML) algorithms have been taught using labeled data sets for training. As a result, the set of data include combining the target variable and the input variables, which is the matching response variable. In the broader setting of EV charging, a simple dataset comprising the vehicle's time of arrival, the city's name, and its departure time may be used as an illustration. The association between the departure time (response variable), city name, and arrival time will be discovered by the ML model. Despite the fact that it is outside the scope of this work to look into every supervised learning method, ML Techniques for EV Charging Behavior: A. The following discussion examines some of the most popular methods for predicting how EVs will charge.

Linear Regression

Linear regression (LR) is a statistical technique for determining the relationship between two variables. Finding the best-fitting line to represent the connection is the goal, which is predicated on the assumption that the independent and dependent variables are linearly related. The line is found by minimizing the sum of the squared differences among the predicted and actual values.

In LR, it is presumed that the connection between the response of the variable and the input characteristics is linear. Equation 1 is a representation of LR:

$$y = b_0 + b_1x_1 + b_2x_2 + \ldots + b_nx_n \tag{1}$$

Here y is the representation of the targeted variable, y-intercept represents b_0, the input features and regression coefficient are represented as follows respectively

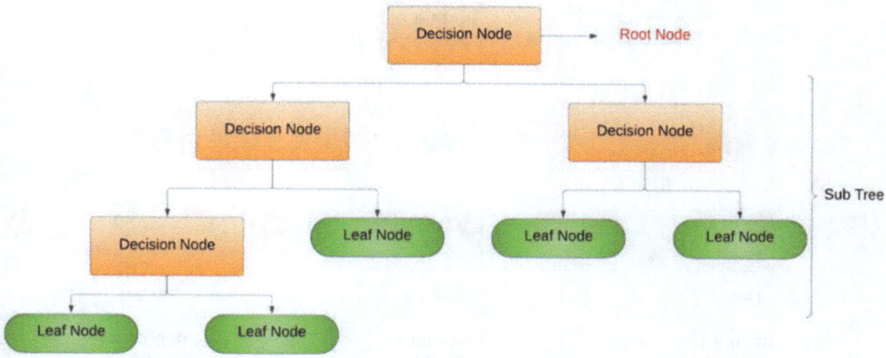

Fig. 4 Decision tree for machine learning

$[x_1; x_{2...}x_n]$ and $[b_1; b_{2...}b_n]$. When the dataset can be separated linearly and the procedure is straightforward to employ, LR is extremely helpful. Overfitting, which happens when a particular model performs throughout the training phase, exceptionally strong by incorporating extraneous input characteristics but it fails to produce generalized predictions, is a significant difficulty in training ML algorithms.

Decision Tree

A supervised learning technique called a Decision Tree (DT) may be used to tackle regression and classification issues, nevertheless, it is most commonly employed to solve problems with classification. It serves as a classifier with a structure resembling a tree, where internal nodes represent the traits of the dataset, branches the process of generating decisions, and each leaf node represents the classification result. A decision tree has two nodes: the one that makes the decision (decision Node) and the branching (Leaf Node). A decision node is used to make decisions since it contains numerous branches, but a Leaf node indicates the outcome of a choice and contains no additional branches.

A decision node is a location where choices are made. The leaf nodes are the places where no further splitting occurs. Figure 4 has shown a demonstration of a Decision tree for machine learning. The average value of every item in the leaf node is used as the prediction value for regression issues. The projected collection of classes is represented by the leaf nodes in classification tasks. DTs are easy to explain, especially when accompanied by a tree diagram that explains how predictions are made. A single DT is, however, frequently unable to make accurate predictions and is vulnerable to overfitting.

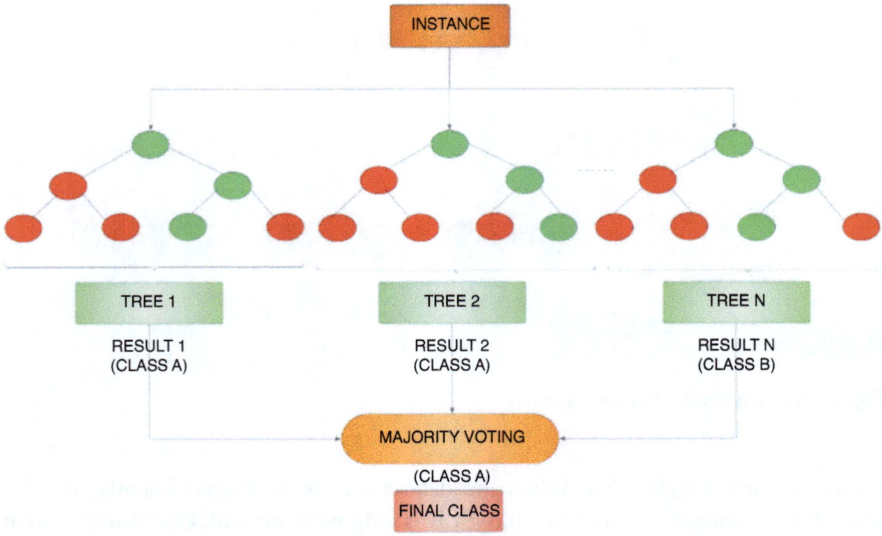

Fig. 5 Random forest

Random Forest

Random Forest (RF) is a prominent classification and regression technique. Given that regression and classification are the two most important aspects of machine learning, we may assert that the algorithm known as Random Forest Algorithm constitutes one of the most important algorithms in the field. It is based on the concept of ensemble learning, which brings together a number of classifiers to tackle a challenging issue and enhance the model's performance. The random forest algorithm, which creates a forest without correlation of decision trees by combining feature randomization with bagging, extends the bagging approach. An example of a random forest is shown in Fig. 5 below.

For regression, the predicted values which are average made by all the trees are aggregated; additionally, the majority vote from all of the trees is combined for categorization. A good example of ensemble machine learning is RF, where numerous machine learning models are first assessed and then combined with an individual model that frequently exhibits greater accuracy in predicting than the individual models. The main distinction between this and a gradient boosting technique, such as XGboost, is that a gradient boosting algorithm creates each tree one after another while taking into account the mistakes made by the prior trees. Comparing these methods to a straightforward DT model, overestimation is significantly reduced.

Support Vector Machine

Support Vector Machine, also known as SVM, is a potent and adaptable technique that is primarily utilized for regression as well as classification issues in supervised learning (ML) methods. A hyperplane in space with multiple dimensions that represents numerous categories is all that an SVM model is. To cut down on errors, SVM will create the hyperplane in an iterative fashion. SVM seeks to identify the maximal marginal hyperplane (MMH) by classifying the datasets.

Important SVM principles include the following:

- *Support Vectors*: The information gathered from points that are close to the hyperplane is known as support vectors. The separation line will be the one determined using these data points.
- *Hyperplane*: A hyperplane is a decision plane as well as space that is divided up between an assortment of different types of objects.
- *Margin*: The margin is the separation between these two lines over the closest data points for each class. It may be calculated using the angle that is perpendicular between the line and the supporting vectors. A high margin is preferred, whereas a small margin is disliked.

K-Nearest Neighbor

K-nearest neighbor or KNN is an algorithm that is used for both regression as well as a classification but mainly it is popular for only classification algorithms. KNN is referred to be a slow or lazy learning algorithm since it needs a dedicated training phase. The prediction of new data points is done by using a Euclidean distance that is used to measure the distance in order to find the k-nearest neighbors. It is assigned to the classes that contain the majority of neighbors.

4.2.2 Unsupervised Learning

In unsupervised learning, there are no labeled output variables; the training dataset only contains input variables. Data labeling is time-consuming and expensive in many real-world applications. The ML model's objective is to discover patterns or structures in the dataset. Unsupervised learning techniques like cluster analysis are frequently used to find groups of things that have certain characteristics. You may use unsupervised learning to identify groups of EV behavioral patterns. Some of the common methods that can be used for EV charging are discussed below.

K-Mean Clustering

To categorize data into k groups, one method for unsupervised learning is known as K-Means clustering based on certain characteristics found in each unique data point. This algorithm is used to analyze our past charging data and produce user behavior hypotheses for scheduling EV charging. Individual data points create k clusters using K-Means clustering, where each point is first randomly allocated to one of the k center points. Based on fresh center calculations, the data points are then redistributed to the nearest center.

Gaussian Mixture Model

In account to several normal distributions of the dataset, the Gaussian mixture model (GMM) serves as a learning model which is probabilistic that may represent sub-populations which are normally distributed. Although GMMs are often employed for unsupervised learning, supervised learning variants do exist. A GMM allows for the fitting of numerous PDFs since it is the weighted sum of m separate Gaussian distributions. As a result, realistic EV charging profiles are generated using a histogram of the data that has been gathered. EV charging profiles are created by using GMM to determine the charging probability of EVs. In several research, GMMs have been used to simulate EV behavior in order to improve the design and operation of future energy networks. GMM is used to represent each charging metric, such as start charging time and initial and ultimate state-of-charge (SOC).

Kernal Density Estimator

In parametric estimating approaches, the probability density function's (PDF) form must be taken for granted. When this is not achievable, by utilizing nonparametric estimation using the kernel function, we can calculate the PDF of a random variable that is continuous. Kernel functions have three requirements: symmetry, non-negation, and a minimum area under the function curve of 1. Gaussian KDE (GKDE) and diffusion-based KDE (DKDE) are two common KDE kernels. In electric vehicles, KDE can be used for smoothing out the density distribution and for matching out other data layers.

5 Energy Trading and Challenges of Electrical Vehicles

5.1 Energy Trading

Given the widespread adoption of EVs in the next years, the transportation sector contribution to greenhouse gas emissions, which mostly result from the extraordinarily high rate of combustion of fossil fuels, will be significantly reduced. However, this level of EV power use might create an additional issue by placing unmanaged charging demand on the power networks. Neglecting this recently brought-up problem might result in serious performance loss in the electricity distribution system, particularly during peak hours. The ability for EVs to construct V2X (Vehicle-to-Anything) interconnections to exchange energy as well as provide reciprocal energy transferring mechanism with the grid indicates how far their cooperation with the grid (charging or discharging mechanism) has progressed. Researchers are increasingly focusing on charging trading choices; therefore, it is crucial to establish a trustworthy network with numerous customers and prosumers to efficiently carry out energy trading activities. Participants can sell power without the help of centralized organizations (like utility corporations) when using the peer-to-peer (P2P) energy trading paradigm. Based on blockchain and smart contract features, the study in [31] offered a power that is secure trade transaction and communications mechanism among EVs.

To more effectively control the energy demand-response balance, they developed Elliptic Curve Cryptography, also known as ECC, which was used to establish an effective agreement protocol that enabled a mutually beneficial power transfer mechanism across EVs and the smart grid to operate. The official test network for Ethereum is the Ropsten dataset, which was used to evaluate its security performance. To maximize the social welfare factor for EV customers, the research in [32] provided a V2V power swapping model using fog computing and blockchain technologies. They improved the Practical Byzantine Fault Tolerance (PBFT) and DPoS consensus protocols to improve the performance of their suggested model, and as a result, they built the DoPSP agreement protocol to operate more effectively. The ability to move vehicles to send energy to or receive it from other moving vehicles presents one of the current problems in energy trading. The first challenge is to outfit EVs with electronics that can securely locate, travel, and transmit power in the vehicular network while taking the dynamic distance into account. Future research should provide alternate options that offer EV charging methods with the least amount of infrastructure needed and EV stop-times, including wireless energy transmission while traveling. Despite the fact that this method can alleviate many current problems, security risks might arise. One can use blockchain technology to advance future technologies that deal with related security energy trade concerns, and use technology that provides excellent individual information protection in a visible, tamper-proof environment.

5.2 Challenges

Despite the fact that the significant progress in the development and evolution of EVs in recent years, there are still certain issues that require investigation in order to provide better solutions. Having a secure and dependable rechargeable battery is the cornerstone of creating an alternative to traditional cars. Technological advances in battery capacity and size, and health might hasten the popularity trend for EVs because the battery capacity is a key factor in determining the driving range. To make recommendations for battery management and protection, additional research on battery life cycles is required. The difficult task is creating an algorithm that takes grid pressure reduction, frequency of power distribution, and expensive restrictions into account. Charger capacity, charging site cluster placements all of these methods, as well as multistage geometry and grid-based heuristic algorithms, may be used to calculate the charging needs. Another sort of infrastructure enhancement is to construct and install certain wireless charging transmitters in specific locations along the driving highways so that passing electric vehicles may receive energy as they pass by to assist them recharge while they are in motion. Since electromagnetic coil alignment across dynamic charging distances to increase transmission power poses a challenging and uncommon problem, this field of study referred to as dynamic energy (wireless) transfer, is rarely studied [33]. The problem arises, though, when several cars are traveling close together and reach a particular transmitter. Future studies might concentrate on analyzing metering and identification tools with charge control units to monitor each EV‘s electricity usage and regulate how much power is delivered from the inductive energy transfer tools to the surrounding cars.

6 Conclusion

The paper began by reviewing the current discussions surrounding the placement of charging stations, scheduling EV charging, and CS recommendations. In this paper, federated learning and blockchain applications are discussed for potential remedies to boost system scalability and security. Applications for energy trading between power companies were also examined. This study included a thorough investigation of EV charging behavior using machine learning. The study included a comparison of numerous works employing supervised and unsupervised EV charging. An extensive investigation of EV batteries and driving behavior is needed further to improve the performance of an EV energy management system.

References

1. A. Almaghrebi, F. Aljuheshi, M. Rafaie, K. James, and M. Alahmad, “Data-Driven Charging Demand Prediction at Public Charging Stations Using Supervised Machine Learning Regression Methods,” Energies, vol. 13, no. 16, p. 4231, Aug. 2020, https://doi.org/10.3390/en13164231.
2. M. Alonso Raposo, M. Grosso, A. Mourtzouchou, J. Krause, A. Duboz, and B. Ciuffo, “Economic implications of a connected and automated mobility in Europe,” Research in Transportation Economics, vol. 92, p. 101072, May 2022, https://doi.org/10.1016/j.retrec.2021.101072.

3. M. R. Khalid, M. S. Alam, A. Sarwar, and M. S. Jamil Asghar, “A Comprehensive review on electric vehicles charging infrastructures and their impacts on power-quality of the utility grid,” eTransportation, vol. 1, p. 100006, Aug. 2019, https://doi.org/10.1016/j.etran.2019.100006.
4. “The State of Freight – Understanding greenhouse gas emissions from goods movement in Canada,” Climate Change and Law Collection, https://doi.org/10.1163/9789004322714_cclc_2017-0028-029.
5. M. O. Metais, O. Jouini, Y. Perez, J. Berrada, and E. Suomalainen, “Too much or not enough? Planning electric vehicle charging infrastructure: A review of modeling options,” Renewable and Sustainable Energy Reviews, vol. 153, p. 111719, Jan. 2022, https://doi.org/10.1016/j.rser.2021.111719.
6. H. Li, A. Ravey, A. N’Diaye, and A. Djerdir, “A Review of Energy Management Strategy for Fuel Cell Hybrid Electric Vehicle,” 2017 IEEE Vehicle Power and Propulsion Conference (VPPC), Dec. 2017, https://doi.org/10.1109/vppc.2017.8330970.
7. J. Wang, J. Deng, Y. Liu and Y. Wang, "Non-intrusive load perception and flexibility evaluation for electric vehicle charging station: a deep learning based approach," 2022 IEEE 5th International Electrical and Energy Conference (CIEEC), Nangjing, China, 2022, pp. 2570–2575, https://doi.org/10.1109/CIEEC54735.2022.9845884.
8. B. Al-Hanahi, I. Ahmad, D. Habibi, and M. A. S. Masoum, “Charging Infrastructure for Commercial Electric Vehicles: Challenges and Future Works,” IEEE Access, vol. 9, pp. 121476–121492, 2021, https://doi.org/10.1109/access.2021.3108817.
9. T.-Y. Ma and S. Xie, “Optimal fast charging station locations for electric ridesharing with vehicle-charging station assignment,” Transportation Research Part D: Transport and Environment, vol. 90, p. 102682, Jan. 2021, https://doi.org/10.1016/j.trd.2020.102682.
10. Y. Huang and K. M. Kockelman, “Electric vehicle charging station locations: Elastic demand, station congestion, and network equilibrium,” Transportation Research Part D: Transport and Environment, vol. 78, p. 102179, Jan. 2020, https://doi.org/10.1016/j.trd.2019.11.008.
11. Q. Feng, H. Li, Y. Zhou, D. Feng, Y. Wang and Y. Su, "Review of electric vehicles’ charging data anomaly detection based on deep learning," 2022 Power System and Green Energy Conference (PSGEC), Shanghai, China, 2022, pp. 337–341, https://doi.org/10.1109/PSGEC54663.2022.9881073.
12. A. Mosavi, M. Salimi, S. Faizollahzadeh Ardabili, T. Rabczuk, S. Shamshirband, and A. Varkonyi-Koczy, “State of the Art of Machine Learning Models in Energy Systems, a Systematic Review,” Energies, vol. 12, no. 7, p. 1301, Apr. 2019, https://doi.org/10.3390/en12071301.
13. P. Boza and T. Evgeniou, “Artificial intelligence to support the integration of variable renewable energy sources to the power system,” Applied Energy, vol. 290, p. 116754, May 2021, https://doi.org/10.1016/j.apenergy.2021.116754.
14. R. Sankaranarayanan, K. S. Umadevi, N. Bhavani, B. M. Jos, A. Haldorai, and D. V. Babu, “Cluster-based attacks prevention algorithm for autonomous vehicles using machine learning algorithms,” Computers and Electrical Engineering, vol. 101, p. 108088, Jul. 2022, https://doi.org/10.1016/j.compeleceng.2022.108088.
15. R. M. P, S. Ponnan, S. Shelly, Md. Z. Hussain, M. Ashraf, and A. Haldorai, “Autonomous navigation system based on a dynamic access control architecture for the internet of vehicles,” Computers and Electrical Engineering, vol. 101, p. 108037, Jul. 2022, https://doi.org/10.1016/j.compeleceng.2022.108037.
16. A. Haldorai, Eswaramoorthy. V, M. Upadhyaya, G. J. Nehru, and D. Kapila, “Self-Learning and Adaptive Networking Protocols and Algorithms for 6G Edge Nodes,” 2023 Fifth International Conference on Electrical, Computer and Communication Technologies (ICECCT), Feb. 2023, https://doi.org/10.1109/icecct56650.2023.10179645.
17. Y. Li, S. He, Y. Li, L. Ge, S. Lou and Z. Zeng, "Probabilistic Charging Power Forecast of EVCS: Reinforcement Learning Assisted Deep Learning Approach," in IEEE Transactions on Intelligent Vehicles, vol. 8, no. 1, pp. 344–357, Jan. 2023, https://doi.org/10.1109/TIV.2022.3168577.

18. Y. Liu, W. Liu, S. Gao, Y. Wang, and Q. Shi, "Fast charging demand forecasting based on the intelligent sensing system of dynamic vehicle under EVs-traffic-distribution coupling," Energy Reports, vol. 8, pp. 1218–1226, Aug. 2022, https://doi.org/10.1016/j.egyr.2022.02.261.
19. C. B. Saner, A. Trivedi, and D. Srinivasan, "A Cooperative Hierarchical Multi-Agent System for EV Charging Scheduling in Presence of Multiple Charging Stations," IEEE Transactions on Smart Grid, vol. 13, no. 3, pp. 2218–2233, May 2022, https://doi.org/10.1109/tsg.2022.3140927.
20. X. Hu, L. Xu, X. Lin, and M. Pecht, "Battery Lifetime Prognostics," Joule, vol. 4, no. 2, pp. 310–346, Feb. 2020, https://doi.org/10.1016/j.joule.2019.11.018.
21. H. J. Wang, B. Wang, C. Fang, W. Li, and H. W. Huang, "Charging Load Forecasting of Electric Vehicle Based on Charging Frequency," IOP Conference Series: Earth and Environmental Science, vol. 237, no. 6, p. 062008, Mar. 2019, https://doi.org/10.1088/1755-1315/237/6/062008.
22. Q. Yan, X. Chen, H. Jian, W. Wei, W. Wang, and H. Wang, "Design of a deep inference framework for required power forecasting and predictive control on a hybrid electric mining truck," Energy, vol. 238, p. 121960, Jan. 2022, https://doi.org/10.1016/j.energy.2021.121960.
23. M. D. Eddine and Y. Shen, "A deep learning based approach for predicting the demand of electric vehicle charge," The Journal of Supercomputing, vol. 78, no. 12, pp. 14072–14095, Mar. 2022, https://doi.org/10.1007/s11227-022-04428-0.
24. M. J. Eagon, D. K. Kindem, H. Panneer Selvam, and W. F. Northrop, "Neural Network-Based Electric Vehicle Range Prediction for Smart Charging Optimization," Journal of Dynamic Systems, Measurement, and Control, vol. 144, no. 1, Jan. 2022, https://doi.org/10.1115/1.4053306.
25. B. Vaidya and H. T. Mouftah, "Smart electric vehicle charging management for smart cities," IET Smart Cities, vol. 2, no. 1, pp. 4–13, Mar. 2020, https://doi.org/10.1049/iet-smc.2019.0076.
26. A. D. Setiawan, A. Hidayatno, B. D. Putra, and I. Rahman, "Selection of Charging Station Technology to Support the Adoption of Electric Vehicles in Indonesia with the AHP-TOPSIS Method," 2020 3rd International Conference on Power and Energy Applications (ICPEA), Oct. 2020, https://doi.org/10.1109/icpea49807.2020.9280125.
27. L. Zhang, S. Wang, and X. Qu, "Optimal electric bus fleet scheduling considering battery degradation and non-linear charging profile," Transportation Research Part E: Logistics and Transportation Review, vol. 154, p. 102445, Oct. 2021, https://doi.org/10.1016/j.tre.2021.102445.
28. G. Wang et al., "Pricing-aware Real-time Charging Scheduling and Charging Station Expansion for Large-scale Electric Buses," ACM Transactions on Intelligent Systems and Technology, vol. 12, no. 1, pp. 1–26, Nov. 2020, https://doi.org/10.1145/3428080.
29. Y. Bie, J. Ji, X. Wang, and X. Qu, "Optimization of electric bus scheduling considering stochastic volatilities in trip travel time and energy consumption," Computer-Aided Civil and Infrastructure Engineering, vol. 36, no. 12, pp. 1530–1548, May 2021, https://doi.org/10.1111/mice.12684.
30. Madeleine Wang Yue Dong, "A Survey on Multi Agent System and Its Applications in Power System Engineering", Journal of Computational Intelligence in Materials Science, vol.1, pp. 001–011, 2023. https://doi.org/10.53759/832X/JCIMS202301001.
31. Ş. Yıldırım and B. Yıldız, "Electric bus fleet composition and scheduling," Transportation Research Part C: Emerging Technologies, vol. 129, p. 103197, Aug. 2021, https://doi.org/10.1016/j.trc.2021.103197.
32. E. Wang et al., "Joint Charging and Relocation Recommendation for E-Taxi Drivers via Multi-Agent Mean Field Hierarchical Reinforcement Learning," IEEE Transactions on Mobile Computing, vol. 21, no. 4, pp. 1274–1290, Apr. 2022, https://doi.org/10.1109/tmc.2020.3022173.
33. Z. Tian et al., "Real-Time Charging Station Recommendation System for Electric-Vehicle Taxis," IEEE Transactions on Intelligent Transportation Systems, vol. 17, no. 11, pp. 3098–3109, Nov. 2016, https://doi.org/10.1109/tits.2016.2539201.

Empowering Smart Cities: AI-Driven Solutions for Urban Computing

1 Introduction

Urban computing is a multidisciplinary field that combines elements of urban planning, computer science, and data analytics. The main goals of this field include the comprehension, modeling, and enhancement of urban settings. There is a growing need to use data-driven approaches and technology to solve urban problems and enhance people's quality of life. The fast growth of metropolitan areas all over the globe is the main cause of this. At its core, urban computing is the collection and analysis of data from a wide variety of urban sources [1]. The urban environment is equipped with sensor networks that monitor several aspects like traffic, energy consumption, and environmental conditions. The data acquired from diverse sources through ongoing gathering can provide researchers with valuable insights into the trends, patterns, and linkages present within the urban environment. This study employs statistical modeling approaches, data analytics, and computer methodologies [2]. Subsequently, it relies significantly on urban sensing and monitoring. Sensor networks provide instantaneous monitoring of several aspects of urban life, such as air quality, noise levels, temperature, and humidity, among other factors. By employing a continuous monitoring technique, cities may promptly detect and resolve environmental issues, thereby enhancing the quality of life for their residents. For instance, if air quality data show a rise in pollution levels, suitable actions could be implemented to mitigate its impact on public health [3]. Consequently, urban planning and construction have encountered challenges related to the computation of urban information systems outcomes. Urban planners take into account several aspects, such as land use, demographics, and transportation patterns, while making decisions about zoning regulations, infrastructure development, and urban growth [4]. The utilization of data in this approach has the capacity to enhance the efficiency, desirability, and eco-friendliness of cities. Moreover, it enables the efficient distribution of resources and proactive resolution of issues. Urban

© European Alliance for Innovation 2024

A. Haldorai et al., *Artificial Intelligence for Sustainable Development*,
EAI/Springer Innovations in Communication and Computing,
https://doi.org/10.1007/978-3-031-53972-5_10

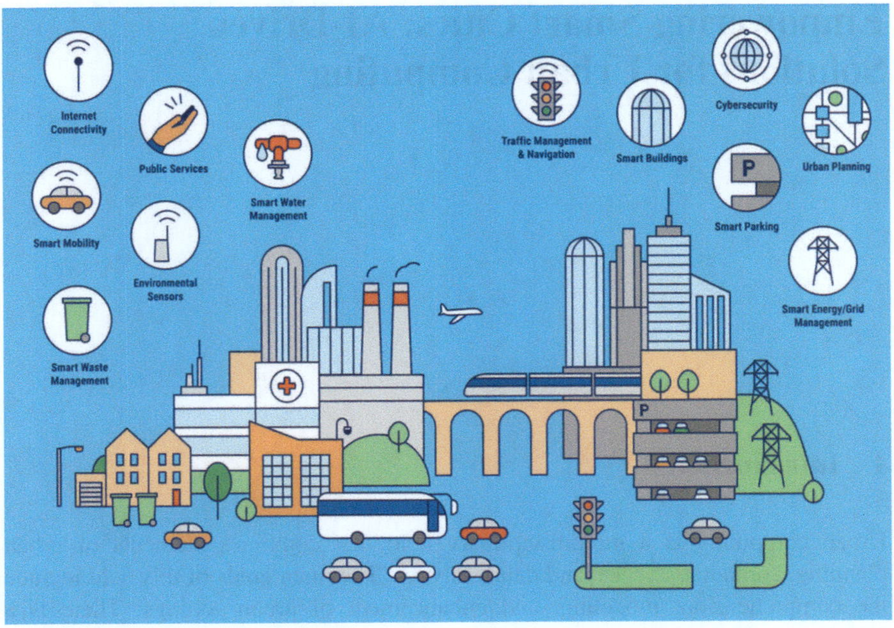

Fig. 1 Elements of urban computing

computing proves to be valuable in the advancement of smart city applications. By integrating computational approaches and technology, cities can harness their potential to construct smart transportation systems, intelligent infrastructure, and efficient energy management systems [5]. These projects enhance the quality of public services, expand the availability of public transportation, and optimize the allocation of resources through the utilization of data and analytics (Fig. 1).

Urban computing focuses on how the residents are actively and passively involved. Active participation by citizens in decision-making is facilitated by social media and smartphone applications, which will enable them to submit data, leave comments, etc. By keeping city dwellers as central figures of its strategy, citizen-centered approach encourages personal responsibility and attachment thereby culminating in more inclusive and responsive urban environments. However, when it comes to Urban Computing, privacy and ethical issues must be considered because data processing and collection are involved. For maintenance of public confidence as well as safeguarding people's privacy, any urban computing system must include personal identity information protection, data anonymization techniques or ways of eliminating algorithmic biases from the outputs. When it comes to data-driven and technology-driven urban planning, design, and administration, urban computing stands out as a powerful paradigm. It improves city life by enhancing citizen participation, promoting sustainable development, gaining insights into urban dynamics, and gathering and evaluating data from a variety of urban sources. Urban computing will be extremely important in determining the future of urban environments as cities continue to expand and face difficult difficulties.

The remaining part is composed as pursues: the overview of urban computing framework is exhibited in Sect. 2. Section 3 gives the Challenges of Urban computing. The significance of AI for urban computing is explored in Sect. 4 and various Urban safety measures are discussed under Sect. 5. Section 6 finishes this discussion with a conclusion.

2 Urban Computing Framework

Urban Computing integrates ubiquitous and nonintrusive sensing technologies with cutting-edge data management, analytical models, and cutting-edge visualization techniques to produce solutions that enhance the urban environment, human life quality, and city operation systems. Figure 2 presents the general structure for

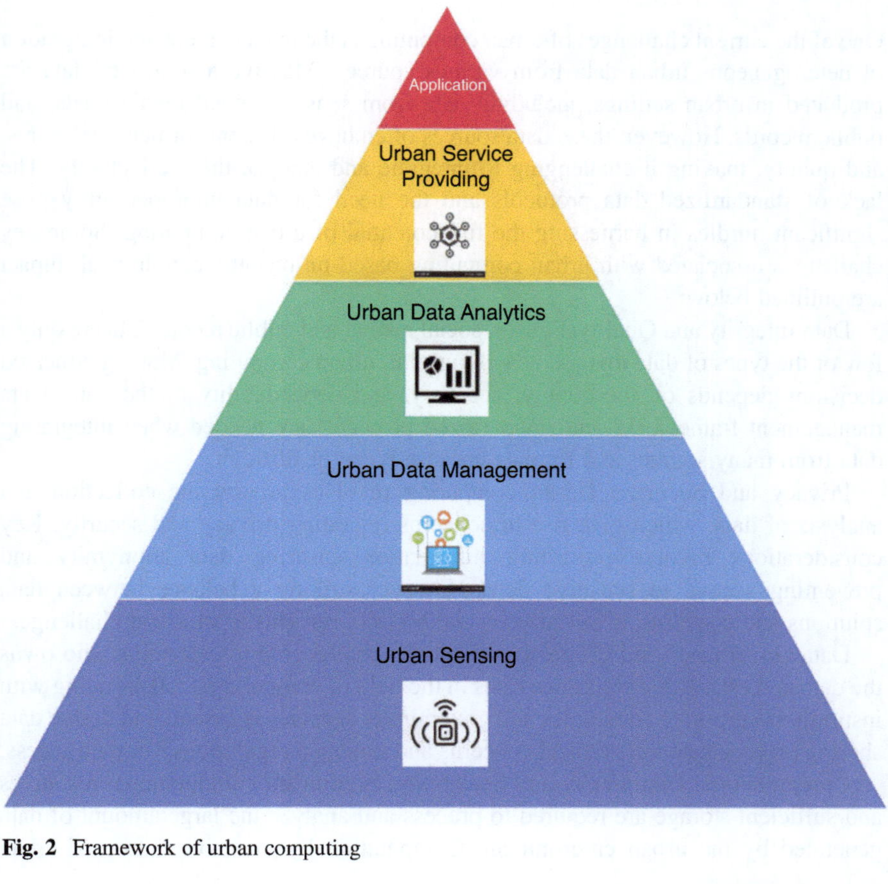

Fig. 2 Framework of urban computing

implementing Urban Computing. Urban Sensing, Urban Data Management, Data Analytics, and Service Providing are the four layers that make up the framework. Each layer has a distinct purpose. Data from urban areas are gathered by the Urban Sensing layer. Different methods, including crowdsensing, mobile sensing, and participatory sensing, can be used to acquire this data.

For effective data analytics, the Urban Data Management layer enables data organization through an indexing structure that includes both textual and spatiotemporal data. The Data Analytics layer uses a variety of methods, including Data Mining, Machine Learning, and Data Visualization, to find patterns in data and extract useful information for further decision-making. The solutions and services offered by the service-providing layer strive to enhance driving experiences while lowering energy use, air pollution, and congestion. For instance, if a traffic anomaly is discovered, the transportation authority will receive this information so it may distribute traffic and identify the abnormality.

3 Challenges of Urban Computing

One of the current challenges of urban computing is the management and integration of heterogeneous urban data from various sources. Massive amounts of data are produced in urban settings, including data from sensors, social media feeds, and public records. However, these data sources often have different formats, structures, and quality, making it challenging to integrate and analyze them effectively. The lack of standardized data protocols and the need for data interoperability pose significant hurdles in harnessing the full potential of urban computing. Some key challenges associated with urban computing based on recent technological impact are outlined below:

Data Integrity and Quality: Sensor, social media, and public record data are only a few of the types of data that are heavily used in urban computing. Making educated decisions depends on the quality, accuracy, and dependability of the data. Data management frameworks and standardized protocols are needed when integrating data from many sources and formats because it can be difficult.

Privacy and Security: Urban computing revolves around the collection and analysis of data, which give rise to concerns regarding privacy and security. Key considerations involve protecting information ensuring data anonymity and preventing access, to sensitive data. However striking a balance between data solutions and concerns, about privacy and security remains a persistent challenge.

Data Governance and Ownership: It can be challenging to determine who owns the data and establish governance rules in the field of computing. Collaborating with institutions corporate companies and government agencies is essential to define data sharing protocols address ethical concerns and establish regulations, for data access.

Computational Complexity and Scalability: Significant computational resources and sufficient storage are required to process and analyze the large amount of data generated by the urban environment. Computationally intensive nature of urban computing has hindered the development of scalable algorithms and infrastructure. The vast amount of urban data require efficient data processing techniques, cloud computing, and distributed computing systems.

Algorithm Bias and Interpretability: Decision-making and data analysis in urban computing are powered by algorithms. But algorithms are not perfect, and they might be biased, which means they can amplify urban inequality. The avoidance of biased results and the establishment of confidence in decision-making processes depend on algorithms that are fair, transparent, and interpretable.

Digital Divide and Inclusion: For data-driven services and technology to be accessible to everyone, urban computing solutions must bridge the digital divide. In order to avoid further inequality and guarantee equitable urban development, it is crucial to bridge the gap between communities that are linked and those who are not.

Citizen Engagement and Participation: Although the goal of urban computing is to include individuals in decision-making, it may be difficult to ensure that their involvement is relevant. In order to encourage participation from all citizens and provide a platform to different perspectives in city planning and development, it is crucial to eliminate obstacles such as low digital literacy, language difficulties, and accessibility concerns.

Ethical and Social Implications: The processes of data collection, usage, and decision-making in urban computing give rise to ethical concerns. To practice responsible urban computing, one must comprehend the societal impacts of data-driven solutions, eradicate bias, and discrimination, and ensure that all individuals benefit from technological advancements.

Information-Driven Decision-Making: AI empowers cities to analyze vast quantities of data obtained from urban sensors, public records, and social media. This data-driven system enhances decision-making in public service, resource allocation, and urban planning.

Intelligent Urban Infrastructure: AI can assist in the development of intelligent infrastructure systems that enhance operational efficiency, reduce energy consumption, and improve transportation effectiveness. It enables the automated surveillance of urban facilities, intelligent transportation systems, and improved energy management..

Traffic Management and Optimization: Minimizing travel times, lowering emissions, and enhancing mobility inside cities are all possible outcomes of traffic pattern analysis, congestion prediction, and optimization by AI algorithms.

Energy Efficiency and Sustainability: Through the analysis of data on energy consumption, demand, and supply, cities may include AI into their energy-saving strategies, grid distribution, and sustainability initiatives.

Urban Planning and Design: AI models aid urban planners in assessing demographic, transportation, and land use data, among other things, to make educated judgments on zoning rules, infrastructure development, and urban growth.

Predictive Maintenance and Safety: AI enables predictive maintenance of urban infrastructure by analyzing data gathered from sensors, thereby detecting and resolving issues before they escalate into disruptive or hazardous situations.

Citizen Engagement and Personalized Services: Residents are able to participate in urban decision-making processes, obtain tailored services, and provide feedback using AI-driven systems that promote citizen participation.

In order to address these challenges, a wide range of artificial intelligence approaches are being explored in different urban computing application scenarios. These techniques include supervised learning, unsupervised learning, semi-supervised learning, matrix factorization, deep learning, visual models, and reinforcement learning.

4 Significance of Artificial Intelligence for Urban Computing

By using data-driven insights and state-of-the-art algorithms, AI has the potential to help cities optimize urban processes, enhance decision-making, and improve the quality of life for residents. Applications, like energy conservation, traffic control, urban planning and proactive maintenance heavily depend on it. Artificial intelligence empowers cities to make decisions operate with efficiency and construct sustainable environments by analyzing large amounts of data related to urban areas. The incorporation of AI into computing has the potential to revolutionize cities making them more intelligent, efficient, and better attuned, to the requirements of their residents. Machine learning algorithms are often employed to process large volumes of data and uncover meaningful patterns. For example, analyzing transportation data can help identify congestion hotspots, predict traffic flow, and optimize transportation systems. Similarly, studying energy consumption patterns can enable more efficient energy management and promote sustainability.

4.1 AI Framework for Urban Computing

A structured approach that integrates various artificial intelligence techniques and methodologies to address urban challenges. It involves the application of machine learning, deep learning, natural language processing, and other AI methods to analyze urban data, extract insights, and make data-driven decisions. Figure 3 outlines the steps involved in applying AI for urban computing.

For any urban computing application, the AI pipeline is incorporated. Initially the relevant and diverse urban data from various sources, such as sensors, IoT devices, social media, public records, and urban databases are collected. The data quality and compatibility for AI analysis are ensured. Then data preprocessing is applied that involves clean, preprocess, and integrate the collected data to ensure consistency, handle missing values, remove outliers, and format. Then in step 2, the process of feature engineering is performed to select and engineer appropriate features from the

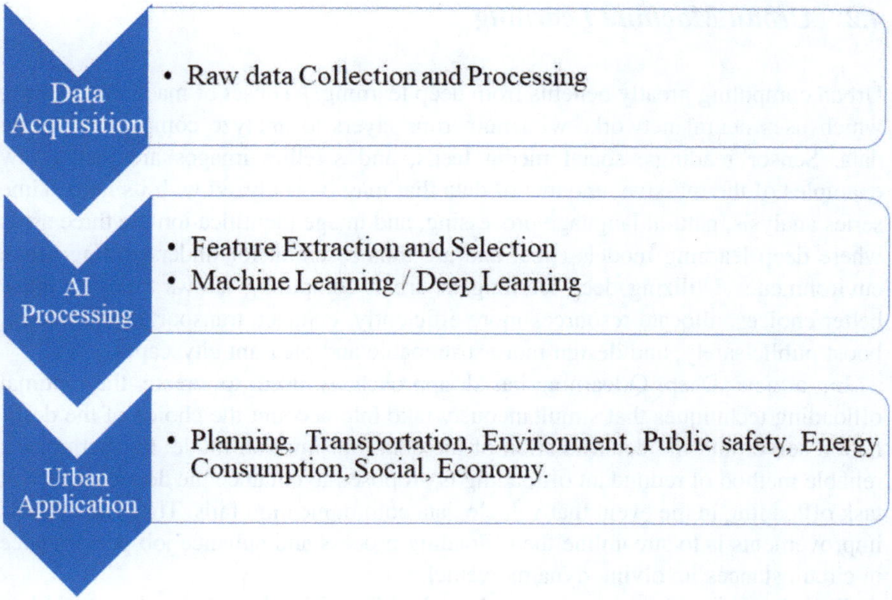

Fig. 3 AI pipeline for urban computing

data that capture the relevant urban characteristics for the specific problem. Feature engineering may involve transforming, aggregating, or creating new features to enhance AI models' performance.

In the next step, an AI Model is selected using appropriate AI techniques and models that align with the problem and data characteristics. This could include machine learning algorithms (e.g., regression, classification, clustering), deep learning models (e.g., neural networks), or other AI approaches (e.g., reinforcement learning, natural language processing). Then the model is properly trained. This involves using labeled data to enable the models to learn patterns, relationships, and predictions relevant to the urban computing problem.

Further the performance of the trained model is assessed using appropriate evaluation metrics and validation techniques. This helps understand how well the models generalize to unseen data and evaluate their effectiveness in addressing the urban computing problem. Then model integration is performed and deployed in real-time or near-real-time environments to provide insights, predictions, or decision support. Regular feedback and data collection allow for constant model performance monitoring. Update or retrain the models periodically to adapt to changing urban conditions, evolving requirements, and improve performance. Finally, stakeholder Engagement and Decision-Making are performed to communicate the insights generated by AI models to stakeholders, including city officials, urban planners, policymakers, and residents. This facilitates informed decision-making and collaborate with stakeholders to address the urban challenges effectively.

4.2 Urban Machine Learning

Urban computing greatly benefits from deep learning, a subset of machine learning, which uses neural networks with numerous layers to analyze complicated urban data. Sensor readings, social media feeds, and satellite images are just a few examples of the massive amounts of data that may be analyzed with its help. Time series analysis, natural language processing, and image identification are three areas where deep learning models excel and are thus essential for understanding urban environments. Utilizing deep learning in urban computing allows cities to make better choices, allocate resources more efficiently, enhance transportation systems, boost public safety, and design more sustainable and pleasant cityscapes.

In, a new Deep Q-learning-based approach is used to create the optimal offloading techniques that simultaneously take into account the choice of the destination server and the determination of the data transmission mode. Furthermore, a reliable method of redundant offloading is proposed to enhance the dependability of task offloading in the event that vehicle data communication fails. The goal of these improvements is to streamline the offloading process and enhance job performance in circumstances involving dynamic vehicles.

Built in [6] is a deep learning model called DeepTRI that can predict the Urban Traffic Revitalization Index (DeepTRI) for major Chinese cities. This model incorporates both the traffic revitalization index and the COVID-19 pandemic data. To construct the graph's topological structure, location data from 29 cities are used. A thorough comprehension of urban traffic issues and revival potential may be achieved by using DeepTRI.

In [7], a novel method for measuring the cleanliness of metropolitan streets is presented, which makes use of deep learning methods in conjunction with mobile edge computing. In particular, the Faster R-CNN is used to precisely classify various types of street trash and quantify the quantity of trash that is there. In order to manage and maintain urban settings efficiently, this new technique offers a strong answer for assessing street cleanliness.

A communication architecture is published in [8] to integrate urban mobile big data with a system for urban environment classification. For the purpose of noise classification, this framework employs a deep learning strategy, more especially the use of a Convolutional Neural Network (CNN). Accurately classifying these disturbances allows for the development of an appropriate noise cancellation algorithm, which in turn reduces the impact of noise during smartphone conversations. Reducing the negative impacts of background noise, this novel solution seeks to improve communication experiences in metropolitan contexts.

In [9], researchers detail their efforts to develop a new method for urban flood forecasting that is both feasible and inexpensive. The FLood Tracking and Prediction System for Urban Implementation is its name. By consistently predicting street-level flood dangers, this platform hopes to provide optimized recommendations to relevant organizations. Extrapolating public early warning signals is another function it does. For accurate and dependable flood predictions, the platform's design employs a hybrid technique that combines Deep Learning and Fuzzy Logic approaches. By leveraging this system, urban areas can better prepare for floods and mitigate their impacts on the community.

4.3 Urban Machine Learning

Urban computing makes interactive visual data analytics, which seamlessly combines visualization techniques with data mining algorithms, even more crucial. It is also a strategy for fusing machine intelligence with human intellect. A real-world problem in cities can be solved by domain experts and data scientists working together thanks to the interactive visual data analytics. This integration of domain expertise (such as urban planning) and data science is made possible by these analytics. To manage, visualize, and comprehend cities and their evolution, sophisticated tools are required.

Public security agencies often employ visualization-assisted crime analysis tools to examine extensive urban areas. These tools typically rely on grid-based heatmaps to uncover the spatial distribution of crimes across entire districts, regions, and neighborhoods. However, Mirante, a crime mapping visualization system, presents a unique approach by enabling spatiotemporal analysis of crime patterns at a more granular street-level scale. The study [10] presents a comprehensive and detailed examination of crime data, facilitating a deeper understanding of crime dynamics and aiding in the development of targeted crime prevention strategies using visualization tools.

The visualization problem caused by the extensive air quality monitoring data from major cities across a country is addressed in research in [11]. To facilitate comprehensible visualization of this massive information, two state-of-the-art visualization tools were developed: timezoom.js for multigranularity time series visualization and symadaptive.js for thematic mapping via the construction of a dynamic symbol declutter map mashup layer. Using these methods, we created a web-based map visualization tool that successfully displays data on urban air quality for all of China's major cities. Users get easier access to and comprehension of the data thanks to this app's simplified data exploration and analysis features.

This project in [12] aims to provide an open and interoperable software framework for urban data administration and analysis, instead of the traditional one-time monolithic software solution. In response to the varied requirements and use cases of its end users, the framework is designed to tackle problems in applied research. The project's goal is to provide a versatile solution that can handle different problems with urban data analysis and management by using this method.

The novel Modal flow system is presented in [13], which offers a notebook-like environment for visualizing flows. To provide new visual encodings and interfaces for modal partitioned flows, it incorporates and tweaks a number of visualization approaches [14]. Ternary plots, scatterplots, flow charts, and origin-destination matrices make up this set. Using the Modalflow technology, researchers may extract valuable insights from urban transportation-related survey and ubiquitous data. As an added bonus, this technology lets you test out a mobile phone network dataset in a specific city, so you can look at urban mobility trends and patterns in detail [15].

5 Empowering Urban Safety Using AI

The goal of empowering urban safety with AI is to improve safety measures, reaction times to incidents, and risk prevention in urban areas by using AI's capabilities. The methods by which AI may help empower urban safety measures are as follows:

- Intelligent Surveillance: With the use of AI, surveillance systems can instantly review live video feeds, flagging any suspicious activity, items, or people of interest, and then notifying the proper authorities. The result is better proactive surveillance and less crime.
- Predictive Analytics: Big data, such as social media feeds, sensor readings, and crime records from years past, may be analyzed by AI systems to identify trends and predict future threats. Due to this, authorities are able to take proactive steps to avoid problems and distribute resources efficiently.
- Emergency Response Optimization: By analyzing real-time data like traffic, event locations, and available resources, AI models can enhance emergency response systems. Government agencies can therefore better deploy emergency services, resulting in faster response times.
- Risk Assessment and Mitigation: Risk assessment and mitigation in urban environments may be aided by AI. In order to mitigate threats, AI models examine data on traffic patterns, infrastructural flaws, and environmental conditions. They then propose solutions to these problems.
- Smart Infrastructure Monitoring: In order to identify structural irregularities or signals of possible breakdowns, sensors and monitoring systems based on AI can continually monitor key infrastructure, including bridges, highways, and buildings. Assuring the security of urban infrastructure is facilitated by this proactive monitoring.
- Safety Prediction and Early Warning: When it comes to safety, AI models may analyze data from a variety of sources, including weather predictions, sensor readings, and event records, to identify potential dangers and provide early warnings. Residents and government may then take preventative measures to guarantee public safety.
- Personal Safety Applications: With the use of AI-powered mobile applications, people may improve their own personal safety in metropolitan areas by accessing features like real-time monitoring, emergency warnings, and safety suggestions.

Cities may use AI to their advantage by anticipating and mitigating threats to public safety, allocating resources wisely, and making overall neighborhoods safer. The use of AI in urban safety solutions allows for a smarter, data-driven strategy, which improves reaction and preventative capacities.

6 Conclusion

Urban computing necessitates the capability to execute computation that is very scalable, rapid, dependable, and adaptable. Nevertheless, in order to efficiently include AI as a crucial element of evidence-based urban planning, developers must have access to extensive datasets, even if they are presently unavailable or need to be expanded, which poses a barrier. To effectively integrate AI into urban data processing, machine learning methods need to be modified to accommodate spatiotemporal data. In addition, it is necessary to implement hybrid learning approaches and evaluate the efficacy of these algorithms utilizing interactive visual analytic tools. This study concludes that AI contributes meaningfully to the analysis and understanding of urban dynamics and facilitates informed decision-making in urban development.

References

1. C. Bayliss, "Machine learning based simulation optimisation for urban routing problems," Applied Soft Computing, vol. 105, p. 107269, Jul. 2021, doi: https://doi.org/10.1016/j.asoc.2021.107269.
2. H. Li and J. Sansalone, "Implementing machine learning to optimize the cost-benefit of urban water clarifier geometrics," Water Research, vol. 220, p. 118685, Jul. 2022, doi: https://doi.org/10.1016/j.watres.2022.118685.
3. A. Haldorai and A. Ramu, "Canonical Correlation Analysis Based Hyper Basis Feedforward Neural Network Classification for Urban Sustainability," Neural Processing Letters, vol. 53, no. 4, pp. 2385–2401, Aug. 2020, doi: https://doi.org/10.1007/s11063-020-10327-3.
4. H. Tanim, C. B. McRae, H. Tavakol-Davani, and E. Goharian, "Flood Detection in Urban Areas Using Satellite Imagery and Machine Learning," Water, vol. 14, no. 7, p. 1140, Apr. 2022, doi: https://doi.org/10.3390/w14071140.
5. L. Wei, W. Tian, E. A. Silva, R. Choudhary, Q. Meng, and S. Yang, "Comparative Study on Machine Learning for Urban Building Energy Analysis," Procedia Engineering, vol. 121, pp. 285–292, 2015, doi: https://doi.org/10.1016/j.proeng.2015.08.1070.
6. K. Nandini and G. Fathima, "Urban Air Quality Analysis and Prediction Using Machine Learning," 2019 1st International Conference on Advanced Technologies in Intelligent Control, Environment, Computing & Communication Engineering (ICATIECE), Mar. 2019, doi: https://doi.org/10.1109/icatiece45860.2019.9063845.
7. A. Roshini and H. Anandakumar, "Hierarchical cost effective leach for heterogeneous wireless sensor networks," 2015 International Conference on Advanced Computing and Communication Systems, Jan. 2015, doi: https://doi.org/10.1109/icaccs.2015.7324082.
8. Z. Lv, J. Li, C. Dong, H. Li, and Z. Xu, "Deep learning in the COVID-19 epidemic: A deep model for urban traffic revitalization index," Data & Knowledge Engineering, vol. 135, p. 101912, Sep. 2021, doi: https://doi.org/10.1016/j.datak.2021.101912.
9. Anandakumar Haldorai, "Key Concepts, Technologies, Current Challenges and Research Areas of Telecommunication Engineering and Neural Communication", Journal of Biomedical and Sustainable Healthcare Applications, vol.3, no.1, pp. 034–044, January 2023. doi: https://doi.org/10.53759/0088/JBSHA202303004.

10. R. M. P, S. Ponnan, S. Shelly, Md. Z. Hussain, M. Ashraf, and A. Haldorai, “Autonomous navigation system based on a dynamic access control architecture for the internet of vehicles,” Computers and Electrical Engineering, vol. 101, p. 108037, Jul. 2022, doi: https://doi.org/10.1016/j.compeleceng.2022.108037.
11. C. Karyotis, T. Maniak, F. Doctor, R. Iqbal, V. Palade and R. Tang, "Deep Learning for Flood Forecasting and Monitoring in Urban Environments," 2019 18th IEEE International Conference On Machine Learning And Applications (ICMLA), Boca Raton, FL, USA, 2019, pp. 1392–1397, doi: https://doi.org/10.1109/ICMLA.2019.00227.
12. G. Garcia-Zanabria et al., "Mirante: A visualization tool for analyzing urban crimes," 2020 33rd SIBGRAPI Conference on Graphics, Patterns and Images (SIBGRAPI), Porto de Galinhas, Brazil, 2020, pp. 148–155, doi: https://doi.org/10.1109/SIBGRAPI51738.2020.00028.
13. J. Samuel et al., “UD-SV: Urban data services and visualization framework for sharing multidisciplinary research,” Transactions in GIS, vol. 27, no. 3, pp. 841–858, Apr. 2023, doi: https://doi.org/10.1111/tgis.13049.
14. Pérez-Messina, E. Graells-Garrido, M. J. Lobo, and C. Hurter, “Modalflow: Cross-Origin Flow Data Visualization for Urban Mobility,” Algorithms, vol. 13, no. 11, p. 298, Nov. 2020, doi: https://doi.org/10.3390/a13110298.
15. Tian Chen and Davin Meng, “A Critical Review of Nanoparticles and Nano Catalyst”, Journal of Computational Intelligence in Materials Science, vol. 1, pp. 012–022, 2023. doi: https://doi.org/10.53759/832X/JCIMS202301002.

AI-Empowered Blockchain Techniques Against Cybersecurity Context in IoT: A Survey

1 Introduction

The term "Internet of Things" (IoT) was first used in 1999 and refers to the process of connecting people and devices via any service, network, or path at any time and location. Communication between objects is critical to achieving the goal of IoT. To ensure data privacy, secrecy, integrity, authentication, and availability, a secure solution is necessary for IoT devices. The emergence of blockchain technology offers a potential solution to many of the security challenges that arise in IoT environments. Blockchain is designed with security in mind and can provide transparency, immutability, data encryption, and auditability to IoT devices. In today's corresponding environment, Modern cyber and defense programmers are advantageous to everyone. Attacks on personal cybersecurity range from identity theft to attempted extortion of important information such as photos of family. All world needs essential infrastructure like Power plants, medical facilities, and financial service companies are examples. Ensuring the security of all these Institutions are important for preservation of the right functioning of society. The work of a team of cyber threat researchers, like his 250-member threat investigation team at Talos, which examines both new and existing dangers and cyberattack tactics, is beneficial to everyone.

They enhance tools that are open-source, identify fresh flaws, and inform people about the important role of cybersecurity [1]. They are trying to increase everyone's internet safety. The practice of cybersecurity involves guarding against harmful breaches into computer networks, servers, PCs, mobile devices, electrical devices, and programs. It is also known as the security of electronic information or information technological security. The expression may be broken down into a few key categories and is used in a range of situations, including business and mobile computing. These attacks frequently seek to disrupt regular corporate operations, extort money from customers via ransomware, or access, modify, or delete important

© European Alliance for Innovation 2024

A. Haldorai et al., *Artificial Intelligence for Sustainable Development*,
EAI/Springer Innovations in Communication and Computing,
https://doi.org/10.1007/978-3-031-53972-5_11

data. Effective cybersecurity safeguards can be difficult to put in place these days since there are more gadgets than humans and attackers have been creating increasingly creative attacks.

The ability of blockchain technology to provide a distributed type of database containing tokenized transactions involving assets that is irreversible and tamper-evident has garnered enormous interest from both industry and academia over the past few years [2]. In addition, the integration of smart contracts, P2P (peer-to-peer) resource allocation, including trade, digital identity management, internet marketplaces for crowdsourced services, and on top of the decentralized consensus are expected to make blockchains a "game changer" in a number of industries, including financial services. Numerous cybersecurity applications already use blockchain and AI. For instance, by monitoring and confirming the authenticity of items and avoiding counterfeiting, blockchain can be used to safeguard the supply chain. By offering a decentralized and open platform, it can also be utilized to protect online transactions, such as financial transactions. By examining network data and spotting unusual activity, AI can be used to identify and stop cyberattacks. Additionally, it can be used to locate and fix software and hardware flaws. Security systems with AI capabilities are better able to recognize and stop cyberattacks because they can learn from previous intrusions and adjust to new dangers. Although AI and blockchain have the potential to revolutionize cybersecurity, there are still obstacles to be solved. One difficulty is the intricacy of these technologies, which can make their implementation and upkeep difficult without specialized knowledge. Interoperability is also a challenge because different blockchain and AI systems might not work together. However, the blockchain and Artificial Intelligence in the cybersecurity appear to have a bright future. We may anticipate more cutting-edge cybersecurity solutions that mix blockchain and AI as the technologies continue to develop and become more widely available.

The creation of machine learning algorithms is one of the main ways that AI is employed in cybersecurity. These algorithms are meant to sift through massive volumes of data and spot trends that could be signs of an impending cyberattack. This enables them to recognize threats far more rapidly and correctly than a person could, allowing them to aid in real-time cyberattack protection for our networks and devices. The creation of chatbots and virtual assistants is another way AI is being applied to cybersecurity. These systems are perfect for offering support and help to people who use technology since they are made to communicate with humans in a natural and intuitive way. They can be utilized to assist individuals with chores like creating new accounts or changing passwords, as well as to inform individuals on the best practices for protecting their data and gadgets. Cybersecurity is also greatly impacted by blockchain technology. A distributed, decentralized ledger known as blockchain enables safe and open record-keeping. This makes it perfect for a range of uses, including cybersecurity. Blockchain technology allows us to build transparent, secure systems that are impervious to hacker attacks and other types of manipulation. Specifically, time-driven and event-driven system models are used to discuss attack and defense methods for cybersecurity reliability and integrity. The survey provides a summary of certain difficulties and unresolved problems.

The contribution of this paper is distilled as following:

1. This survey lists a few communication methods that best meet the requirements of IoT devices and some of the security issues in IoT are discussed.
2. It is examined whether merging blockchain with IoT is effective.
3. This article addresses the potential benefits of AI-blockchain technology for IoT devices in terms of security.
4. The combining new technologies like blockchain with IoT are analyzed successfully and finally, challenges of IoT and blockchain integration are provided.

The novel communication methods that could improve IoT device performance are introduced and The security issues with IoT are examined in Sect. 2. The connection of blockchain and the IoT is covered in Sect. 3. The security improvements and issues of integrating blockchain and IoT are discussed. In Sect. 4, introduces the research gap and directions. Section 5 brings the survey paper to conclude.

2 IoT and Its Communication Technologies

One provider network permits a wireless call to be placed from one person to another. Even so, it qualifies as an IoT when the same phone notifies you that someone is at your door because multiple networks are involved in the second scenario. IoT devices collect, exchange, and distribute data in real-time among other gadgets, programmers, and systems via embedded sensors. IoT (Internet of Things) is an amalgamation of specifically created technologies that connects internet-capable physical items and enables them to share information with one another through a wireless network. Networks in the Internet of Things (IoT) enable smart devices to relay and receive data and information, monitor activities, and control equipment remotely without the need for human resources' involvement or control. There are numerous communication systems that can accomplish a few requirements for a variety of IoT applications. As a result, no single communication method can satisfy the needs of the various IoT applications currently in use [3]. For instance, ZigBee can be used to build low-cost device-to-device networks because it doesn't need a centralized infrastructure. Additionally, BLE enables the use of a button cell battery to power IoT devices.

2.1 IoT Devices and Its Communication

IoT (Internet of Things) devices are tiny, microcontroller-based hardware components, such as sensors, actuators, gadgets, appliances, equipment, or machines, that are programmed to send data and carry out specific activities through the internet or other wireless networks. They may be included into autos, machinery,

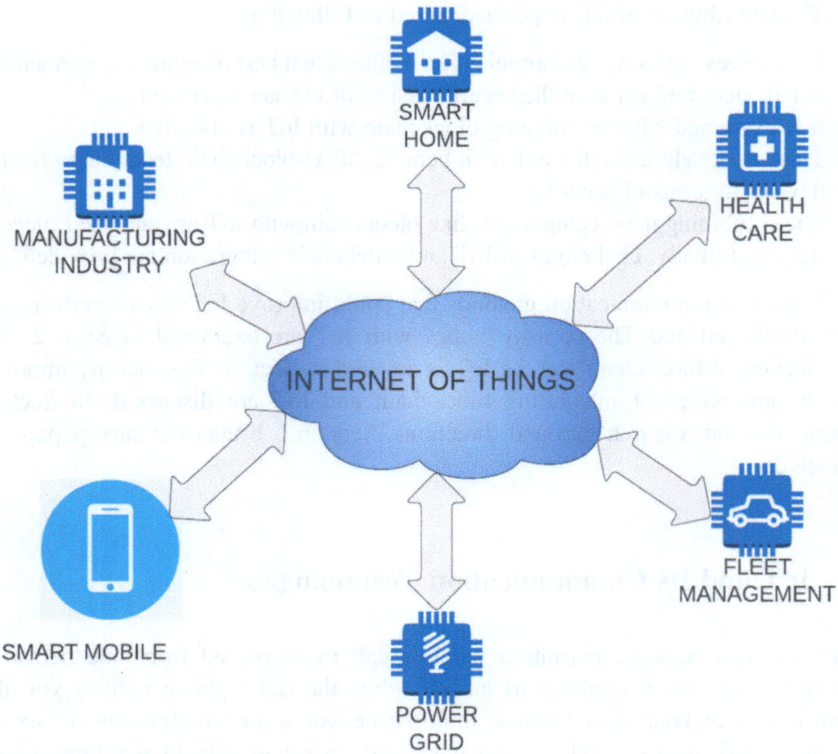

Fig. 1 IoT and its applications

environmental sensors, medical devices, home appliances including fans, lighting, air conditioners, and refrigerators, as well as other things. The majority of IoT devices use AI, machine learning, and other deep-tech tools to incorporate intelligence and autonomy into physical assets, systems, and processes, including autonomous vehicles, smart manufacturing, medical equipment, and home automation. Data processing must be done on the IoT endpoint rather than in a cloud environment due to network capacity restrictions, concerns about data privacy, and user experience—more of an edge computing approach. Communication between diverse devices using various systems and networks is made possible by common protocols. A device that talks with another through a wireless network often goes through seven layers of protocols [4]. According to how it works, information is transmitted from a physical device via the network, entered into a session, then shown where it is needed, and eventually reached the application protocol where it will be used, altered, and retransmitted.

The infrastructure and standards required to transfer data from devices to practical applications are provided by the IoT ecosystem. It does, however, have a number of "layers." Applications of IoT where IoT devices are connected are shown in Fig. 1.

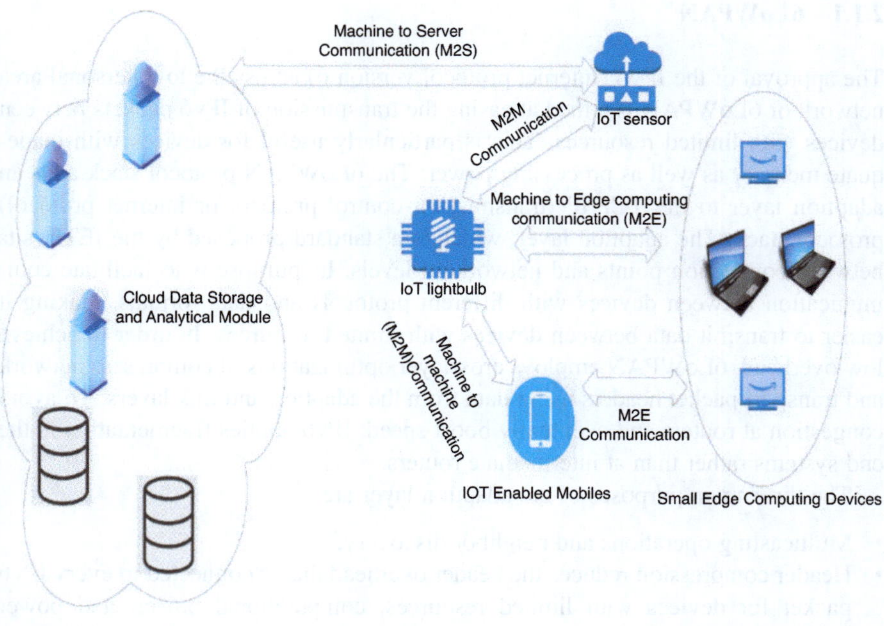

Fig. 2 IoT communication architecture

As an alternative, the sensors can push data (information) to specialized gateway devices, which would then transfer it over GSM or WI-FI to the cloud. When this happens, services of cloud send the outcome generated to the mobile device in order to update or upgrade the user on the latest developments. One wearable infant movement surveillance option is Mimobaby [5]. It collects data via baby clothes attached with sensors. The data are then sent to a nearby custom gateway, where it is used to transfer data to a cloud service using in-home Wi-Fi. The cloud services then notify the parents' smartphones in real-time. A handful of the most popular communication methods utilized in IoT applications are shown in Fig. 2. The cloud may be used to process the data gathered by IoT solutions further, store historical data, or identify patterns. Users of mobile devices are able to act right away or complete actuation chores. In these situations, short-range communication protocols, such Bluetooth, are used for mobile device to hardware communication, and WI-FI or GSM is used for long-range communication duties. Although IoT platforms based on cloud technologies are attempting to create an existence of their own ecosystems by promoting the creation and distribution of third-party extensions (also known as plugins) through app stores. Such tendencies have been observed time and time again in both the PC and smartphone sectors.

2.1.1 6LoWPAN

The approval of the IPv6 (Internet protocol version 6) across the low personal area network or 6LoWPAN is aimed at easing the transmission of IPv6 packets between devices with limited resources. This is particularly useful for devices with inadequate memory as well as processing power. The 6LoWPAN protocol stack adds an adaption layer to the TCP/IP (transmission control protocol or internet protocol) protocol stack. The adaption layer, which is a standard proposed by the IETF, sits between connection points and networking levels. Its purpose is to facilitate communication between devices with different protocols and data formats, making it easier to transmit data between devices with limited resources. In order to achieve low overhead, 6LoWPAN employs crosslayer optimizations; it compresses network and transport packet headers using data from the adaption and link layers. To avoid congestion at routers and eventually boost speed, IPv6 applies fragmentation at the end systems rather than at intermediate routers.

The three main purposes of the adaption layer are:

- Multicasting operations and neighbor discovery.
- Header compression reduces the header overhead that is connected to every IPv6 packet for devices with limited resources, computational power, and power consumption.
- Layer 2 forwarding, sometimes referred to as data link layer forwarding, transmits network layer datagrams over a number of radio hops [6].

With the aid of one particular or multiple edge routers route Internet Protocol datagrams in between various media, two 6LoWPAN networks can be linked to one or more IP networks. Any random link, including Ethernet, Wi-Fi, or 3G/4G, can offer connectivity to other IP networks. While 6LoWPAN only describes how IPv6 will be implemented over the standard established by IEEE 802.15.4, routers at the edge may additionally provide the IPv6 protocol conversion techniques for connecting 6LoWPAN networks with IPv4 networks, which include NAT64 as mentioned in RFC 6146. The 6LoWPAN nodes are not required to implement IPv4 completely or partially in order to use these IPv6 transition techniques (Fig. 3).

2.1.2 ZigBee

A standard called ZigBee was created to allow two-way wireless communication between wireless devices. ZigBee‘s extremely low-power consumption makes it suited for devices with limited resources [7]. ZigBee IP, a new iteration of the standard, has been approved. For D2D communications with IoT devices, ZigBee IP is intended to be used. Without a central hub, devices employing ZigBee IP can join the network, talk with one another, and share data. In the ZigBee IP topology, there are three different types of devices: ZC, ZR, and ZED. ZC coordinates the network’s formation and management. Relaying data provided by ZEDs and, if

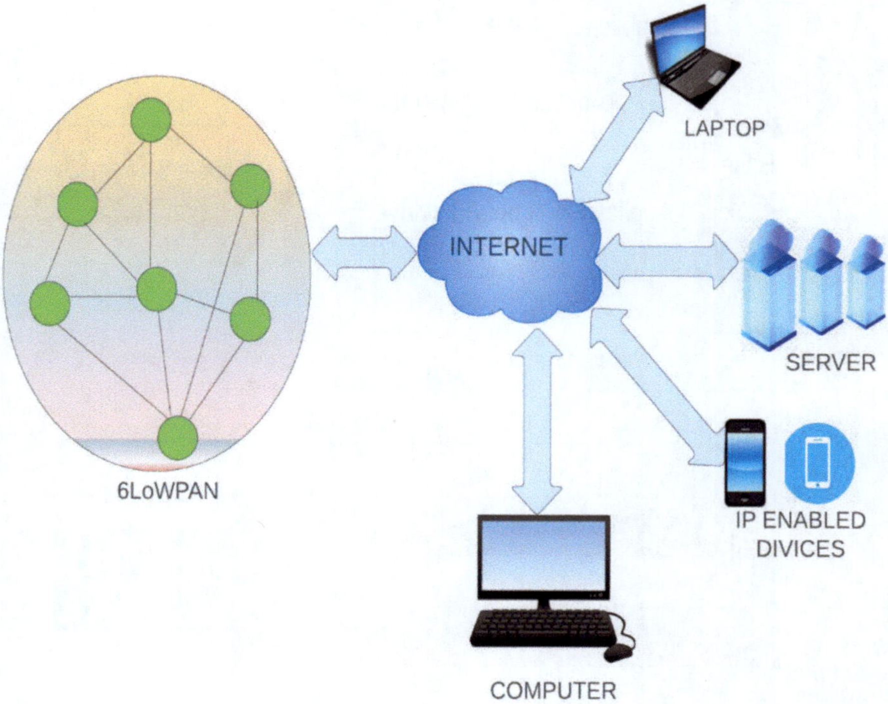

Fig. 3 Architecture of 6LoWPAN

required, growing the ZigBee internet protocol network are the responsibilities of ZRs. It is the responsibility of ZRs to extend the ZigBee IP network as necessary and relay traffic sent by ZEDs. A wireless device called ZED is controlled by ZC, and ZR relays its traffic. The Architecture of Zigbee is shown in Fig. 4. The coordinator, which is located in the system's core, is an immensely significant component of this network's structure. There is only one coordinator for each network, who is employed for various functions. Researchers choose an appropriate channel, that can scan channels to find the most effective one with the lowest possible level of interference, and give each device a distinct ID and address in order to transmit messages as well as instructions throughout the network. Zigbee routers serve as intermediaries, enabling data transfer between them alongside additional devices. In order to conserve battery power, end devices can only connect with parent nodes to a limited extent. The total number of routers, coordinators, and end devices will vary depending on the kind of network, such as star, tree, or mesh networks [8]. The Zigbee protocol architecture, which consists of a stack of extra layers, including the network and application layers exclusive to Zigbee, is defined by the IEEE 802.15.4 specification at the physical and MAC levels.

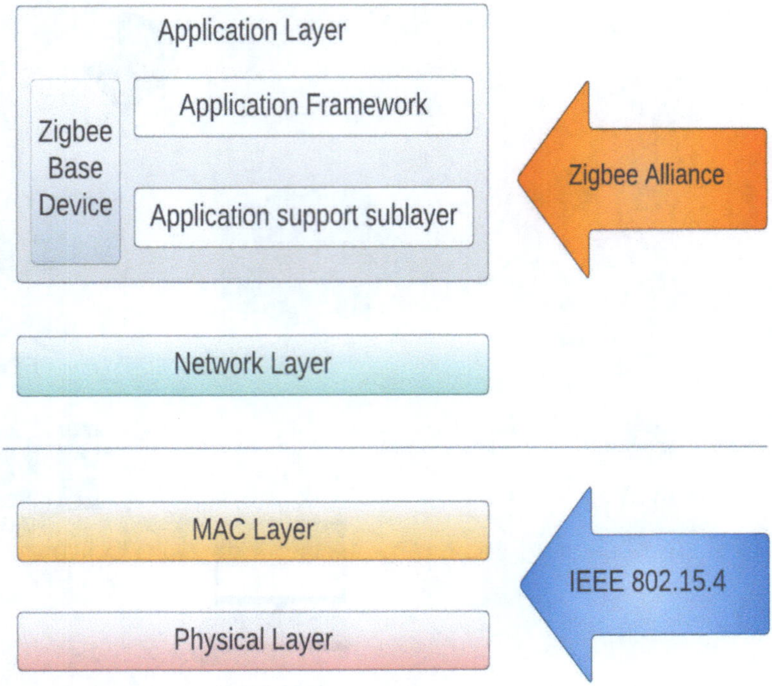

Fig. 4 Zigbee architecture

- Physical Layer: This layer performs both demodulation and modulation operations on signals, respectively, during transmission and reception.
- MAC Layer: By connecting carrier sensing multiple access collision avoidances (CSMA), this layer ensures the reliable delivery of data. Additionally, beacon frames are transmitted in order to synchronize communication.
- Network Layer: The network layer takes responsibility for all network-related operations, including network setup, connection and separation of the device from the internet, routing, device settings, etc.
- Application Support Sublayer: The data management services required for communication between Zigbee device application objects and the network layers. This layer is in charge of comparing two devices based on their services.
- Application Framework: Key-value pair and general message services are the two distinct kinds of data services offered by this framework. The key-value pair is used to get properties from application objects, whereas the generic message represents a structure that the developer specifies. In Zigbee devices, ZDO acts as an interface between application information (data) and the application protocol stack (APS) layer. It is in charge of finding, turning on, and establishing connections with other network devices.

2.1.3 Bluetooth Low Energy (BLE)

Bluetooth's low energy protocol stacking (sometimes referred to as the protocol stack) is made up of the host and the controller. This controller and host separation has its roots in the creation of conventional devices with Bluetooth BR/EDR (Basic Rate/Enhanced Data Rate), where the two components are carried out independently. Any applications along with identities are stacked underneath the GATT and GAP layers of the protocol stack and offer many APIs for dealing with the capabilities of the Bluetooth connectivity Low-energy Protocol Stack. The stack project and associated files are used to implement the Bluetooth Low-Energy Protocol Stack Activity [9]. This task, which is the most important in the system and is shown in Fig. 5, uses the Bluetooth energy-efficient protocol stack.

In order to enable low-cost devices to operate with extremely minimal power, BLE was developed to enhance the wireless communication offered by the original Bluetooth standard [10]. BLE is the best option for devices that broadcast and

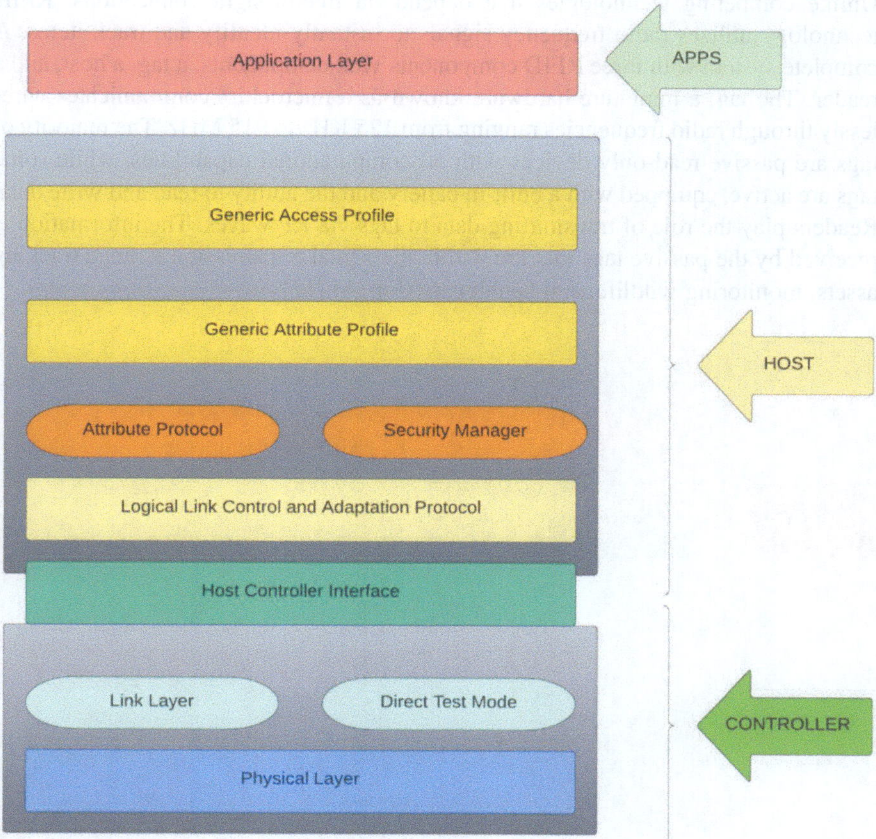

Fig. 5 Bluetooth low-energy stack and protocol

receive little amounts of data over short distances and at low data rates [11]. The advertising channel and the piconet channel are the two physical layer channels that BLE uses. Devices must select one channel at a time; they cannot use both channels concurrently. The function of the piconet channel is to enable communication between devices linked to a particular piconet and other devices linked to the same piconet. The advertising channel, on the other hand, is in charge of transmitting broadcast frames to detach devices or linking them together, as the name suggests. In the BLE, the repeat request and acknowledgment are two distinct activities performed by the link layer. Error control and multiplexing, as well as connection-oriented channel support, are two additional duties of the L2CAP. Once a link layer connection has been established using IPSS, BLE can transfer data and find IP-enabled devices.

2.1.4 Radio Frequency Identification (RFID)

Unlike competing technologies that depend on line-of-sight connections, RFID technology utilizes radio frequency signals to instantly identify and track items. A complete system with three RFID components vital components: a tag, a host, and a reader. The tag, a miniature hardware known as a microchip, communicates wirelessly through radio frequencies ranging from 125 KHz to 915 MHz. The majority of tags are passive read-only devices with no computational capabilities, while some tags are active, equipped with a built-in battery and the ability to read and write data. Readers play the role of transmitting data to tags via RF waves. The information is received by the passive tags that are within the signal transmission's range tracking assets, monitoring wildlife, and health care (Fig. 6) [12].

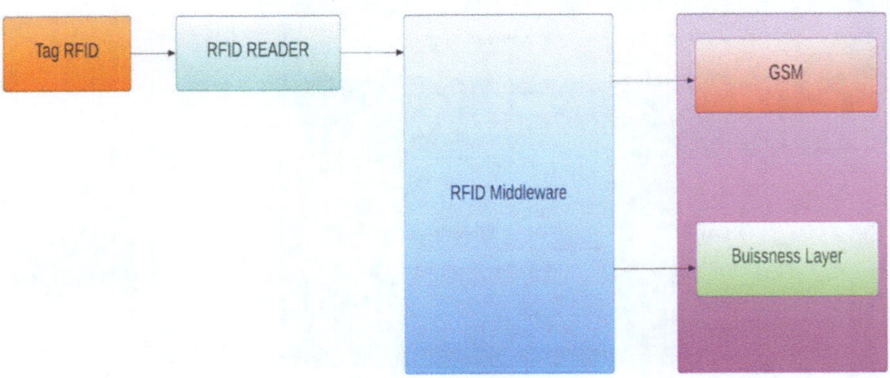

Fig. 6 RFID architecture

RFID can be utilized in a number of situations, including:

- In clothes retailers or fashion-related companies, the RFID tag is affixed to garments or other things. The store's entrance alarm will go off once the item exits without being deactivated.
- The tags on animals or cars can also be followed by the reader.

2.1.5 Wireless Local Area Network (WLAN)

Wireless Local Area Network (WLAN) is the prevailing wireless technology employed for indoor broadband wireless access. Wireless routers and modems have become commonplace purchases for both inexperienced consumers and businesses. The purpose of WLAN is to expand wired networks and offer mobility to laptops and other portable devices. In Europe, there are two widely adopted standards for WLAN: HIPER-LAN and IEEE 802.11 (commonly known as Wi-Fi). This article will focus on the former. The IEEE 802.11 standard introduced two network configuration categories known as Independent Basic Service Set (IBSS) and Infrastructure BSS (IBSS). In an IBSS, an Access Point (AP) assumes the responsibility of coordinating with other wireless devices or stations (STAs) within its coverage area. Thus, the AP acts as a bidirectional bridge connecting the wired network (e.g., Ethernet) with wireless devices. On the other hand, an STA is a mobile device that combines one or more wireless network interfaces with another mobile device like a laptop or mobile phone [13]. Prior to establishing a connection, the wireless device engages in packet exchange with the AP. APs within range broadcast their presence through beacon frames, notifying wireless devices. When a wireless device wishes to connect, it sends a probe request frame to the nearest AP it desires to associate with. Upon receiving the probe request, the AP responds with a probe response, requesting additional details and allocating resources.

Fig. 7 Wireless fidelity (Wi-Fi)

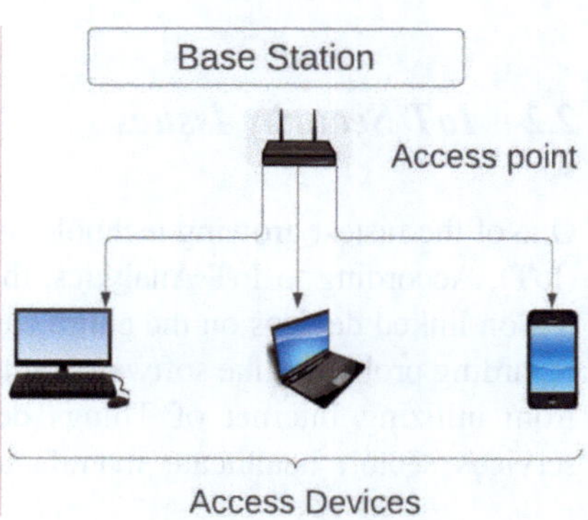

2.1.6 Wireless Fidelity (Wi-Fi)

Figure 7 shows the operation of a Wi-Fi system, where the Access Point enables data transmission between wireless devices within the same IBSS, as well as between the wired network and wireless devices. This facilitates seamless communication among wireless devices and connectivity to the wired network.

2.1.7 Light Fidelity (Li-Fi)

Modern wireless communication technology called Li-Fi, utilizes LED lights to transmit data. Data transmission is an essential task in our increasingly connected world. However, existing wireless networks that connect our devices to the internet often suffer from slow speeds when multiple devices are connected. Additionally, the limited availability of bandwidth poses challenges in achieving high data transfer rates and establishing secure networks as the number of internet-connected devices increase. Traditional wireless communication technologies rely on a small fraction of the electromagnetic spectrum or radio waves for data transmission. In contrast, Li-Fi offers a significantly broader transmission spectrum [14]. The fundamental concept behind Li-Fi is the ability to transmit data through LED light at speeds imperceptible to the human eye. Instead of using gigahertz radio waves like traditional methods, Li-Fi harnesses an underutilized portion of the electromagnetic spectrum known as the visible spectrum.

Li-Fi boasts strong security measures since light, unlike radio waves, cannot penetrate walls easily, making it more challenging for potential eavesdroppers to intercept wireless device traffic. Additionally, Li-Fi is less harmful to the human body compared to Wi-Fi, as light does not have any adverse effects, unlike radio waves used in Wi-Fi that can penetrate the human body. Similar to Wi-Fi, Li-Fi utilizes electromagnetic communication, but in the form of light, whereas Wi-Fi relies on radio waves that coexist with Bluetooth, microwaves, and other technologies within the frequency spectrum [15–18]. This includes natural light and conventional lightbulbs, apart from LED lamps. Figure 8 illustrates the architecture of Li-Fi, showcasing its components and functionality.

2.2 *IoT Security Issues*

One of the fastest growing technological innovations today is the Internet of Things (IoT). According to IoT Analytics, there will probably be more than twenty-seven billion linked devices on the entire earth by 2025. However, rising security worries regarding problems like software vulnerabilities and hackers may deter some people from utilizing Internet of Things devices. Companies engaged in the financial services sector, healthcare manufacturing, logistics, the retail sector, and other

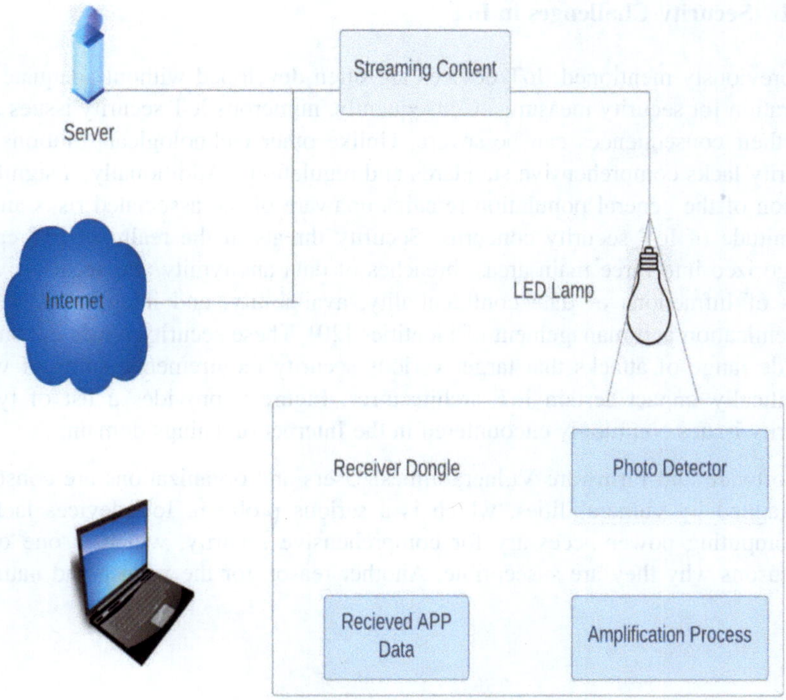

Fig. 8 Li-Fi architecture

industries that are already in the process of deploying devices should pay particular attention to these IoT security concerns. Organizations should pay special attention to system security because IoT systems are widely used [19]. Any weakness can lead to system crashes or hacks, which will then affect hundreds or thousands of people. There are several security concerns as a result of the numerous security vulnerabilities that are usually present in IoT devices. The IoT security that is strong can cover all aspects of protection, such as improving the component, monitoring, updating the firmware, controlling access, identifying threats, and fixing vulnerabilities. IoT security is essential because of the pervasiveness, insecurity, and prime target status of these systems. It is feasible to stop IoT devices from leaking crucial data or functioning as a backdoor for different parts of the network by securing them against unauthorized access. IoT security problems exist in everything between smart grids to vehicles, watches, and smart home appliances. IoT security's two main objectives are to:

- Assure that every piece of information is collected, processed, maintained, and transmitted in a secure manner.
- Recognize and take action on IoT component vulnerabilities.

2.2.1 Security Challenges in IoT

As previously mentioned, IoT devices are often developed without adequate consideration for security measures. Consequently, numerous IoT security issues arise, and their consequences can be severe. Unlike other technological solutions, IoT security lacks comprehensive standards and regulations. Additionally, a significant portion of the general population remains unaware of the associated risks and the magnitude of IoT security concerns. Security threats in the realm of IoT can be categorized into three main areas: breaches of data anonymity and security, violations of Infractions of data confidentiality, availability, and integrity as well as authentication and management of identities [20]. These security issues encompass a wide range of attacks that target various security requirements, some of which specifically impact certain IoT architectures. Figure 9 provides a list of typical security issues commonly encountered in the Internet of Things domain.

1. Software and Firmware Vulnerabilities: Users and organizations are constantly plagued by vulnerabilities, which is a serious problem. IoT devices lack the computing power necessary for comprehensive security, which is one of the reasons why they are susceptible. Another reason for the widespread nature of

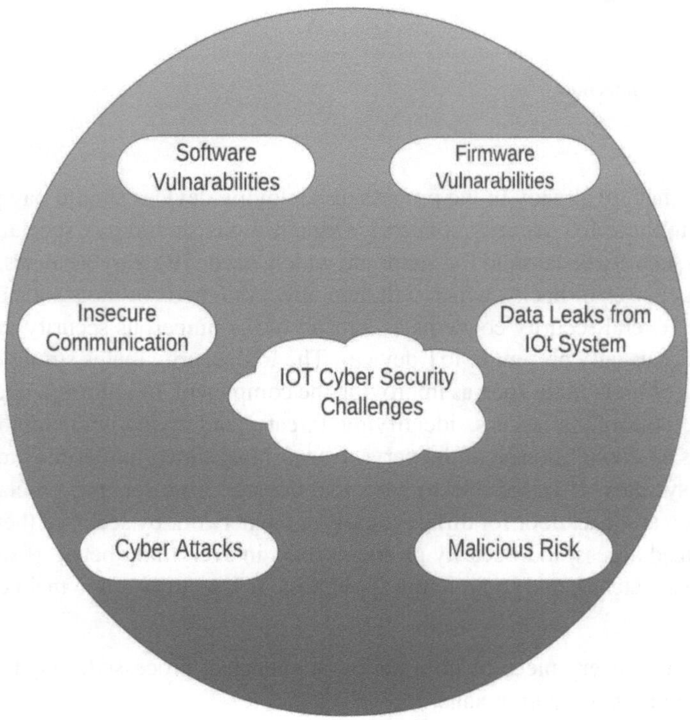

Fig. 9 IoT cybersecurity challenges

vulnerabilities is the insufficient budget for creating and evaluating secure firmware, which is spurred by the price of the gadgets and their incredibly quick development cycles. Firmware is a frequently exposed attack surface that attackers utilize to break into networks, steal data, or even take over the devices themselves to do damage or cause disruption. Bottom line: An IoT device with insecure firmware is also insecure.

2. Data Leaks from IoT System: All data are transmitted, stored, and used by cloud-based services, which are also exposed to threats from the outside. As a result, the actual devices as well as environments in the cloud may experience data loss. An additional potential cause of breaches of information in your IoT system is third-party services.
3. Insecure Communication: One of the most significant risks of inadequate interaction is the possibility of man-in-the-middle assault (MitM). If your device does not deploy reliable secure authentication and encryption protocols; otherwise, hackers might readily conduct MitM attacks to compromise the control the updating process and manage our gadget. Even attackers can install malware and modify functionality [21]. Even if the device you are using is not among the victims of a MitM attack, hackers can still access the data it sends to other systems and devices via unencrypted messaging.
4. Cyberattacks: The devices of IoT lack security protocols strictly because they are created to meet the board of an organization. Attackers used this advantage to gain access to the company's systems through one of the vulnerable IoT devices. The attacks are cyberattacks that use any types of IoT device to gain access to sensitive consumer data. The attackers often damage a gadget, installing a malware on it, or gain access to additional company personal data or information. Figure 10 shows a list of the most common types of IoT device attacks.
5. Malicious Risk: Despite the minimal computing capability of the majority of IoT devices, malware may nevertheless infiltrate them. The functionality of the IoT system can alter private information collected, and other attacks can be carried out if an attacker manages to introduce malware into the system. In addition, some utilities may be preinfected with viruses if their manufacturers do not take appropriate software security precautions.

3 AI and Blockchain for IoT

Blockchain is the name of the technology that enables a collection of untrustworthy parties to validate transactions. Blockchain enables the auditability of transactions in a distributed, an encrypted, irreversible, and transparent ledger [22]. Blockchain employs a peer-to-peer network (decentralized style) to store and process data as opposed to a client-server architecture (centralized style), which is how it differs from cloud computing. In a blockchain, any entity at any moment has access to every transaction that occurs, starting with the very first one in the system. This access can

Fig. 10 Common attacks on IoT system

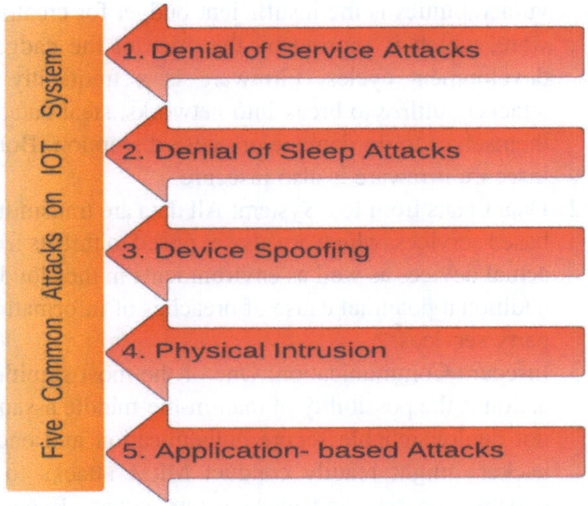

be checked, gathered, or aggregated. The blockchain protocol aims to create data that take the kind of structure of a chain of blocks.

Since the introduction of Bitcoin, the first widely adopted cryptocurrency, in 2009, the concept of blockchain has gained significant traction. The blockchain serves as the foundational data structure for Bitcoin, ensuring its functionality as a trustless electronic cash system by preventing double-spending. A Bitcoin transaction record includes information such as the value, sender, receiver, and signature. Multiple or different transactions can be included in a single type of block; hence, the term "blockchain" specifically refers to the data structure employed by Bitcoin. Each block, except for the genesis block, which is the initial block, is linked to the preceding block through its hash. The interconnection of blocks ensures the integrity of the blockchain. Additionally, each block contains the Merkle root hash of the transactions to safeguard against tampering. Figure 11 showcases an AI-enabled blockchain.

As previously mentioned, the peer-to-peer (P2P) architecture used in blockchain technology necessitates that every peer in the network fulfills four essential Services such as storage spaces. wallets, also known routing, as well as mining [23]. In a blockchain network, various nodes serve different purposes. For transaction and block propagation to occur, each node in the P2P network of the cryptocurrency must possess routing capabilities. Storage functionality is required for some nodes to maintain a copy of the entire chain (full nodes), while others may store only a portion of the chain (light nodes).

As a result, prior to adding a block to the blockchain, its validity is verified by the network's peers. Since blocks in the network are created simultaneously by multiple miners, the blockchain can have several branches. To address this lack of comparability, a solution is implemented by considering the longest branch as the valid one [25]. This concurrent and intensive block generation process leads to the creation of a unique distributed trustless consensus mechanism. The complexity of block

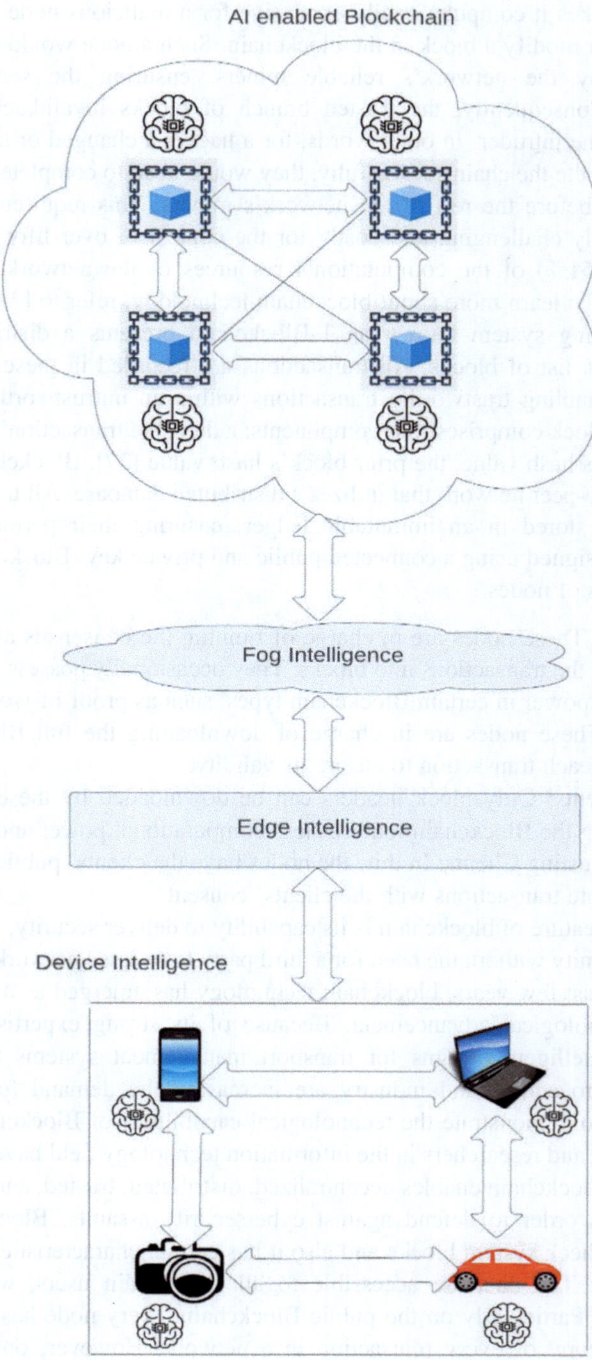

Fig. 11 AI-enabled blockchain architecture

generation makes it computationally expensive for a malicious node attempting to tamper with or modify a block in the blockchain. Such a node would eventually be outmatched by the network's reliable miners, ensuring the security of the blockchain. Consequently, the trusted branch of blocks invalidates any blocks produced by the intruder. In other words, for a hacker a changed or tampered with block to complete the chain successfully, they would need to complete the Proof-of-Work (PoW) before the rest of the network's miners. This requirement makes it computationally challenging and costly for the control of over fifty percent (or a minimum of 51%) of the computational resources of the network by only one organization. To learn more about blockchain technology, refer to [24] and [26].

An upcoming system known as T-Blockchain presents a distributed ledger consisting of a list of blocks. All transactions are recorded in these blocks using encryption, enabling trustworthy transactions within an untrustworthy participant group. Each block comprises four components: a date, the transaction's number, the current block's hash value, the prior block's hash value [27]. Blockchain is specifically a peer-to-peer network that utilizes a distributed database. All transactions are gathered and stored in an immutable ledger, ensuring their permanence. Each transaction is signed using a connected public and private key. Blockchain contains different kinds of nodes:

Miner Nodes: These nodes are in charge of running the consensus algorithms and assembling the transactions into blocks. They occasionally possess the maximum computing power in certain Blockchain types, such as proof-of-work consensus.

Full Nodes: These nodes are in charge of downloading the full Blockchain and examining each transaction to ensure its validity.

The Thin Clients: Only block headers can be downloaded by these nodes. As a result, using the Blockchain requires less computational power and storage.

The Server-Trusting Clients: In this, the nodes have the clients' public keys and are able to create transactions with the clients' consent.

The primary feature of blockchain is its capability to deliver security, data integrity, and anonymity without the need for a third party to oversee network transactions. Over the past few years, blockchain technology has emerged as a highly promising technological advancement. Because of its strong expertise in creating parallel, intelligent designs for transport management systems [28]. Modern business procedures and industry are increasing the demand for Blockchain systems. To demonstrate the technological capabilities of Blockchain, financial institutions and researchers in the information technology field have done a lot of research. Blockchain enables decentralized, distributed, trusted, and secured data sharing. In order to defend against cybersecurity assaults, Blockchain nodes regularly check system blocks and also it has certain characteristics as follows:

Transparency: The data are accessible to all Blockchain users, which lends it credibility. Particularly on the public Blockchain, every node has access to the complete copy of every transaction in a network. However, only permission nodes were allowed access in the private and consortium blockchains [29].

Programmable: Blockchain may feature smart contracts, which support contract negotiation and execution. A contract that is smart is a piece of code that has the ability to run, be validated, or be enforced independently. Smart contracts can be used to create black-lists and white-lists of IP addresses. The most well-known open Blockchain with smart contract capabilities is the Ethereum platform.

Immutability: The Blockchain works by connecting every block with the header hash from the previous block. As a result, the Blockchain records are unchangeable and not revocable.

Decentralization: The Blockchain's database is a distributed one that has a lot of related data. It creates a connection between the current and prior blocks. Furthermore, peer-to-peer networks without a reliable third party are used to build the Blockchain [30].

Pseudonymous Address: To prevent identity disclosure without sacrificing the transparency characteristic is the major motivation.

Traceability: The Blockchain assigns a timestamp to each transaction in the distributed ledger. Any transaction between two parties can be verified and tracked by all participants as a result.

Autonomy: Utilizing public/private key pairs, each node has the ability to send and receive transactions without the aid of a reliable third party.

3.1 *Integration of AI-Enabled Blockchain with IoT*

Attacks on cybersecurity have increased recently, and safeguarding IoT from these attacks remains a top security priority. These attacks currently have many-to-one dimensions, making them challenging to counter. The available literature research focused on strengthening IoT security by combining it with other technologies to close the gap between IoT constraints and security requirements. Due to its limited resource availability, IoT has weak security methods, which is the primary driver for this integration. IoT networks require a third party to govern and monitor the traffic because they are unable to use an advanced strategy [31]. IoT and AI integration is a popular technology that uses AI to analyze large amounts of data from various IoT devices. These methods rely on creating a single point of detection that has strong analytical skills and can identify attack patterns from network traffic [32]. AI-based techniques may have sophisticated security features, but because they are based on a centralized server/client model, they have a number of drawbacks. Furthermore, they made the incorrect assumption that the third party is always a reliable source. As a result, AI-based models have grown incompatible with IoT's nature and are no longer able to offer a solid defense against the expansion of IoT networks.

These centralized or isolated defense systems are simple to get around for attackers. As a result, countering cybersecurity assaults requires a distributed and collaborative detection technique, making the ideal choice to create a decentralized architecture. Blockchain technology has received a lot of interest recently for offering decentralized defense mechanisms. Through the provision of this technology can improve the security of Internet of Things (IoT) networks since it has a decentralized infrastructure and safe data exchange across untrusted nodes. Additionally, it allows different IoT devices to take part in the detecting process without the help of a reliable third party [33].

However, given that Blockchain is a new technology, the recently put out solutions based on Blockchain encounter significant challenges. They lack a thorough understanding of IoT networks to identify cybersecurity attacks. Additionally, there are several drawbacks to Blockchain techniques, including scalability and efficiency. Additionally, Blockchain-based models do not possess the intelligence necessary to meet the adaptive criteria necessary to counter newer and more sophisticated cybersecurity attack variants. Given IoT limitations and the present situation of the aforementioned models, IoT networks demand more efficient and intelligent decentralized defense solutions based upon additional promising technologies at the same time to combat cybersecurity attacks. In this context, blockchain and AI technologies have shown promise as allies, combining strengths like blockchain's decentralization, and AI's analytical capabilities. The core idea behind these methods is to address IoT risks by incorporating AI and blockchain technologies, such as Blockchain's decentralization feature. The biologically inspired methods have also demonstrated their capacity to improve IoT security. Empirical studies that examine IoT networks' resilience to cybersecurity assaults are the basis of our knowledge on how to improve their security. While the majority of research suggest a generic defense mechanism, many of them focus on preventing, detecting, and mitigating a particular kind of attack.

The IoT revolution is altering the world through the automation and optimization of manual processes as well as the generation and acquisition of vast amounts of data that enhance people's lives by supplying important knowledge. By automating services, this useful information can be used to create clever applications that enhance management and lifestyles in general. For the past few years, cloud computing has been the sole choice for giving IoT devices the services they require to collect, analyze, and transform data into real-time services. There are several chances to share and access information as a result of the expansion of IoT devices. The incorporation of cutting-edge technology, such as the Internet of Things and cloud computing, has the potential to significantly enhance quality of life. The potential for IoT and blockchain integration has been noted by certain experts. Blockchain will advance Internet of Things (IoT) technologies by offering reliable and traceable shared resources. The Benefits of Using Blockchain with AI are mentioned in the above Fig. 12. Data sources may always be specified securely and kept constant throughout time. Many Internet of Things (IoT) devices provide safe data sharing capabilities.

Traceability would reveal any data leak in the chain, preventing fraud and enabling the hunt for infections that can endanger human life [34]. In various fields, including smart homes and smart cities, new actors can enter the ecosystem in search of data dependability to enhance their services. As a result, the IoT and blockchain integration might deliver trustworthy and secure data to IoT applications. It was argued that blockchain will significantly improve the IoT ecosystem by addressing a

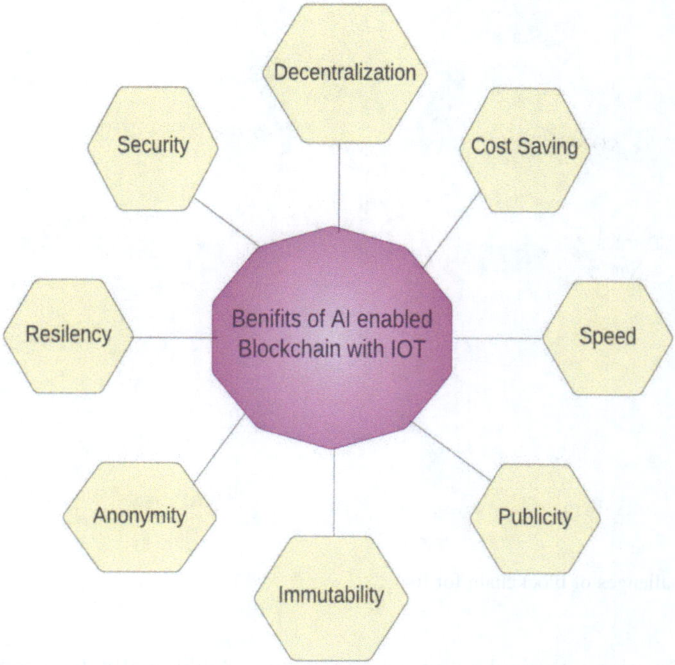

Fig. 12 Benefits of AI and blockchain with IoT

number of difficulties with dependability, scalability, and privacy. Some of the challenges that can be faced by integrating these technologies are shown in the Fig. 13.

IoT technologies have various difficulties, including problems with device performance, capacity, and heterogeneity. However, one of the biggest difficulties is security. IoT devices have a number of security vulnerabilities; we will quickly discuss them in this paragraph, and any interested readers can learn more about them in [35]. Since IoT technology has been the target of numerous attacks (some of which are brought on by integration with cloud computing), it is vital to link IoT with another technology to increase its security. As numerous academics have noted blockchain may hold the key to strengthening the security of the IoT paradigm. However, the validity of the data produced by IoT devices is a possible problem when merging IoT with blockchain.

In order to train a Support Vector Machine (SVM) classifier [36] suggested a Blockchain-based training strategy termed secure SVM. The major objective of the suggested model is to protect the privacy of IoT data in applications for smart cities. Without the use of a third-party auditor, blockchain technologies were used to securely distribute the data among several IoT data providers. IoT data providers can use Blockchain to encrypt the data using their private keys and deposit it on the

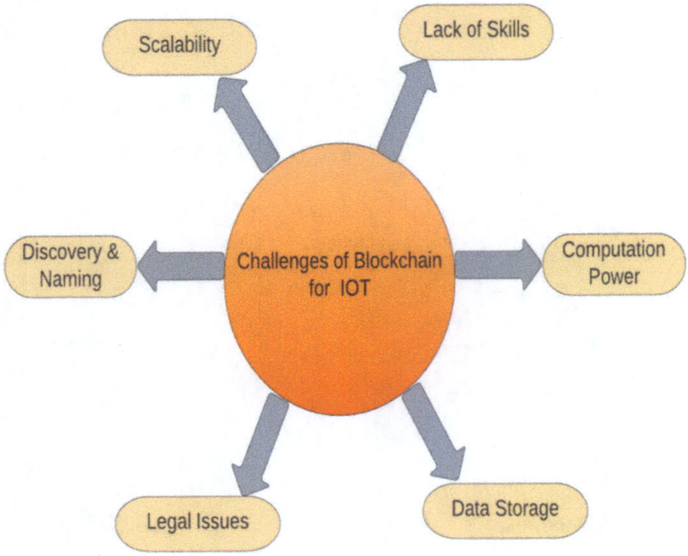

Fig. 13 Challenges of blockchain for IoT

Blockchain using a particular transaction structure. Additionally, they combined the Paillier cryptosystem with Blockchain to address the issues of data confidentiality, integrity, and ownership during the training process [37]. The suggested method includes IoT devices, IoT data sources, a single shared Blockchain ledger, and a single the Internet of Things data analyst who trains the classifier. IoT devices first sense the data and communicate it to IoT data providers. The data are then encrypted by the IoT data provider and kept in a public Blockchain ledger. Then, utilizing training algorithms like secure comparison and secure polynomial addition, the IoT data analyst can access the encrypted data that are stored on the shared Blockchain ledger. Using two real-world datasets, the accuracy and effectiveness of the proposed model were assessed.

4 Research Gap and Future Direction

IoT networks face numerous security difficulties despite recent emphasis being paid to securing IoT networks using AI and Blockchain approaches, as most current solutions concentrate on integrating IoT with AI or Blockchain independently. In order to emphasis that each technology has incomparable advantages and that blockchain technology combined with artificial intelligence can result in solutions that can withstand cybersecurity threats. Blockchain and AI complement one other and help each other overcome obstacles. By combining these, we can create an explainable AI, a distributed and decentralized IoT framework, digitally signed transactions, safe data exchange, and an immutable IoT. Consequently, utilizing each technology can shield the Internet of Things from cybersecurity threats.

Collaborative Attack Mitigation Model: Some of the current defense strategies are underequipped to handle cybersecurity attacks on their own. Additionally, using the resources of other businesses could be a productive strategy to divide the workload and lessen attacks. One of the most important pieces of information that businesses may give to lessen threats is the IP address. Recently, the researchers used the Blockchain to promote IP addresses that were blacklisted or whitelisted, specifically to thwart DDoS attacks. Although using Blockchain to promote black- or whitelisted IP addresses has advantages, this strategy serves as a backup. As a result, further research is necessary before it can be coupled with current AI-based solutions without altering them. Additionally, these integrations require additional study when dealing with a dynamic IP address.

Create a Trusted Signature Database: Using a network of IoT nodes connected to a Blockchain, researchers have used Blockchain to create a trusted database containing attack signatures. These nodes work together to build the database, and each node uses its private key to sign each transaction. Each node keeps an eye on the traffic in order to regularly define signatures using an analytical process. These studies, however, have a number of drawbacks, are still in the early stages of study, and require further examination because they do not take the insider attack into account.

5 Conclusion

In this study, for protecting IoT networks by fusing AI and Blockchain are systematically reviewed. We shed light on the architecture of the Internet of Things, its security needs, its difficulties and weaknesses, and its cybersecurity threats. We also emphasized the types of blockchain and their properties. The majority of studies centered on combining IoT with blockchain or AI. In this article, about how to secure the IoT's network by merging AI and Blockchain technology. The findings indicate that combining IoT with AI and blockchain is a viable strategy that can give rise to new, potent technologies that could improve IoT security in a variety of ways and open up new avenues for digitalization. Blockchain and AI complement one other and help each other overcome obstacles. Consequently, utilizing each technology can shield the Internet of Things from cybersecurity threats. A decentralized and distributed IoT structure, digitally signed transactions, safe data sharing, and an immutable IoT can all be created by integrating these technologies. As a result, it might trigger the following revolution in business and other fields.

References

1. T. T. A. Dinh, R. Liu, M. Zhang, G. Chen, B. C. Ooi, and J. Wang, "Untangling Blockchain: A Data Processing View of Blockchain Systems," IEEE Transactions on Knowledge and Data Engineering, vol. 30, no. 7, pp. 1366–1385, Jul. 2018, https://doi.org/10.1109/tkde.2017.2781227.
2. J. Zou, B. Ye, L. Qu, Y. Wang, M. A. Orgun, and L. Li, "A Proof-of-Trust Consensus Protocol for Enhancing Accountability in Crowdsourcing Services," IEEE Transactions on Services Computing, vol. 12, no. 3, pp. 429–445, May 2019, https://doi.org/10.1109/tsc.2018.2823705.
3. A. Reyes-Yanes, S. Gelio, P. Martinez, and R. Ahmad, "Wireless Sensing Module for IoT Aquaponics Database Construction," International Journal of Electronics and Electrical Engineering, vol. 9, no. 2, pp. 43–47, Jun. 2021, https://doi.org/10.18178/ijeee.9.2.43-47.
4. S. Alharbi, A. Attiah, and D. Alghazzawi, "Integrating Blockchain with Artificial Intelligence to Secure IoT Networks: Future Trends," Sustainability, vol. 14, no. 23, p. 16002, Nov. 2022, https://doi.org/10.3390/su142316002.
5. S. Aldhaheri, D. Alghazzawi, L. Cheng, A. Barnawi, and B. A. Alzahrani, "Artificial Immune Systems approaches to secure the internet of things: A systematic review of the literature and recommendations for future research," Journal of Network and Computer Applications, vol. 157, p. 102537, May 2020, https://doi.org/10.1016/j.jnca.2020.102537.
6. M. Shen, X. Tang, L. Zhu, X. Du, and M. Guizani, "Privacy-Preserving Support Vector Machine Training Over Blockchain-Based Encrypted IoT Data in Smart Cities," IEEE Internet of Things Journal, vol. 6, no. 5, pp. 7702–7712, Oct. 2019, https://doi.org/10.1109/jiot.2019.2901840.
7. Y. Liu, F. R. Yu, X. Li, H. Ji, and V. C. M. Leung, "Blockchain and Machine Learning for Communications and Networking Systems," IEEE Communications Surveys & Tutorials, vol. 22, no. 2, pp. 1392–1431, 2020, https://doi.org/10.1109/comst.2020.2975911.
8. S. K. Singh, S. Rathore, and J. H. Park, "BlockIoTIntelligence: A Blockchain-enabled Intelligent IoT Architecture with Artificial Intelligence," Future Generation Computer Systems, vol. 110, pp. 721–743, Sep. 2020, https://doi.org/10.1016/j.future.2019.09.002.
9. X. Han, R. Zhang, X. Liu, and F. Jiang, "Biologically Inspired Smart Contract: A Blockchain-Based DDoS Detection System," 2020 IEEE International Conference on Networking, Sensing and Control (ICNSC), Oct. 2020, https://doi.org/10.1109/icnsc48988.2020.9238104.
10. M. Kowalski, Z. W. Y. Lee, and T. K. H. Chan, "Blockchain technology and trust relationships in trade finance," Technological Forecasting and Social Change, vol. 166, p. 120641, May 2021, https://doi.org/10.1016/j.techfore.2021.120641.
11. P. Sandner, J. Gross, and R. Richter, "Convergence of Blockchain, IoT, and AI," Frontiers in Blockchain, vol. 3, Sep. 2020, https://doi.org/10.3389/fbloc.2020.522600.
12. R. Vishwakarma and A. K. Jain, "A survey of DDoS attacking techniques and defence mechanisms in the IoT network," Telecommunication Systems, vol. 73, no. 1, pp. 3–25, Jul. 2019, https://doi.org/10.1007/s11235-019-00599-z.
13. S. Singh, P. K. Sharma, B. Yoon, M. Shojafar, G. H. Cho, and I.-H. Ra, "Convergence of blockchain and artificial intelligence in IoT network for the sustainable smart city," Sustainable Cities and Society, vol. 63, p. 102364, Dec. 2020, https://doi.org/10.1016/j.scs.2020.102364.
14. B. D. Deebak and F. AL-Turjman, "Privacy-preserving in smart contracts using blockchain and artificial intelligence for cyber risk measurements," Journal of Information Security and Applications, vol. 58, p. 102749, May 2021, https://doi.org/10.1016/j.jisa.2021.102749.
15. K. R. Ozyilmaz and A. Yurdakul, "Designing a Blockchain-Based IoT With Ethereum, Swarm, and LoRa: The Software Solution to Create High Availability With Minimal Security Risks," IEEE Consumer Electronics Magazine, vol. 8, no. 2, pp. 28–34, Mar. 2019, https://doi.org/10.1109/mce.2018.2880806.
16. T. A. Ahanger, "Defense Scheme to Protect IoT from Cyber Attacks using AI Principles," International Journal of Computers Communications & Control, vol. 13, no. 6, pp. 915–926, Nov. 2018, https://doi.org/10.15837/ijccc.2018.6.3356.

17. H. F. Atlam, R. J. Walters, and G. B. Wills, "Intelligence of Things: Opportunities & Challenges," 2018 3rd Cloudification of the Internet of Things (CIoT), Jul. 2018, https://doi.org/10.1109/ciot.2018.8627114.
18. Y. Qian et al., "Towards decentralized IoT security enhancement: A blockchain approach," Computers & Electrical Engineering, vol. 72, pp. 266–273, Nov. 2018, https://doi.org/10.1016/j.compeleceng.2018.08.021.
19. J. Nieminen et al., "Networking solutions for connecting bluetooth low energy enabled machines to the internet of things," IEEE Network, vol. 28, no. 6, pp. 83–90, Nov. 2014, https://doi.org/10.1109/mnet.2014.6963809.
20. K. E. Jeon, J. She, P. Soonsawad, and P. C. Ng, "BLE Beacons for Internet of Things Applications: Survey, Challenges, and Opportunities," IEEE Internet of Things Journal, vol. 5, no. 2, pp. 811–828, Apr. 2018, https://doi.org/10.1109/jiot.2017.2788449.
21. C. Wohlin, E. Mendes, K. R. Felizardo, and M. Kalinowski, "Guidelines for the search strategy to update systematic literature reviews in software engineering," Information and Software Technology, vol. 127, p. 106366, Nov. 2020, https://doi.org/10.1016/j.infsof.2020.106366.
22. B. Wu, Q. Li, K. Xu, R. Li, and Z. Liu, "SmartRetro: Blockchain-Based Incentives for Distributed IoT Retrospective Detection," 2018 IEEE 15th International Conference on Mobile Ad Hoc and Sensor Systems (MASS), Oct. 2018, https://doi.org/10.1109/mass.2018.00053.
23. S. R and A. H, "Improved EPOA clustering protocol for lifetime longevity in wireless sensor network," Sensors International, vol. 3, p. 100199, 2022, https://doi.org/10.1016/j.sintl.2022.100199.
24. Evans Mwiti Ochieng, "A Study of the History, Functions, Roles, and Challenges of Human Resources Management", Journal of Enterprise and Business Intelligence, vol.3, no.1, pp. 054–064, January 2023. https://doi.org/10.53759/5181/JEBI202303006.
25. A. Haldorai, J. Sivaraj, M. Nagabushanam, and M. Kingston Roberts, "Cognitive Wireless Networks Based Spectrum Sensing Strategies: A Comparative Analysis," Applied Computational Intelligence and Soft Computing, vol. 2022, pp. 1–14, Oct. 2022, https://doi.org/10.1155/2022/6988847.
26. W. Li, S. Tug, W. Meng, and Y. Wang, "Designing collaborative blockchained signature-based intrusion detection in IoT environments," Future Generation Computer Systems, vol. 96, pp. 481–489, Jul. 2019, https://doi.org/10.1016/j.future.2019.02.064.
27. G. Spathoulas, N. Giachoudis, G.-P. Damiris, and G. Theodoridis, "Collaborative Blockchain-Based Detection of Distributed Denial of Service Attacks Based on Internet of Things Botnets," Future Internet, vol. 11, no. 11, p. 226, Oct. 2019, https://doi.org/10.3390/fi11110226.
28. I. Dutt, S. Borah, and I. K. Maitra, "Immune System Based Intrusion Detection System (IS-IDS): A Proposed Model," IEEE Access, vol. 8, pp. 34929–34941, 2020, https://doi.org/10.1109/access.2020.2973608.
29. M. A. Cheema, H. Khaliq Qureshi, C. Chrysostomou, and M. Lestas, "Utilizing Blockchain for Distributed Machine Learning based Intrusion Detection in Internet of Things," 2020 16th International Conference on Distributed Computing in Sensor Systems (DCOSS), May 2020, https://doi.org/10.1109/dcoss49796.2020.00074.
30. S. Aldhaheri, D. Alghazzawi, L. Cheng, B. Alzahrani, and A. Al-Barakati, "DeepDCA: Novel Network-Based Detection of IoT Attacks Using Artificial Immune System," Applied Sciences, vol. 10, no. 6, p. 1909, Mar. 2020, https://doi.org/10.3390/app10061909.
31. P. Nespoli, F. G. Marmol, and J. M. Vidal, "A Bio-Inspired Reaction Against Cyberattacks: AIS-Powered Optimal Countermeasures Selection," IEEE Access, vol. 9, pp. 60971–60996, 2021, https://doi.org/10.1109/access.2021.3074021.
32. J. Ashraf et al., "IoTBoT-IDS: A novel statistical learning-enabled botnet detection framework for protecting networks of smart cities," Sustainable Cities and Society, vol. 72, p. 103041, Sep. 2021, https://doi.org/10.1016/j.scs.2021.103041.
33. A. A. Ghali, R. Ahmad, and H. Alhussian, "A Framework for Mitigating DDoS and DOS Attacks in IoT Environment Using Hybrid Approach," Electronics, vol. 10, no. 11, p. 1282, May 2021, https://doi.org/10.3390/electronics10111282.

34. P. Kumar, R. Kumar, G. P. Gupta, and R. Tripathi, "A Distributed framework for detecting DDoS attacks in smart contract-based Blockchain-IoT Systems by leveraging Fog computing," Transactions on Emerging Telecommunications Technologies, vol. 32, no. 6, Sep. 2020, https://doi.org/10.1002/ett.4112.
35. Dua, D.; Graff, C. Machine Learning Repository. University of California, Irvine, School of Information and Computer Sciences. 2017. Available online: http://archive.ics.uci.edu/ml (accessed on 16 January 2022).
36. X. Qu et al., "A Survey on the Development of Self-Organizing Maps for Unsupervised Intrusion Detection," Mobile Networks and Applications, vol. 26, no. 2, pp. 808–829, Oct. 2019, https://doi.org/10.1007/s11036-019-01353-0.
37. S. Walling and S. Lodh, "Performance Evaluation of Supervised Machine Learning Based Intrusion Detection with Univariate Feature Selection on NSL KDD Dataset," Feb. 2023, https://doi.org/10.21203/rs.3.rs-2537820/v1.

Artificial Intelligence for Intelligent Spectrum Management Toward Futuristic Communication

1 Introduction

The predicted launch of 6G mobile networks in 2030 will meet the advanced communication needs of an impending data-centric, hyperconnected civilization. These networks will offer ultra-high-speed connectivity, ultra-low-latency communication, and highly scalable low-latency machine-type communication, resulting in remarkable network capabilities. By lowering latency and boosting machine-to-machine (M2M) communication, 6G will revolutionize networks and boost overall network performance. Although its focus is mostly on human subscribers, 6G also intends to improve the network experience for robots. As a replacement for 5G, the following-generation wireless communication network for machines will use cutting-edge innovations including artificial intelligence (AI) and machine learning [1].

The optimization of communication networks in a variety of configurations has long been a key function of AI. Machine learning has developed as a formidable AI tool for boosting the effectiveness and adaptability of wireless communication, enabling robots to demonstrate intelligence comparable to that of humans. Through AI, the introduction of 6G will further transform wireless communication. In contrast to 5G, 6G is developed to fully exploit useful AI applications.

Telecommunications businesses (telcos) will be able to increase operational efficiency in a variety of ways thanks to the integration of AI [2], including by placing base stations in the best possible locations, using less energy, and automating the maintenance of intricate networks. Telcos will be able to use massive volumes of real-time data gathered from linked devices and human communication channels to effortlessly integrate AI into a variety of wireless networks and services. Thus, the arrival of 6G will unleash the revolutionary potential of AI in wireless communication, bringing about new heights of effectiveness and intelligence.

© European Alliance for Innovation 2024

A. Haldorai et al., *Artificial Intelligence for Sustainable Development*,
EAI/Springer Innovations in Communication and Computing,
https://doi.org/10.1007/978-3-031-53972-5_12

The remainder is made up of the following: Sect. 2 presents machine learning and its importance for GDM; Sect. 3 presents the system design of the suggested work. Various performance metrics and the findings from Sect. 5 are described in Sect. 4. Lastly, a conclusion wraps up this debate in Sect. 6.

2 Features of 6G Communication Networks

The availability of improved services from the perspectives of network data availability, mobile data rate, and seamless ubiquitous connectivity is predicted to exceed the bar already established by 5G communications and surpass what is currently considered the norm in communications [4]. Figure 1 lists a summary of the features that 6G is anticipated to support. Terahertz (THz) Frequencies: 6G is expected to explore the utilization of higher frequency bands in the terahertz range, beyond the frequency bands used in previous generations. Terahertz frequencies offer wider bandwidth and the potential for extremely high data rates, enabling unprecedented speed and capacity for data transmission [6].

- Extremely High Data Rates: 6G is anticipated to deliver even faster data rates than its predecessors, aiming for multiterabit per second (Tbps) speeds. This will enable lightning-fast downloads, ultra-high-definition streaming, and support for bandwidth-intensive applications like holographic communication and virtual reality.
- Ultra-Low Latency: The data transmission time should be drastically reduced thanks to 6G's ultra-low latency. Real-time communication will be made possible as a result, supporting programs like remote surgery, self-driving cars, and industrial automation that need quick responses.
- Massive Connectivity: 6G is envisioned to support massive connectivity, accommodating an exponentially larger number of connected devices compared to previous generations. It aims to provide seamless connectivity for billions of devices, including Internet of Things (IoT) devices, sensors, and wearables, fostering the development of smart cities, smart homes, and industrial IoT applications.
- Intelligent and Cognitive Networks: The network infrastructure for 6G is anticipated to use cutting-edge machine learning (ML) and artificial intelligence (AI) technology. This will enable intelligent and self-adaptive networks that can autonomously optimize resource allocation, manage network traffic, and dynamically adapt to changing conditions for improved performance and efficiency.
- Heterogeneous Network Architecture: 6G may adopt a heterogeneous network architecture, integrating various access technologies, such as satellite, terrestrial, aerial, and underwater networks, to provide ubiquitous and seamless connectivity across different environments and scenarios.

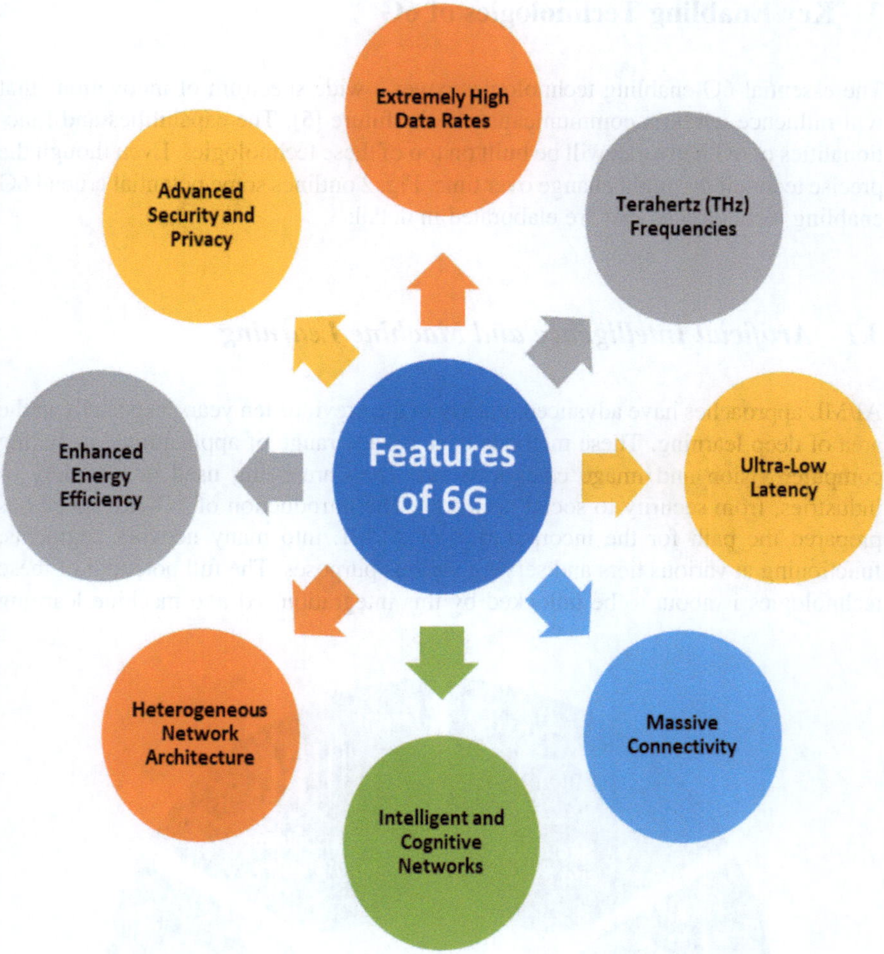

Fig. 1 Features of 6G communication networks

- Enhanced Energy Efficiency: 6G aims to prioritize energy efficiency and sustainability by optimizing network infrastructure and reducing power consumption. It may explore energy-efficient transmission technologies, intelligent power management, and eco-friendly network design to minimize environmental impact.
- Advanced Security and Privacy: 6G networks are expected to incorporate enhanced security and privacy mechanisms to address the evolving threats and concerns. This may include advanced encryption techniques, secure authentication protocols, and privacy-preserving technologies to protect user data and ensure secure communication.

3 Key Enabling Technologies of 6G

The essential 6G enabling technologies cover a wide spectrum of innovations that will influence wireless communication in the future [5]. The capabilities and functionalities of 6G networks will be built on top of these technologies. Even though the precise technology might change over time, Fig. 2 outlines some potential crucial 6G enabling technologies that are elaborated in detail:

3.1 Artificial Intelligence and Machine Learning

AI/ML approaches have advanced quickly in the previous ten years, especially in the area of deep learning. These methods have a wide range of applications, including computer vision and image categorization. They are being used in a variety of industries, from security to social networks. The introduction of 5G-Advanced has prepared the path for the incorporation of AI/ML into many network segments, functioning at various tiers and serving various purposes. The full potential of these technologies is about to be unlocked by this integration. AI and machine learning

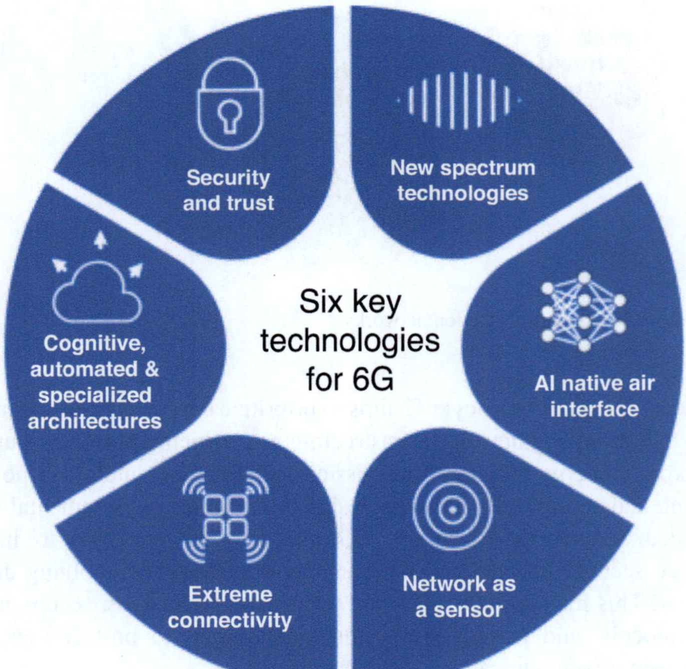

Fig. 2 Key technologies of 6G

will revolutionize various facets of network operations with 5G. Self-optimizing networks use AI/ML techniques, for instance, to schedule cell sites more effectively. Similar to this, beam formation optimization in the radio layer makes use of AI/ML algorithms to improve performance while reducing complexity. These developments in AI/ML have the potential to significantly increase network performance across many different roles [3].

3.2 Spectrum Bands

Spectrum allotment has a significant impact on radio connection. Every time a new generation of mobile technology is released, it becomes increasingly important to obtain new spectrum resources so that the potential of the most recent developments can be fully realized. It is also crucial to switch from outdated to newer technologies in the mobile communication spectrum that is currently in use. The distribution of pioneer spectrum blocks will be crucial in the development of 6G. These blocks should operate at low frequencies between 460 and 694 MHz, providing wide coverage. Furthermore, peak data speeds exceeding 100 Gbps will be attained by using sub-terahertz frequencies. Additionally, mid-bands between 7 and 20 GHz will support urban outdoor cells, enabling extreme MIMO (Multiple-Input Multiple-Output) technology to increase capacity.

While 5G-Advanced will expand the capabilities of 5G beyond simple data connection, 6G will reportedly advance localization by using a large spectrum and investigating new spectral ranges up to terahertz frequencies. It will significantly improve positioning accuracy, especially in situations like subterranean and interior facilities without satellite signals. By enabling centimeter-level precision, this advancement will make 5G more applicable in a variety of settings.

3.3 Network as Sensor

The distinguishing feature of 6G is its extraordinary capacity to observe and comprehend the immediate environment, people, and objects. The network can learn important details about different objects, including their type, shape, relative location, velocity, and possibly even material qualities, by gathering and analyzing signals that reflect off of them. Because of this capability, the network can be a useful source of situational awareness. This perceptual technique aids in the development of a "mirror" or digital image of the real world when paired with other sense modalities. Every point of interaction with the network is effectively extended by it, extending our sensory experiences. This vast amount of data is used with AI/ML techniques to make the network more intelligent and to provide new perspectives and insights based on the real world. Our view of and engagement with the world around us will be revolutionized by 6G thanks to the convergence of perception and intelligent processing.

3.4 Extreme Connectivity

Significant enhancements and advancements will be made in 6G to meet the most demanding connectivity requirements, including ultra-low latency and the Ultra-Reliable Low-Latency Communication (URLLC) service, which was introduced with 5G. Network stability could be improved by improvements such simultaneous transmission, multihop wireless communication, device-to-device connectivity, and the incorporation of AI/ML capabilities. The implementation of upgraded mobile broadband will significantly improve real-time video communications, holographic engagements, and even the updating of digital twin models using video sensors. It will also result in lower latency and higher reliability. Significant developments in these fields will occur throughout the 6G era, offering users a higher standard of service [7].

Researchers predict the introduction of novel network scenarios in the 6G environment, where networks will comprise subnetworks with different needs. Networks will act as endpoints in a network-of-networks that will result from this. Examples include circumstances where hundreds of sensors are dispersed across an area smaller than 100 meters, such as a body or automotive area network. To function properly, these sensor networks require continuous connectivity within 100 microseconds and exceptionally high reliability. A new era for equipment designers will begin with the removal of the necessity for lengthy and troublesome cable connections, especially in industries like robots and autos where completely wireless networks will become commonplace. The design and functionality of such equipment will be made more flexible and innovative thanks to the wireless paradigm, creating new opportunities and revolutionizing numerous industries.

3.5 Novel Network Architectures

As network needs and strain rise, 5G, the first technology intended to replace wired connectivity in commercial and industrial environments, requires even more complex architecture. While the RAN is included in some cloud-native implementations, the core of 5G uses a services-based architecture. Utilized are heterogeneous cloud environments, which combine private, public, and hybrid clouds. Consolidating functions can save costs when the RAN core gets more decentralized and higher layers become more centralized. New network and service orchestration solutions provide an unparalleled level of network automation, lowering operational costs, by leveraging AI/ML breakthroughs. These developments provide increased flexibility, specialization, and efficiency in connectivity solutions to meet the expanding needs of industries.

3.6 *Security and Trust*

All types of networks are increasingly at danger of cyberattacks in the current environment. To effectively combat constantly evolving threats, it is crucial to establish strong security protocols. The forthcoming advent of 6G networks offers a distinct opportunity to actively address diverse difficulties, such as the menace of jamming. Ensuring the availability and integrity of the network is of utmost importance in this endeavor. Furthermore, with the ongoing progress of cutting-edge technologies like mixed reality, which seamlessly integrate virtual and real-world components, safeguarding user privacy will become increasingly imperative. It is crucial to meticulously tackle these privacy challenges and maintain user privacy in these fully immersive experiences. In anticipation of upcoming cyber threats, 6G networks will prioritize the implementation of strong security measures, with a particular emphasis on protecting user privacy.

4 Challenges of Futuristic Communication

As we anticipate the future of 6G technology, it is evident that there is great potential for remarkable progress. However, for this groundbreaking technology to thrive, it is crucial to tackle substantial challenges that arise during its development and integration [10]. Extensive investigation has discovered multiple crucial obstacles that must be effectively overcome by the year 2030, employing cutting-edge technologies.

- Spectrum Availability: Identifying appropriate frequency channels for the development of 6G technology is a significant obstacle. To fully investigate the capabilities of higher frequencies, like the untapped terahertz spectrum, one must navigate a complex network of technical and regulatory challenges. Extensive collaboration and study among governments, regulatory agencies, and major industry participants will be required to allocate and regulate these frequency bands.
- Technology and Infrastructure: The development of fundamental techniques and infrastructure for 6G networks entails various challenges. The deployment of 6G will need the development of novel components, including terahertz-capable antennas, state-of-the-art signal processing techniques, and high-speed data transmission systems.
- Energy Consumption: Anticipated advancements in 6G networks are projected to result in enhanced data speeds and connection, which may therefore result in an increase in energy usage. Striking an equilibrium between high-performance systems and efficient use of energy is a challenging endeavor. In order to address this dilemma, it is imperative to shift our focus toward developing energy-efficient technologies, enhancing network configurations, and adopting

intelligent power management strategies. These techniques are crucial in reducing the adverse effects of 6G on the environment. By placing energy conservation as a top priority, 6G technology has the capacity to encourage sustainable practices while ensuring high-quality network performance, ultimately leading the way for a responsible and environmentally aware era of wireless communication.

- Security and Privacy: The increasing complexity and interdependence of 6G networks necessitate the implementation of strong security and privacy protocols. The primary objectives are on safeguarding user data, enhancing network infrastructure, and mitigating cyberattacks. In order to accomplish this, it is essential to integrate sophisticated security measures, encryption protocols, authentication methods, and privacy-preserving technology. Through the efficient implementation of these approaches, we can establish a strong sense of confidence and guarantee the dependability of 6G networks. 6G prioritizes the security and privacy of users, enabling customers to confidently utilize the full potential of the upcoming wireless communication age.
- Standardization: Ensuring uniformity is crucial to guarantee the compatibility, reliability, and widespread adoption of 6G networks. Accomplishing this task is challenging and necessitates cooperation among diverse entities, such as industry participants, research organizations, and standards entities. Having a universally accepted set of standards can greatly support the global implementation and seamless integration of 6G technology. These standards will serve as the building blocks for creating a cohesive and streamlined 6G ecosystem, enabling effective communication, fostering innovation, and maximizing the full potential of 6G networks on a global scale.
- Ethical and Social Implications: The emergence of 6G technology has sparked numerous ethical and social considerations, much like any other transformative advancement. Its impact will undoubtedly reach all levels of society, from accessibility to employment opportunities, privacy to addressing the digital divide. If we hope for 6G to be widely embraced, it is crucial that we address these challenges head-on. This means promoting fairness and equal opportunities, being conscious of how we use technology, bridging the digital gap, protecting user privacy, fostering inclusive employment opportunities, and implementing necessary regulations for the responsible implementation and utilization of 6G. Failure to do so could hinder its potential and cause harm to individuals and society as a whole. By adhering to these principles, the implementation of 6G technology can yield favorable societal outcomes while mitigating potential challenges.
- Regulatory and Policy Frameworks: Establishing and implementing 6G will necessitate the presence of accommodating legislative and policy frameworks. Governments and regulatory bodies will need to address spectrum allocation, licensing, privacy regulations, security guidelines, and competition policies. Establishing flexible and adaptive frameworks that foster innovation while protecting public interests is essential for the successful rollout of 6G.

- Research and Development: 6G is still in the early research phase, and significant advancements are needed to overcome technical challenges and realize its full potential. Investing in research and development activities, fostering collaborations between academia, industry, and government, and supporting innovation will be crucial to drive the progress of 6G technology.

5 AI for Advanced Computing

The world's information is currently doubling every 12 hours, and the real-time data that are used to extract new knowledge are expanding even faster. Communication networks are used to transfer all of that information. Future applications like fiber-like fixed wireless, 3D digital twinning, and immersive digital world experiences will require a lot of bandwidth. As a result, there is an enormous need for additional capacity because the demands on our networks have increased dramatically and will continue to do so. To put the situation in perspective, a 35% annual rise in demand would require a 2000% increase in capacity over the course of ten years. Therefore, 6G must be built to offer at least 20 times the wide-area capacity of 5G [8].

Futuristic computing encompasses several aspects that are anticipated to shape the future of computing technology. These aspects explore advancements in various areas, including hardware, software, and the overall computing ecosystem. Machine learning is a crucial tool for wireless communications since it has made considerable strides in fields including speech recognition, computer vision, and natural language processing. Wireless communication issues (signal recognition, detection, characterization, channel estimates, optimal network resource allocation, error correction coding schemes, and other physical layer applications) have been addressed using deep learning, a subset of machine learning [9].

The development of 6G will be largely dependent on deep unfolded techniques that allow a variety of receiver activities, including signal estimation and identification, self-interference (SI) cancellation, and improved error correction. Additionally, new methods for managing network slices and dynamic security are being studied. These initiatives are all geared toward creating an intelligent ecology. Further developments in AI/ML will also be made possible by 6G technology itself, such as the ability to utilize the data that are local to the 6G sensor by effectively transmitting the (AI/ML) algorithms. Even though it is still in the early stages of development, 6G has significant evolutionary changes that have the ability to improve user experiences and open up new use cases (Table Table 1).

Table 1 AI model support for 6G

AI model	6G feature
Convolutional neural networks	Intelligent video streaming, high-resolution imaging, and object recognition
Recurrent neural networks	Predicting user mobility patterns, modeling temporal dynamics in network traffic, and enabling context-aware services
Generative adversarial networks	Generating realistic network traffic patterns
Reinforcement learning	Autonomous network management, adaptive resource allocation, and dynamic spectrum management to optimize network performance in real-time.
Support vector machines	User behavior analysis, network anomaly detection, and intrusion detection to enhance network security.
Decision trees and random forests	Network optimization, user profiling, and quality of service prediction.
Deep reinforcement learning	Intelligent network orchestration, dynamic network slicing, and adaptive resource allocation.
Bayesian networks	Spectrum prediction, interference modeling, and quality of experience estimation.

6 Case Study—Cognitive Radio for Intelligent Spectrum Management During Disaster Recovery

6.1 Cognitive Radio for Emergency Communication Support

Disasters are unforeseen events that not only result in major property damage or fatalities but also could destroy the current communication networks. The telecommunications networks suffer significant damage during natural disasters, which prevents people in the afflicted area from communicating urgently. Disaster recovery is a critical process that involves restoring and maintaining communication networks and services during and after a disaster [11]. When compared to the spatial and temporal domains, the frequency domain analysis shows a relatively low level of spectrum utilization. This necessitates the use of an intelligent device that is aware of its environment and capable of dynamically adjusting to the current radio frequency (RF) environment by sharing spectral resources and using spectrum more effectively. Figure 3 outlines various features supported by CR for intelligent spectrum management.

CR technology enables intelligent and adaptive use of the available RF spectrum, can play a role in disaster recovery by providing flexible and resilient communication capabilities. CRs can intelligently switch their spectrum for communication to use spectrum optimally, there act as a resilient communication network that are ideally suited for disaster management [12].

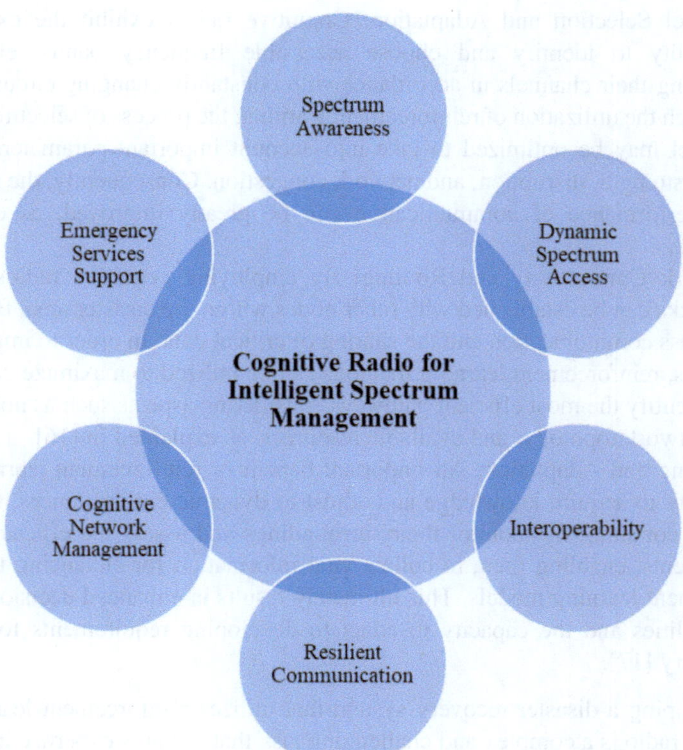

Fig. 3 CR for intelligent spectrum management

6.2 *Reinforcement Learning and CR for Disaster Management*

Reinforcement learning, a branch of machine learning, can be employed to optimize the operation of cognitive radios in disaster scenarios. The process of establishing emergency communication support during disaster using cognitive radio with reinforcement learning is outlined as follows.

- Disaster Scenario Assessment: When a disaster occurs, the first step is to assess the communication infrastructure damage and identify the available radio resources. This could involve analyzing the status of base stations, mobile networks, and other communication systems [13].
- Spectrum Sensing and Allocation: Cognitive radios utilize spectrum sensing techniques to detect available frequency bands that are not being used or are underutilized. By leveraging reinforcement learning techniques, one can effectively learn and dynamically select the most advantageous frequency bands, adapting to ever-changing radio conditions and fulfilling specific quality of service requirements [14].

- Channel Selection and Adaptation: Cognitive radios exhibit the exceptional capability to identify and choose accessible frequency bands, effortlessly adjusting their channels in accordance with constantly changing circumstances. Through the utilization of reinforcement learning, the process of selecting the best channel may be optimized to take into account important parameters such as signal strength, disruption, and network congestion. Consequently, the reliability and performance of communication can be greatly improved, as evidenced in [15].
- Network Connectivity and Routing: By employing cognitive radios, ad hoc networks can be established with other nodes within the disaster area, facilitating seamless communication and the sharing of critical data. In order to improve this process, reinforcement learning methods can be utilized to maximize connection and identify the most efficient pathways considering aspects such as node mobility, network topology, and available resources, as explained in [16].
- Learning and Adaptation: An important benefit of reinforcement learning is its capacity to acquire knowledge and adjust to dynamic circumstances. Cognitive radios continuously monitor their surroundings and assess the efficacy of their judgments, enabling them to collect vital information for enhancing their reinforcement learning models. This ultimately results in enhanced decision-making capabilities and the capacity to adapt to developing requirements for disaster recovery [17].

Developing a disaster recovery system that utilizes reinforcement learning and cognitive radio is a complex and challenging task that requires expertise in wireless communication, networking protocols, and machine learning [18]. Furthermore, the efficiency of such a system would rely on several factors, such as the accessibility of appropriate hardware, the precision of sensing methods, and the capacity to incorporate cognitive radios into current communication infrastructure, which can be facilitated by integrating advanced AI algorithms.

7 Conclusion

The paper provides an overview of the significant technological breakthroughs in 6G wireless communication, including its features, technologies, and problems. Also, we discuss the advances in AI for supporting various aspects of 6G for advanced computing. Further to highlight the significance of AI toward intelligent decision-making and spectrum management, a case study on disaster management using cognitive radio and reinforcement learning is presented. Researchers can boldly assert that successfully implementing AI-powered 6G communication networks will revolutionize the digital era.

References

1. H. Kim, “6G Wireless Communications and Networks Systems,” Artificial Intelligence for 6G, pp. 307–339, 2022, doi: https://doi.org/10.1007/978-3-030-95041-5_7.
2. T. S. Kumar and N. Kavitha, “Artificial Intelligence Techniques for 6G,” Future Trends in 5G and 6G, pp. 305–319, Nov. 2021, doi: 10.1201/9781003175155-17.
3. A. Haldorai and A. Ramu, “Security and channel noise management in cognitive radio networks,” Computers & Electrical Engineering, vol. 87, p. 106784, Oct. 2020, doi: https://doi.org/10.1016/j.compeleceng.2020.106784.
4. A. Haldorai, A. Ramu, and M. Suriya, “Organization Internet of Things (IoTs): Supervised, Unsupervised, and Reinforcement Learning,” EAI/Springer Innovations in Communication and Computing, pp. 27–53, 2020, doi: https://doi.org/10.1007/978-3-030-44407-5_2.
5. M. I. S. Mamode and T. P. Fowdur, “Machine Learning-Based Scheduling in 5G/6G Communication Systems,” Driving 5G Mobile Communications with Artificial Intelligence towards 6G, pp. 277–308, Feb. 2023, doi: 10.1201/9781003205494-9.
6. R. Thakur and D. Rane, “Machine Learning and Deep Learning for Intelligent and Smart Applications,” Future Trends in 5G and 6G, pp. 95–113, Nov. 2021, doi: 10.1201/9781003175155-5.
7. S. R and A. H, “Improved EPOA clustering protocol for lifetime longevity in wireless sensor network,” Sensors International, vol. 3, p. 100199, 2022, doi: https://doi.org/10.1016/j.sintl.2022.100199.
8. G. D. Vignesh, A. Ramu, J. T. Raja, P. Ponmurugan, A. Haldorai, and G. Senthilkumar, “Sensor data fusion techniques in the construction of generalized VORONOI graph for on-line motion planning in robot navigation,” International Journal of System Assurance Engineering and Management, Sep. 2022, doi: https://doi.org/10.1007/s13198-022-01773-4.
9. Evans Mwiti Ochieng, “A Study of the History, Functions, Roles, and Challenges of Human Resources Management”, Journal of Enterprise and Business Intelligence, vol.3, no.1, pp. 054–064, January 2023. doi: https://doi.org/10.53759/5181/JEBI202303006.
10. A. Haldorai, J. Sivaraj, M. Nagabushanam, and M. Kingston Roberts, “Cognitive Wireless Networks Based Spectrum Sensing Strategies: A Comparative Analysis,” Applied Computational Intelligence and Soft Computing, vol. 2022, pp. 1–14, Oct. 2022, doi: https://doi.org/10.1155/2022/6988847.
11. S. Sarkar, “Machine Learning in 6G: The Future of Wireless Communication,” 2022 6th International Conference on Intelligent Computing and Control Systems (ICICCS), May 2022, doi: https://doi.org/10.1109/iciccs53718.2022.9788210.
12. T. Simeonova, “Features of communication services, requirements and applications for 6G,” Yearbook Telecommunications, vol. 8, pp. 23–33, Oct. 2021, doi: https://doi.org/10.33919/ytelecomm.21.8.3.
13. S. M, “Machine learning and quantum computing for 5G/6G communication networks – A survey,” International Journal of Intelligent Networks, vol. 3, pp. 197–203, 2022, doi: https://doi.org/10.1016/j.ijin.2022.11.004.
14. R. Devnikar and V. Hendre, “Applications of Machine Learning in Wireless Communication: 5G and Beyond,” Future Trends in 5G and 6G, pp. 149–162, Nov. 2021, doi: 10.1201/9781003175155-7.
15. Anandakumar H, “Nano-Antibacterial Materials as an Alternative Antimicrobial Strategy”, Journal of Computational Intelligence in Materials Science, vol. 1, pp. 045–055, 2023. doi: https://doi.org/10.53759/832X/JCIMS202301005.
16. S. Singh, A. Anand, S. Sharma, T. Choudhury, and S. Mukherjee, “Futuristic Communication Technologies,” Smart and Sustainable Intelligent Systems, pp. 491–510, Mar. 2021, doi: https://doi.org/10.1002/9781119752134.ch35.

17. S. Murugan and M. Sumithra, "Energy Efficient Cognitive Radio Spectrum Sensing for 5G Networks – A Survey," EAI Endorsed Transactions on Energy Web, p. 169169, Jul. 2018, doi: https://doi.org/10.4108/eai.29-3-2021.169169.
18. M. Suriya and M. G. Sumithra, "Enhancing Cooperative Spectrum Sensing in Flying Cell Towers for Disaster Management Using Convolutional Neural Networks," EAI/Springer Innovations in Communication and Computing, pp. 181–190, Oct. 2019, doi: https://doi.org/10.1007/978-3-030-19562-5_18.

A Review on Smart Navigation Techniques for Automated Vehicle

1 Introduction

It is believed that automobile technology is subordinate to smart city services. Congestion caused by the growing number of automobiles on the road is unavoidable. Congestion will undoubtedly result in fuel waste, pollution of the environment, time waste, fatal accidents, etc. [1]. Using intelligent transportation can help drivers escape gridlock. The use of dedicated short-range communications (DSRC), which were developed by the IEEE Community, in vehicular ad hoc networks (VANETs) [2] which enable vehicles to establish connections with RSUs [4] installed along the roads—is growing in popularity recently. In industrial applications, NAVIGATION is a critical capability of autonomous vehicles [1]. However, because there are not any satellite location signals within, conventional navigation methods won't work [2]. While current positioning techniques frequently struggle to offer a reliable solution for centimeter-level inside localization, indoor autonomous cars must build their own navigation systems by outfitting a few extra sensors [3]. As a result, the manufacturing sector has become interested in AV technology to ensure operations are not interrupted while simultaneously maintaining the quality of the company [4, 5]. For tasks like moving goods from the assembling area to the manufacturing line, from the manufacturing line to the packing area, and vice versa, technology like this can also step in for a human. Additionally, the advancement of this technology is not only concentrated on the manufacturing sector but also has the potential to solve the global pandemic problem in other sectors like as hospitals, food delivery services from restaurants, and civil construction [6, 7]. Although learning is a form of intelligence, it only shows itself when it is applied to or used to govern an agent. Any machine, robot, or application having sensors and actuators could be an agent, whether it is real or artificial. Unmanned Aerial Vehicles (UAVs) have become the most important robots to be controlled out of all the different types of robots since they can fly, allowing them to move to places

© European Alliance for Innovation 2024

A. Haldorai et al., *Artificial Intelligence for Sustainable Development*,
EAI/Springer Innovations in Communication and Computing,
https://doi.org/10.1007/978-3-031-53972-5_13

and situations that Unmanned Ground Vehicles (UGV) as well as Unmanned Surface Vehicles (USV) cannot. The navigation and control algorithm of a UAV is one of the most important factors to take into account when piloting it.

A human driver must have talent, knowledge, and presence of mind in order to operate a vehicle. When attempting to include artificial intelligence into automatic driving schemes, the driving environment presents a challenging task. Combining reinforcement learning and deep learning is seen to be the most effective way to give machines the ability to drive cars at the human level. In Atari video games, this combination has already demonstrated its viability [38, 39, 41], and [43]. Planning is the responsibility of RL, and learning is the responsibility of DL [13]. These notifications' availability and accuracy are dependent on the system's general architecture and failures impacting the navigation sensors. The onboard sensors' ability to work in tandem with one another increases robustness against individual sensor malfunctions and dropouts, certain environmental conditions, which restrict their capacity to adjust to novel, uncharted surroundings. Deep Learning (DL) and Reinforcement Learning (RL), two intelligent machine learning alternatives, effectively address this, but their utility is constrained when applied alone. An innovative approach called Deep (DRL) combines DL and RL algorithms to more effectively address the planning of paths and trajectory generating issues. To increase its performance, resilience, and adaptability, the robot must learn from its prior experiences and observations and make generalizations [8]. For the robot to learn features, models, and policies from data and apply them to new tasks or situations, deep learning, transfer learning, and meta-learning are necessary [18, 19]. These methods, though, are time- and resource-intensive and are susceptible to overfitting, generalization error, and sample efficiency issues [33–35].

The researchers and engineers would struggle to find an appropriate indoor navigation solution for their specific use cases among the many navigation techniques. The reason for this is that there are not any studies that compare how well various navigation strategies perform in actual application situations. In this study, we examine the various interior navigation techniques used by autonomous vehicles and compare them using the same dataset. We review the outcomes of the comparative approaches and come to some conclusions that will help us design a navigation system that is appropriate for indoor localization.

Main contribution of this paper is as follows:

1. The algorithms like oracle-guided independent parameter sharing TD3, approaches like LiDAR Slam, and generalized advantage estimation in navigation systems are discussed [28].
2. Different navigation strategies are discussed like data-driven inertial navigation, Automotive Dead Reckoning, and Deep indoor vehicle positioning [36].
3. Deep reinforcement learning techniques are examined and some evaluation findings are given in this paper [42].

The rest of this paper is divided into the following sections.

The traditional smart navigation strategies are covered in Sect. 2. In Sect. 3, some common localization algorithms are described. The experimental conditions, the evaluation findings, and some discussions are introduced in Sect. 4. Our task is completed in Sect. 5.

2 Approaches and Algorithms in Navigation

A reinforcement learning algorithm's approach for calculating each action's advantage is utilized to update the policy [24]. The issue of excessive variance in conventional reinforcement learning systems, which can result in sluggish learning and subpar performance, was addressed by its development. GAE computes the advantage using a linear combination of the value function and the reward, which aids in lowering variance and accelerating learning [24]. The researchers and engineers would be perplexed in choosing an appropriate solution for indoor navigation for their particular usage scenarios from among the different LiDAR SLAM approaches. This is due to the paucity of studies comparing the effectiveness of various SLAMs in actual application environments using the same assessment parameters.

2.1 LiDAR SLAM Approaches

The three methodologies listed below serve as the fundamental concepts for determining the robot's current condition. They can be used in a variety of ways, which result in numerous SLAM Methods. The Gaussian filter-based approach makes the assumption that the robot's predicted state complies with the multivariate Gaussian distribution [10, 25]. The robot's pose is estimated using a set of particles in the particle filter-based technique. Each particle represents a potential reality-based state for the robot. The particle filter-based technique calculates the current state of the robot x_t based on the prior state x_{t-1} recursively, much like the Gaussian filter. The fundamental principle of the Bayesian filter is followed in the recursive estimation of the robot state. It is important to note that while particle filters are frequently employed in both LiDAR and visual SLAM, Gaussian filters and graph optimization are more frequently used in the latter. We give a quick introduction to the key LiDAR-based SLAM approaches, including Gmapping, KartoSLAM, HectorSLAM, LagoSLAM, CoreSLAM, LOAM, and Cartographer. The Robot Operating System (ROS) is used to run these solutions. The characteristics of the various approaches are compiled in Table 1.

The proposed system can be a suitable replacement to safeguard the privacy of the car users, who can securely utilize the navigation service for a hassle-free drive in a smart city, when the computing, communication, and security needs are taken into account.

Table 1 Comparison of various LiDAR SLAM

Methods	Open source	Real time	Fuse IMU or Visual	Algorithm	Map type	Localization performance
Gmapping	√	√		PF	Grid map	Good
CoreSLAM	√	√		PF	Grid map	Poor
KartoSLAM	√	√		Graph	Grid map	Good
LagoSLAM	√	√		EKF	Grid map	Medium
HectorSLAM	√	√	IMU		Grid map	Good
LOAM	√	√	IMU		Point cloud map	Excellent
v-LOAM		√	Visual		Point cloud map	Excellent
Cartographer	√	√	IMU	Graph	Grid map	Good

2.2 Generalized Advantage Estimation

Autonomous vehicles have the potential to completely transform transportation, but their actual implementation will depend on how well they can negotiate complex and dynamic situations. The difficulty stems from the fact that traditional autonomous navigation techniques sometimes rely on preexisting maps and guidelines, which might not fully take into account the complexity and unpredictability of real-world events. When faced with unforeseen obstacles, this reliance on preexisting data can result in inferior performance and occasionally even accidents. Generalized Advantage Estimation (GAE) in reinforcement learning has been proposed as a possible option for autonomous cars to tackle these problems. As a case study, GAE-based reinforcement learning has been applied to small robots, specifically the TurtleBot.

In this method, an algorithm based on Python was created to allow the robot to interact with its surroundings and navigate them in order to maximize rewards. The robot could walk around and interact with obstacles in a 2D world that was simplified to a grid of cells. and the navigating between neighboring cells in the grid, it may go left, right, up, or down. The robot's state was initialized and the environment was reset at the beginning of every training session. Behaviors were carried out in the environment, and the actor and critic networks were crucial in predicting values and behaviors at each time step, respectively. Buffers were also employed to store values and incentives while the students were studying.

The payouts and losses for the actor and critic networks were calculated by the GAE algorithm once an episode was finished. Following that, backpropagation was employed to update the networks using these losses. As training went on, the robot's average reward increased, demonstrating that it could successfully navigate its environment and arrive at the intended condition. TurtleBot robot navigation tasks in a simulated environment were successfully demonstrated by this work using a Deep Reinforcement Learning (Deep-RL) algorithm in conjunction with GAE. These results imply that artificial intelligence can learn quickly and perform very well in navigation tasks when paired with GAE. By adding more complicated and realistic surroundings and investigating different decision-making strategies, the algorithm's performance may be further improved as research progresses. This study confirms the possibility of future widespread adoption of autonomous cars and lays the groundwork for future advancements in the field of autonomous navigation.

2.3 Oracle-Guided Independent Parameter-Sharing TD3

The deep neural network in the approach discussed is trained using TD3 (Twin Delayed Deep Deterministic), which is acknowledged as a state-of-the-art Reinforcement Learning (RL) algorithm [26, 37]. The decision to go with TD3 highlights both its sophisticated features and efficiency when applied to reinforcement learning tasks. Once TD3 has been chosen, the technique is then used to train an independent learner parameter-sharing training scheme. The purpose of this system is to facilitate collaborative information transfer among independent learners and improve the learning process. The research paper also suggests an improved iteration of this training plan known as oracle-guided independent parameter-sharing (TD3). This version is specially designed to be scalable, with the goal of resolving issues related to the growth of the learning system. The focus on scalability demonstrates a forward-thinking approach in the creation of RL techniques, indicating an understanding of the algorithm's capacity to manage larger and more complicated task.

Algorithm 1: Oracle-guided training

Input: The structure of the actor and critic networks.

Output: The weights associated with the actor and critic networks.

Initialization

1. Randomly initialize the parameters φ and θ for the actor and critic networks across all Unmanned Aerial Vehicles (UAVs).

2. Set up target actor and critic networks for all UAVs.

3. Establish a shared replay buffer, denoted as B, for all UAVs.

4. for episode = 1, 2, . . . do.

5. for t = 1, 2, . . ., T do.

6. for UAV u = 1, . . ., N do.

7. Training Loop.

8. For each episode from 1 to an arbitrary number.

9. For each time step, from 1 to a specified maximum T.

10. For each UAV from 1 to N.

11. Select an action with exploration noise $a_b^u \sim \pi\left(s_t^u\right) + N(0, \sigma^2)$and observe the corresponding reward $\left(R_t^u\right)$ and the new state $\left(s_{t+1}^u\right)$.

12. Obtain oracle information from the simulator and replace the last two dimensions of $\left(s_{t+1}^u\right)$.

13. end for.

14. Remove low-value transitions, select a transition randomly, and save it in the replay buffer B.

15. Sample a batch of M transitions (s, a, r, s') from B.

(continued)

16. Introduce random noise $\in \sim N(0, 0.1^2)$.
17. Update the actor's action ã according to a policy $\tilde{a} \leftarrow \pi_{\varphi'}(s') + \in$.
18. Compute the Q-target for the state-action pair (s, a) using the sampled transition and the updated actor's action ã $Q_{\text{target}}(s, a) \leftarrow r + \gamma \min_{i=1,2} Q_{\theta_i'}(s', \tilde{a})$.
19. Minimize the Mean Squared Error (MSE) between the current critic network's Q-value and the computed Q-target $Q_\theta(s, a)$ and $Q_{\text{target}}(s, a)$.
20. If the current time step t is even ($t\%2 = 0$).
21. Then update the actor by performing gradient ascent based on a specific equation.
22. Update the target networks according to a specified equation.
23. end if.
24. end for.
25. end for.

Training a Reinforcement Learning (RL) algorithm for flocking behavior is difficult because UAVs (unmanned aerial vehicles) must simultaneously forecast the flock center and modify their behavior. There is a chance that this dual goal will become unstable while being trained. Using the oracle-guided Multiagent Reinforcement Learning (MARL), training process is the suggested answer to this problem. This method works under the presumption that every UAV can see the flock center and can ask the simulator directly for this information. Learning superior policies is possible by utilizing the oracle's help. But there's a significant drawback when it comes to execution because there's no way to reach the flock core. As such, flocking and navigation in the real environment cannot be directly facilitated by the learned policy. The Oracle-Guided Training Scheme of TD3, or Twin Delayed Deep Deterministic, is contained in Algorithm 1. TD3 efficiently learns a control strategy by integrating the data from the flock center with an independent learning and parameter-sharing training technique [27]. A policy that uses the precise flock center and local observations as inputs is used to gather data from the simulated environment during the training phase. An estimator is then trained using this data. The flock center is predicted by the estimator using the selected policy and local observations. The importance of this method is demonstrated by its ability to solve the flocking and navigation problem for up to 32 UAVs, even in situations where communication is inadequate. An efficient solution is jointly produced by the estimator, which is produced through supervised learning, and the generated policy, which is aided by reinforcement learning [29–31].

3 Smart Navigation Strategies in Deep Learning

Autonomous vehicles powered by artificial intelligence have previously investigated its existence in various levels of autonomous driving scenarios. The relevant autonomous driving projects completed by several simulators are shown below. The navigation system of autonomous cars is briefly introduced in this section, followed by a summary of the methods employed in various forms of navigation, and lastly a discussion of the difficulties encountered when operating in various situations.

3.1 Data-Driven Inertial Navigation

Recent years have seen a noteworthy upsurge in research and development for data-driven inertial navigation algorithms [9, 10, 12], indicating their increasing demand. One notable example of an innovative inertial navigation program is the Remote Inertial Data Integration (RIDI) [14]. RIDI provides a new method for inertial navigation by using support vector machines (SVMs) to regress an individual's velocity. A study [11] that uses a convolutional neural network (CNN) to calculate a user's immediate velocity highlights a similar improvement in inertial navigation. After that, an Inertial Navigation System (INS) is constrained by this velocity data [50]. Although a number of deep learning-based inertial positioning and navigation systems have been created, it is important to keep in mind that these systems are frequently designed with human users in mind and might not be immediately useful for resolving the issues surrounding autonomous vehicle positioning [54].

Particular attention has been paid to vehicle localization using deep learning techniques in the context of autonomous vehicles, as demonstrated by the work in [15]. Specifically, the use of deep learning techniques for precise vehicle positioning is the focus of this research. Furthermore, a comparable method outlined in [16] uses an inertial measurement sequence to estimate vehicle movement characteristics, such as direction and velocity deviations, using a Temporal Convolutional Network (TCN). The development of inertial navigation methods specifically suited to the demands of self-driving vehicles has been further emphasized by this strategy.

An inertial positioning system typically consists of a 3-axis gyroscope and 3-axis accelerometer, as shown in Fig. 1. Newton's classical mechanics forms the foundation of the theory guiding inertial navigation devices. Newton's laws state that unless an external force acts upon an object in motion, causing a proportionate acceleration, the object will continue to travel in a uniform rectilinear path. When continuously integrated over time, accelerometers, which are sensitive to acceleration, allow one to determine changes in velocity and, by extension, changes in location [17]. However, tracking the direction of the acceleration vector and breaking it down is crucial to achieving precise localization and navigation inside the inertial coordinate system. This procedure guarantees an in-depth understanding of the motion of the vehicle. Maps, weather, and speed patterns are examples of auxiliary data that Vetorch, a deep learning framework, needs but may not always have access to in the field of inertial sequence learning [20]. A smartphone-based vehicle location system is suggested as a solution to this problem, doing away with the requirement for additional data. It is remarkable that the deep learning-based method enables indoor vehicle navigation even in the absence of extra infrastructure assistance, as it is able to directly learn the motion state of the vehicle from raw sensor input data obtained from several sources. Motion parameter regression is used in the learning process, and it is essentially independent. A trained lightweight neural network is then used to assess the direction and speed of the vehicle, which helps to restore its route without requiring more contextual data. The new approach shows how deep learning

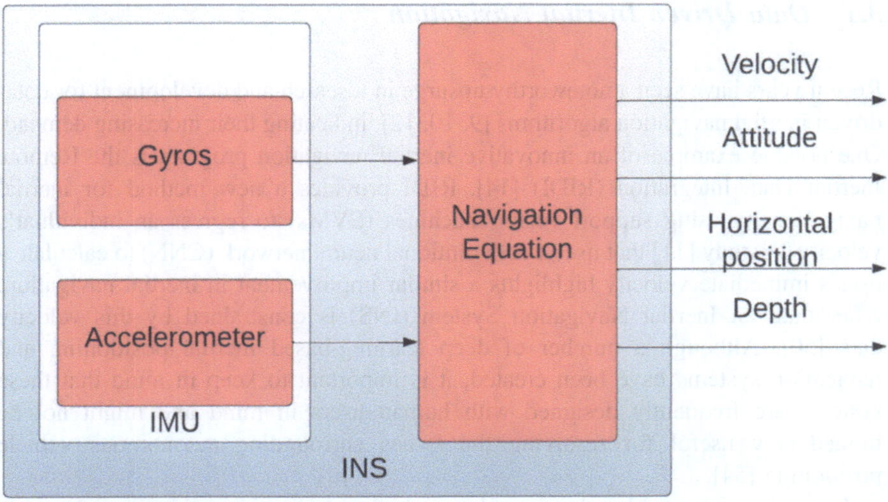

Fig. 1 Inertial navigation system

algorithms can directly extract useful information from unprocessed sensor data for efficient vehicle navigation.

3.1.1 Automotive Dead Reckoning

Analyzing and dissecting the direction of the acceleration vector is an important stage in the localization and navigation process within the inertial coordinate system. This is accomplished by multiplying the direction cosine matrix c_b^i, as given in Eq. (1), by the acceleration vector, represented as f, on the left. This procedure results in the acceleration vector being transformed from the coordinates of the Inertial Measurement Unit (IMU) to the inertial coordinate system, which permits its decomposition. The relative position of the IMU with regard to the inertial coordinate system is represented mathematically by the direction cosine matrix c_b^i. The acceleration vector is efficiently aligned with the inertial coordinate system using this matrix multiplication, yielding a breakdown of its constituent parts. Understanding the various contributions of the acceleration vector in various directions inside the inertial coordinate system depends on this decomposition.

$$f^i = c_b^i f^b \tag{1}$$

$$c_b^i(t + \Delta t) = c_b^i(t)A(t) \tag{2}$$

$$A(t) = c_{bt-1}^{bt} = I + \begin{matrix} 0 & -w_z & w_y \\ w_z & 0 & -w_x \\ -w_y & w_x & 0 \end{matrix} \tag{3}$$

Here, the directional cosine matrices with respect to the inertial coordinate system before and after the carrier rotation are indicated by the symbols $c_b^i(t)$ and $c_b^i(t+\Delta t)$ respectively. These matrices describe the carrier's orientation in terms of directional cosines both before and after a rotation that takes place during a period of time Δt. In the carrier coordinate system, the directional cosine matrix from instant t to moment $(t + \Delta t)$ is represented as *A(t)*. Equation (3) describes this matrix *A(t)* in later attitude updates, where t approaches 0 and the rotation angle is minimal. Notably, this form makes use of a 3 × 3-unit matrix, indicated by I. As shown in (4) and (5), the rotation of the carrier coordinate system about its roll (*x*-axis), pitch (*y*-axis), and yaw (*z*-axis) during the time interval Δt is characterized by the angular velocities w_x, w_y, and $w_z$respectively. The carrier's change in orientation throughout the given time period is determined by these angular velocities. Equations (4) and (5) show that by repeatedly integrating the decomposed acceleration values, the trajectory of the vehicle serving as the carrier may be found. This integration procedure helps determine the trajectory of the vehicle by making it easier to calculate its position over time.

$$v(t+\Delta t) = v(t) + \int_t^{t+\Delta t} \left(c_b^i(t) f^i(t) - g(t)\right) dt \tag{4}$$

$$D(t+\Delta t) = D(t) + \int_t^{t+\Delta t} v(t) dt \tag{5}$$

The accelerations in the carrier coordinate system and inertial coordinate system are represented in this context as *fi* and *fb* respectively. The rates of change of velocity inside each of these coordinate systems are represented by these accelerations. In particular, acceleration in the inertial coordinate system is represented by fi but acceleration in the carrier coordinate system is represented by *fb*. Measurements of *w* taken with a gyroscope reveal information about the angular velocity within the carrier frame. The carrier coordinate system's rotation rate is represented by this angular velocity, *w*. The system obtains insights into the carrier's orientation changes over time by integrating gyroscope measurements. Other variables include *D*, which represents displacement or change in position within the inertial coordinate system, and *v*, which represents velocity inside it. In this context, the term G refers to the Earth's gravitational pull [21, 44, 46]. For the navigation equations to effectively simulate the dynamics of the system, especially in terrestrial circumstances, gravity must be taken into account. Gravity is a significant component that determines the motion and behavior of the system.

3.1.2 Deep Indoor Vehicle Positioning

To achieve optimal placement accuracy and computing efficiency, the suggested method uses two different deep neural networks, DeepVIP-L and DeepVIP-M, to adjust to various parameters. These networks' design decisions are based on particular needs, guaranteeing flexibility in a range of situations. Long Short-Term Memory Networks (LSTM), a kind of recurrent neural network renowned for its capacity to capture long-term dependencies in sequential data, are used in the implementation of DeepVIP-L. Because of this, DeepVIP-L works especially effectively in situations requiring a high level of positional precision. In applications where keeping a comprehensive history of positional data is essential for precise predictions, LSTM networks' innate capacity to store and use information over lengthy sequences is useful. Conversely, MobileNet-V3, a thin and effective neural network architecture, is used by DeepVIP-M [32, 47–49]. Finding a balance between computing cost and accuracy is crucial, and this decision is particularly relevant in situations where computational resources are limited. Because of its computationally efficient design, MobileNet-V3 is appropriate for real-time applications and scenarios where hardware constraints must be taken into account.

- DeepVIP-L: Long Short-Term Memory (LSTM) networks are deliberately used in the architecture of DeepVIP-L to produce significant features for indoor vehicle placement. Using the temporal relationships present in sensor signals, this is accomplished. Notably, by using two layers of LSTM as the shared bottom layer, the gradient explosion and disappearing issues that are frequently experienced in conventional neural networks are lessened. This design decision guarantees the extraction of pertinent characteristics while addressing training-related issues with deep networks [22, 32]. The LSTM layers are excellent at capturing the possible temporal features buried in the time series data because they preserve local concealed states within a temporal window. The architecture then uses two completely connected layers to regress velocity and heading. Reducing the discrepancy between the actual values obtained from the ground truth trajectory and the projected motion parameters is the training objective. The optimization method uses this discrepancy as the loss function. The Adam optimizer is used to speed up the optimization process and find the best network settings. By helping to reduce the loss values iteratively, this optimizer enables the network to adjust and enhance its predictions. The speed v and heading fluctuation Δh are predicted within a certain time interval by applying Eq. (6). The linkages between the temporal dependencies recorded by the LSTM layers, the output parameters denoting vehicle motion, and the input sensor signals are all summed up in this equation. With the use of LSTM networks and a rigorous training schedule, DeepVIP-L's entire architecture is designed to efficiently utilize temporal information for precise and significant vehicle locations.

$$(v, \Delta h) = \text{LSTM}(\{(a_i, w_i, m_i, g_i)\}) \quad (6)$$

$$\begin{cases} x_t = x_{t-1} + v_t dt * \cos(h_{t-1} + \Delta h) \\ y_t = y_{t-1} + v_t dt * \sin(h_{t-1} + \Delta h) \end{cases} \quad (7)$$

- DeepVIP-M: Using deep learning innovations specifically optimized for mobile platforms, it aims to solve the problems related to indoor vehicle location under resource constraints. By combining 1x1 convolutions with squeeze and expand modules, DeepVIP-M drastically lowers the number of parameters by leveraging the efficiency of SqueezeNet [32], an early yet powerful deep network. Because of this architectural decision, the model is guaranteed to be computationally efficient and suitable for mobile platform deployment. The first stage of DeepVIP-M is to use a one-dimensional Convolutional Neural Network (CNN) layer to extract relevant properties from sensor input. This layer is essential for gathering pertinent data that help with precise indoor vehicle placement. The importance of each feature channel in the network model is then illustrated using an improved squeeze-and-excitation module. This module contributes to an optimum representation of the sensor data by promoting beneficial elements and suppressing less relevant ones according to their relative value. DeepVIP-M uses the hard form of the swish activation function (h-swish), which was inspired by MobileNet-V3 [23, 51], to further increase efficiency. By lowering the interconstraint of different tasks throughout the optimization learning process, this choice of activation function helps improve the model's placement accuracy.

3.2 *Deep Reinforcement Learning for Automated Vehicle Navigation*

The vehicle agent, observation/state, action, and reward system are the four essential parts of the Deep Reinforcement Learning (DRL) system [25]. The primary decision-maker is the vehicle agent, which uses information obtained from the surrounding traffic environment. With the aim of optimizing the ultimate benefit, its primary function is to produce a suggested plan of action based on these findings. In order to minimize the amount of time needed to get there, the ultimate prize is established. Essentially, how the DRL system works is that the vehicle agent assesses the condition of the surroundings, suggests courses of action, and gets feedback in the form of incentives or penalties depending on how well those courses of action work.

- *Vehicle Agent*: Based on data collected from traffic observations, the dynamic neural network (NN) model known as the Vehicle Agent is intended to be self-evolving and generate decisions that may be put into effect. This neural network's architecture makes use of a five-layer multilayer perceptron (MLP). The agent exhibits a high pace of exploration in the early phases of its training. The agent can extensively study alternate routes toward its objective destination because of its high exploration rate, which motivates the agent to display more adventurous and varied behaviors [26, 30, 32]. The agent uses exploration rate decay (ERD) as a training approach. The purpose of this method is to progressively lower the exploration rate over time. The agent shifts to performing more deterministic activities when the exploration rate drops. Through this change, the agent should be able to develop a more deterministic strategy π:S→A that will drastically lessen the unpredictability in its decision-making process. This deterministic strategy is designed to maximize the agent's traveling time expectations, especially under comparable travel circumstances. To put it simply, the Vehicle Agent's learning procedure is meant to provide a compromise between discovery and utilization, encouraging flexibility in the early learning phases and progressively moving toward a more optimal and deterministic decision-making policy as training goes on.

- *Action:* The vehicle agent's navigational decisions are referred to as Action in the system's context. Choosing one of these options entails figuring out how to navigate to m connected edges from the vehicle's current position. Within this framework, actions are discrete values that signify certain choices the vehicle agent has made on how to navigate the environment [52, 53]. Turning left, turning right, or continuing straight from the present edge are the options when $m = 3$. Every single one of these acts is a choice that affects how the car moves through its surroundings. The number of connected edges (represented by the symbol m) from which the vehicle can select its next course determine the particular set of actions.
- State: State is a useful illustration of the flow of traffic at the moment. Multiple parameters characterizing the circumstances in the world's urban transport network are included in the representation variables to accurately capture the dynamics' complexity [55]. Here, the state is described as a vector with the elements $[m_e, n_e, l_e, x_v, y_v]$, where $[m_e]$ is the number of cars in the edges, $[n_e]$ is the average driving speed in the edges, $[l_e]$ is the length of the roads in the edges, and $[x_v, y_v]$ is the agent's starting point and ending point. The calculation is as follows:

$$t_e = \begin{cases} \frac{l_e}{v_e}, & \text{if } n_e > 0 \\ \frac{l_e}{m_e}, & \text{if } n_e = 0 \end{cases}$$

where $[n_e]$ is the number of cars on the road, $[m_e]$ is the posted speed limit, $[l_e]$ is the length of the road, and $[v_e]$ is the average driving speed. A segment called Decision Zone is compartmentalized in each edge to show the ideal time for acquiring a state because the vehicle agent only needs the most recent state to make decisions. The decision zone d_v is written as follows:

$$d_v = \min\left[l_{\text{edge}}, v_{\max} + v_{\max} b(v)\tau_v\right]$$

Due to the safety considerations of the urban transport network, the decision zone is influenced by both the kind of vehicle and the driver's reaction time.

- Reward: The most crucial element in the DRL system is reward since it influences each agent's convergence to an optimal strategy by π_0 rewarding positive function approximations. Our work's overarching reward-setting principle is to maximize projected future discounted returns. The projected benefit is described as follows:

$$R = \sum_{K=0}^{T} \gamma^{k} r_{S+K}$$

The lengthy delay problem affects the conventional method of allocating rewards for each action in a sequential decision-making process. In comparison to a discounted incentive system, the proposed reward scheme greatly improves the convergence performance of our DRL approach.

4 Evaluation Finding in Deep Reinforcement Learning

S is the state space in the reinforcement learning paradigm, and *A* is the action space. The agent, which is our intelligent model, receives state *s* and reward *r* and outputs action a for TSC. The environment represents all of the traffic states associated with an intersection. It acquires an action $a \in A$ for TSC and outputs state $s \in S$ and reward r which is equivalent to s and a. To maximize the cumulative expected rewards, we need to learn the objective of policy $\pi*$. If at the start time slot *t,* if the agent receives state *st* at *t* and generates an action *at*, and obtains time slot t with sequence of reward as *rm*, where *m* = *t, t + 1, t +* 2..., then we can get the expected cumulative rewards as a *Q* value. Here $\gamma \in [0, 1]$ is to calculate the present value of future rewards which is a discount factor.

$$Q(s_t, a_t) = E\left[\sum_{m=t}^{\infty} \gamma^{m-t} r_m\right]$$

$$= E\left[r_t + \gamma^{Q(s_{t+1}, a_{t+1})}\right]$$

By the Bellman optimality equation, we can express the maximum $Q(s_t, a_t)$, which corresponds to the optimal action policy at the time slot t, i.e., $\pi * t = \text{argmax}_{\pi t}\ Q_{\pi t}(s_t, a_t)$ shown as below:

Table 2 Parameters for intelligent vehicle

Parameters	Value
1. Learning rate	0.001
2. Episode	10,000
3. Exploration	1.0 → 0.05
4. Target update of network (per learning step)	3000
5. Discount factor	0.99
6. Mini-batch for update	32
7. Prioritization important sampling	0.4 → 1.0
8. Prioritization exponent	0.6

$$Q^{\pi_t^*}(s_t, a_t) = E(r_t(s_t, a_t) + \gamma \max_{a_{t+1} \epsilon A} Q(s_{t+1}, a_{t+1})) \tag{8}$$

Through dynamic programming, we can solve Eq. [8]. About 10,000 episodes are run in SUMO with a 0.001 rate of learning to train the vehicle agent, as shown in Table 2. The greedy rate, commonly referred to as the exploration rate, dropped from 1 to 0.05. Every 3000 learning steps, the network of target parameters will be changed. The reward's discount percentage is 0.99 [32, 56]. The mini-batch to be trained is 32 and the overall replay storage capacity for storing each transaction is 10,000. Prioritization important sampling is raised from 0.4 to 1 and the prioritization exponent is set to 0.6.

The deep neural network architectures and training parameters utilized in the suggested framework are presented in this subsection. The input level, two hidden layers with a dueling structure, and the output layer are the five completely connected layers of the neural network. There are 150 neurons in the first concealed layer and 100 neurons in the second hidden layer. RELU [40] is used as an activation mechanism in both hidden layers. The advantage stream along with the value stream are two streams of completely connected layers that emerge from a dueling network. The advantage stream has as many outputs as actions, which is the greatest number of connected roads in the urban network, but the value stream has just one output. The TensorFlow library is used to build the neural network [45, 57].

In the second example (see Fig. 2), the network was trained over a long time span of 100,000 episodes using the Deep Deterministic Policy Gradient (DDPG) Reinforcement Learning (RL) algorithm. Interestingly, during this training phase, no changes were made to the network architecture or its parameters. The agent's coverage increased to its maximum value throughout this time, peaking at an average reward of 50 following the training episodes. This moving average of order 50 was used to calculate the average reward value, which gave a smoothed picture of the learning process over time. A histogram is used in Fig. 3 to visualize accelerometer and rotation vector datapoints on the test set. This visual aid enables a perceptive analysis of the distribution and properties of the collected data points. In addition, the plot shows the average episode reward in relation to the number of episodes, which makes it easy to comprehend how the RL-trained model learns and performs in this specific setting. The graphical display facilitates the understanding of the evolution of the agent's performance throughout training, demonstrating the convergence and

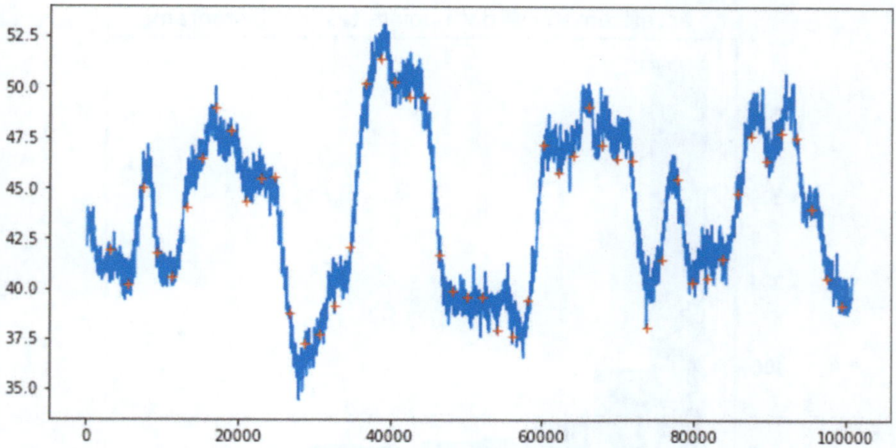

Fig. 2 Training parameters

efficacy of the DDPG RL algorithm in optimizing the cumulative rewards within the given scenario. However, they struggle to keep up the same performance when demand increases in traffic. Another justification for this is that if there are multiple ways when traveling to a location of similar distance, making a poor navigational choice in a car has little to no effect on trip time because both routes are more likely to see comparable traffic when demand is lower. When the city map is larger, the improvement also becomes more noticeable.

When the environment is large and complicated, the mean reward first rises quickly and reaches a maximum in the nearby neighborhood. This is explained by the fact that, during the initial phases of training, the agent systematically searches the map, concentrating on regions that are near its starting location. This exploration technique causes a significant increase in the average reward since it enables the agent to rapidly understand the local features of the environment. However, as the training goes over 20,000 episodes, there is a noticeable change in the mean reward. The average payment has decreased, suggesting that the agent is beginning to investigate new areas on the map. This behavioral shift indicates that the agent has progressed past its first exploration stage and is now exploring uncharted territory, running into fresh obstacles and environmental dynamics. This behavior indicates that the agent is continuously improving its perception of the surroundings and its process for making decisions. The agent continues to explore its surroundings, learning new things with every episode. At the end of the training period, the average reward eventually converges to a global maximum as a result of this repeated learning process. The aforementioned pattern illustrates the agent's ability to adjust and refine its approach over time, ultimately reaching a course of action that optimizes cumulative rewards within the intricate and vast surroundings.

A graphical representation of the route of waypoints generated by our technique is given in Fig. 4. Three location icons—red, green, and blue—are used in this representation to indicate the navigation task and show the path from the beginning

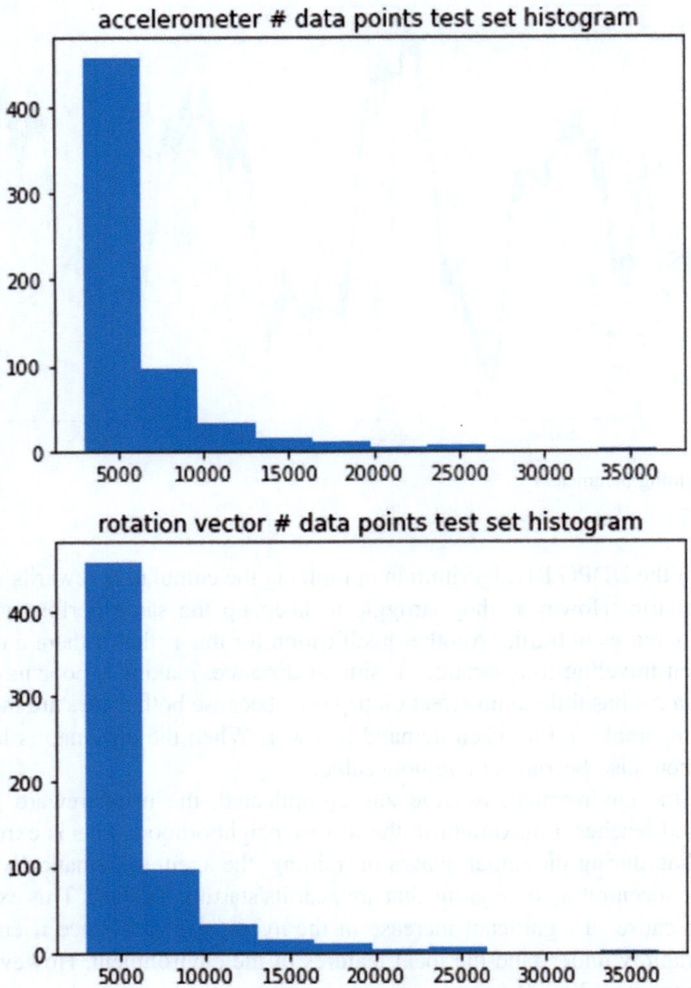

Fig. 3 Training set of accelerometer and rotation vector

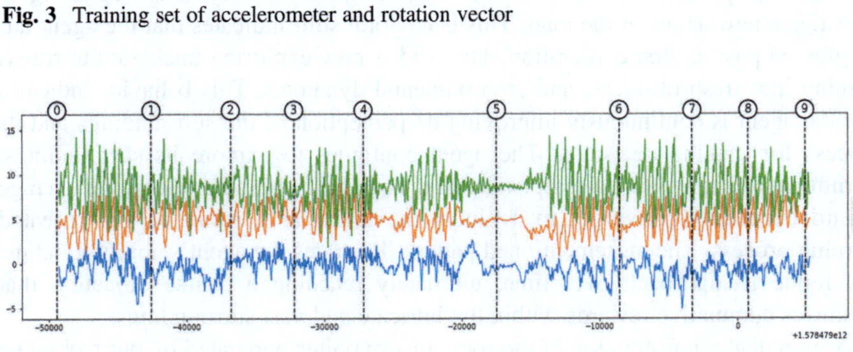

Fig. 4 Map for navigation task

point to the destination. This graphic makes it easier to understand how the algorithm creates waypoints for drone navigation in situations with plenty of obstacles. Three different vehicle kinds that we added to the maps using the SUMO tool are used in our experimental setup to add realism. These cars replicate real-world traffic situations for sporadic and dual-route travel. Different vehicle types have different characteristics when it comes to speed, acceleration, and deceleration. At the end of the training sessions, it exhibited an impressive success rate in successfully navigating and performing duties for each episode, even with the difficult conditions and different vehicle characteristics. Given that the agent had no prior knowledge of the particular aspects of the environment, this achievement is remarkable and demonstrates the efficacy and versatility of our algorithm in maneuvering through intricate and ever-changing settings.

5 Discussion and Conclusion

A significant modern problem in many densely populated cities is traffic congestion. As a result, numerous automobile navigation systems have been developed over the past few decades to help cars get where they're going as soon as possible when traffic is heavy. Finding the best answer in a complicated urban context is a difficult task, though. Pathways with regular shapes and irregular shapes are utilized to test the comparison methods in order to make fair comparisons. To make the evaluation situations more realistic, we also drive the autonomous cars at low speeds as well as elevated speeds, especially small areas and vast areas, in addition to a thorough evaluation with various paths. With the help of three different autonomous cars, we evaluate the localization effectiveness of different algorithms and techniques. Both of these autonomous cars used in the experiment have McLam wheels. Because the McLam wheel can move within any direction and has gained popularity in the market, we want to deploy this sort of vehicle. To create a more sustainable transport network, the proposed frameworks must also focus on vehicle emissions in addition to travel time. Additionally, because the model was developed using simulated traffic data, further fine-tuning will be necessary before it can be used with actual traffic data.

This paper discusses about the different navigation techniques, algorithms, and deep reinforcement learning. According to the findings, Deep RL is a competent algorithm for learning to steer autonomous vehicles and find its way over perilous terrain in order to arrive at a destination. Additionally, the strategies like deep indoor positioning automotive dead reckoning and data-driven inertial navigation are explained. While other algorithms which are not based upon machine learning techniques can also do the same tasks, many of them rely on localization techniques that have been shown to be computationally more demanding than deep RL, such as SLAM. Along with various frameworks, a comparison and distinctions between reinforcement learning-based autonomous driving simulators have to be examined further.

References

1. B. Zhou et al., “DeepVIP: Deep Learning-Based Vehicle Indoor Positioning Using Smartphones,” IEEE Transactions on Vehicular Technology, vol. 71, no. 12, pp. 13299–13309, Dec. 2022, https://doi.org/10.1109/tvt.2022.3199507.
2. F. Shen, J. W. Cheong, and A. G. Dempster, “A DSRC Doppler/IMU/GNSS Tightly-coupled Cooperative Positioning Method for Relative Positioning in VANETs,” Journal of Navigation, vol. 70, no. 1, pp. 120–136, Jul. 2016, https://doi.org/10.1017/s0373463316000436.
3. L. Chen, W. Zhan, W. Tian, Y. He, and Q. Zou, “Deep Integration: A Multi-Label Architecture for Road Scene Recognition,” IEEE Transactions on Image Processing, vol. 28, no. 10, pp. 4883–4898, Oct. 2019, https://doi.org/10.1109/tip.2019.2913079.
4. A. A. Rahmani Hosseinabadi, A. Slowik, M. Sadeghilalimi, M. Farokhzad, M. Babazadeh shareh, and A. K. Sangaiah, “An Ameliorative Hybrid Algorithm for Solving the Capacitated Vehicle Routing Problem,” IEEE Access, vol. 7, pp. 175454–175465, 2019, https://doi.org/10.1109/access.2019.2957722.
5. L. F. Galindres-Guancha, E. M. Toro-Ocampo, and R. A. Gallego- Rendón, “Multi-objective MDVRP solution considering route balance and cost using the ILS metaheuristic,” International Journal of Industrial Engineering Computations, pp. 33–46, 2018, https://doi.org/10.5267/j.ijiec.2017.5.002.
6. J. Wu et al., “Autonomous Cooperative Flocking for Heterogeneous Unmanned Aerial Vehicle Group,” IEEE Transactions on Vehicular Technology, vol. 70, no. 12, pp. 12477–12490, Dec. 2021, https://doi.org/10.1109/tvt.2021.3124898.
7. S. Padmini, S. S. Dash, S. Chandrasekhar, and P. Vedula, “Fuzzy Logiccontrol Of Differential Protection For Large Power Transformer,” i-manager’s Journal on Circuits and Systems, vol. 1, no. 1, pp. 10–15, Feb. 2013, https://doi.org/10.26634/jcir.1.1.2194.
8. J. C. Jesus, J. A. Bottega, M. A. S. L. Cuadros, and D. F. T. Gamarra, “Deep Deterministic Policy Gradient for Navigation of Mobile Robots in Simulated Environments,” 2019 19th International Conference on Advanced Robotics (ICAR), Dec. 2019, https://doi.org/10.1109/icar46387.2019.8981638.
9. J. Graeter, A. Wilczynski, and M. Lauer, “LIMO: Lidar-Monocular Visual Odometry,” 2018 IEEE/RSJ International Conference on Intelligent Robots and Systems (IROS), Oct. 2018, https://doi.org/10.1109/iros.2018.8594394.
10. Q. Zou, Q. Sun, L. Chen, B. Nie, and Q. Li, “A Comparative Analysis of LiDAR SLAM-Based Indoor Navigation for Autonomous Vehicles,” IEEE Transactions on Intelligent Transportation Systems, vol. 23, no. 7, pp. 6907–6921, Jul. 2022, https://doi.org/10.1109/tits.2021.3063477.
11. J. M. Santos, D. Portugal, and R. P. Rocha, “An evaluation of 2D SLAM techniques available in Robot Operating System,” 2013 IEEE International Symposium on Safety, Security, and Rescue Robotics (SSRR), Oct. 2013, https://doi.org/10.1109/ssrr.2013.6719348.
12. C. Wang, J. Wang, and X. Zhang, “A Deep Reinforcement Learning Approach To Flocking And Navigation Of Uavs In Large-Scale Complex EnvironmentS,” 2018 IEEE Global Conference on Signal and Information Processing (GlobalSIP), Nov. 2018, https://doi.org/10.1109/globalsip.2018.8646428.
13. C. Yan, C. Wang, X. Xiang, Z. Lan, and Y. Jiang, “Deep Reinforcement Learning of Collision-Free Flocking Policies for Multiple Fixed-Wing UAVs Using Local Situation Maps,” IEEE Transactions on Industrial Informatics, vol. 18, no. 2, pp. 1260–1270, Feb. 2022, https://doi.org/10.1109/tii.2021.3094207.
14. S. Bijjahalli and R. Sabatini, “A High-Integrity and Low-Cost Navigation System for Autonomous Vehicles,” IEEE Transactions on Intelligent Transportation Systems, vol. 22, no. 1, pp. 356–369, Jan. 2021, https://doi.org/10.1109/tits.2019.2957876.
15. T. Wang, R. Qin, Y. Chen, H. Snoussi, and C. Choi, “A reinforcement learning approach for UAV target searching and tracking,” Multimedia Tools and Applications, vol. 78, no. 4, pp. 4347–4364, Feb. 2018, https://doi.org/10.1007/s11042-018-5739-5.

16. B. Baruah and S. Dhal, "A Security and Privacy Preserved Intelligent Vehicle Navigation System," IEEE Transactions on Dependable and Secure Computing, vol. 20, no. 2, pp. 944–959, Mar. 2023, https://doi.org/10.1109/tdsc.2022.3145649.
17. M. Bosse and R. Zlot, "Continuous 3D scan-matching with a spinning 2D laser," 2009 IEEE International Conference on Robotics and Automation, May 2009, https://doi.org/10.1109/robot.2009.5152851.
18. T. Y. Tang, D. J. Yoon, F. Pomerleau, and T. D. Barfoot, "Learning a Bias Correction for Lidar-Only Motion Estimation," 2018 15th Conference on Computer and Robot Vision (CRV), May 2018, https://doi.org/10.1109/crv.2018.00032.
19. F. Chanier, P. Checchin, C. Blanc, and L. Trassoudaine, "Comparison of EKF and PEKF in a SLAM context," 2008 11th International IEEE Conference on Intelligent Transportation Systems, Oct. 2008, https://doi.org/10.1109/itsc.2008.4732631.
20. J. Xiong, J. W. Cheong, Z. Xiong, A. G. Dempster, S. Tian, and R. Wang, "Adaptive Hybrid Robust Filter for Multi-Sensor Relative Navigation System," IEEE Transactions on Intelligent Transportation Systems, vol. 23, no. 8, pp. 11026–11040, Aug. 2022, https://doi.org/10.1109/tits.2021.3098739.
21. N. Kwak, I.-K. Kim, H.-C. Lee, and B.-H. Lee, "Analysis of Resampling Process for the Particle Depletion Problem in FastSLAM," RO-MAN 2007 - The 16th IEEE International Symposium on Robot and Human Interactive Communication, 2007, https://doi.org/10.1109/roman.2007.4415080.
22. S. Thrun and M. Montemerlo, "The Graph SLAM Algorithm with Applications to Large-Scale Mapping of Urban Structures," The International Journal of Robotics Research, vol. 25, no. 5–6, pp. 403–429, May 2006, https://doi.org/10.1177/0278364906065387.
23. N. Kbayer and M. Sahmoudi, "3D-mapping-aided gnss localization for integrity monitoring in urban environments," 2017 14th International Multi-Conference on Systems, Signals & Devices (SSD), Mar. 2017, https://doi.org/10.1109/ssd.2017.8167014.
24. R. Sabatini, T. Moore, and S. Ramasamy, "Global navigation satellite systems performance analysis and augmentation strategies in aviation," Progress in Aerospace Sciences, vol. 95, pp. 45–98, Nov. 2017, https://doi.org/10.1016/j.paerosci.2017.10.002.
25. M. Cao, V. O. K. Li, and Q. Shuai, "Book Your Green Wave: Exploiting Navigation Information for Intelligent Traffic Signal Control," IEEE Transactions on Vehicular Technology, vol. 71, no. 8, pp. 8225–8236, Aug. 2022, https://doi.org/10.1109/tvt.2022.3176620.
26. S. Sivashangaran and M. Zheng, "Intelligent Autonomous Navigation of Car-Like Unmanned Ground Vehicle via Deep Reinforcement Learning," IFAC-PapersOnLine, vol. 54, no. 20, pp. 218–225, 2021, https://doi.org/10.1016/j.ifacol.2021.11.178.
27. J. W. Betz and K. R. Kolodziejski, "Generalized Theory of Code Tracking with an Early-Late Discriminator Part II: Noncoherent Processing and Numerical Results," IEEE Transactions on Aerospace and Electronic Systems, vol. 45, no. 4, pp. 1557–1564, Oct. 2009, https://doi.org/10.1109/taes.2009.5310317.
28. Y. Ni, J. He, L. Cai, J. Pan, and Y. Bo, "Joint Roadside Unit Deployment and Service Task Assignment for Internet of Vehicles (IoV)," IEEE Internet of Things Journal, vol. 6, no. 2, pp. 3271–3283, Apr. 2019, https://doi.org/10.1109/jiot.2018.2882436.
29. Z. Wei, J. Li, X. Wang, and C.-Z. Gao, "A Lightweight Privacy-Preserving Protocol for VANETs Based on Secure Outsourcing Computing," IEEE Access, vol. 7, pp. 62785–62793, 2019, https://doi.org/10.1109/access.2019.2915794.
30. T. Qian, C. Shao, X. Wang, and M. Shahidehpour, "Deep Reinforcement Learning for EV Charging Navigation by Coordinating Smart Grid and Intelligent Transportation System," IEEE Transactions on Smart Grid, vol. 11, no. 2, pp. 1714–1723, Mar. 2020, https://doi.org/10.1109/tsg.2019.2942593.
31. C. Sur, Y. Park, and K. H. Rhee, "An efficient and secure navigation protocol based on vehicular cloud," International Journal of Computer Mathematics, vol. 93, no. 2, pp. 325–344, Jul. 2014, https://doi.org/10.1080/00207160.2014.934685.

32. B. Zhou et al., "DeepVIP: Deep Learning-Based Vehicle Indoor Positioning Using Smartphones," IEEE Transactions on Vehicular Technology, vol. 71, no. 12, pp. 13299–13309, Dec. 2022, https://doi.org/10.1109/tvt.2022.3199507.
33. R. Ercek, P. De Doncker, and F. Grenez, "NLOS-multipath effects on Pseudo-Range estimation in urban canyons for GNSS applications," 2006 First European Conference on Antennas and Propagation, Nov. 2006, https://doi.org/10.1109/eucap.2006.4584889.
34. N. Alam, A. Kealy, and A. G. Dempster, "An INS-Aided Tight Integration Approach for Relative Positioning Enhancement in VANETs," IEEE Transactions on Intelligent Transportation Systems, vol. 14, no. 4, pp. 1992–1996, Dec. 2013, https://doi.org/10.1109/tits.2013.2265235.
35. A. Ramezani Dooraki and D.-J. Lee, "A Multi-Objective Reinforcement Learning Based Controller for Autonomous Navigation in Challenging Environments," Machines, vol. 10, no. 7, p. 500, Jun. 2022, https://doi.org/10.3390/machines10070500.
36. X. Wang, N. Cui, and J. Guo, "INS/VisNav/GPS relative navigation system for UAV," Aerospace Science and Technology, vol. 28, no. 1, pp. 242–248, Jul. 2013, https://doi.org/10.1016/j.ast.2012.11.004.
37. W. Wang, L. Wang, J. Wu, X. Tao, and H. Wu, "Oracle-Guided Deep Reinforcement Learning for Large-Scale Multi-UAVs Flocking and Navigation," IEEE Transactions on Vehicular Technology, vol. 71, no. 10, pp. 10280–10292, Oct. 2022, https://doi.org/10.1109/tvt.2022.3184043.
38. J. N. Gross, Y. Gu, and M. B. Rhudy, "Robust UAV Relative Navigation With DGPS, INS, and Peer-to-Peer Radio Ranging," IEEE Transactions on Automation Science and Engineering, vol. 12, no. 3, pp. 935–944, Jul. 2015, https://doi.org/10.1109/tase.2014.2383357.
39. M Senthil Vadivu, M Purushotham Reddy, Kantilal Rane, Narendra Kumar, A. Karthikayen and Nitesh Behare, "An IoT-Based System for Managing and Monitoring Smart Irrigation through Mobile Integration", Journal of Machine and Computing, vol.3, no.3, pp. 196-205, July 2023. https://doi.org/10.53759/7669/jmc202303018.
40. D. K. Mishra, A. Thomas, J. Kuruvilla, P. Kalyanasundaram, K. R. Prasad, and A. Haldorai, "Design of mobile robot navigation controller using neuro-fuzzy logic system," Computers and Electrical Engineering, vol. 101, p. 108044, Jul. 2022, https://doi.org/10.1016/j.compeleceng.2022.108044.
41. Dong Jo Kim, "A Discussion of Key Aspects and Trends in Self Driving Vehicle Technology", Journal of Machine and Computing, vol.3, no.4, pp. 556-565, October 2023. https://doi.org/10.53759/7669/jmc202303047.
42. A. A. Agarkar, Md. Z. Hussain, J. T. Raja, A. Haldorai, S. Selvakanmani, and M. Thangamani, "Analysis of taskable mobile IoT sensing systems for coverage and throughput," International Journal of System Assurance Engineering and Management, Feb. 2023, https://doi.org/10.1007/s13198-023-01872-w.
43. N. Kumar, S. S. Rahman, and N. Dhakad, "Fuzzy Inference Enabled Deep Reinforcement Learning-Based Traffic Light Control for Intelligent Transportation System," IEEE Transactions on Intelligent Transportation Systems, vol. 22, no. 8, pp. 4919–4928, Aug. 2021, https://doi.org/10.1109/tits.2020.2984033.
44. H. Wei, G. Zheng, V. Gayah, and Z. Li, "Recent Advances in Reinforcement Learning for Traffic Signal Control," ACM SIGKDD Explorations Newsletter, vol. 22, no. 2, pp. 12–18, Jan. 2021, https://doi.org/10.1145/3447556.3447565.
45. S. Koh et al., "Real-time deep reinforcement learning based vehicle navigation," Applied Soft Computing, vol. 96, p. 106694, Nov. 2020, https://doi.org/10.1016/j.asoc.2020.106694.
46. P. Palos and A. Huszak, "Comparison of Q-Learning based Traffic Light Control Methods and Objective Functions," 2020 International Conference on Software, Telecommunications and Computer Networks (SoftCOM), Sep. 2020, https://doi.org/10.23919/softcom50211.2020.9238290.
47. "Traffic signal timing via deep reinforcement learning," IEEE/CAA Journal of Automatica Sinica, vol. 3, no. 3, pp. 247–254, Jul. 2016, https://doi.org/10.1109/jas.2016.7508798.

48. P. A. Lopez et al., "Microscopic Traffic Simulation using SUMO," 2018 21st International Conference on Intelligent Transportation Systems (ITSC), Nov. 2018, https://doi.org/10.1109/itsc.2018.8569938.
49. Y. Liu, W. Zhang, F. Chen, and J. Li, "Path planning based on improved Deep Deterministic Policy Gradient algorithm," 2019 IEEE 3rd Information Technology, Networking, Electronic and Automation Control Conference (ITNEC), Mar. 2019, https://doi.org/10.1109/itnec.2019.8729369.
50. P. Polack, F. Altche, B. d'Andrea-Novel, and A. de La Fortelle, "The kinematic bicycle model: A consistent model for planning feasible trajectories for autonomous vehicles?," 2017 IEEE Intelligent Vehicles Symposium (IV), Jun. 2017, https://doi.org/10.1109/ivs.2017.7995816.
51. Q. Guo, S. Xin, H. Sun, Z. Li, and B. Zhang, "Rapid-Charging Navigation of Electric Vehicles Based on Real-Time Power Systems and Traffic Data," IEEE Transactions on Smart Grid, vol. 5, no. 4, pp. 1969–1979, Jul. 2014, https://doi.org/10.1109/tsg.2014.2309961.
52. R. Gao, F. Ye, G. Luo, and J. Cong, "Smartphone-Based Real-Time Vehicle Tracking in Indoor Parking Structures," SpringerBriefs in Computer Science, pp. 81–109, 2018, https://doi.org/10.1007/978-981-10-8378-5_5.
53. Q. Wang, Y. Gu, J. Liu, and S. Kamijo, "DeepSpeedometer: Vehicle speed estimation from accelerometer and gyroscope using LSTM model," 2017 IEEE 20th International Conference on Intelligent Transportation Systems (ITSC), Oct. 2017, https://doi.org/10.1109/itsc.2017.8317935.
54. Gong, Goldman, and Lach, "Deepmotion: a deep convolutional neural network on inertial body sensors for gait assessment in multiple sclerosis," 2016 IEEE Wireless Health (WH), Oct. 2016, https://doi.org/10.1109/wh.2016.7764572.
55. J. Chen et al., "A Data-Driven Inertial Navigation/Bluetooth Fusion Algorithm for Indoor Localization," IEEE Sensors Journal, vol. 22, no. 6, pp. 5288–5301, Mar. 2022, https://doi.org/10.1109/jsen.2021.3089516.
56. M. Gadaleta and M. Rossi, "IDNet: Smartphone-based gait recognition with convolutional neural networks," Pattern Recognition, vol. 74, pp. 25–37, Feb. 2018, https://doi.org/10.1016/j.patcog.2017.09.005.
57. Y. Singh, M. Bibuli, E. Zereik, S. Sharma, A. Khan, and R. Sutton, "A Novel Double Layered Hybrid Multi-Robot Framework for Guidance and Navigation of Unmanned Surface Vehicles in a Practical Maritime Environment," Journal of Marine Science and Engineering, vol. 8, no. 9, p. 624, Aug. 2020, https://doi.org/10.3390/jmse8090624.

AI-Based Effective Communication in Software-Defined VANET: A Study

1 Introduction

Artificial intelligence (AI) has become a favor for automobile industries to make a feasible future as well as strengthen the safety and experience of human beings. Intelligent Transport System (ITS) is part of AI which promises an enhanced level of transportation and safety. ITS is arresting technology that has progressed applications to lay out innovative services for transportation environment in this diligent world. Some countries like United Kingdom, South Korea, Australia, Singapore, Japan, United States are using ITS for traffic updates and navigations [1]. Every successful deployment using ITS is to secure an efficient intelligent transportation. In order to secure our lives while traveling in this engaged world, utilizing the benefits of technology is a greater advancement. VANET (Vehicular Ad Hoc Networks) is popular for connected technology in vehicles. In the Past and recent research shows an extensive interest in the field of VANET. The consideration of VANET in ITS is due to its importance in solving traffic and safety problems [2]. The system followed by VANET is like communication between vehicles where vehicles aim to be intelligent transport that can collect the information and transfer the information between Vehicle to Vehicle (V2V) and Vehicle to Infrastructure (V2I) [3, 4]. The effective implementation of VANET technology in ITS can reduce accidents and save many lives.

However, in VANET, mobility of vehicle is high and random due to the increase in number of vehicles in cities and urban environment. In VANET due to high mobility, it suffers obstacles in terms of data transmission which requires abounding computing resources as well as communication. The issues in routing like stability of network, scalability, and availability are affecting many services in VANET [5, 6]. Therefore, traffic efficiency can be provided by low-latency service and high data transmission for a better user experience in VANET. Though energy consumption is also a concern in vehicular communication to realize sustainable or green city [7].

© European Alliance for Innovation 2024

A. Haldorai et al., *Artificial Intelligence for Sustainable Development*,
EAI/Springer Innovations in Communication and Computing,
https://doi.org/10.1007/978-3-031-53972-5_14

The traditional methods for overcoming the difficulties in vehicular communication entail the use of well-established technologies such dedicated short-range communication (DSRC) and cellular communication based on long-term evolution (LTE) [8–10]. Operating in accordance with the IEEE 802.11 standard, DSRC effectively facilitates information sharing between nearby cars. However, because to limitations on resilience and communication coverage, its usefulness to widespread vehicular communication is still rather limited. Numerous features of vehicle ad hoc networks (VANETs) are currently the subject of substantial ongoing study, including routing, security, and energy efficiency [11]. However, there are still issues that need to be resolved, including: (1) Network administration, (2) choosing the best routing option, (3) energy management for large-scale vehicular communication, and (4) assuring reliability and scalability. In order to develop a strong communication framework between cars that includes all relevant parameters and has the ultimate goal of reducing traffic accidents, it is crucial to overcome these impediments.

The diminishing of issues in vehicular communication were done by some common approaches like deployment of more access points, base stations (BS) or relay. Although, the manipulation of these methods will be a reason for high cost, increase in energy consumption, and unbending network management in VANET. Because of high cost, it is hard to deploy in short time. AI technologies are promising technology in every field especially in ITS and next-generation network. AI-based Intelligent Reflecting Surface and a Software-Defined Network (SDN) can provide better communication, flexibility, and network configuration which can proffer energy efficiency and in addition security. IRS can alter signal propagation environment because it consists of shielding layer, interface & communication layer, computing layer, metamaterial layer, and sensing & actuation layer [12–14]. SDN is based on Road Side Unit (RSU) and on-board unit in which RSU will be connected to SDN controller to obtain network information [15, 16]. In VANET, SDN provides the following advantages: (1) it simplifies network management and ensures VANET elasticity [17, 20]; and (2) the network issues will be reduced and provide efficient routing decision [18]. The advantages of IRS are less expensive, nearly passive, and have high energy efficiency which can be used in the RSU because of its thin and light nature which can assist in transmission of data [20].

In this paper, our contribution to overcome the issues in VANET is as follows:

1. The analysis of AI-based IRS and SDN is done in order to improve the vehicle communication, in this the planes of the framework are discussed.
2. Secure data transmission that is transmission and management of data in AI-based IRS and vehicle to vehicle communication is briefly discussed.
3. A small study on collaborative learning (CL) and Deep Reinforcement Learning (DRL) to determine the transmission power is contributed in this paper.
4. Cluster head selection in vehicle group with short explanation of particle swarm optimization and density peak clustering algorithm is given.

The rest of this article is organized as follows. Section 2 discusses the literature related to VANET security solutions. The AI-based IRS and SDN vehicular framework with overview and its preliminaries are presented in Sect. 3. In Sect. 4, the Collaborative Learning and DRL-based learning and transmission are detailed. In Sect. 5, vehicle to vehicle communication is detailed and discussed. In Sect. 6, empirical findings are detailed, and concludes this chapter.

2 Related Work

2.1 *Artificial Intelligence in Communication*

Artificial Intelligence (AI) is been used in many application domains because it has a capability of enhancing conventional data-driven methods. In VANET, the collecting of data is in a frequent way from various sources [21]. The data that are collected from the sources are used for various purposes which include awareness of the driver, routing predicting mobility so that we can improve the comfort, safety, and hazardous situations in roads [22]. According to [23], the author has created a mechanism to safeguard data and preserve location privacy using a VANET service location protocol. They combined trusted node distance verification, received RF with strength-based transmitter node range approximation, and distance estimation methods based on artificial intelligence for modifying the broadcast range [23]. To augment VANET, Different techniques can be used like machine learning, deep learning, metaheuristic search algorithms, and optimization techniques. For security, the intrusion detection system (IDS) is used in addition to prevent security attacks [24].

Predominantly, we can say that AI can increase the idea of drivers and can avoid road accidents. But even we have more exploration in this field of ad hoc networks, still AI adoption is ignored for management of communication in vehicles.

2.2 *Data Transmission in VANETs*

The investigation of various solutions has been prompted by the need to address data transmission issues in Vehicular Ad Hoc Networks (VANETs). Dedicated Short-Range Communication (DSRC), which permits information transmission among close-by cars, is a notable example [8]. However, the DSRC's coverage restrictions make it unsuitable for reliable vehicle access and data transmission. Due to its insufficient transmission efficiency, DSRC's effectiveness in enabling large amounts of data transfer remains limited. LTE-based cellular vehicular communication, which has better data transmission capabilities and user accessibility, is the next technology in the field of VANETs following DSRC [9]. However, LTE has poor energy efficiency and high-transmission latency, which are issues that are especially important given the demands of latency-sensitive services. In several research [26, 27], DSRC and LTE are combined with cooperative communication models [19]. A software-defined cooperative data sharing plan for 5G-VANETs is one example [26] that integrates cellular and DSRC communication [11]. Another strategy [25] uses a fuzzy logic system to optimize next hop node selection, employing inputs such as speed, distance, and Link Residual Time (LRT) to direct hop selection. Researchers in [28] also develop an Internet of Vehicles (IoV) system that makes use of unmanned aerial vehicles (UAVs) for content distribution in order to address data transmission problems. While this fixes the data transmission problems, there are now further worries about cost and energy use.

2.3 Intelligent Reflecting Surface

IRS is an emerging and revolutionary technology to program the communication of wireless environment via countless and reconfiguring independently adjusting reflecting elements [29]. In paper [30], they combined IRS and BS beamforming with machine learning to consolidate the optimization problem. It has shown improvement in learning convergence rate and efficiency. In the previous wireless communication networks in order to improve the performance of the network, they used transceivers. In [31], IRS-assisted multiple-input single-output (MISO) transmission scheme has been proposed by optimizing phase shift and beamforming vector. The research on IRS is continuing in order to explore the benefits of IRS. In wireless communication network, IRS helps in controllability and programmability. The authors in [32] investigation were done for IRS-assisted multiuser communication by channel estimation, where the algorithm, analysis, and framework are included. To magnify the ground user communication, article [33] has proposed an integrated UAV-IRS relaying system. The authors of the paper [35] introduce the idea of incorporating Intelligent Reflecting Surfaces (IRS) within a mobile edge computing system. This novel method seeks to improve data offloading effectiveness and decrease processing latency. The system deliberately manipulates wireless signal propagation by utilizing IRS technology, which optimizes data transmission between mobile devices and edge computing resources. By efficiently managing data traffic and accelerating processing operations, this integration has the potential to greatly increase the system's overall performance, creating a more streamlined and responsive computing environment. In the article [34] to improve the system performance and to allocate system resources, they have proposed an IRS-aided full-duplex cognitive radio systems. The combining of VANET with IRS-AI is not much popular in other researches. IRS is an emerging technology that can be used in wireless communication. However, IRS with AI can improve the communication performance in VANET.

2.4 SDN Networking in VANET

Software-defined networking (SDN) is an emerging technology in wireless networking architecture with some promising properties that can manage the weaknesses of traditional networks. SDN has the capability of forwarding decisions by decoupling the control plane from data plane that forwards the data [36]. In the existing, SDN was integrated with fog computing and edge computing but the pervasive technology of SDN can be used in VANET communication like location-based services, content sharing services [20], etc. The authors in [37] have proposed a software wireless-defined network (SWDN), they have used a Detection and Discovery Control in Handoff (DISpatCH) approach to manage handoff process IEEE 802.11 Networks. Software-Defined Vehicular Network will provide energy efficiency for authentication mechanism and also make the system more secure against attacks as it uses the intrusion detection system.

3 AI-Based IRS in Software-Defined Vehicular Network

A futuristic architectural strategy that gives greater flexibility and adaptability in network design is software-defined networking (SDN). This method adds an abstraction layer that successfully captures the complex hierarchies and observable strata present in a network. SDN is designed with the overarching goal of improving network administration and control, allowing businesses and service providers to quickly and successfully adapt to the constantly changing demands of their industry. Routing decisions and data forwarding are the two independent but related functional modules that form the basis of SDN. SDN presents a revolutionary method for routing choices that enables more dynamic and context-aware traffic management. On the other hand, the data forwarding component makes it possible for data packets to be transmitted within the network in an effective manner. Contrary to traditional network architecture, that relies on individual network devices to choose how to route data according to the configuration of their routing tables, this process uses a centralized routing table. Ten years of SDN involvement in networking have resulted in numerous networking advancements. In SDNs, which normally carry out procedures on fixed infrastructure/roadside units (RSU) instead of mobile devices, mobile devices are employed to transfer the data in line with the policies defined centrally. Open Flow is a well-known centralized management protocol. Figure 1 depicts an SDN-based VANET design.

Vehicular Ad Hoc Networks Support for QoS and SDN-Based Mobility Management. SDN is founded on the central control principle, and it can be challenging to consider SDN for VANETs because of their distinctive characteristics, which have led to the division of VANETs into distributed and self-contained networks. However, only a few elements of vehicle networks can be combined with SDN concepts. Vehicles follow a predictable topology, for instance, when using a GPS

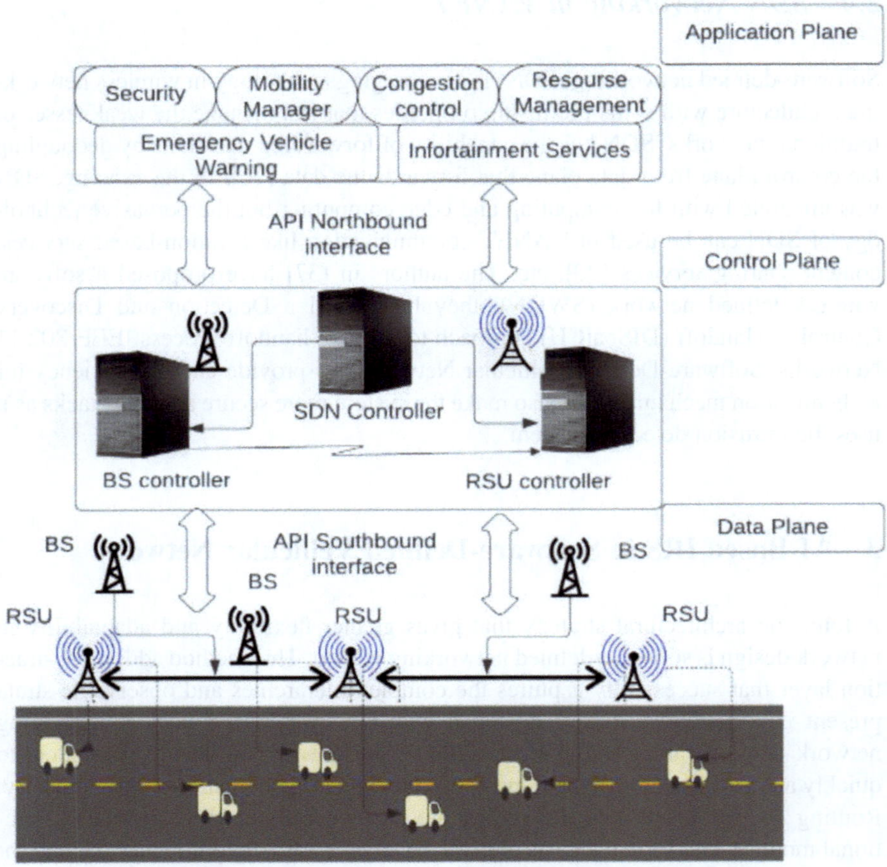

Fig. 1 SDN-based VANET design

tool for road maps. Permitting traffic optimization using a global network perspective there are extra difficulties when taking SDN into account for VANETs. When dealing with the issue of traffic congestion in the VANET, it is important to determine the route and divert traffic according to the density of the cars along the route in order to reduce congestion and delay. To address the congestion issue in VANET, we require a global perspective of the network. Due to the fact that traffic congestion increases queuing delays, lengthens travel times, and reduces vehicle speed, the SD-VANET design may therefore be essential in reducing traffic congestion. The control plane (controller) for devices that convey data is decoupled by the logically centralized SDN architecture, as was already mentioned. As a result, SDN enables programmability on routers, switches, and RSU. To the control plane, sometimes referred to as the SDN controller, these forwarding devices transmit data.

The continual changes in link and channel conditions, as well as the heavy processing needs for determining centralized routes, make it difficult to apply Software-Defined Networking (SDN) principles to vehicle networks. The constantly changing dynamics of vehicle communication, where the standard of connections between cars is subject to regular changes, is what gives rise to this complexity. Getting vehicles to migrate smoothly across several network domains is a major challenge. Due to the necessity to maintain continuous connectivity while moving vehicles change from one coverage area to another, this task is extremely challenging. To deal with the density or sparsity of dynamic topologies, more rules are needed. Multiple coexisting wireless access technologies' erratic network traffic adds to the difficulty. A wireless media is used to transmit information between different vehicles and the RSU. In order to inform drivers and improve road safety, a variety of data are communicated using this method. Communication occurs between the control plane and the data plane in the SDN-based VANET system. The traffic forwarding hardware is contained in the data plane, whereas the control plane manages communication flow.

The IRS-based SDN-based VANET's vehicles move very quickly. Each vehicle has rules for interacting with the data it receives. The RSU or similar device that is directly connected to the SDN controller with IRS assists the nodes or the cars in the case of any new data flow. It is necessary to position the controller close to a busy location to make it easier for all nodes to access it and prevent latency or waiting for a response. The RSU should also be positioned close to a lot of traffic to lower communication costs overall. Centralized control and distributed control are the two methods available for managing network functionality. In centralized control, the system's functionality can be managed by a central controller, but if it fails or malfunctions, it is impossible to manage the system's functionality until the control gain is restored to its initial state. The distributed controller, which makes use of multiple controllers to regulate the system's operation, offers a solution to this issue. The system's complexity is reduced because these controllers are connected via a wireless means. It is possible to govern the actions of the other controllers by utilizing a single controller as the parent controller. The network OS is used to regulate the general performance of the network in an IRS-based SDN VANET. For instance, the OS in conventional computer systems offers abstractions in the manner of APIs to offer access to common resources. The network OS uses SDN programming languages to implement the required system capabilities.

This section puts forth the AI-based IRS-driven wireless communication framework for Software-Defined VANETs shown in Fig. 2. We begin by outlining the structure and component parts of the suggested architecture for vehicular communication. Then, data transmission with AI-IRS assistance is explained, along with safety, energy efficiency, and security in SDVN. There are generally three planes in SDN, they are Data plane, Control plane, and Application plane. In the proposed AI-based IRS in software-defined vehicular network framework, we have vehicle plane, control plane with IRS-aided layer and application layer. The details of these three planes are illustrated.

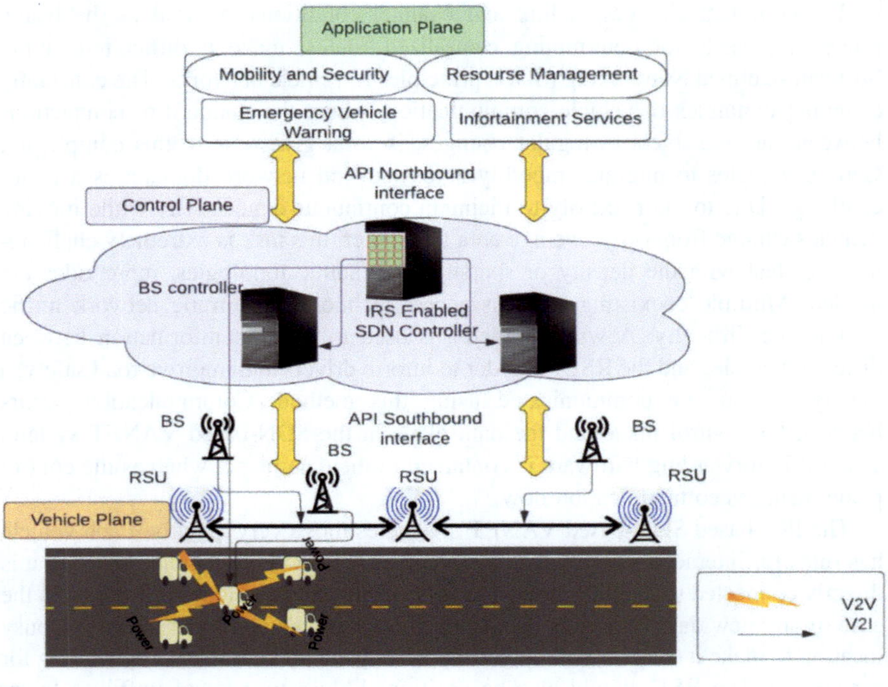

Fig. 2 AI-based IRS SDVN framework

- *Control Plane with IRS*: The control plane with Intelligent Reflective Surfaces (IRS) comprises intelligent reflective surfaces that are placed in each allocated cell in a manner reminiscent to surfaces on buildings. The main goal of this integration is to increase the communication network's capacity for data transfer. The base station (BS) can efficiently manage the communication and channel programming capabilities that are built into the IRS technology. In this situation, the primary objective of IRS technology is to enable seamless data flow between automobiles and the BS. By utilizing its communication and channel programming skills, this is accomplished. The management of these capabilities falls to the BS, which makes sure that the IRS technology is best set up for the specific communication scenario. A typical computer system's operating system (OS) is in charge of overseeing all of the system's resources. This function is, however, handled by the SDN controller in the VANET that is SDN-based. The SDN system's second layer, sometimes known as the network's brain, has a network controller. Each device that is directly connected to the network controller requires a controller to install, update, and delete the rules on each device. The SDN controller's definition of the services is not subject to any set rules. The features that the SDN controller offers are referred to as the services. The part of the network architecture known as the control plane determines the topology and traffic flow of the network. Based on parameters provided by the Control Plane, the Data Plane is the layer of the network architecture that actually manages the

Fig. 3 CHV and CMV

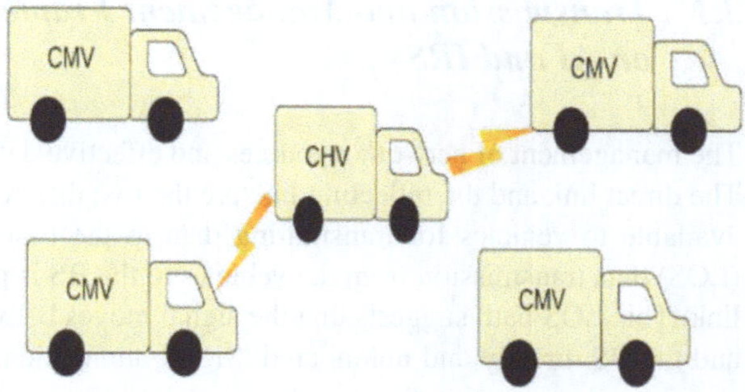

traffic. The implementation of new policies and procedures to improve the system's performance involves numerous difficulties, and it is challenging to apply this attribute of the SDN controller. In addition, the system's centralization may result in problems including security risks, dependability problems, performance problems, and system instability. Regarding suitable positioning the controller, there are numerous research issues in this layer.

- *Vehicle Plane*: This layer is made up of a number of dispersed, data sensing, data storage, data preprocessing, and data transmission-capable underlying vehicles. The cars on this aircraft are divided into several clusters according to their geographical positions, and each cluster consists of two different types of vehicle users: one cluster head vehicle (CHV) and a large number of cluster member vehicles (CMV) as shown in Fig. 3. The member vehicles gather data on the road environment and transmit it to their CHV. Information regarding head vehicles and CMVs is gathered. CHVs occasionally communicate the local data they have collected to the neighboring BS for additional processing due to the limited energy, cache, and computer resources.

- *Application Layer*: On a network, the Application Plane is the layer where services and applications are located. These layers, which include the Control Plane and the Data Plane, are where requests are made for network functions. Devices or programs that manage security, operations, and monitoring are found on this layer on conventional networks. The controller API is where applications are written directly, although this layer helps with portability, code reuse, and network abstraction. The high-level network abstraction is covered first. We overlook the low-level operation of the programming language by employing the high-level language for development. The program should be developed taking into consideration high-level policies. Each device physically connected to the controller will receive the OpenFlow rules that the compiler automatically generates and installs. The network compiler also makes the code more reusable. The system has to execute two different programs in order for it to function properly. Both programs measure the amount of web traffic coming into the system while one program transfers the packet from a single interface to another interface. It is possible to construct a single program that performs both functions since the network compiler makes it easier to reuse resources.

3.1 Transmission and Management Framework Based on AI and IRS

The management of network resources and effective data transmission are designed. The direct link and the reflecting link are the two different propagation links that are available to vehicles for transmitting data to the base station (BS). Line-of-sight (LOS) data transmission from the vehicles to the BS is part of the direct propagation link. This LOS path suggests that the signal moves between the transmitting vehicle and the BS directly and unhindered. Signal attenuation is the process by which the signal weakens as the distance that separates the transmitting vehicle and the BS widens. In essence, the physical distance between the two endpoints of this direct propagation link affects the signal's attenuation. Obstacles (such as buildings) may hinder the immediate pathway of data transmission; this is a regular occurrence in metropolitan settings. In addition, the route loss will significantly reduce the strength of the electromagnetic signals, the BS receives from those cars that are at the cell's perimeter.

All of these occurrences cause a dramatic decline in the signal-to-interference-plus-noise ratios (SINR) at the base station, which has an impact on the speed and quality of transmission. The IRS technology is used during the reflecting propagating mode to speed up data transfer from CHVs to the BS. The CHV-IRS link, IRS reflection, & IRS-BS link are the three components of the IRS-aided reflecting path. The IRS, which is often placed at the cell's edge, receives the vehicle signal first. The incident signal's phase and amplitude are then modified by the IRS by modifying the reflection coefficient of every component. In order to accomplish smooth three-dimension beamforming for directed signal transmission, the IRS-reflection elements then cooperate. The VANET communication system participants, including all vehicles, BS and IRS-aided channels are included in the intended AI-driven management mechanism. The suggested AI-driven method determines the settings of the cars, the IRS, and the BS, and sends those settings to the associated vehicles and the IRS controller via specific channels, respectively. Data transfer with help from the IRS and resource allocation guided by AI.

IRS-Aided Data Transmission: The two different propagating links that are available to cars for data transmission to the base station (BS) are the direct link and the reflecting link. The direct propagation link includes line-of-sight (LOS) data transmission from the vehicles to the BS. Figure 4 shows an illustration of IRS-aided data transmission. This LOS path implies that the signal travels directly and unobstructed between the transmitting vehicle and the BS. The mechanism by which the signal deteriorates when the gap between the transmitting vehicle and the BS expands is known as signal attenuation. In essence, the attenuation of the signal depends on the physical separation of the two endpoints that constitute the direct propagation link. The construction of a communication network for automotive devices supported by IRS technology, specifically designed for automotive Ad Hoc Networks (VANETs), is the main topic of this particular segment. The main goal of this network configuration is to simultaneously increase data transfer rates

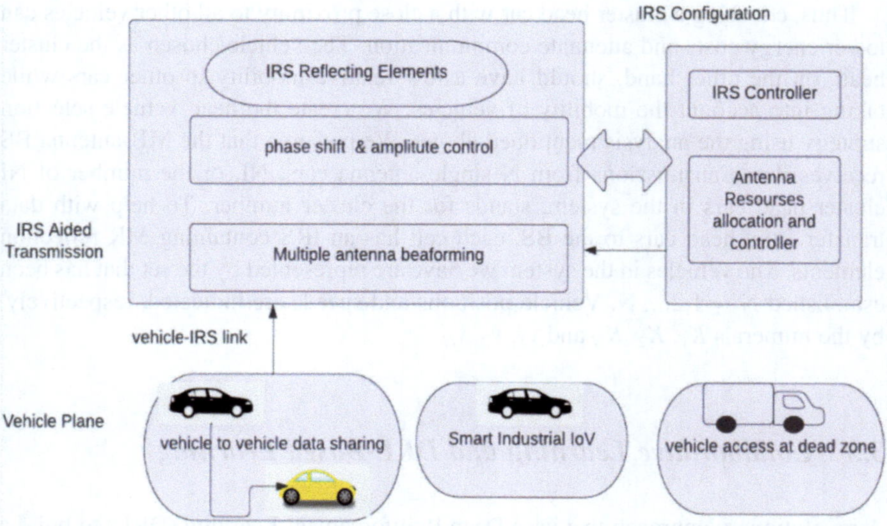

Fig. 4 IRS-aided data transmission

while reducing energy usage. For seamless vehicle-to-vehicle (V2V) and vehicle-to-infrastructure (V2I) connections, each vehicle included in our proposed system is outfitted with both cellular and IEEE 802.11 interfaces. By enabling vehicles to connect successfully with one another and with infrastructure components, this comprehensive interface design helps create an ecosystem for vehicular communications that is both optimized and effective.

3.2 *Vehicle-to-Vehicle Communication*

With the use of vehicle-to-vehicle (V2V) communication, automobiles can exchange digital signals on their location, speed, and direction. With the use of this technology, automobiles can send and receive signals omnidirectionally up to 10 times per second, giving them a 360-degree awareness of all other vehicles in their vicinity. These signals are used by equipped vehicles to spot potential accident risks and inform the driver via an auditory, tactile, or visual alarm so they can take preventative action. V2V transmissions, which have a range of more than 300 meters, can reveal concealed threats brought on by weather or topography. V2V communication improves collision prevention rather than merely mitigation in current accident-avoidance systems. The role of cluster head vehicles in data transmission to the base station (BS) is crucial. It develops a head car selection approach based on vehicle authority and velocity to enable efficient vehicular communication. During the V2V stage, communication attenuation is influenced by the separation between a member vehicle and its cluster head.

Thus, choosing a cluster head car with a close proximity to all other vehicles can lower energy costs and attenuate communication. The vehicle chosen as the cluster head, on the other hand, should have a low relative mobility to other cars while taking into account the mobility of vehicles. We create the head vehicle selection strategy using the analysis mentioned above. We presume that the MB-antenna BS receives data transmissions from N single-antenna cars. NI, or the number of NI cluster head cars in the system, stands for the cluster number. To help with data transfer from head cars to the BS, each cell has an IRS containing MR reflection elements. The vehicles in the system we have are represented by the set that has been established N = 1,2..., N. Vehicle positions and speeds are indicated, respectively, by the numerals K_1, K_2 ,K_N and v_1, v_2 ,v_N.

3.3 Collaborative Learning and DRL-Based Learning

A revolutionary approach that uses Deep Reinforcement Learning (DRL) to build a complete system for managing and allocating network resources. The main goal is to create a framework for resource allocation and control that smoothly incorporates DRL concepts while taking into account the communication requirements of the suggested system. The DRL-driven technique for regulating and allocating network resources is the cornerstone of the hybrid vehicular communication system. This system is based on the ideas of a base station (BS) detection matrix, intelligent reflective surfaces (IRS) layouts for signal reflection, and maximizing vehicle transmission power. The total effectiveness of IRS-facilitated data transfer is greatly increased by coordinating communication resources among cars, IRS, and BS based on the DRL-driven scheme. The collaborative optimization of resources across vehicles, BS, and IRS is prioritized. The main objective of this cooperative optimization is to increase system-wide energy efficiency while maintaining predetermined latency constraints. This cutting-edge IRS-driven vehicle communication system aims to increase transmission rates while reducing energy and time expenses. One of the most effective learning strategies, Reinforcement Learning (RL), a potent artificial intelligence paradigm, takes center stage and closely resembles how humans learn. Earlier in this section, the foundational ideas of Markov decision processes and reinforcement learning are laid forth. It also explores deep learning techniques, a key branch of machine learning. Deep Reinforcement Learning (DRL), a combination of deep learning and reinforcement learning, is the next step in the exploration. DRL is particularly well-suited for complicated scenarios with high-dimensional and infinite-state settings because it is skilled at producing precise and reliable function approximations.

Deep Reinforcement Learning (DRL)-based network resource management is a practical solution to the complex and dynamic nature of vehicular communication systems. DRL, a well-known area of artificial intelligence, is particularly good at improving decision-making in environments with frequent changes and interactions. Its ability to skillfully manage the allocation and control of variable resources inside Vehicular Ad Hoc Networks (VANETs) originates from the merging of two potent approaches, namely Reinforcement Learning (RL) and deep learning. The capacity of DRL to optimize the efficiency of the entire network using CSI and network conditions in real-time. The improvement of energy efficiency is the main goal of this strategy. The transmission power of vehicles, the design of Intelligent Reflective Surfaces (IRS) for signal reflection, and the coefficients utilized for Base Station (BS) identification are all cooperatively optimized to achieve. This is challenging, though, because of how dynamic and unpredictable network condition is. Network entities may learn about the networks they are a part of and accumulate knowledge about them thanks to learning algorithms like RL/DRL, which enables them to make the best judgments possible. Due to its poor learning convergence speed and inability to handle continuous state space, Q Learning is not an effective learning method. Although the policy gradient technique can handle infinite state-action spaces, it could end up with a less than ideal outcome. Additionally, the optimization problem under a high-dimensional input state space is impossible for Q-learning and policies gradient methods to solve. Even though DQN excels at learning policies in high-dimensional state spaces, it is nonlinear.

4 Cluster Head Selection

Router efficiency is mostly determined by cluster stability, reliability, and the choice of cluster heads. The routing system depends on the presence of cluster heads (CHs), which oversee data transmission and cluster maintenance. Although urban traffic networks, which are characterized by high car density, need taking link dependability into consideration, distance usually shapes clusters. The connection reliability parameter is a new one introduced by this study during cluster formation. Due to the increasing vehicle mobility in VANET networks, clusters could lose their structure. Regular maintenance intervals are advised in order to adjust. Vehicle density in urban areas affects the stability of clusters and the efficiency of routing. Car distribution and, hence, stable CH selection are shaped by variables such as direction, surrounding cars, speed, and position. Reliable links and connectivity are ensured via a data flow between clusters, which is facilitated by suitable CHs. The difficult VANET clustering job is CH selection. It takes into account three factors: alignment with cluster direction, similar position and speed, and appropriate cluster placement for density and connectedness. Cluster Head selection is using Particle Swarm Optimization (PSO) and taking advantages from the Density Peak Clustering Algorithm.

4.1 Particle Swarm Optimization

The Particle Swarm Optimization (PSO) method, which belongs to the class of swarm intelligence methods, is frequently used to address optimization issues. The PSO algorithm uses a population-based architecture and was inspired by the group behavior of birds in a flock. Each person, seen as a particle, represents a potential answer. With slight adjustments to their placements and speeds, these particles move through a multidimensional search space. Particles dynamically change their positions and velocities by inferring information from their individual flying experiences and the sociopsychological interactions among the swarm. Each particle's mobility is controlled by this adaptation, which is impacted by their prior knowledge and experience. Notably, particles keep track of the best answer found thus far in the search space. By using this approach, the search solution is discovered in order to go back to the search space's top positions. The PSO approach has a number of benefits, including: its understanding and application are straightforward [38].

- There are not many variables to change.
- There is not enough computing time needed for the technique.
- The method is based not just on position but also on velocity.
- It can swiftly converge.

The position of each particle represents a possible answer to the problem. These particles collectively refine their placements as the process goes along to arrive at the best results. This is accomplished by a mechanism in which each particle dynamically modifies its position and velocity which is denoted as "x" and "v," in order to explore innovative possibilities and effectiveness. The position and velocity formula:

$$v_{id}(t+1) = w \times v_{id}(t) + c_1 r_1 \left[P_{ij}(t) - x_{ij}(t)\right] + c_2 r_2 \left[P_{gj}(t) - x_{ij}(t)\right] \quad (1)$$

$$x_{ij}(t+1) = x_{ij}(t) + v_{ij}(t+1) \quad (2)$$

where the inertia weight is indicated by $j = 1, 2 \ldots, D$; and $i = 1, 2 \ldots, m$; $P_{ij}(t)$ is the current optimal position that particle i has been searching for, and $P_{gj}(t)$ is the current optimal position of the particle swarm as a whole. $v_{id}(t)$ is the velocity of particle i at iteration j; $x_{ij}(t)$ is the position of particle i at iteration j; c_1 and c_2 are the learning factors; and r_1 and r_2 are random values between 0 and 1. Three components make up the velocity update formula: The particle's level of trust in its current movement is expressed in the rest section. The particle's current velocity multiplied by its inertial weight is what it represents. This portion is therefore determined using the inertial motion of the initial velocity. The particle's assessment of its own past is shown in the second section, which deals with the condition of personal influence. A social influence is demonstrated in the third section. The exchange of knowledge and collaboration between the particles is the subject. Figure 5 shows the flow chart of PSO algorithm.

The PSO algorithm generates a population of evenly spaced-out particles and determines where each one should be placed in the search space. Based on their individual endpoints and the previous ideal position, the particles then update their positions and velocities. By transferring information amongst the particles, the algorithm ultimately discovers the best answer. Because the process is iterative, it runs repeatedly until the halting conditions are met. The algorithm's overall convergence is strengthened by this.

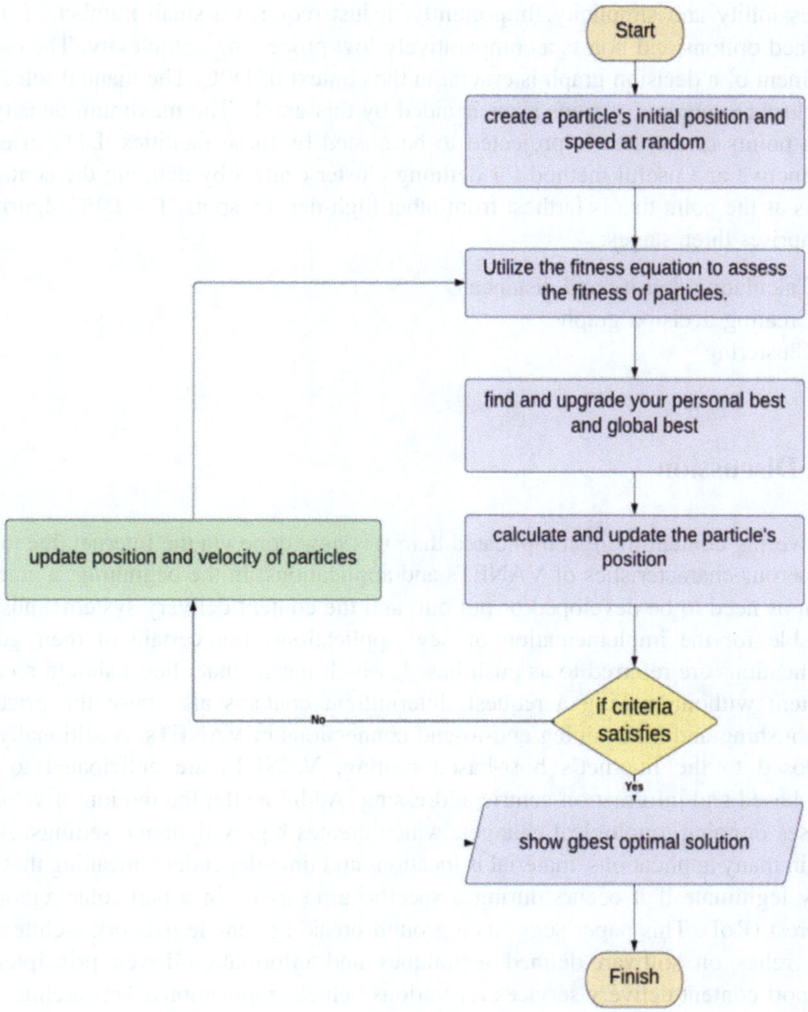

Fig. 5 Flow chart of particle swarm optimization

4.2 Density Peak Clustering Algorithm

Density-based clustering methods have been used extensively in many different fields, proving their broad applicability. By effectively locating clusters where closely connected data points congregate, these approaches are intended to locate areas of high density in the data space. The Density Peak Clustering (DPC) algorithm stands out as a leading candidate among these density-based clustering algorithms. This technique can be used for a variety of applications because of its accessibility and simplicity. Importantly, it just requires a small number of user-defined options and boasts a comparatively low-processing complexity. The establishment of a decision graph is crucial in the context of DPC. The manual selection of cluster centers of various sizes is aided by this graph. The maximum density of data points in the area is projected to be hosted by these facilities. DPC offers a distinctive and useful method for defining cluster centers by defining the center of mass as the point that is farthest from other high-density spots. The DPC algorithm comprises three stages:

- Calculating density and distance
- Creating decision graphs
- Clustering

5 Discussion

Delivering content more complicated than it is now done via the Internet due to the numerous characteristics of VANETs and applications. In the beginning, a number of apps need to be developed or put out, and the content delivery systems must be reliable for the implementation of new applications. For certain of their goals, applications are referred to as push-based, which means that clients should receive content without making a request. Intermittent contacts also raise the price of establishing and maintaining end-to-end connections in VANETs. Additionally, as opposed to the Internet's host-based routing, VANETs are anticipated to use geo-based and information-centric addressing. Additionally, the motion of vehicles causes ongoing topological changes, which creates highly dynamic settings. After all, in many applications, material is location- and time-dependent, meaning that it is only legitimate if it occurs during a specific time frame in a particular region of interest (RoI). This paper suggests a ground-breaking vehicle network architecture that relies on software-defined techniques and information-driven principles to support content delivery services for various vehicle applications. This architecture has been painstakingly created to seamlessly integrate the communication protocol used in information-centric networks (ICN) with the principles of software-defined networking (SDN), as well as to extend existing concepts to fit the unique characteristics of vehicular ad hoc networks (VANETs).

An architecture for content distribution that is fundamentally driven by the flow of information is produced by this unusual fusion. This architecture's SDN components have the capacity to dynamically modify the rules that control various operations so that they conform to the most recent network state. Due to the lack of global network information, it is challenging for decentralized VANET routing protocols to determine a route's validity with any degree of accuracy. As a result, the predicted paths might not be reliable until the packets pass, which would probably cause the packet to be dropped. When sending packets takes a long time, either because they are sent in succession or because a long route is chosen, the problem becomes serious because the loopholes grow in both scenarios. Lower packet delivery ratios

(PDR) occur from the likelihood that the packets won't make it to the destination at the conclusion of the sequence due to a route break. As a result, the system is frequently compelled to take different paths, which increases latencies. The algorithms that take into account various traffic densities during CH lifespan are inevitably influenced by traffic density. When there are more automobiles, the lifetime lengthens. This results from the network topology frequently changing as the number of vehicles increases. Studies for an effective routing protocol in vehicular ad hoc networks that satisfies the demand of the contemporary intelligent transportation system may be found by the researcher. In order to reach this milestone, a quick survey of VANETs was carried out. We evaluated each protocol's benefits and drawbacks and determined the reasons for any shortcomings. The studied operating situation, performance metrics, simulation tool, number of nodes, and speed, density, and simulation duration of the suggested protocols are summarized in a table. It was discovered that the primary causes of inconsistent performance in a dynamic environment of VANETs are the realistic mobility model and the design of protocols to function in a given operational situation.

6 Conclusion

This work aims to enhance comprehension in the realm of SDN-based VANET systems by studying their intricacies. The article delves into the application of SDN-based VANETs with AI-driven Intelligent Reflective Surfaces (IRS), utilizing technologies like Deep Reinforcement Learning (DRL) and collaborative learning. These SDN-based VANETs offer superior communication compared to conventional VANETs. Moreover, by integrating SDN controllers with IRS, security concerns can be managed, enabling long-range vehicle connections. The article explores the PSO and DPC algorithms, along with the selection of Cluster Heads (CHs) in vehicles. Future suggestions encompass a joint latency-energy optimization communication system for VANETs, prioritizing data transmission efficiency. Addressing privacy, the article acknowledges the security challenge in vehicular communication. It proposes the potential development of a hybrid clustering algorithm, amalgamating the strengths of PSO and DPC, to enhance routing in SDN-based VANETs.

References

1. G. Singh, D. Bansal, and S. Sofat, "Intelligent Transportation System for Developing Countries A Survey," International Journal of Computer Applications, vol. 85, no. 3, pp. 34–38, Jan. 2014, doi: https://doi.org/10.5120/14824-3058.
2. Y. Tang, N. Cheng, W. Wu, M. Wang, Y. Dai, and X. Shen, "Delay-Minimization Routing for Heterogeneous VANETs With Machine Learning Based Mobility Prediction," IEEE Transactions on Vehicular Technology, vol. 68, no. 4, pp. 3967–3979, Apr. 2019, doi: https://doi.org/10.1109/tvt.2019.2899627.

3. M. Afrashteh and S. Babaie, “A Route Segmented Broadcast Protocol Based on RFID for Emergency Message Dissemination in Vehicular Ad-Hoc Networks,” IEEE Transactions on Vehicular Technology, vol. 69, no. 12, pp. 16017–16026, Dec. 2020, doi: https://doi.org/10.1109/tvt.2020.3041754.
4. K. Kandali and H. Bennis, “An Efficient Routing Protocol Using an Improved Distance-Based Broadcasting and Fuzzy Logic System for VANET,” International Journal of Intelligent Engineering and Systems, vol. 13, no. 6, pp. 81–93, Dec. 2020, doi: https://doi.org/10.22266/ijies2020.1231.08.
5. L. Wu et al., “An Efficient Privacy-Preserving Mutual Authentication Scheme for Secure V2V Communication in Vehicular Ad Hoc Network,” IEEE Access, vol. 7, pp. 55050–55063, 2019, doi: https://doi.org/10.1109/access.2019.2911924.
6. T. S. J. Darwish, K. Abu Bakar, and K. Haseeb, “Reliable Intersection-Based Traffic Aware Routing Protocol for Urban Areas Vehicular Ad Hoc Networks,” IEEE Intelligent Transportation Systems Magazine, vol. 10, no. 1, pp. 60–73, 2018, doi: https://doi.org/10.1109/mits.2017.2776161.
7. T. Zhang, Z. Shen, J. Jin, A. Tagami, X. Zheng, and Y. Yang, “ESDA: An Energy-Saving Data Analytics Fog Service Platform,” Lecture Notes in Computer Science, pp. 171–185, 2019, doi: https://doi.org/10.1007/978-3-030-33702-5_13.
8. F. Yang, J. Han, X. Ding, Z. Wei, and X. Bi, “Spectral Efficiency Optimization and Interference Management for Multi-Hop D2D Communications in VANETs,” IEEE Transactions on Vehicular Technology, vol. 69, no. 6, pp. 6422–6436, Jun. 2020, doi: https://doi.org/10.1109/tvt.2020.2987526.
9. A. Al-Dulaimi, S. Al-Rubaye, and Q. Ni, “Energy Efficiency Using Cloud Management of LTE Networks Employing Fronthaul and Virtualized Baseband Processing Pool,” IEEE Transactions on Cloud Computing, vol. 7, no. 2, pp. 403–414, Apr. 2019, doi: https://doi.org/10.1109/tcc.2016.2564405.
10. S. Al-Rubaye, A. Al-Dulaimi, J. Cosmas, and A. Anpalagan, “Call Admission Control for Non-Standalone 5G Ultra-Dense Networks,” IEEE Communications Letters, vol. 22, no. 5, pp. 1058–1061, May 2018, doi: https://doi.org/10.1109/lcomm.2018.2813360.
11. Q. Pan, J. Wu, J. Nebhen, A. K. Bashir, Y. Su, and J. Li, “Artificial Intelligence-Based Energy Efficient Communication System for Intelligent Reflecting Surface-Driven VANETs,” IEEE Transactions on Intelligent Transportation Systems, vol. 23, no. 10, pp. 19714–19726, Oct. 2022, doi: https://doi.org/10.1109/tits.2022.3152677.
12. H. Han, J. Zhao, D. Niyato, M. D. Renzo, and Q.-V. Pham, “Intelligent Reflecting Surface Aided Network: Power Control for Physical-Layer Broadcasting,” ICC 2020 - 2020 IEEE International Conference on Communications (ICC), Jun. 2020, doi: https://doi.org/10.1109/icc40277.2020.9148827.
13. Y. Cai, Z. Wei, S. Hu, D. W. K. Ng, and J. Yuan, “Resource Allocation for Power-Efficient IRS-Assisted UAV Communications,” 2020 IEEE International Conference on Communications Workshops (ICC Workshops), Jun. 2020, doi: https://doi.org/10.1109/iccworkshops49005.2020.9145224.
14. C. Huang et al., “Holographic MIMO Surfaces for 6G Wireless Networks: Opportunities, Challenges, and Trends,” IEEE Wireless Communications, vol. 27, no. 5, pp. 118–125, Oct. 2020, doi: https://doi.org/10.1109/mwc.001.1900534.
15. G. Raja, A. Ganapathisubramaniyan, S. Anbalagan, S. B. M. Baskaran, K. Raja, and A. K. Bashir, “Intelligent Reward-Based Data Offloading in Next-Generation Vehicular Networks,” IEEE Internet of Things Journal, vol. 7, no. 5, pp. 3747–3758, May 2020, doi: https://doi.org/10.1109/jiot.2020.2974631.
16. S. Anbalagan, D. Kumar, G. Raja, and A. Balaji, “SDN assisted Stackelberg Game model for LTE-WiFi offloading in 5G networks,” Digital Communications and Networks, vol. 5, no. 4, pp. 268–275, Nov. 2019, doi: https://doi.org/10.1016/j.dcan.2019.10.006.
17. M. Liyanage, M. Dananjaya, J. Okwuibe, and M. Ylianttila, “SDN based operator assisted offloading platform for multi-controller 5G networks,” 2017 IEEE International Symposium on Local and Metropolitan Area Networks (LANMAN), Jun. 2017, doi: https://doi.org/10.1109/lanman.2017.7972134.

18. S. Anbalagan, D. Kumar, D. Ghosal, G. Raja, and M. V, "SDN-Assisted Learning Approach for Data Offloading in 5G HetNets," Mobile Networks and Applications, vol. 22, no. 4, pp. 771–782, Mar. 2017, doi: https://doi.org/10.1007/s11036-017-0838-5.
19. Q. Wu and R. Zhang, "Towards Smart and Reconfigurable Environment: Intelligent Reflecting Surface Aided Wireless Network," IEEE Communications Magazine, vol. 58, no. 1, pp. 106–112, Jan. 2020, doi: https://doi.org/10.1109/mcom.001.1900107.
20. G. Raja, S. Anbalagan, G. Vijayaraghavan, P. Dhanasekaran, Y. D. Al-Otaibi, and A. K. Bashir, "Energy-Efficient End-to-End Security for Software-Defined Vehicular Networks," IEEE Transactions on Industrial Informatics, vol. 17, no. 8, pp. 5730–5737, Aug. 2021, doi: https://doi.org/10.1109/tii.2020.3012166.
21. Mchergui, T. Moulahi, and S. Zeadally, "Survey on Artificial Intelligence (AI) techniques for Vehicular Ad-hoc Networks (VANETs)," Vehicular Communications, vol. 34, p. 100403, Apr. 2022, doi: https://doi.org/10.1016/j.vehcom.2021.100403.
22. M. Arif, G. Wang, M. Zakirul Alam Bhuiyan, T. Wang, and J. Chen, "A survey on security attacks in VANETs: Communication, applications and challenges," Vehicular Communications, vol. 19, p. 100179, Oct. 2019, doi: https://doi.org/10.1016/j.vehcom.2019.100179.
23. S. Kaushik, R. C. Poonia, S. K. Khatri, D. Samanta, and P. Chakraborty, "Transmit Range Adjustment Using Artificial Intelligence for Enhancement of Location Privacy and Data Security in Service Location Protocol of VANET," Wireless Communications and Mobile Computing, vol. 2022, pp. 1–13, Sep. 2022, doi: https://doi.org/10.1155/2022/9642774.
24. Ullah Khan, M. Abul Hassan, A. Abdollahi, and B. Baig, "AI-enabled communication protocols for vehicular ad hoc networks (VANETs): fundamentals and future challenges," Artificial Intelligence of Things in Smart Environments, pp. 19–34, Feb. 2022, doi: https://doi.org/10.1515/9783110760477-002.
25. B. S. and D. J. Aravindhar, "Optimizing Data Transmission in VANET using Fuzzy Logic System with Link Residual Time," 2019 Third International conference on I-SMAC (IoT in Social, Mobile, Analytics and Cloud) (I-SMAC), Dec. 2019, doi: https://doi.org/10.1109/i-smac47947.2019.9032497.
26. P. Wang, B. Di, H. Zhang, K. Bian, and L. Song, "Cellular V2X Communications in Unlicensed Spectrum: Harmonious Coexistence With VANET in 5G Systems," IEEE Transactions on Wireless Communications, vol. 17, no. 8, pp. 5212–5224, Aug. 2018, doi: https://doi.org/10.1109/twc.2018.2839183.
27. G. Luo et al., "Software-Defined Cooperative Data Sharing in Edge Computing Assisted 5G-VANET," IEEE Transactions on Mobile Computing, vol. 20, no. 3, pp. 1212–1229, Mar. 2021, doi: https://doi.org/10.1109/tmc.2019.2953163.
28. Z. Su, M. Dai, Q. Xu, R. Li, and H. Zhang, "UAV Enabled Content Distribution for Internet of Connected Vehicles in 5G Heterogeneous Networks," IEEE Transactions on Intelligent Transportation Systems, vol. 22, no. 8, pp. 5091–5102, Aug. 2021, doi: https://doi.org/10.1109/tits.2020.3043351.
29. Q. Wu and R. Zhang, "Intelligent Reflecting Surface Enhanced Wireless Network: Joint Active and Passive Beamforming Design," 2018 IEEE Global Communications Conference (GLOBECOM), Dec. 2018, doi: https://doi.org/10.1109/glocom.2018.8647620.
30. J. Kang and S. Kim, "Deep Reinforcement Learning based Secure Wireless Communication with Self-powered Intelligent Reflecting Surface," Jul. 2022, doi: https://doi.org/10.22541/au.165873784.43044471/v1.
31. C. Pan et al., "Intelligent Reflecting Surface Aided MIMO Broadcasting for Simultaneous Wireless Information and Power Transfer," IEEE Journal on Selected Areas in Communications, vol. 38, no. 8, pp. 1719–1734, Aug. 2020, doi: https://doi.org/10.1109/jsac.2020.3000802.
32. Z. Wang, L. Liu, and S. Cui, "Channel Estimation for Intelligent Reflecting Surface Assisted Multiuser Communications: Framework, Algorithms, and Analysis," IEEE Transactions on Wireless Communications, vol. 19, no. 10, pp. 6607–6620, Oct. 2020, doi: https://doi.org/10.1109/twc.2020.3004330.

33. T. Shafique, H. Tabassum, and E. Hossain, “Optimization of Wireless Relaying With Flexible UAV-Borne Reflecting Surfaces,” IEEE Transactions on Communications, vol. 69, no. 1, pp. 309–325, Jan. 2021, doi: https://doi.org/10.1109/tcomm.2020.3032700.
34. D. Xu, X. Yu, Y. Sun, D. W. K. Ng, and R. Schober, “Resource Allocation for IRS-Assisted Full-Duplex Cognitive Radio Systems,” IEEE Transactions on Communications, vol. 68, no. 12, pp. 7376–7394, Dec. 2020, doi: https://doi.org/10.1109/tcomm.2020.3020838.
35. T. Bai, C. Pan, Y. Deng, M. Elkashlan, A. Nallanathan, and L. Hanzo, “Latency Minimization for Intelligent Reflecting Surface Aided Mobile Edge Computing,” IEEE Journal on Selected Areas in Communications, vol. 38, no. 11, pp. 2666–2682, Nov. 2020, doi: https://doi.org/10.1109/jsac.2020.3007035.
36. A. K. Sangaiah, A. Javadpour, C.-C. Hsu, A. Haldorai, and A. Zeynivand, “Investigating Routing in the VANET Network: Review and Classification of Approaches,” Algorithms, vol. 16, no. 8, p. 381, Aug. 2023, doi: https://doi.org/10.3390/a16080381.
37. N. Gayathri, H. Anandakumar, S. Gowri, J. Keerthika, and S. Nithyapriya, “An analysis of Network Security Attacks and their mitigation for Cognitive Radio Communication,” 2023 9th International Conference on Advanced Computing and Communication Systems (ICACCS), Mar. 2023, doi: https://doi.org/10.1109/icaccs57279.2023.10112905.
38. Kandali, L. Bennis, O. E. Bannay, and H. Bennis, “An Intelligent Machine Learning Based Routing Scheme for VANET,” IEEE Access, vol. 10, pp. 74318–74333, 2022, doi: https://doi.org/10.1109/access.2022.3190964.

Part IV
Intelligent AI Prediction Techniques for Advanced Applications

Bi-Model Emotional AI for Audio-Visual Human Emotion Detection Using Hybrid Deep Learning Model

1 Introduction

The rise of online business transactions in the current digital era makes it difficult for organizations to comprehend the feelings of their employees and clients because there is no face-to-face interaction. However, the integration of emotional AI into the process has the potential to revolutionize this dynamic. Emotional AI can effectively differentiate between various moods, providing valuable insights. Consequently, it is justifiable for businesses and governments to employ emotional AI to survey individuals, detect their emotions, and respond appropriately [1].Text, video, and audio are the three main subcategories of emotional artificial intelligence. Opinion mining or sentiment analysis are common terms used to describe the process of extracting emotions from text. This procedure comprises analyzing text samples using machine learning and natural language processing techniques to determine and classify the prevailing sentiment, be it neutral, positive, or negative. This approach is used in a variety of fields, such as assessing customer sentiment to filter reviews for products, hotels, and movies. The ability of a single word to elicit a wide range of emotions presents a significant difficulty for text-based emotion identification.

Emotional AI utilizes multiple criteria such as intonation, tone, pitch, pace of voice, and extended pauses to assess a person's sentiment based on their speech. This cutting-edge technology has the ability to detect hidden tension, even in situations where someone uses dry humor or expresses sentiments that contradict their true feelings. It can effectively handle challenges related to sarcasm, exaggeration, and other verbal habits using sophisticated approaches [2]. Voice-based emotional artificial intelligence finds significant application in identifying dissatisfied customers in call centers and directing them to the appropriate personnel who can address their concerns and improve their experience. In customer service, audio-based emotion recognition plays a practical role in predicting reviews based on call

© European Alliance for Innovation 2024

A. Haldorai et al., *Artificial Intelligence for Sustainable Development*,
EAI/Springer Innovations in Communication and Computing,
https://doi.org/10.1007/978-3-031-53972-5_15

interactions. Differentiating between intense anger and excitement can be challenging since both emotions may be conveyed through a high-pitched tone. Additionally, one of the most complex hurdles lies in accurately identifying vocal modulation, which adds another layer of difficulty to the process.

AI face emotion detection leverages facial recognition and computer vision technologies to analyze facial expressions in videos and ascertain the dominant sentiment [3, 4]. It proves particularly useful in discerning between instances of lying and truth-telling. In practical applications, this form of emotional artificial intelligence can be employed during interviews to gauge candidates' emotional states, such as nervousness, confidence, or sincerity. Financial organizations can utilize it as a screening tool for loan applications, among other use cases. However, there are inherent challenges associated with capturing clear facial images in dynamic settings where lighting and background conditions may vary. These variations can hinder the accuracy of emotion detection. Additionally, some individuals are naturally less expressive, which presents an added challenge in detecting subtle emotional cues. Despite these complexities, ongoing advancements in technology are aimed at mitigating these limitations and improving the reliability of AI face emotion detection systems.

Emotion detection based on AI is used in advertising marketing, healthcare, and customer service. AI emotion detection algorithms assist marketers in determining which ads resonate most with their target audience and what characteristics to incorporate in their videos to achieve the best results [5]. Affectiva, a Boston-based emotional AI business, provides one example. It created a system to assist marketers in gauging client reactions to various adverts. The program monitors users' body language, facial expressions, and voice with their permission to comprehend their emotional state [6]. Emotional support AI is employed in the healthcare sector to aid doctors and patients alike. Within this domain, medical professionals utilize this technology in conjunction with voice assistants to determine stress levels and provide appropriate responses. It actively monitors patients' moods within assessment rooms and even during surgical procedures. In call centers, companies leverage emotional recognition AI to enhance the customer experience. These solutions offer several benefits [7]. They enable the selection of the most suitable agent for each client, offer real-time feedback to agents, and alert them when their composure begins to falter, enabling them to respond appropriately to dissatisfied customers.

Recognizing various emotions still remains a skill dominated by humans, but artificial intelligence (AI) is quickly catching up due to its ability to analyze vast amounts of data. Emotional AI encompasses a range of approaches. It has the capability to process human speech and identify vocal intonations that signify emotions like anger, stress, and more. Moreover, this technology can detect subtle facial micro-expressions that occur fleetingly, too quick for the human eye to perceive. Emotion recognition entails machines attempting to analyze, interpret, and categorize human emotions based on facial features. It is considered one of the most challenging tasks in high-level visual processing due to the inherent gap between low-level image pixels and complex emotions [8, 9]. Nevertheless, visual emotion analysis shows great promise as it plays a crucial role in advancing powerful artificial intelligence [10].

This chapter focuses on the fundamental processes that underpin emotion recognition with AI. The following steps constitute the core of the proposed approach: Step 1: Obtain an image frame from a camera feed, which can originate from various sources such as IP cameras, CCTV systems, or USB cameras. Step 2: Perform image preprocessing tasks, including cropping, resizing, rotating, and color correction, to enhance the quality and consistency of the image. Step 3: Utilize a Convolutional Neural Network (CNN) model to extract essential features from the preprocessed image. Step 4: Employ the extracted features to identify and categorize emotions exhibited in the image. This chapter's primary contributions lie in presenting these sequential processes as the foundation for emotion recognition with AI. By outlining these steps, the chapter aims to provide a comprehensive framework for effectively analyzing emotions through artificial intelligence techniques.

- Recognizing and understanding the various opinions the customer mind thoughts from the audio-based data.
- Identifying the emotions of the human from the image or video with AI algorithm.
- Implementing a hybrid deep learning architecture for bi-model emotion detection.

Section 1 explains the basic details and introduction about the emotion detection. The works related to emotion detection is detailly discussed in Sect. 2. Similarly, Sect. 3 and Sect. 4 present the proposed work and experimental setup clearly. Section 5 delves into the discussion and analysis of the obtained results, while Sect. 6 encapsulates the conclusion and outlines avenues for future work.

2 Literature Review

In their study, Chen et al. [11] employed auto-encoder architecture to recognize facemask emotions during humanoid–robot interactions. Their approach utilized the SRDSAN method to decrease falsification and enhance training efficacy and dimensional difficulty. Additionally, the DSAN technique facilitated exact feature selection, while soft-max regression aided in classifying the input signals. An emotive recognition system that combines facial expressions and electroencephalogram (EEG) data was developed and implemented by Hassouneh et al. [12]. Their approach makes use of deep neural network and machine learning techniques to accurately and instantly recognize and analyze emotions. Their work involved the utilization of the optical flow technique to detect facial regression using virtual markers. This technique proved beneficial for physically challenging individuals due to its ability to recognize fewer computational complexities. In their research, Ekundayo et al. [13] implemented a multilabel Convolutional Neural Network

(CNN) for facemask expression acknowledgment and ordinal intensity valuation. While several existing emotive recognition practices, such as FER, function effectively, none of them are suitable for achieving faultless multiclass emotive classification. To address this, Ekundayo et al. employed a multi-label convolutional neural network in their paper. Jayanthi et al. [14] developed an integrated emotional classification system that combined speech and static images using deep classifier fusion. Feeling identification plays a crucial role in determining an individual's stress level. The system focused on two key characteristics: emotion detection and speech modulation.

Babajee et al. [15] used a deep learning technique in their study to identify face cues from human expressions. This chapter presents seven categories for facial emotion recognition, including happy and sad. As such, the chapter provides a database of 32,398 examples for gathering different types of emotional identification with the help of the Facial Action Coding System (FACS). Notably, this study does not serve as an optimization technique; rather, it only concentrates on the identification strategy. In contrast, Tan et al. [16] implemented short-term emotion recognition and comprehension via neuro-sensing. To examine spatiotemporal EEG patterns, they used a spiking neural network model. This study introduces the novel use of the SNN approach, contributing to the understanding of brain system functionality. Hossain and Muhammad [17] developed an acoustic-pictorial emotion recognition system using deep learning and a cognitive wireless architecture in the context of Internet-based medical care. This system recognizes patients' emotions on its own and adds a great deal of value to the advancement of online healthcare. The researchers conducted tests to demonstrate the system's efficacy within the framework.

Yadahalli et al. [18] employed a deep-learning method to detect facial micro-expressions in their study. They collected a dataset comprising eight layers, encompassing six different emotions: happiness, sorrow, rage, fear, neutrality, and startle. The gathered dataset facilitated the development of the FER model, indicating that the use of a convolutional neural network enhances accuracy. The study achieved multimodal face expression analysis using a single algorithm, simplifying the output process. Li et al. [19] conducted a study on deep facial expression recognition, highlighting the challenges posed by facial expression tracking in network systems. The lack of training groups and variations in facial appearances were identified as major challenges. To address these issues, the researchers organized the dataset using a neural pipeline technique, reducing the complexity associated with FER. Khalil et al. [20] summarized research on deep learning technology for speech-based emotion recognition and highlighted its capability for speech-based sentiment recognition. Their simulation of multimodal sentiment recognition demonstrated efficient data input through audio-visual and other modalities. Yang et al. [21] employed a Stacked Autoencoder. They utilized EEG signals and discrete entropy computation methods for signal measurement. Overall, deep learning systems exhibited favorable performance across various emotional recognition classes.

In their research, Ben et al. [22] proposed a method based on video analysis to examine micro-expressions on the face. Deep learning algorithms have been extensively studied and applied in various domains, such as voice-based image recognition, demonstrating promising potential for speech and image analysis. A dual neural network was created by Ntalampiras [23] to recognize emotions through speech. For improved performance, this model blends with learning analogies. The study showcased the effectiveness of their framework under dynamic conditions, and analyzing the activation graph at each layer provided valuable insights into the model's predictions. Other notable studies have also contributed to emotion recognition and similarity measurement. In their research, Bejani et al. [24] utilized various audio features.

Elmadany et al. [25] introduced MLPCCA (Multi-set Locality-Preserving CCA) for audio-visual emotion recognition in multi-view learning settings. Zhang et al. [26] employed two pretrained deep convolutional neural networks (DCNN) on audio and visual features. Traditional classifiers like SVM were utilized for audio-visual recognition. Recent advances in deep learning for automatic facial emotion recognition were examined by Mellouk and Handouzi [27]. Their examination showed a focused effort on the part of researchers to apply technology for the deciphering and analysis of facial expressions, extracting information that would improve computer-generated predictions. The results of the study demonstrated how well deep learning algorithms recognize emotions. Please refer to Table 1 for an extensive summary of the studies that have been done in the area of emotion detection.

Table 1 Emotion detection

Reference	Dataset description	Predicted emotion
MultiPie [28]	750,000 images captured with 19 illumination	Rage, revulsion, indifferent, joyful, narrowed eyes, shriek, astonishment
SFEW [29]	700 images	Six emotions
CK+ [30]	593 videos for various expressions	Encompasses six fundamental emotions
GEMEP FERA [31]	289 images in various pose	Anger, fear, sadness, relief, happy
FER2013 [32]	35,887 grayscale images	Encompasses six fundamental emotions
BU-3DFE [33]	500 3D facial images at −45°, +45°	Six emotions
CASME II [34]	247 micro-expressions	Happy, surprise, regression and others
Oulu-CASIA [35]	Different illumination	Encompasses six fundamental emotions and neutral
AffectNet [36]	440,000 samples	Encompasses six fundamental emotions and neutral
RAFD-DB [37]	30,000 samples	Encompasses six fundamental emotions and neutral
RaFD [38]	8040 samples with different face poses	Encompasses six emotions and neutral

3 Proposed Model

Using a deep learning system that can distinguish between various emotions based on facial and aural cues is the main goal of this chapter. We will investigate a novel hybrid CNN model in this study to identify emotions from audio and visual inputs. The parameter sharing method of CNN, which greatly lowers the number of operating parameters, makes it a popular choice for image processing [39]. Getting data input is the primary job of the input layer of a CNN. Before being sent to the neural network, speech or picture data is usually encoded into feature vectors.

The input from the previous layer is subjected to the convolution process in the following convolution layer. By establishing local and global connections, this process allows CNN to reduce the number of factors involved and improve learning efficiency [40, 41]. The linear rectification and pooling layers use a multilayer convolution approach to down sample the data once the low-level features have been recovered from the data. This pooling not only significantly reduces the network's training parameters but also somewhat improves the model's fit. Finally, the input data is passed through fully connected layers to the neurons, and the final result is obtained from the output layer. Figure 1 illustrates the proposed approach.

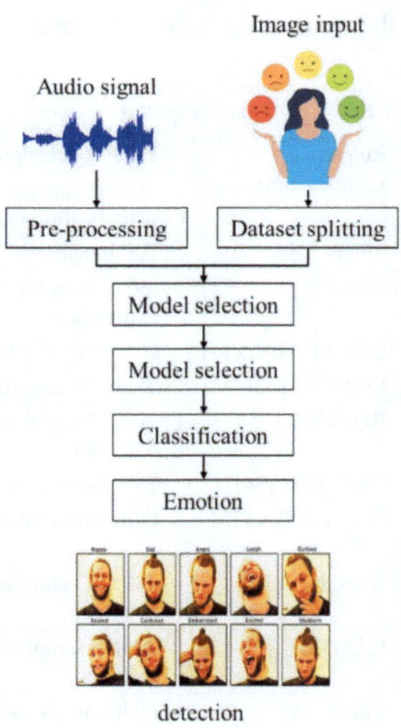

Fig. 1 Image/audio emotion detector

An overview of the proposed method describes the first image preprocessing phase that lowers image noise. Kaggle datasets are selected based on their preprocessing and pre-classification. Then, using both visual and auditory inputs, a hybrid CNN model is used to identify emotions in these previously processed images. Convolutional Neural Networks (CNNs), which are specifically engineered for image classification tasks, are the deep learning algorithm utilized. The CNN uses convolution to process the input picture through the RELU layer, then the max-pooling algorithm. In order to extract significant features, this sequence is repeated several times. After that, the output image is sent to the fully linked layer, which improves output accuracy and aids in image classification. The RELU layer plays a crucial role in performing the multilayer function of the CNN. The convolutional layer, commonly utilized in image classification, assists in identifying different facial expression patterns through the RELU activation function. The data is then passed through the maximum pooling layer, which helps in reducing overfitting, dimensionality of the image, and analyzing erroneous data.

In this study, the emotion detector employs a hybrid CNN model to recognize emotion from image/audio. The suggested emotion detector employing the CNN-LSTM model is a hybrid of the CNN and LSTM architectures. The basic CNN architecture is displayed in Fig. 2. An input layer, convolution layer, LSTM layers, three fully connected (FC) layers, and a softmax output layer comprise the proposed CNN-LSTM model. The convolution layer collects low-level input features, while the LSTM layer extracts high-level input features. To successfully extract features from the dataset, the convolution layer is provided. In addition, the extracted features are continuously processed by the LSTM layers to extract high-level features of Tamil handwritten characters. The output of the LSTM layer is passed into the fully connected layer for final Tamil character identification. Four FC layers are envisaged in the proposed design. The output layer with the Softmax function is the last FC layer. The Adam optimizer optimizes the remaining three FC layers.

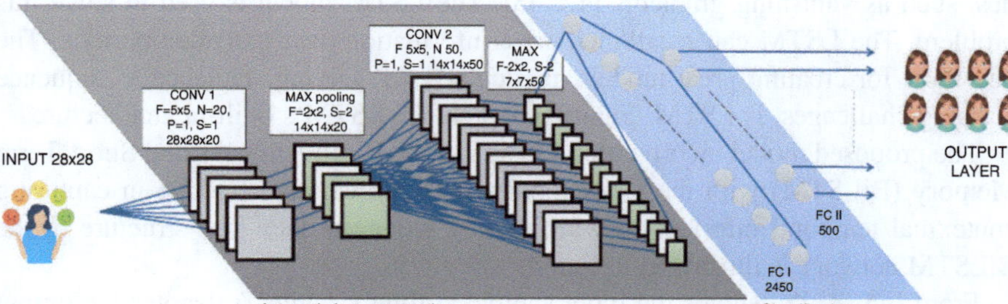

Fig. 2 Traditional CNN architecture

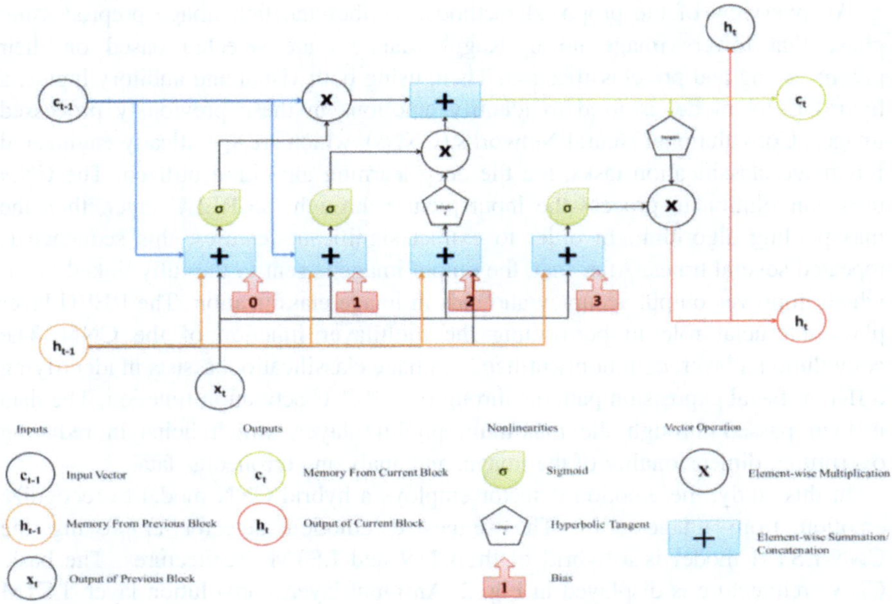

Fig. 3 LSTM model

One of the RNN architectures is LSTM. The RNN model and sequential data processing are the ideal architecture for remembering previous information like a human brain. The typical deep learning models have inputs such as the hidden and output units, which process the input data independently of the previous one. As a result of the model's inability to memorize, this standard model comprises a variety of weights and biases. However, the RNN uses the same weights throughout the process because it can remember. The fundamental CNN model has some difficulties, such as vanishing gradients [42, 43]. The LSTM model is used to solve this problem. The LSTM can recall/memorize information over extended periods. The key layer for creating great models, particularly for solving sequence to sequence learning challenges, is LSTM. Figure 3 depicts the LSTM's built-in architecture.

The proposed model incorporates a fusion of the Bi-directional Long Short-Term Memory (BiLSTM) with the CNN model. It is particularly effective in capturing contextual temporal information within voice sequence data. The structure of the BiLSTM network is illustrated in Fig. 4.

From Fig. 3, *Xt* denotes the input sample at time *t*, while *Yt* denotes the output sample at time *t*. The update state of unit information can be expressed as Eq. (1) using the forgetting gate circuit.

$$f_t = \sigma\left(W_f . [h_{t-1}, x_t] + b_f \right. \quad (1)$$

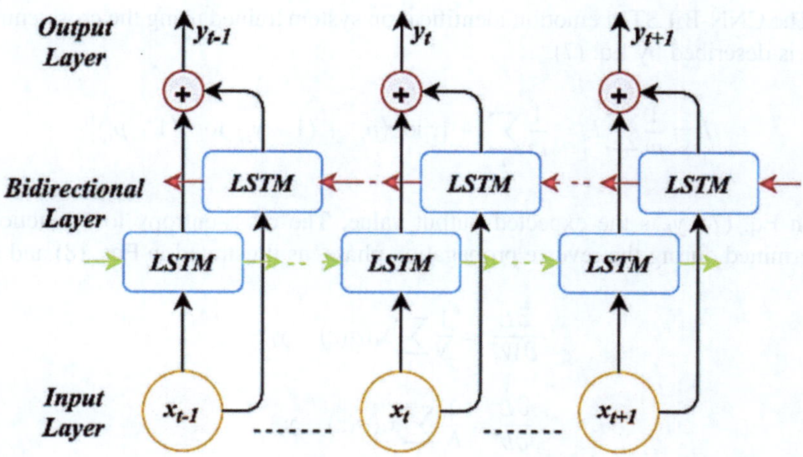

Fig. 4 BiLsTM architecture

The sigmoid activation function, represented by σ, is represented by the weights and bias settings for the forgetting gate, respectively, in Eq. (1) above. The unit's status must be updated by the input gate controller. After processing the current input data (*xt*) and the hidden state data (*ht*–1), the tan *h* system layer compresses the output and creates the state vector *C*, which can be appended. To identify which state has to be changed, the sigmoid activation function allows the sigmoid network to assign a value to each input that falls between 0 and 1. Equations (2) and (3), where W_i and W_c are the input gate connection weights, b_i and b_c are the bias.

$$i_t = \sigma(W_i.[h_{t-1}, x_t] + b_i \tag{2}$$

$$C_t = \tan\, h\, (W_c.[h_{t-1}, x_t] + b_c \tag{3}$$

As demonstrated in Eq. (4), the C_t is the sum of the previous state C_{t-1} and the updated value $i_t * C_t$.

$$C_t = f_t * C_{t-1} + i_t * C_t \tag{4}$$

The cells change under the influence of the tan *h* network layer to take on a new state depending on the information that was previously input. The decision of which part of the cell state to output is mostly determined by the sigmoid network layer, as shown by Eqs. (5) and (6), where W*l* and b*l* denote the weights and bias.

$$o_t = \sigma(W_o.[h_{t-1}, x_t] + b_o \tag{5}$$

$$h_t = o_t * \tan\, h\, (C_t) \tag{6}$$

The CNN-BiLSTM emotion identification system trained using the cross-entropy loss is described by Eq. (7).

$$L = \frac{1}{m}\sum_i L_i = \frac{1}{m}\sum_i - \ [y_i \log(p_i) + (1 - y_i)\ \log\ (1 - p_i)] \tag{7}$$

In Eq. (7), y_i is the expected output value. The cross-entropy loss function is determined during the reverse propagation phase, as illustrated in Eqs. (8) and (9).

$$\frac{\partial L}{\partial W_i} = \frac{1}{N}\sum_x x_i(\sigma(z) - y_i) \tag{8}$$

$$\frac{\partial L}{\partial b} = \frac{1}{N}\sum_x x_i(\sigma(z) - y_i) \tag{9}$$

Obviously, when the error $\sigma(z) - y_i$ is large, the weight update is fast, and when the error $\sigma(z) - y_i$ is small, the weight update is slow. The BiLSTM network's forward propagation can be expressed as the following Eqs. (10, 11, 12, 13, and 14).

$$r_t = \sigma(W_r.[h_{t-1}, x_t]) \tag{10}$$

$$Z_t = \sigma(W_Z.[h_{t-1}, x_t] \tag{11}$$

$$h_t = \tan h\ (W_h.[r_t * h_{t-1}, x_t] \tag{12}$$

$$h_t = (1 - Z_t) * h_{t-1} + Z_t * h_t \tag{13}$$

$$y_t = \sigma(W_0.h_t) \tag{14}$$

The connection between two vectors is shown by the symbol * in the equation [*ht* – *0*, *xt*]. Furthermore, the input data at the moment is represented by *xt*, the candidate set from the previous iteration is represented by *ht* – 0, and the connection weights for the update and reset gates are *Wz* and *WZ*. After determining the final output, Eq. (15) determines the loss (Fig. 5).

$$E_t = \frac{1}{2}\left(y_d - y_t^0\right)^2 \tag{15}$$

3.1 Transfer Learning

The CNN-Bi-LSTM hybrid design is illustrated in Fig. 4. In order to expedite the processing time, we employed a technique called transfer learning for emotion recognition. Transfer learning is a machine learning technique that exploits the

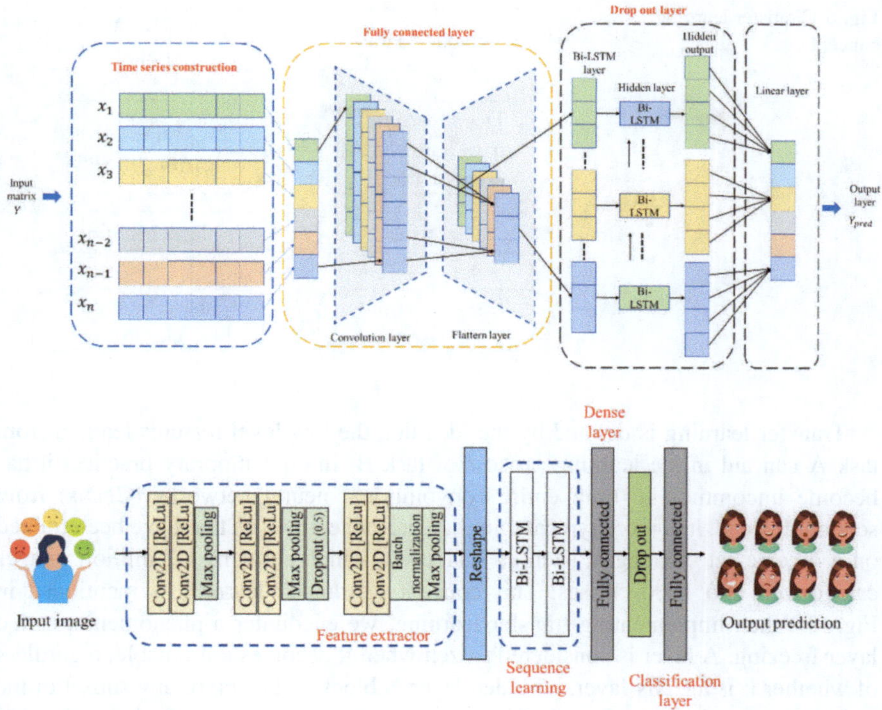

Fig. 5 (**a**) CNN-Bi-LsTM. (**b**) CNN-Bi-LsTM layer architecture

knowledge acquired by a model during training on one task to facilitate the training of a different but related task. This method is particularly useful when the second task shares similarities with the first task or when there is a scarcity of data available for the second task. By utilizing the learned features from the initial task as a starting point, the model can enhance its learning efficiency and effectiveness when tackling the subsequent task. Moreover, transfer learning aids in mitigating overfitting since the model has already captured general characteristics that are applicable to the new task. An interesting observation in deep neural networks trained on image data is that the early layers of the network tend to learn fundamental features such as detecting edges, recognizing colors, and capturing intensity variations. These low-level features serve as a foundation for higher-level representations. These features appear to be independent of any particular dataset or task, as they are present regardless of whether the image being processed involves detecting a lion or a car. Thus, it is crucial to identify and incorporate these low-level traits in both scenarios.

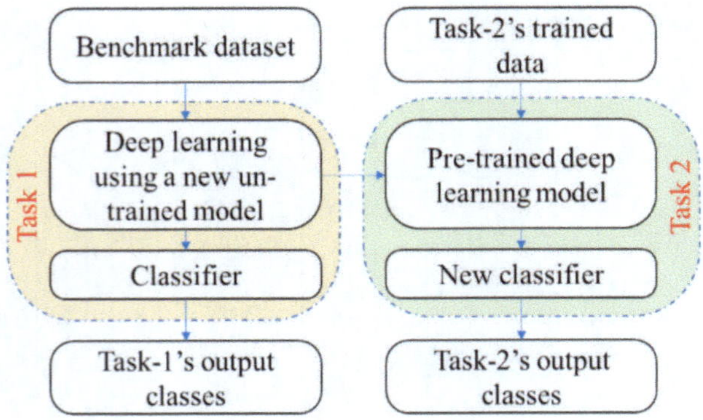

Fig. 6 Transfer learning concept

Transfer learning is defined by the idea that the low-level features learned from task A can aid in the learning process of task B. In contemporary practice, it has become uncommon to train entire convolutional neural networks (CNNs) from scratch. Instead, it is more typical to utilize pretrained models that have been trained on a diverse set of images, such as ImageNet, which contains 1.2 million images categorized into 1000 classes. This concept of transfer learning is mentioned in Fig. 6. When implementing transfer learning, we encounter a phenomenon called layer freezing. A layer is considered frozen when it becomes untrainable, regardless of whether it is a CNN layer, a hidden layer, a block of layers, or any subset of the overall layer set. Consequently, the weights of frozen layers remain unchanged during the training process, while nonfrozen layers undergo regular training procedures. In the context of solving a problem using transfer learning, we initiate the process with a pretrained model as our starting point.

There are two approaches for leveraging the knowledge of a pretrained model. The first method involves freezing a portion of the pretrained model's layers and training only the remaining layers on a new dataset specific to the task at hand. This allows modification of the trainable layers to suit the characteristics of the new dataset while retaining valuable learned features from the pretrained model. The second method entails constructing a new model that incorporates selected features extracted from the layers of the pretrained model, utilizing these extracted features in the newly created model. In both cases, certain learned characteristics of the pretrained model are discarded, and the remaining components are fine-tuned using the new dataset. During this training process, the model is adjusted to accommodate the features of the fresh dataset, ensuring that shared features across tasks are incorporated and the transfer learning technique is effectively employed. In the context of the suggested study on emotion identification, the appropriate strategy will be applied to optimize the model's performance.

4 Experimental Setups

Whether the input data is auditory or visual, each convolutional layer in the network executes convolutional operations to change it. Each layer's output is then transferred to the following layer for additional processing. The purpose of these convolutional layers is to identify and extract different patterns from the input. In the case of the initial CNN used for backdrop removal, the input image typically contains not only the face but also shapes, edges, textures, and objects. Four filters – a corner detector, a circle detector, and an edge detector – are applied in the first convolutional layer to process the input image. These filters are used to highlight and identify particular patterns within the image. After face detection, the next part of the CNN applies more filtering methods, like median and Gaussian noise filters inside its filters. The eyes, ears, lips, nose, cheeks, and other distinguishing aspects of the face can all be identified with the use of these procedures. By employing these filters, the CNN can focus on extracting relevant information related to facial characteristics and enhancing the accuracy of facial feature recognition.

To diminish the data space's dimensionality, the chapter employs kernel filters. Face feature extraction, a crucial method in emotional recognition, is approached through geometric, hybrid, comprehensive, and template-based methods. Following a preprocessing step, the feature extraction technique is applied, and the resulting features are input into a convolutional neural network (CNN). The research aligns with how the specified feature extraction technique enhances the CNN model for facial emotion identification. These extracted features are subsequently input into the Bi-LSTM architecture. Figure 7 illustrates the proposed architecture, incorporating transfer learning techniques. The first dataset encompasses diverse facial

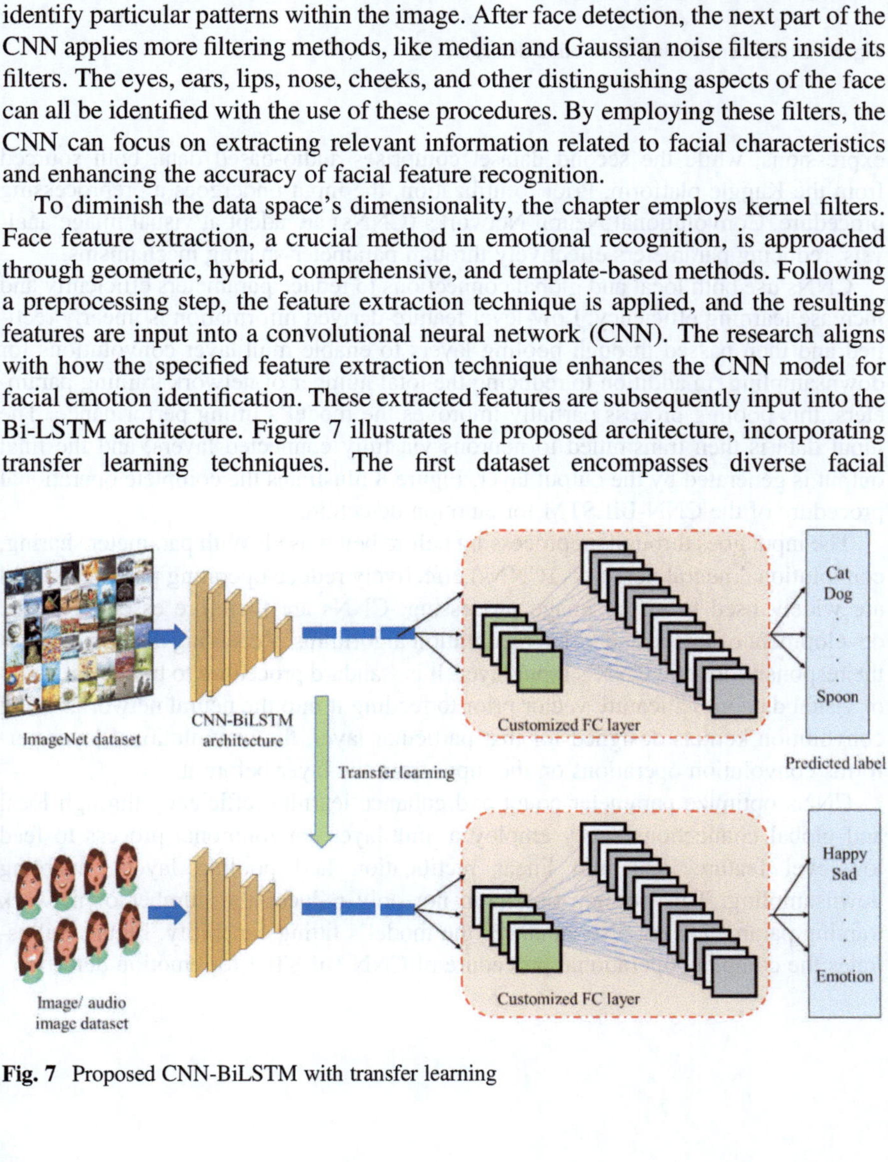

Fig. 7 Proposed CNN-BiLSTM with transfer learning

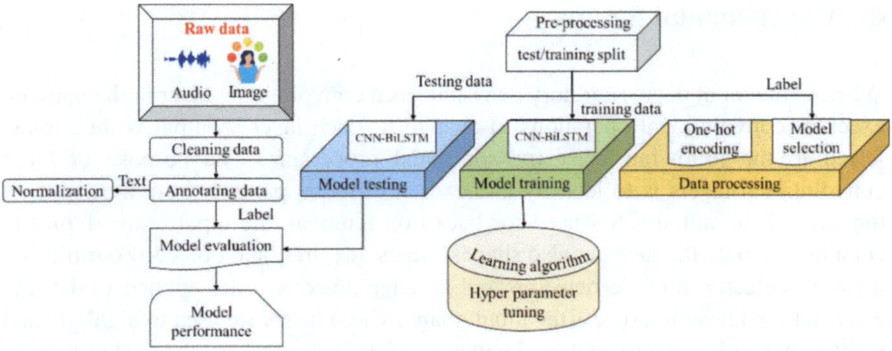

Fig. 8 Experimental setup of emotion detection

expressions, while the second dataset comprises audio-based data, both sourced from the Kaggle platform. Prior to utilization, the input undergoes a preprocessing procedure. Convolutional Neural Networks (CNNs) are adept at visual image analysis, reducing parameters effectively through parameter-sharing mechanisms.

CNNs use both local and global connections to reduce parameters efficiently and increase learning efficiency. Low-level feature-derived information is linearly rectified and then passed through pooling layers to enable multilayer convolutions for downsampling. In addition to reducing the total number of network training parameters, this pooling process partially improves the model's fitting performance. The input data is then transmitted to neurons via fully connected layers, and the final output is generated by the output layer. Figure 8 illustrates the complete operational procedure of the CNN-BiLSTM for emotion detection.

The input goes through preprocessing before being used. With parameter sharing, convolutional neural networks (CNNs) effectively reduce operating parameters and are widely used in visual image processing. CNNs are therefore essential to the development of image and video recognition algorithms. Receiving the data input is the responsibility of a CNN's input layer. It is standard procedure to transform audio or visual data into a feature vector prior to feeding it into the neural network. Using convolution kernels designed for that particular layer, the convolutional layer performs convolution operations on the input from the layer before it.

CNNs optimize parameter count and enhance learning efficiency through local and global connections. They employ a multilayer convolutional process to feed low-level feature data into linear rectification and pooling layers, enabling downsampling. This pooling operation not only reduces the number of network training parameters but also enhances the model's fitting capability. Figure 8 illustrates the complete operational procedure of CNN-BiLSTM for emotion detection.

5 Result and Discussion

This research uses the prominent dataset Speech Emotion Recognition (en), which contains 7442 original clips from 91 performers of various ages, races, and ethnicities. The actors spoke a series of 12 statements, each accompanied by one of 6 emotions (anger, disgust, fear, happiness, neutral, and sadness). Crema, Ravdess, Savee, and Tess are the four most common datasets in English in Speech Emotion Recognition (en). Each one includes audio in.wav format with some key labels.

The dataset consists of seven primary emotions: happy, fearful, angry, disgusted, surprised, sad, or neutral. SAD – sadness; ANG – anger; DIS – disgust; FEA – fear; HAP – happiness; and NEU – neutral are all referenced in Crema. Tess is very similar to Crema in that the file name includes the emotion label. The audio files in this dataset are called in Savee in such a way that the prefix letters explain the emotion classes as follows: “a” = “anger,” “d” = “disgust,” “f” = “fear,” “h” = “happiness,” “n” = “neutral,” ““sa” = “sadness,”” “su” = “surprise.” Emotions are described in Ravdess as 01 = neutral, 02 = calm, 03 = pleased, 04 = sad, 05 = furious, 06 = afraid, 07 = disgusted, and 08 = surprised. The dataset sample counts are displayed in Fig. 9. The dataset samples are displayed in as wave plot and spectrogram in Figs. 10.

The audio dataset is passed through the preprocessing steps like audio dataset sample, noise audio, stretched audio, shifted audio, and pitched audio. The preprocessing steps are presented in Fig. 11.

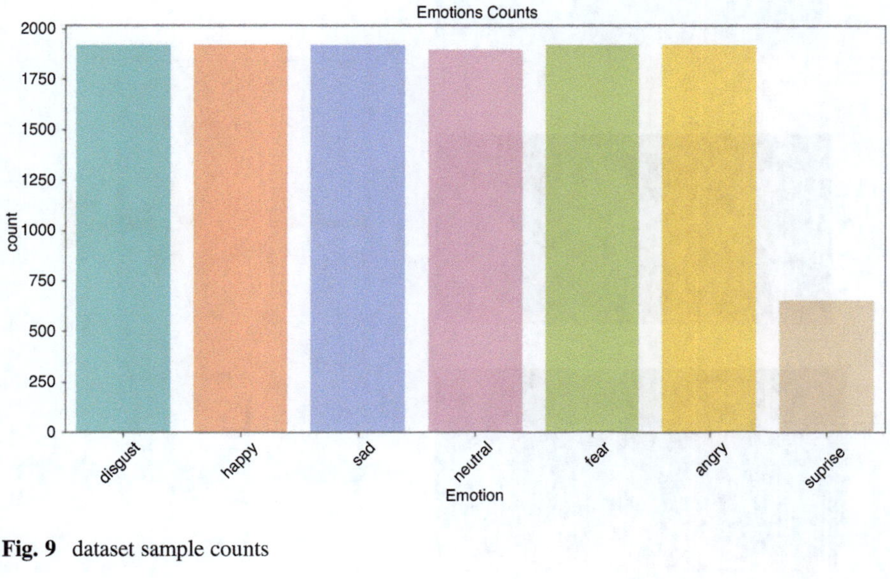

Fig. 9 dataset sample counts

The image-based dataset is considered from FER – 2013 dataset with seven emotion types. The dataset comprises 35,685 instances of 48 × 48 pixel grayscale facial images, which are divided into training and testing sets. These images are classified according to the emotional expressions displayed, including happiness, neutrality, sadness, anger, surprise, disgust, and fear. The sample image count and images shown in Fig. 12 and Fig. 13.

Eighty percent of the data is input for the training phases, whereas 20% is considered for the testing phase. As a result, the preprocessing procedure aids in the reduction of noise in the filter. The feature selection aids in reducing the filter's high dimensionality. To detect emotion, this work employs a hybrid CNN-BiLSTM with transfer learning approach. As a result, the system generates output images with precise dimensional quality and purity. Concurrent feature extraction and data classification are steps in the process. Features from both the testing and training phases are transmitted after the categorization. Figure 14 shows the training plot that demonstrates this procedure.

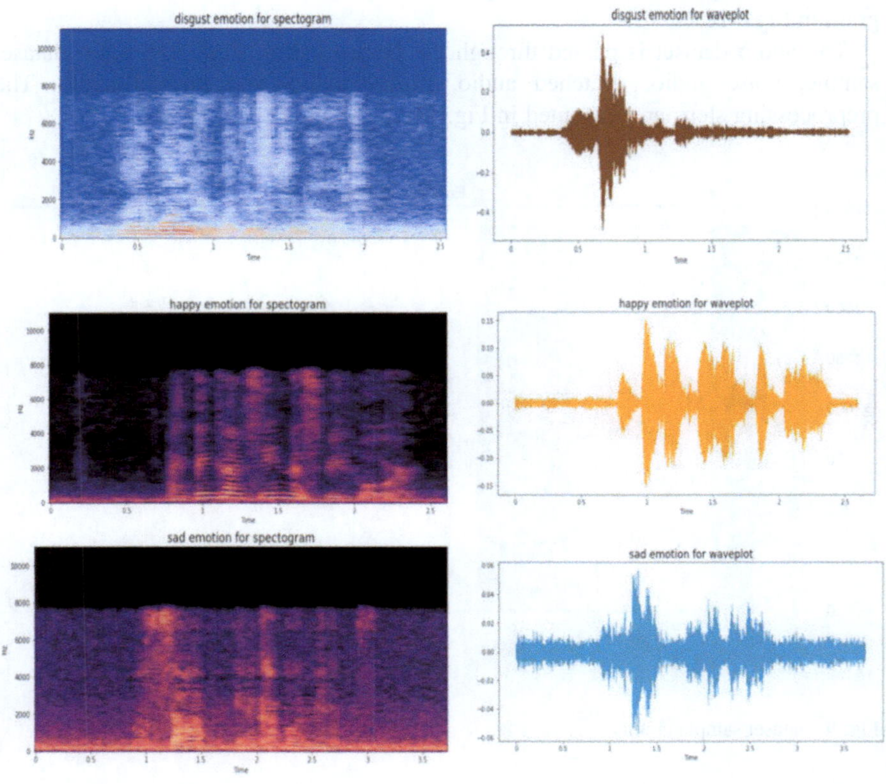

Fig. 10 Visualization of audio dataset samples

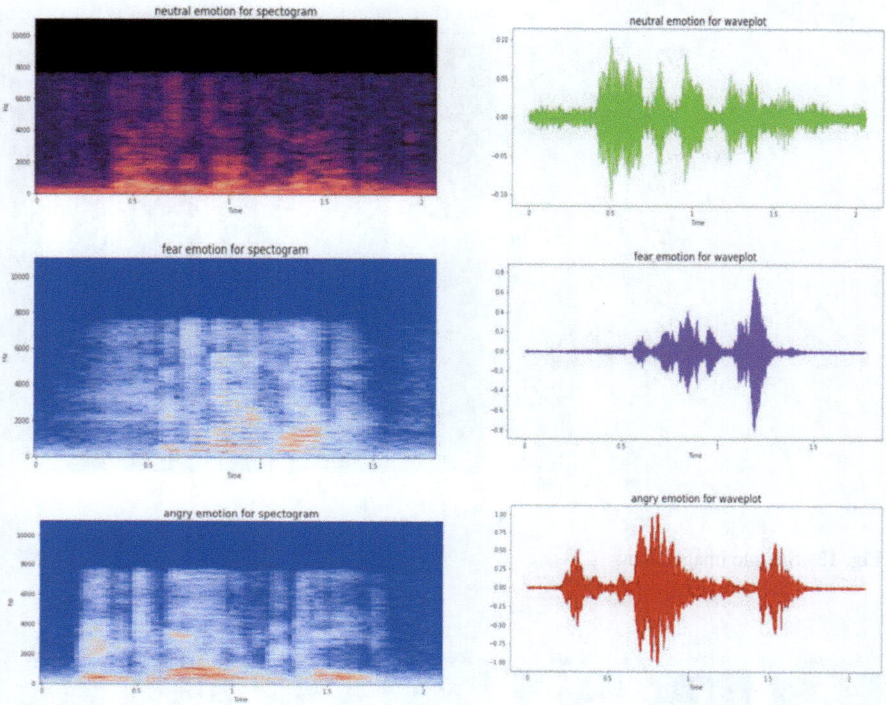

Fig. 10 (continued)

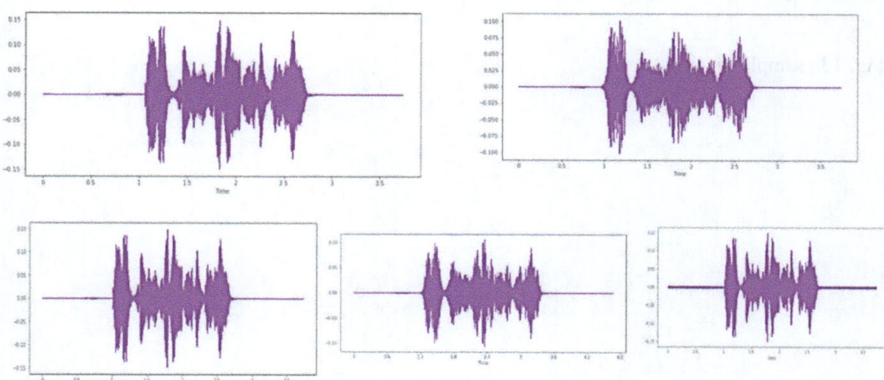

Fig. 11 Preprocessing of audio signals (**a**) audio dataset sample (**b**) noise audio (**c**) stretched audio (**d**) shifted audio (**e**) pitched audio

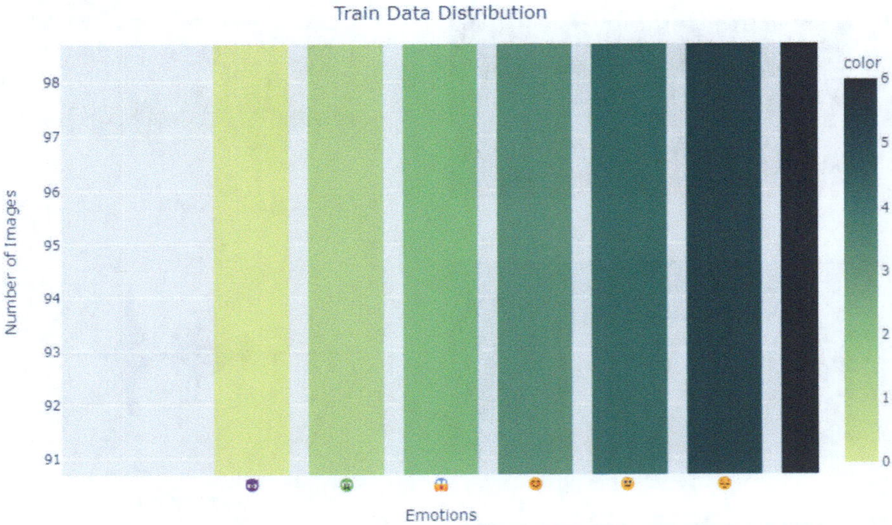

Fig. 12 Sample image count

Fig. 13 sample images

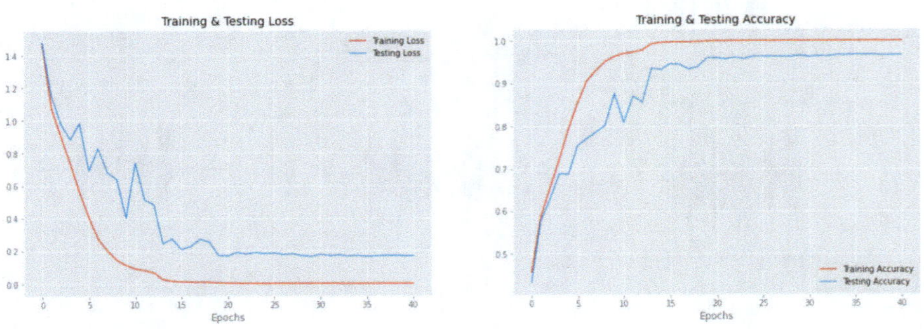

Fig. 14 Accuracy versus loss for emotion detection

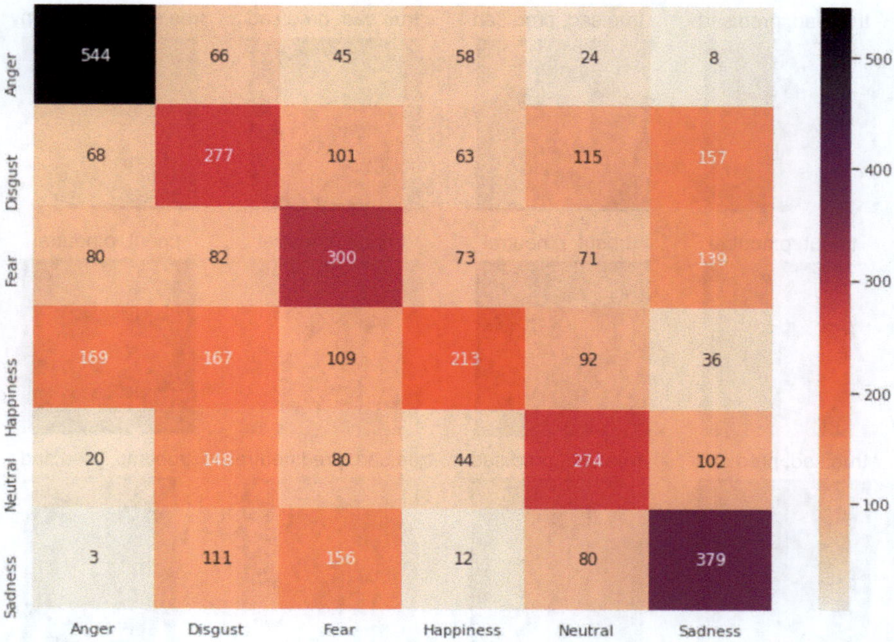

Fig. 15 Confusion matrix for emotion detection

A confusion matrix is a table that is used to define a classification algorithm's performance. A confusion matrix visualizes and summarizes a classification algorithm's performance. Figure 15 depicts a confusion matrix. The matrix shows how many true positives (TP), true negatives (TN), false positives (FP), and false negatives (FN) the model produced on the test data. The query image testing is presented in Fig. 16.

6 Conclusion and Future Work

In this study, a transfer learning technique was employed to develop an emotional detection system using image and audio data. The deep learning method and a modified CNN were applied. The results showed improved categorization and precision compared to previous techniques. The simulation findings, which showed successful emotion identification, showed that the suggested technique beat the existing system. Additionally, the suggested encoding module outperformed alternative classifiers, such as a straightforward CNN, in terms of performance. In conclusion, this part offers a succinct comparison of the proposed strategy with earlier works, demonstrating the suggested strategy's superior performance in this field. The tested designs may need to be modified in the future, or new transformer models may be used to improve visual representations.

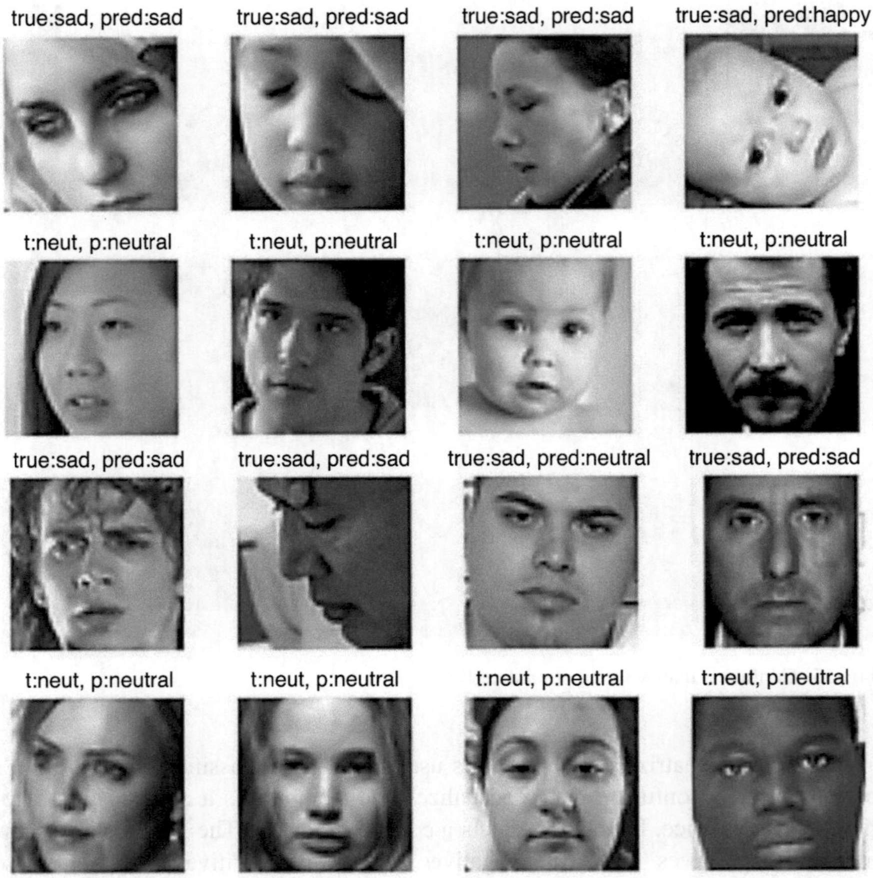

Fig. 16 Query image testing

References

1. J. Deng and F. Ren, "Multi-Label Emotion Detection via Emotion-Specified Feature Extraction and Emotion Correlation Learning," in IEEE Transactions on Affective Computing, vol. 14, no. 1, pp. 475–486, 1 Jan.-March 2023, doi: https://doi.org/10.1109/TAFFC.2020.3034215.
2. H. Kim, J. Ben-othman, L. Mokdad and P. Bellavista, "A Virtual Emotion Detection Architecture With Two-Way Enabled Delay Bound toward Evolutional Emotion-Based IoT Services," in IEEE Transactions on Mobile Computing, vol. 21, no. 4, pp. 1172–1181, 1 April 2022, doi: https://doi.org/10.1109/TMC.2020.3024059.
3. Y. Yu et al., "Cloud-Edge Collaborative Depression Detection Using Negative Emotion Recognition and Cross-Scale Facial Feature Analysis," in IEEE Transactions on Industrial Informatics, vol. 19, no. 3, pp. 3088–3098, March 2023, doi: https://doi.org/10.1109/TII.2022.3163512.

4. F. Anzum and M. L. Gavrilova, "Emotion Detection From Micro-Blogs Using Novel Input Representation," in IEEE Access, vol. 11, pp. 19512–19522, 2023, doi: https://doi.org/10.1109/ACCESS.2023.3248506.
5. C. Suman, R. Chaudhari, S. Saha, S. Kumar and P. Bhattacharyya, "Investigations in Emotion Aware Multimodal Gender Prediction Systems From Social Media Data," in IEEE Transactions on Computational Social Systems, vol. 10, no. 2, pp. 470-479, April 2023, doi: https://doi.org/10.1109/TCSS.2022.3158605.
6. Z. Xu and S. Wang, "Emotional Attention Detection and Correlation Exploration for Image Emotion Distribution Learning," in IEEE Transactions on Affective Computing, vol. 14, no. 1, pp. 357-369, 1 Jan.-March 2023, doi: https://doi.org/10.1109/TAFFC.2021.3071131.
7. N. Aslam, F. Rustam, E. Lee, P. B. Washington and I. Ashraf, "Sentiment Analysis and Emotion Detection on Cryptocurrency Related Tweets Using Ensemble LSTM-GRU Model," in IEEE Access, vol. 10, pp. 39313–39324, 2022, doi: https://doi.org/10.1109/ACCESS.2022.3165621.
8. F. Alrowais et al., "Modified Earthworm Optimization With Deep Learning Assisted Emotion Recognition for Human Computer Interface," in IEEE Access, vol. 11, pp. 35089-35096, 2023, doi: https://doi.org/10.1109/ACCESS.2023.3264260.
9. E. A. Veltmeijer, C. Gerritsen and K. V. Hindriks, "Automatic Emotion Recognition for Groups: A Review," in IEEE Transactions on Affective Computing, vol. 14, no. 1, pp. 89–107, 1 Jan.–March 2023, doi: https://doi.org/10.1109/TAFFC.2021.3065726.
10. S. Lee, S. Lee and H. Kim, "Differential Security Barriers for Virtual Emotion Detection in Maritime Transportation Stations With Cooperative Mobile Robots and UAVs," in IEEE Transactions on Intelligent Transportation Systems, vol. 24, no. 2, pp. 2461–2471, Feb. 2023, doi: https://doi.org/10.1109/TITS.2022.3172668.
11. L. Chen, M. Zhou, W. Su, M. Wu, J. She, and K. Hirota, “Softmax regression based deep sparse autoencoder network for facial emotion recognition in human-robot interaction,” Information Sciences, vol. 428, pp. 49–61, 2018.
12. A. Hassouneh, A. M. Mutawa, and M. Murugappan, “Development of a real-time emotion recognition system using facial expressions and EEG based on machine learning and deep neural network methods,” Informatics in Medicine Unlocked, vol. 20, Article ID 100372, 2020.
13. O. Ekundayo and S. Viriri, “Multilabel convolution neural network for facial expression recognition and ordinal intensity estimation,” PeerJ Computer Science, vol. 7, p. e736, 2021.
14. K. Jayanthi and S. Mohan, “An integrated framework for emotion recognition using speech and static images with deep classifier fusion approach,” International Journal of Information Technology, pp. 1–11, 2022.
15. P. Babajee, G. Suddul, S. Armoogum, and R. Foogooa, “Identifying human emotions from facial expressions with deep learning,” in Proceedings of the 2020 Zooming Innovation in Consumer Technologies Conference (ZINC), pp. 36–39, IEEE, Novi Sad, Serbia, May 2020.
16. C. Tan, M. Šarlija, and N. Kasabov, “NeuroSense: short-term emotion recognition and understanding based on spiking neural network modelling of spatio-temporal EEG patterns,” Neurocomputing, vol. 434, pp. 137–148, 2021.
17. M. S. Hossain and G. Muhammad, “An Audio-Visual Emotion Recognition System Using Deep Learning Fusion for a Cognitive Wireless Framework,” IEEE Wireless Communications, vol. 26, no. 3, pp. 62–68, Jun. 2019, doi: https://doi.org/10.1109/mwc.2019.1800419.
18. Yining Chen, “The Past, Present, and Future of Enterprise Resource Planning”, Journal of Enterprise and Business Intelligence, vol.3, no.2, pp. 065–074, April 2023. doi: https://doi.org/10.53759/5181/JEBI202303007.
19. B. Jackson, S. R, B. Balachander, A. Haldorai, V. S., and A. N. A. Sahaya, “Modified Design Structure Of A Metamaterial Microstrip Patch Array Antenna For Rf Energy Optimization,” Materiali in tehnologije, vol. 56, no. 2, Apr. 2022, doi: 10.17222/mit.2022.384
20. Sim Sze Yin and Yoni Danieli, “An Evaluation of Smart Livestock Feeding Strategies”, Journal of Robotics Spectrum, vol.1, pp. 066–077, 2023. doi: https://doi.org/10.53759/9852/JRS202301007.

21. A. Kumar, K. Abhishek, X. Liu, and A. Haldorai, "An Efficient Privacy-Preserving ID Centric Authentication in IoT Based Cloud Servers for Sustainable Smart Cities," Wireless Personal Communications, vol. 117, no. 4, pp. 3229–3253, Nov. 2020, doi: https://doi.org/10.1007/s11277-020-07979-8.
22. Kazuo Adachi and Ryo Sato, "Customer Relationship Management in an Era of Digital Disruption and Business Model Evolution", Journal of Enterprise and Business Intelligence, vol.3, no.1, pp. 001–011, January 2023. doi: https://doi.org/10.53759/5181/JEBI202303001.
23. S. Ntalampiras, "Speech emotion recognition via learning analogies," Pattern Recognition Letters, vol. 144, pp. 21–26, Apr. 2021, doi: https://doi.org/10.1016/j.patrec.2021.01.018.
24. M. Bejani, D. Gharavian, and N. M. Charkari, "Audiovisual EmotionRecognition using ANOVA Feature Selection Method and Multi-ClassifierNeural Networks," Neural Computing and Applications, vol. 24, no. 2,pp. 399–412, 2014.
25. N. E. D. Elmadany, Y. He, and L. Guan, "Multiview Learning via DeepDiscriminative Canonical Correlation Analysis," in IEEE InternationalConference on Acoustics, Speech and Signal Processing. IEEE, 2016,pp. 2409–2413.
26. S. Zhang, S. Zhang, T. Huang, and W. Gao, "Multimodal DeepConvolutional Neural Network for Audio-Visual Emotion Recognition,"in ACM on International Conference on Multimedia Retrieval. ACM,2016, pp. 281–284.
27. W. Mellouk and W. Handouzi, "Facial emotion recognition using deep learning: review and insights," Procedia Computer Science, vol. 175, pp. 689–694, 2020, doi: https://doi.org/10.1016/j.procs.2020.07.101.
28. R. Gross, S. Baker, I. Matthews, and T. Kanade, "Face Recognition Across Pose and Illumination," Handbook of Face Recognition, pp. 193–216, doi: 10.1007/0-387-27257-7_10.
29. A. Dhall, R. Goecke, S. Lucey, and T. Gedeon, "Static facial expression analysis in tough conditions: Data, evaluation protocol and benchmark," 2011 IEEE International Conference on Computer Vision Workshops (ICCV Workshops), Nov. 2011, doi: https://doi.org/10.1109/iccvw.2011.6130508.
30. P. Lucey, J. F. Cohn, T. Kanade, J. Saragih, Z. Ambadar, and I. Matthews, "The Extended Cohn-Kanade Dataset (CK+): A complete dataset for action unit and emotion-specified expression," 2010 IEEE Computer Society Conference on Computer Vision and Pattern Recognition – Workshops, Jun. 2010, doi: https://doi.org/10.1109/cvprw.2010.5543262
31. M. F. Valstar, B. Jiang, M. Mehu, M. Pantic, and K. Scherer, "The first facial expression recognition and analysis challenge," Face and Gesture 2011, Mar. 2011, doi: https://doi.org/10.1109/fg.2011.5771374.
32. I. J. Goodfellow et al., "Challenges in representation learning: A report on three machine learning contests," Neural Networks, vol. 64, pp. 59–63, Apr. 2015, doi: https://doi.org/10.1016/j.neunet.2014.09.005.
33. Lijun Yin, Xiaozhou Wei, Yi Sun, Jun Wang, and M. J. Rosato, "A 3D Facial Expression Database For Facial Behavior Research," 7th International Conference on Automatic Face and Gesture Recognition (FGR06), doi: 10.1109/fgr.2006.6.
34. W.-J. Yan et al., "CASME II: An Improved Spontaneous Micro-Expression Database and the Baseline Evaluation," PLoS ONE, vol. 9, no. 1, p. e86041, Jan. 2014, doi: https://doi.org/10.1371/journal.pone.0086041.
35. G. Zhao, X. Huang, M. Taini, S. Z. Li, and M. Pietikäinen, "Facial expression recognition from near-infrared videos," Image and Vision Computing, vol. 29, no. 9, pp. 607–619, Aug. 2011, doi: https://doi.org/10.1016/j.imavis.2011.07.002.
36. A. Mollahosseini, B. Hasani, and M. H. Mahoor, "AffectNet: A Database for Facial Expression, Valence, and Arousal Computing in the Wild," IEEE Transactions on Affective Computing, vol. 10, no. 1, pp. 18–31, Jan. 2019, doi: https://doi.org/10.1109/taffc.2017.2740923.
37. S. Li, W. Deng, and J. Du, "Reliable Crowdsourcing and Deep Locality-Preserving Learning for Expression Recognition in the Wild," 2017 IEEE Conference on Computer Vision and Pattern Recognition (CVPR), Jul. 2017, doi: https://doi.org/10.1109/cvpr.2017.277.

38. O. Langner, R. Dotsch, G. Bijlstra, D. H. J. Wigboldus, S. T. Hawk, and A. van Knippenberg, “Presentation and validation of the Radboud Faces Database,” Cognition & Emotion, vol. 24, no. 8, pp. 1377–1388, Dec. 2010, doi: https://doi.org/10.1080/02699930903485076.
39. A. V. Savchenko, L. V. Savchenko and I. Makarov, "Classifying Emotions and Engagement in Online Learning Based on a Single Facial Expression Recognition Neural Network," in IEEE Transactions on Affective Computing, vol. 13, no. 4, pp. 2132–2143, 1 Oct.-Dec. 2022, doi: https://doi.org/10.1109/TAFFC.2022.3188390.
40. T. Dar, A. Javed, S. Bourouis, H. S. Hussein and H. Alshazly, "Efficient-SwishNet Based System for Facial Emotion Recognition," in IEEE Access, vol. 10, pp. 71311–71328, 2022, doi: https://doi.org/10.1109/ACCESS.2022.3188730.
41. F. Huang, X. Li, C. Yuan, S. Zhang, J. Zhang and S. Qiao, "Attention-Emotion-Enhanced Convolutional LSTM for Sentiment Analysis," in IEEE Transactions on Neural Networks and Learning Systems, vol. 33, no. 9, pp. 4332–4345, Sept. 2022, doi: https://doi.org/10.1109/TNNLS.2021.3056664.
42. B. García-Martínez, A. Fernández-Caballero, R. Alcaraz and A. Martínez-Rodrigo, "Application of Dispersion Entropy for the Detection of Emotions With Electroencephalographic Signals," in IEEE Transactions on Cognitive and Developmental Systems, vol. 14, no. 3, pp. 1179–1187, Sept. 2022, doi: https://doi.org/10.1109/TCDS.2021.3099344.
43. S. Kakuba, A. Poulose and D. S. Han, "Deep Learning-Based Speech Emotion Recognition Using Multi-Level Fusion of Concurrent Features," in IEEE Access, vol. 10, pp. 125538–125551, 2022, doi: https://doi.org/10.1109/ACCESS.2022.3225684.

An End-to-End Offline Handwritten Tamil Text Identification Using Modified RAdam Optimizer with Effective Post-processing Techniques

1 Introduction

To make a paperless world, digital libraries are essential. Due to the advancements in the electronic kit, such as high-end scanners, digital cameras, and mobile scanning and capturing the paper document in image format is becoming the most common in the digital library scenario. But storing the text in image format has some drawbacks; extracting and retrieving the present information from these stored images is a very tedious task [1]. Optical character recognition/reader (OCR) is used to solve this issue, which converts the image-based document into a text-based format. Text-based files are easy to handle in the digital world, like searching, analyzing, and formatting than image-based data. OCR helps the machine-based automatic text identification system to print or handwrite documents [2]. Further, the OCR is categorized into two categories, such as online and offline identifications for printed/handwritten texts. Printed text identification in the OCR is the ability of a technique to predict the character label from the document accurately; similarly, the handwritten OCR is the ability to predict correct names (labels) of character present in the scanned handwritten document [3].

Though the OCR systems are available for many scripts, even for printed Tamil OCR, not much research has been done in the Offline Handwritten Tamil Text (OHTT) identification [4]. The present work concentrates on the OHTT identification. Usually, Tamil script has a massive quantity of symbols. Hence, recognition of Tamil text is a challenging task [5]. In recent times, deep learning classification models are a successful approach to get the best accuracy level [6]. A successful deep learning classification model means it will consider proper dataset and processing time. For this proposed research work, the deep learning classification model considers the Inception-v3 architecture with the transfer learning technique.

© European Alliance for Innovation 2024

A. Haldorai et al., *Artificial Intelligence for Sustainable Development*, EAI/Springer Innovations in Communication and Computing,
https://doi.org/10.1007/978-3-031-53972-5_16

Nowadays, the transfer learning technique is the keyword in the deep learning systems to reduce the processing time efficiently [7]. The main contributions of this study are:

- Experimental compile on the handwritten Tamil HP (Hewlett-Packard Lab: hpl-tamil-iso-char-offline-1.0.) dataset with the proposed Inception-v3 model (transfer learning-based).
- Moreover, this research study concentrates on the optimization area and the segmentation phase. In this study, the Tamil text recognition uses a new modified RAdam optimizer for weight updates and new segmentation algorithm, since optimization and segmentation steps play an active role in the identification system.
- The training model uses the rectification term with Adam optimizer, which is called RAdam, to reduce the convergence issue. In this RAdam optimizer, the recommended model considers the quasi-hyperbolic (QH) momentum instead of standard momentum for acceleration value, to improve the speed of convergence.
- Similarly, for a successful identification system, the post-processing steps like spell detection and correction are also necessary. So, the proposed research work is going to concentrate on error correction. The planned work focuses on Tamil to English translation based on modified Google translation and printing the recognized handwritten Tamil, translated English text in an editable page.

The introduction part is given in Sect. 1. Section 2 gives the details about literature review, Sect. 3 explains about the background work, deep learning models. Similarly, the transfer learning techniques are presented in Sect. 3. The proposed models are explained in the Sect. 4, very deeply. The experimental setup and result analysis are also presented in the following sections.

2 Literature Review

The achievement of identification systems of printed type numerals led to research on online recognition for handwritten alphabets. The satisfactory improvements of this online recognition opened the doors to the generation of offline handwritten recognition [8, 9]. For Latin and various Asian scripts, different new research methods and advanced algorithms have been suggested. High recognition rates and top solutions were conveyed by the researchers. The proposed THCR system can't use all these existing ideas; since this proposed THCR system has its own identity and needs. But this THCR needs some useful ideas and tricks from existing research works to improve the performance and recognition rate of the proposed model. The main value of surveying existing works is choosing the best way for the proposed THCR system concerning overall accuracy, processing time, and error rate, since every stage of the recognition system includes several processes and algorithms. This review process helps to improve the knowledge about the recognition systems, methods, and special techniques.

According to Vijayalakshmi, S et al. [10], Tamil has a low identification rate for the optical character recognition system because of its varied writing styles and large number of characters. This system's objective is to identify a digital format for written Tamil documents. With the use of CRNN, the suggested strategy aims to reach a 96% accuracy rate. Despite the fact that several studies have been planned for Tamil written language, it is still the most challenging assignment. The deep learning algorithm rule has been employed in this domain's trending analysis to increase the input document's accuracy. A thorough study on the recognition of handwritten Tamil characters utilizing locational and directional approaches combined with various combinations of zone and quad procedures is presented by Antony Robert Raj et al. in [11]. In order to pinpoint the location of constructions within the characters, locational elements are introduced. Each zone is then divided into four quads, and features are used to address unused sections and loops using the pixel availability from these quads. In order to designate distinct forms, directional features are extracted from the upper (three columns one row) and lower (three columns one row) zones. Their related location features are then pooled. For character classification, the labeled forms from the upper and lower zones are mixed with the directional features from the intermediate zones (three columns one row) and their respective locational features. To categorize the character classes, a suitable machine learning algorithm is chosen.

In their work, Gnanasivam P [12] proposed a handwritten Tamil character recognition system using a deep learning algorithm. The primary objective of this project is to convert character images into text and provide an audio output using the Google Cloud Translation API. By employing a deep learning algorithm, the system successfully transforms undigitized Tamil letters into a readable format. The training process involved 121 characters, which included 12 vowels (soul letters), 18 consonants (body letters), 90 soul-body letters, and one special character. A dataset comprising approximately 40,000 image samples was collected from students and the Kaggle website for training purposes. The results achieved were highly accurate (ranging from 99% to 100%) for 114 out of the 121 characters. The remaining seven characters can achieve accuracy if captured under appropriate lighting conditions. Overall, the model exhibited an accuracy of 95%. Additionally, the system is capable of recognizing words that consist of various combinations of the aforementioned letters.

One of the main challenges in Indian handwritten character recognition is the presence of overlapping shapes between characters. These overlapping shapes make it difficult to accurately identify the characters, leading to a lower recognition rate. These factors contribute to the complexity of handwritten character recognition. In reference [13], a new approach is proposed to address the recognition of handwritten characters in the Telugu language using Deep Learning (DL) techniques. The proposed approach aims to enhance the recognition rate of individual characters. The overall accuracy achieved by this approach is 94%. Furthermore, in paper [14], a novel dataset of Arabic letters called ""Hijja"" is introduced. This dataset consists of 47,434 characters written exclusively by children aged 7–12. The authors also propose an automatic handwriting recognition model based on convolutional neural networks (CNN). The model is trained on both the Hijja dataset and the Arabic

Handwritten Character Dataset (AHCD). The experimental results demonstrate promising performance, with accuracies of 97% on the AHCD dataset and 88% on the Hijja dataset. This performance outperforms existing models in the literature for Arabic handwritten character recognition.

3 Background

The General planning design of OCR is demonstrated in Fig. 1. Automatic type of machine-based text identification contains a group of sub-processes such as text preprocessing, text segmentation, feature extraction from text, classification, and finally, post-processing [10]. The feature extraction phase is optional for Convolutional Neural Network (CNN)-based deep learning classification system since CNN can do extract the features automatically [11].

Anyhow, general recognition of handwritten text is complicated due to different kinds of writing ways of various writers; especially handwritten Tamil text is more complex, since it has curves, loops in character, and similar structure character. The research follows offline identification, so scanning or image capturing is the primary step. During these steps, maybe the input image may become noisy due to low-quality scanner/mobile. Also, the writers may write text with slope and slant error. To escape from these issues, proper preprocessing is essential. Further, the inappropriate segmentation may lead to less recognition accuracy rate. So, concentration on segmentation is also one of the needful steps.

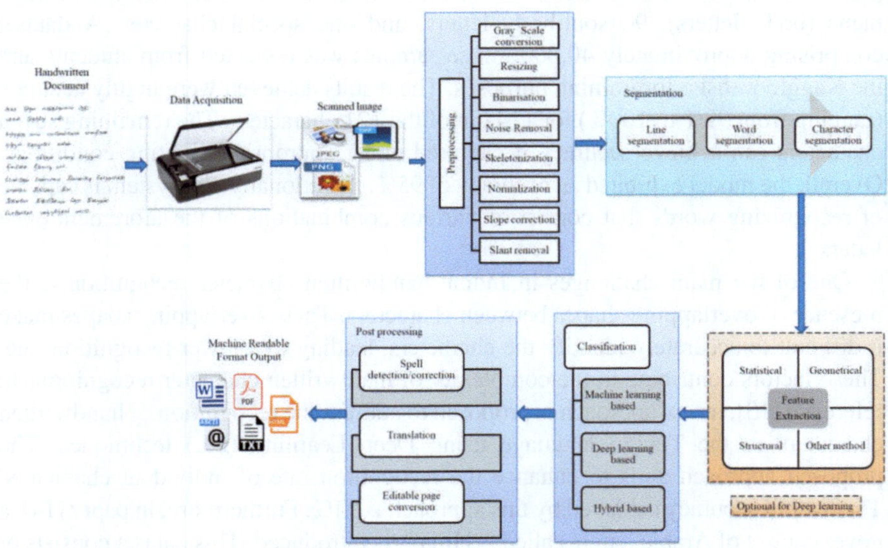

Fig. 1 The general design of OCR

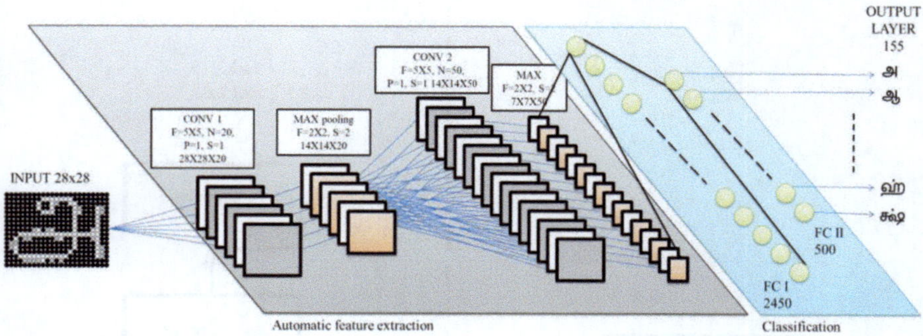

Fig. 2 OHTT using CNN model

3.1 Pretrained Models

The deep models perform the classification/identification task on image, text, and sound-based inputs [12]. The deep learning architectures can obtain high-level accuracy, even sometimes equal to human performance [13]. Many researchers have achieved success by using deep learning models like CNN, AlexNet, GoogleNet, RNN (Recurrent Neural Network), and so on [14]. CNN is a standard neural network-based design, which uses shared weights. Since the Inception-v3 attains the best accuracy in the various research fields, the planned work uses the deep learning classification model inception-v3. When using gigantic deep learning classification models, the processing time is very high, depending on the number of hidden layers. So, to reduce the processing time, the planned work takes advantage of the transfer learning approach. Transfer learning is already a successful approach in various researches, like image recognition and traffic light identification. Figure 2 shows the Tamil text identification using the CNN model.

3.2 Transfer Learning

Generally, the transfer learning concept is learning the knowledge from the natural-based image dataset called IMAGENET, along with the corresponding pretrained model weights. The performance evaluation on these two image classification tasks is that the transfer learning concept results in less processing time with simple, lightweight model training without training from scratch. When reusing the pretrained model, the reused (base) layers are frozen, and the classification layers get fine-tuned to avoid over-parameterization. During the training process, the fine-tuned classification layers only retrain for parameter updates. In this paper, the concept explores the properties of transfer learning for Tamil handwritten character imaging recognition. In the Tamil script identification, Inception-v3 [15] pretrained

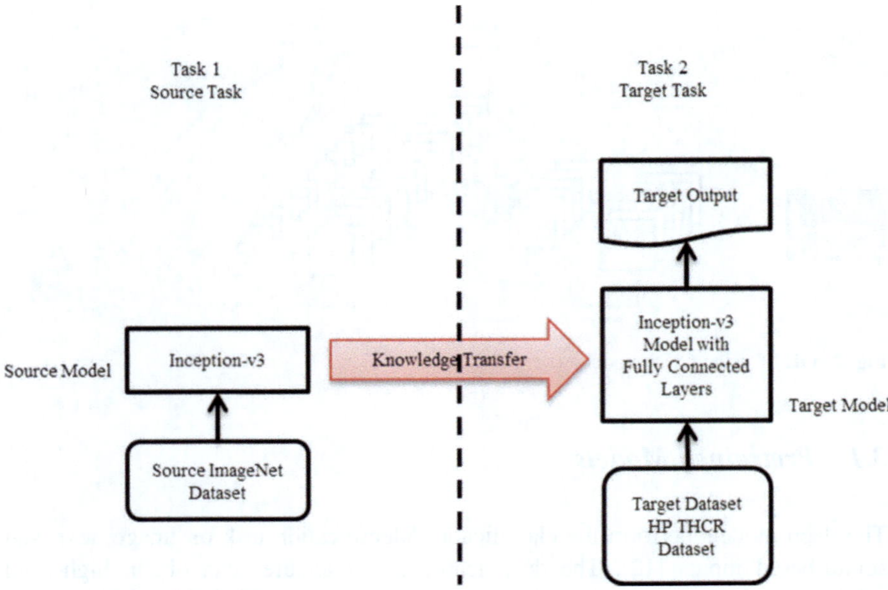

Fig. 3 Transfer learning concept

model based on IMAGENET; it is considered a source task, and the Tamil script recognition regarded as the target task. The research uses the transfer learning method of reusing the pretrained system's information on IMAGENET for new script identification tasks. While transferring the information or knowledge between two disparate domains, the parameters would be varying. Fine-tuning is one of the approaches to resolve these parameter variations, in which the topmost layer classifications of the Inception-v3 pre-trained model is removed. New classification layers depend on the Tamil dataset, which is used to replace the removed layers. The transfer learning concept demonstrates in Fig. 3.

3.3 RAdam Optimizer

Deep learning is an extremely iterative procedure. In the environment of deep learning, optimization algorithms used to train the network model and choose the best and most suitable hyper-parameter values of the model [16]. The best optimizers are giving a way to reduce the loss value during training. But the optimization process should handle the problems such as local minima and vanishing gradients. Optimizers are processes or approaches used to regulate the hyper-parameter elements of the deep learning-based model, such as weights and learning rates, to decrease the error function.

Usually, the deep learning-based models use the heuristic-based Gradient Descent (GD), Stochastic GD (SGD), Adam optimizer, Adadelta for training, and also weight tuning. GD is one of the supreme underlying optimizers, but the most used optimization due to secure computations. It's used heavily in linear regression and classification algorithms. Stochastic Gradient Descent attempts to update the model's parameters more frequently. Therefore, it converges in less time. When using Adagrad optimizer, the learning rate value changes for each training parameter. This Adam optimization method is excessively fast, so it will converge quickly. It rectifies the vanishing learning rate, high variance. Most commonly, Adam optimizers are widely used in deep architectures. When using Adam optimizer for the parameter update, the results may suffer from severe convergence due to the significant variance in adaptive learning rates. But the proper warm-up may resolve this issue and boost the accuracy level little more. For this purpose, a new advance optimizer called RAdam is used in this planned work for weight update.

When using RAdam [17], the model's performance will not change for various learning rates, since it is insensitive depending on the learning rate like SGDO and adaptive-based optimizers. Creating an accurate deep learning model with a high generalized form and trying to finish the training phase in fewer epochs are the main aims behind the RAdam optimizer. Nowadays, RAdam becomes a new trademark optimizer in deep learning models, acting as the best replacer for standard Adam optimizer by correcting generalization problems. The RAdam optimizer can directly replace the usual Adam optimizer without any changes in setting, but some fine-tuning has to done to achieve better performance. The suggested model uses this RAdam optimizer for OHTT in a modified way in terms of considering the previous value of the update gradient with the gradient value.

4 Proposed OHTT Identification System

The architecture for the proposed OHTT identification uses the modified RAdam optimizer shown in Fig. 4. The HP Tamil database images are loaded and segregated as training, validation, and test set. Subsequently, the data augmentation stage, the research work, considers the Inception-v3 pretrained architecture for training. The first important idea of this research is transfer learning since the parameter values of the Inception-v3 architecture on standard ImageNet is used. The second concentration area of the proposed work is optimization, which means that the weight update is part of the training. Specifically, the planned work utilizes a new modified RAdam optimizer instead of the standard and famous Adam optimizer. After the retraining process, the trained model is saved.

The scanned Tamil handwritten document query image acts as an input image. The input query image preprocessed to convert it into a suitable form for secure processing. Usually, the preprocessing stage includes dataset image reshaping, noise removing, color conversion, slant/slope correction, detection, and array conversion [18]. The third concentration of the planned work is segmentation. The Tamil

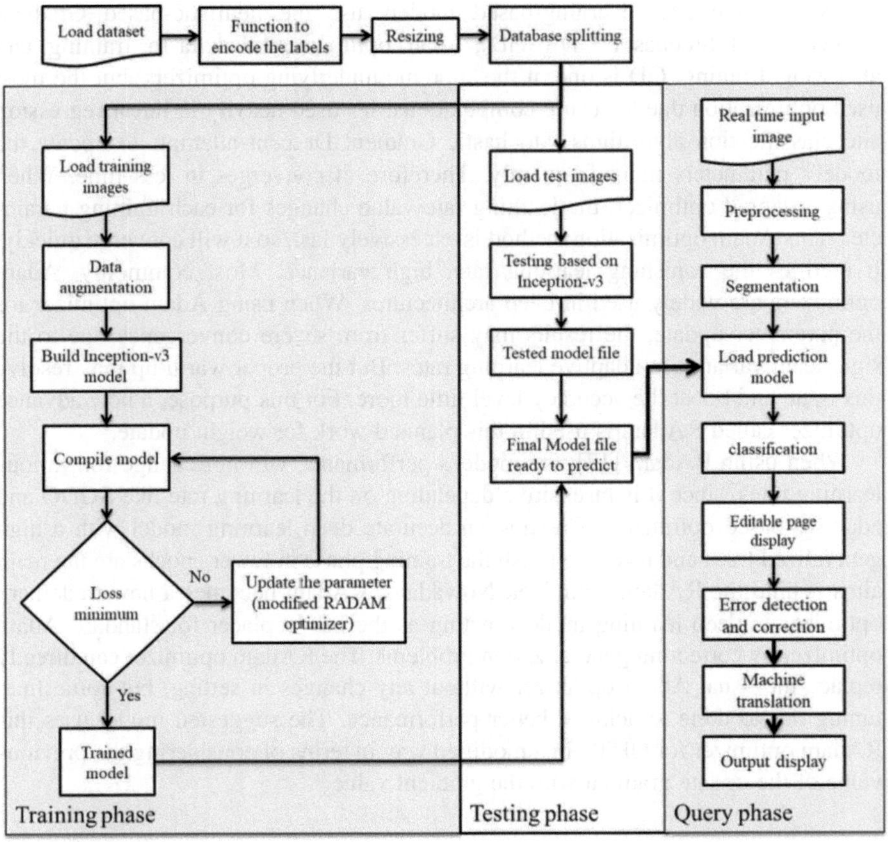

Fig. 4 The architecture of OHTT identification

characters have loops, overlapping, and slope/slant errors, due to which the segmentation will be a tedious process. So, the proper concentration on this area may lead to excellent accuracy at the end, undoubtedly. The most significant part of this research is the post-processing like editable page display, spell correction, and translation. Converting the scanned handwritten Tamil image into an editable page is essential to search, edit, and analyze efficiently. In this chapter, the concept explores the properties of transfer learning for Tamil handwritten character imaging recognition. In the Tamil script identification Inception-v3 pretrained model based on IMAGENET [19], it is considered a source task, and the Tamil script recognition regarded as the target task. The inception-v3 architecture [20] is shown in Fig. 5.

The research uses the transfer learning method of reusing the pretrained system's information on IMAGENET for new script identification tasks. While transferring the information or knowledge between two disparate dominions, the parameters would be varying [21]. Fine-tuning is one of the approaches to resolve these

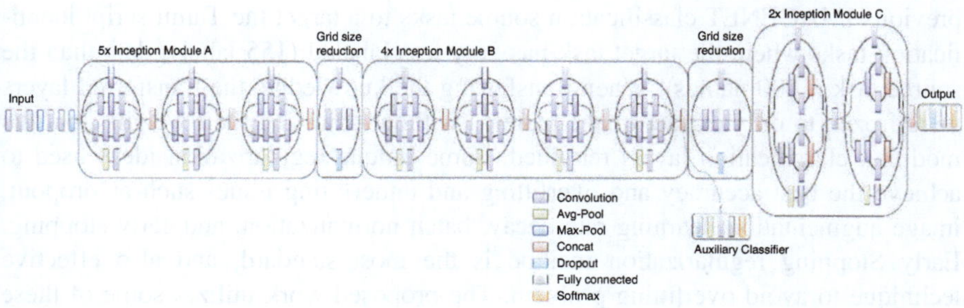

Fig. 5 Inception-v3 model

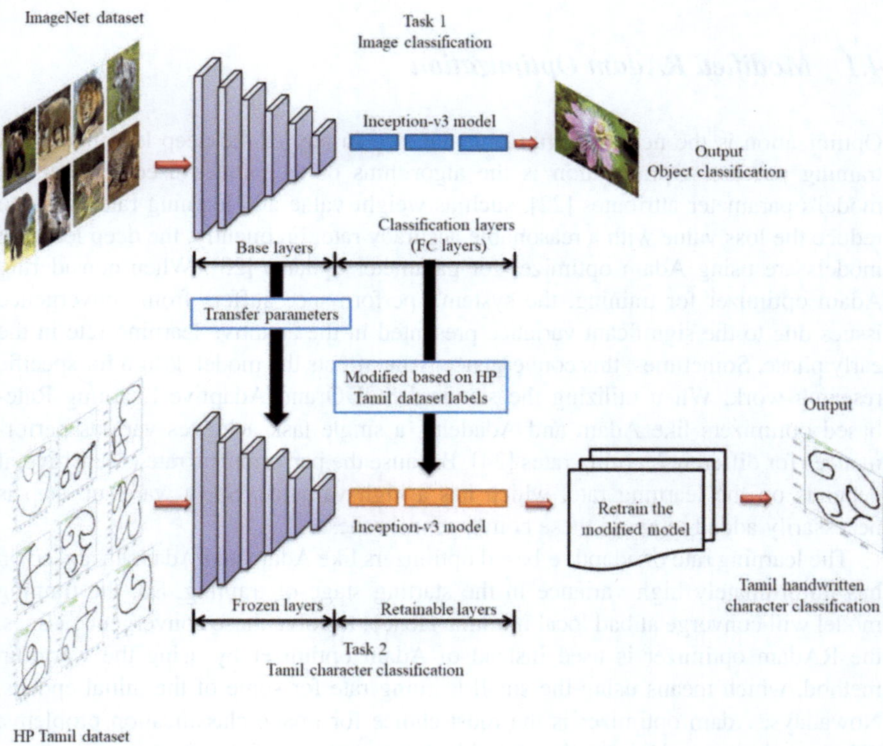

Fig. 6 OHTT identification system using transfer learning

parameter variations in which the topmost layer classifications of the Inception-v3 pretrained model is removed [15]. New classification layers depend on the Tamil dataset, which used to replace the removed layers.

Figure 6 shows the transfer learning concept between the processes of IMAGENET image classification and the Tamil handwritten script identification. In that, the planned pretrained model tries to extract the knowledge from the previous IMAGENET classification source tasks to a target the Tamil script identification tasks where the target task has very less labeled (155 labels) data than the source task (1000 names). When transferring the knowledge, the transferred layers were frozen to only extract image patterns without adjusting their weights, but the modified classification layers retrained. Some unique regularization ideas used to achieve the best accuracy and overfitting and underfitting issues such as dropout, image augmentation, learning rate decay, batch normalization, and early stopping. Early Stopping regularization method is the most standard, and also effective technique to avoid overfitting problem. The proposed work utilizes some of these regularization ideas, such as image augmentation technique, dropout technique, and early stopping technique to get the best training rate without overfitting issues.

4.1 Modified RAdam Optimization

Optimization is the necessary fundamental step in any of the deep learning-based training process. Optimization is the algorithms or techniques used to alter the model's parameter attributes [22], such as weight value and learning rate value, to reduce the loss value with a reasonable accuracy rate. Frequently, the deep learning models are using Adam optimizer for parameter updates [23]. When considering Adam optimizer for training, the system's performance suffers from convergence issues due to the significant variance presented in the adaptive learning rate in the early phase. Sometimes, this convergence issue affects the model design for specific research work. When utilizing the standard SGDO and Adaptive Learning Rate-based optimizers like Adam and Adadelta, a single task achieves various performances for different learning rates [24]. Because the performance rate is sensitive, it depends on the learning rate, which has a high variance. So, a warm-up step is necessarily added to avoid these convergence issues.

The learning rate of adaptive-based optimizers like Adam and Adadelta and so on has unfortunately high variance in the starting stage of training. So, the training model will converge at bad local minima. Hence, to solve these convergence issues, the RAdam optimizer is used instead of Adam optimizer by using the warm-up method, which means using the small learning rate for some of the initial epochs. Nowadays, Adam optimizer is the most choice for image classification problems. Adam optimizer calculates the individual learning rate for different parameters. Adam uses EMA (Exponential Moving Average) function to estimate the adaptive learning rate on gradient value. The moving averages on the gradient for standard Adam optimizer is given in Eqs. 1 and 2. The parameter updates value for Adam optimizer is revealed in Eq. 3.

$$m_t = \beta_1 m_{t-1} + (1 - \beta_1) g_t \tag{1}$$

$$v_t = \beta_2 v_{t-1} + (1-\beta_2) g_t^2 \tag{2}$$

$$\theta_t = \theta_{t-1} + \alpha_t m_t l_t \tag{3}$$

where, β_1 is 0.9 and β_2 is 0.999, g_t means gradient value. α_t, l_t is the step size and learning rate value.

The moving averages on the gradient for RAdam optimizer are shown in Eqs. 4 and 5. The parameter updates value for the RAdam optimizer is revealed in Eq. 6.

$$m_t = \beta_1 m_{t-1} + (1-\beta_1) g_t \tag{4}$$

$$v_t = 1/\beta_2 v_{t-1} + (1-\beta_2) g_t^2 \tag{5}$$

$$\theta_t = \theta_{t-1} + \alpha_t r_t \widehat{m}_t l_t \tag{6}$$

where $\widehat{m}_t$ the bias is corrected moving average. "r_t" is the rectification term. To reduce the problem of convergence, the rectification term is added with standard Adam optimizer. The rectification term is shown in Eq. 7, where, "ρ"is the degree of freedom.

$$r_t = \sqrt{\frac{(\rho_t - 4)(\rho_t - 2)\rho_\alpha}{(\rho_\alpha - 4)(\rho_\alpha - 2)\rho_t}} \tag{7}$$

When using the RAdam optimizer, the convergence problem will get reduced, but the training performance is equal to Adam optimizer. The best optimizer means updating and tuning the parameters in the direction toward less loss value in a sequence of step size. In a much recent work, momentum is the key point to increase the performance of the deep models by reducing the convergence issues. So, the main focus of the planned work is on the momentum and how it can boost the accuracy of the model with effective modification of the momentum value. The QH momentum with standard Adam optimizer achieved the best result on performance. This planned model also uses the QH momentum instead of simple for parameter tuning process on the RAdam optimizer. The RAdam optimizer is also updating the parameter based on standard momentum. Based on the QH momentum, parameter updates are shown in Eqs. 8 and 9.

$$g_{t+1} = \beta g_t + (1-\beta).\widehat{\nabla L_t \theta_t} \tag{8}$$

$$\theta_{t+1} = \theta_t - \alpha\left[(1-v).\widehat{\nabla L_t \theta_t} + v.g_{t+1}\right] \tag{9}$$

The QH momentum method is simple to apply since they are cheap on a computational basis. This modified momentum can act as the best substitution for standard momentum. The QH momentum using the exponential moving average value is based on immediate decaying factor. This algorithm is recovering the performance issues of Adam optimizer in the variability of setting implementation.

The implementation of standard Adam optimizer based on the QH momentum is displayed in Eq. 10.

$$\theta_{t+1} = \theta_t - \alpha \left[\frac{(1 - v_1).\widehat{\nabla L_t}(\theta_t) + v_1.g'_{t+1}}{\sqrt{(1 - v_2).\left(\widehat{\nabla L_t}(\theta_t)\right)^2 + v_2.s'_{t+1} + \epsilon}} \right] \quad (10)$$

Here, ϵ is a small number to avoid any division by zero. The variance problem can reduce by using the warm-up function and increasing the value of "ϵ," where,

$$g_{t+1} = \beta_1 g_t + (1 - \beta_1).\widehat{\nabla L_t}\theta_t \quad (11)$$

$$s_{t+1} = \beta_2 g_t + (1 - \beta_2).\left(\widehat{\nabla L_t}(\theta_t)\right)^2 \quad (12)$$

$$g'_{t+1} = \left(1 - \beta_1^{t+1}\right)^{-1}.g_{t+1} \quad (13)$$

$$s'_{t+1} = \left(1 - \beta_2^{t+1}\right)^{-1}.v_{t+1} \quad (14)$$

In this study, the research presents a modest and spontaneous method to revise the RAdam optimizer and make it more efficient. Thus, the planned research moves the usual standard Adam optimizer and modern RAdam optimizer to the next level by improving generalization with fewer convergence issues. The parameter updates of the modified RAdam based on the quasi-hyperbolic momentum algorithm is revealed in Eq. 15.

$$\theta_{t+1} = \theta_t - \alpha r_t \left[\frac{(1 - v_1).\widehat{\nabla L_t}(\theta_t) + v_1.g'_{t+1}}{\sqrt{(1 - v_2).\left(\widehat{\nabla L_t}(\theta_t)\right)^2 + v_2.s'_{t+1} + \epsilon}} \right] \quad (15)$$

4.2 Segmentation

The subdivision of handwritten Tamil characters is a very crucial task [24, 25, 26]. Generally, the segmentation process starts with a line [27]. The separation of a touched or skewed line is complicated [28, 29]; similarly, separating the touching character from the text is problematic [30]. The proposed work will use the data clustering-based line segmentation, and graph distance theory [31, 32], based on the word and character separation used to solve those difficulties. Extra care is needed for the Tamil handwritten script line segmentation, due to the standard loops and curves. This chapter concentrates on the line segmentation based on the data clustering algorithm. The midpoint dictates lines present in the content with the assistance of a piecewise projection profile. The competitive learning algorithm is applied to change the midpoints depending on the geometrical information. The geometrical information comprises three kinds of feature information. The middle pixel of the symbol is used to form the first two features, and the neighborhood statistics of text pixels formed the third feature. This planned line separation approach is well suited for touched line and skewed line separation.

However, it remains a challenge in word and symbol separation tasks, because separating the joint type of individual symbols from the words is another tedious task. After applying the preprocessing steps like thinning operation on the input text, the usual projection profile approach is applied to the words to get the initial separation path. The distance association conditions are used to get the correct separation path. The overlapped symbols are separated based on the graph distance theory. In this way of separation too, three features are used concerning the structural property—the first feature is considered by distances between the center pixel and the individual symbol pixels. In the second type, feature cut vectors are generated in eight diverse ways. For creating the third feature, the amount of eight neighbors of every individual symbol pixel is computed. Thus, in this chapter, two categories of algorithms are combined and used to achieve the best segmentation rates for the overlapped and skewed kind of characters.

4.3 Post-processing Techniques

Next, the proposed Tamil identification system concentrates on post-processing, such as the editable page display, automatic spell detection/correction [33, 34], and English translation. The editable page is necessary to edit, search, and save the recognized handwritten Tamil document. Such operations are not possible in an image-based scanned document. After identifying the query image in the editable display, the proposed work takes the spell checking as a new goal. Even though the planned Radam-based Inception-v3 OHTT system performs well, there are some spelling errors present in the identification result. These kinds of errors occur due to improper handwriting style and similar shapes of characters. An automatic spell detection/correction algorithm is essential [35, 36]; otherwise, such errors will undoubtedly affect the recognition rate. To correct the spelling errors of the recognized Tamil text, it is recommended to use an adaptable spell checking method, which identifies the spelling errors and recommends context-sensitive corrections. When correcting the spelling errors, giving suggestions creates comfort for handlers in order to take a suggested correction without manual typing [37]. Bigram Language Model is used to make suggestions for the revision. After the successful error correction, Google's Neural Machine-based Translation (GNMT) approach is used for automatic language translation [38, 39]. Based on the error correction rate, the translation rate also varies, since the spelling correction is essential for the machine translation.

5 Experimental Setup

The experimental setup for the Tamil text recognition is displayed in Fig. 7. The arranged work utilizes the Python deep learning structure to compile the model. All image samples for training and testing the models are taken from the HP MILE Lab database. The total available training samples are divided into mini-batches of 32 for training the network. The dataset contains images of varying dimensions; hence, they are resized before feeding it into the grid during the training and testing phases. The network parameters are initialized by the transfer learning approach of Inception-v3 deep learning systems. And the cross-entropy-based loss was considered to measure the misclassification error of the network. Dropout regularization with a 0.2 dropout probability and an early stopping technique are also used to save the model from overfitting issues. The classification rate accuracy of the proposed OHTT model is approximately 94.2% on the given HP MILE Lab database.

The standard adaptive learning-based optimizers have very high variance at the starting stage of the training. To solve this issue, a warm-up with a less initial learning rate should apply for the starting stage of a training session or should turn off the momentum at the starting stage. Generally, 0.001 is the learning rate for Adam optimizer, but the RAdam optimizer uses the minimum learning rate as 0.00001. While giving the random images of the test database, the images are

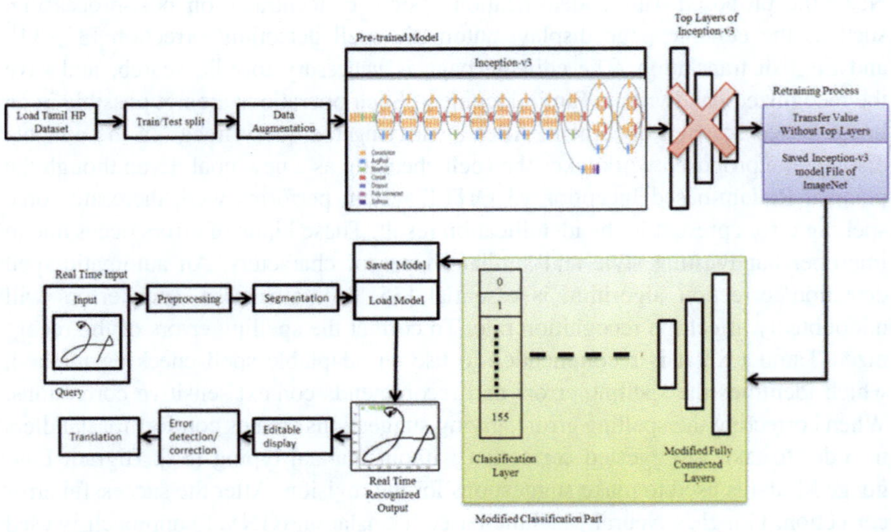

Fig. 7 Experimental setup for Tamil text recognition

compared with the saved trained model. The editable display of recognized text is needed to modify or analyze the text, so the final post-processing stage concentrates on the editable page display. The wrong prediction of text will decrease the performance of the OCR system. Solving the issue of the erroneous prediction spell error detection and correction based on the dictionary will help the system boost efficiency. Moreover, the planned work extends toward the machine translation. Finally, the proposed work proves in terms of precision rate and processing time analysis.

6 Result Analysis

Figure 8 shows the variation of accuracy and the cross-entropy based loss on the Tamil handwritten dataset. The red mark symbolizes the training set, and the blue line represents the validation set. The training model and validation model accuracy show the learning ability rate of the training, validation database set, based on epochs. The validation database means the randomly chosen images from the training database. Cross-entropy is used as a loss function in the training process, which is an error function that indicates the progress of the learning process. The proposed experiments observed that the RAdam achieves consistent improvements over the vanilla Adam with stable results. The research shows that the training of the suggested model is extra stable with less fluctuation. From the validation, the curve can be the view that the validation curve is much more likely or near accurately tracks the training curve due to the RAdam optimizer performance. The comparison between different models based on various optimizer is revealed in Fig. 9 and Table 1.

From Fig. 9 and Table 1, one is able to understand that the RAdam gives an excellent performance since the heuristic learning rate works well to smooth the training process and improve generalization for OHTT. The planned modified RAdam decreases the significant variance with a warm-up step by analytics, so the

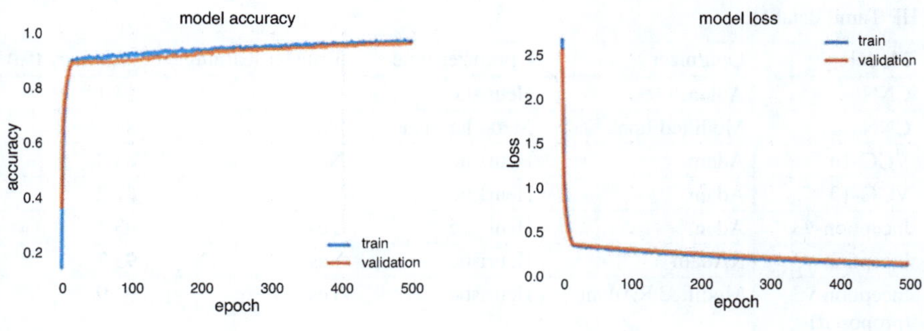

Fig. 8 The variation of (**a**) accuracy and (**b**) loss using modified RAdam

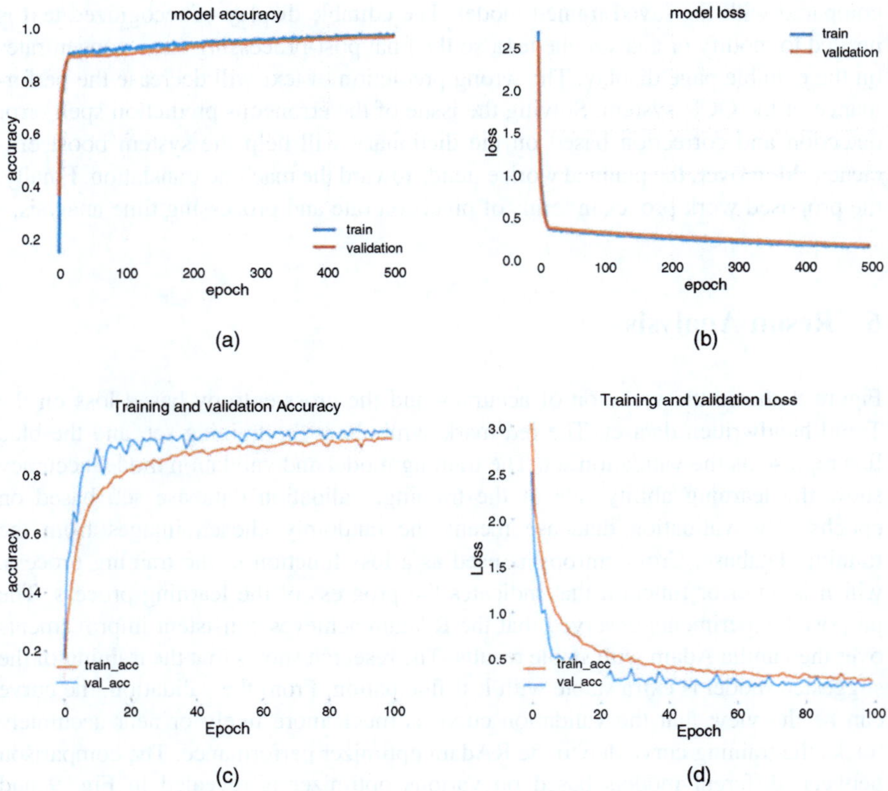

Fig. 9 Result variation between different model using Inception-v3 model with different optimizers for HP Tamil database: (**a**) Accuracy variation with Radam (**b**) Loss variation with RAdam (**c**) Accuracy variation with Adam, (**d**) Loss variation with Adam

Table 1 Comparison table result variation between different model using different optimizers for HP Tamil database

Model	Optimizer	Optimizer type	Transfer learning	Accuracy (%)
CNN	Adam	Heuristic	No	83.1
CNN	Modified lion	Meta- heuristic	No	84
VGG-16	Adam	Heuristic	No	85.3
VGG-19	Adam	Heuristic	No	87.2
Inception-v3	Adam	Heuristic	Yes	93.1
Inception-v3	RAdam	Heuristic	Yes	93.7
Inception-v3 (proposed)	Modified RAdam	Heuristic	Yes	94.9

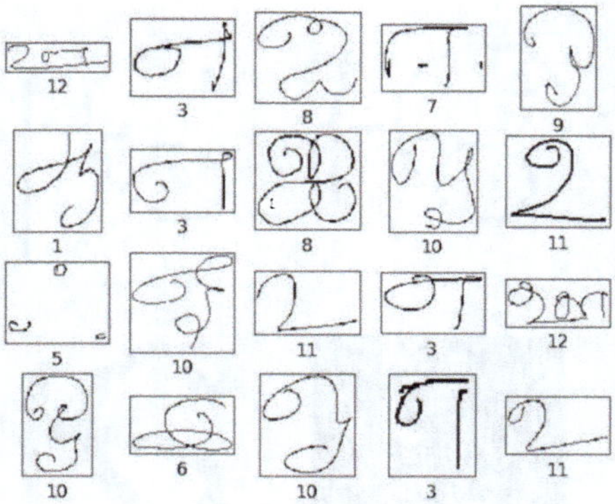

(a) Dataset image

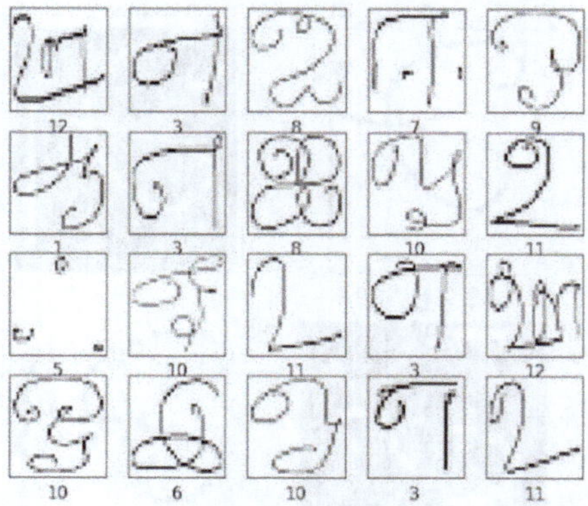

(b) Resized image

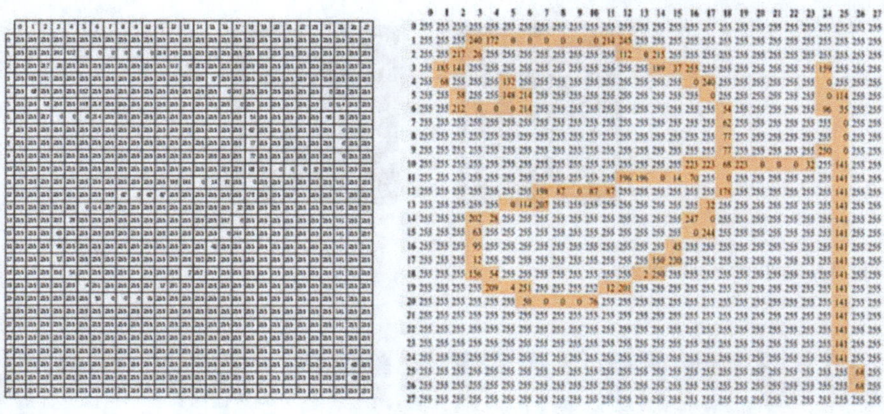

(c) Array image

Fig. 10 Dataset samples of HP Lab Tamil dataset

convergence problem gets reduced. While continuing the training process, the variance value will stabilize, and the rate of learning will further increase. Using the transfer learning concept with Inception-v3 and Adam optimizer combination will reduce the processing time than the previous model; it will also increase the training accuracy to 93.1%. But in those models, the fluctuation value is high. The RAdam with the Inception-v3 model helps to reduce these fluctuations in result, which produces the 93.07% training accuracy. There is no significant difference between the performance rates. But the modified RAdam optimizer helps to improve the performance to 94.9%. Thus, the suggested model is better than the previous model based on accuracy, processing time, and stability with less convergence issue.

The dataset image samples, resized samples, and array conversion are displayed in Fig. 10. The preprocessed output of the input samples is presented in Fig. 11.

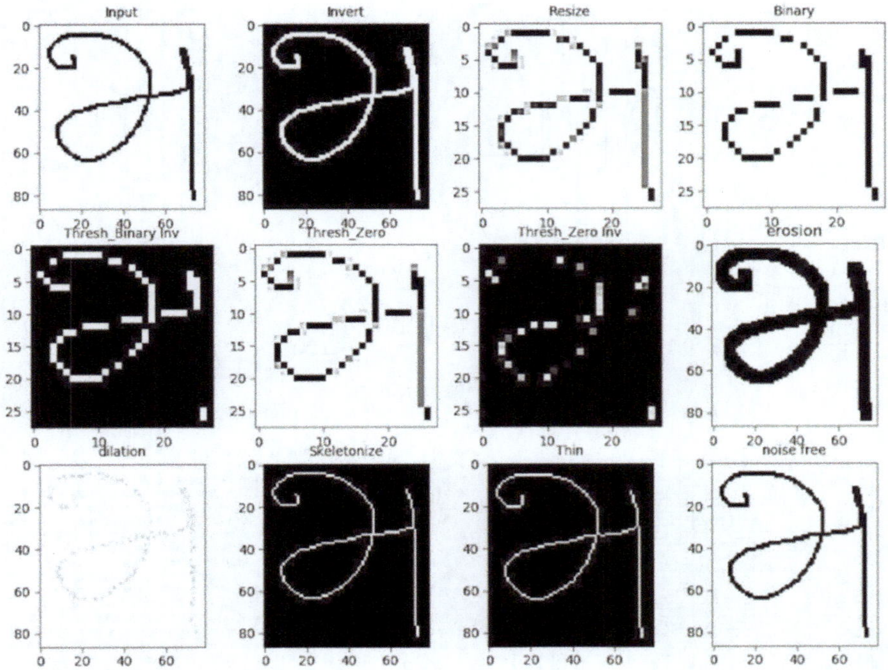

Fig. 11 Preprocessed images

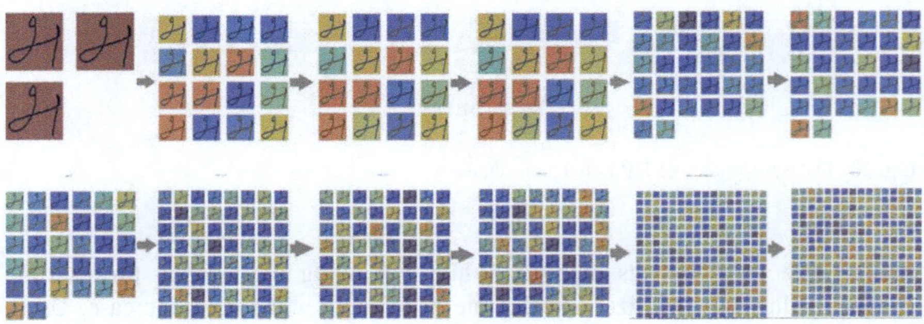

Fig. 12 Feature map visualization CNN model for THCR

Figure 12 shows the visualized feature maps. Some of the test image identification outputs based on Radam-based Inception-v3 model is displayed in Fig. 13.

When giving the query image, after the process of the preprocessing stage, the input query image passes through the segmentation step. The segmented result is compared with the saved proposed trained model, which gives the recognized output in an editable page. The perceived output is passed through the spell detection/correction system, which will detect and correct the spelling errors present in the

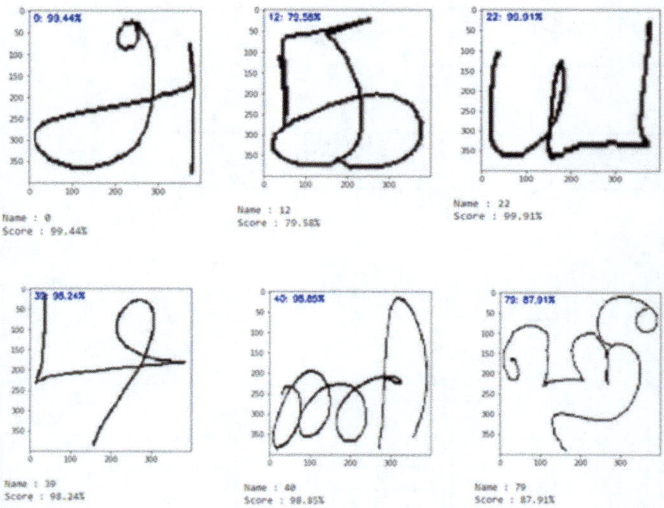

Name	Symbol
0	அ
12	க
22	ய
39	டி
40	ணி
79	ஹீ

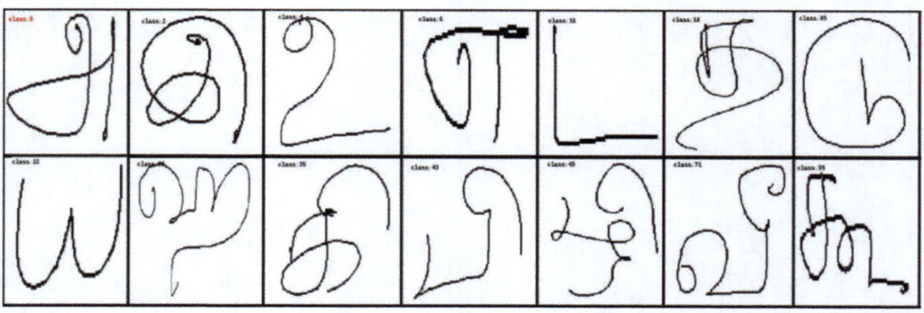

Fig. 13 Test image prediction

recognized production. Any wrong prediction will affect the recognition rate, so spelling correction is essential to improve recognition accuracy. The corrected text is translated into the English language with the help of a translator. Figure 14 and Table 2 reveals the query image analysis.

After the successful segmentation process, the segmented Tamil text characters are identified with the help of the saved Inception-v3 trained model. In Fig. 14, "தமதுthamathu" and "பறனவ) paranava" are wrongly predicted. After the error correction process, one error is corrected. "பறனவ) paranava" is altered to "பறவைparavai." But "தமதுthamathu" is not fixed. The correct word for that is "நமது) namathu." Due to its similar shape, this error occurred. Already, in the Tamil language "தமதுthamathu" word is available. So it is not corrected in the error correction process also. The corrected Tamil text translated into the English version.

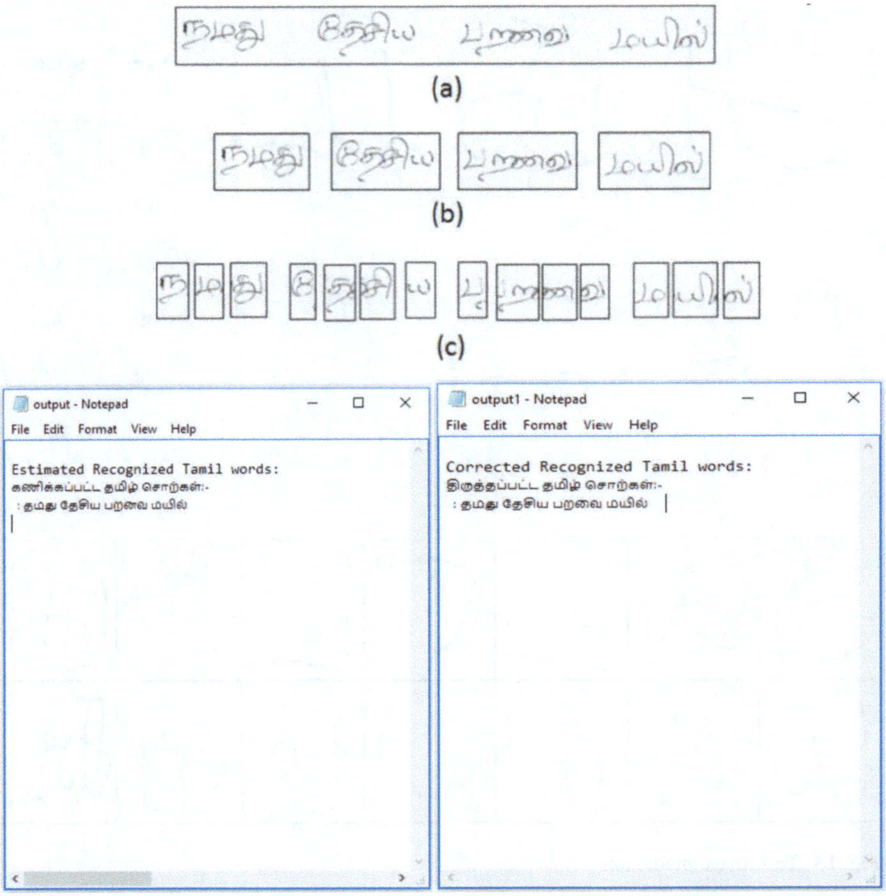

Fig. 14 Query image analysis (**a**) input query image (**b**) word segmentation output (**c**) character segmentation (**d**) character recognition in editable page (**e**) after error correction output (**f**) translated output

7 Conclusion

The suggested deep learning model of Inception-v3 with the transfer learning technique, the modified RAdam optimizer, is executed significantly well on the OHTT task. The planned model has achieved 94.7% accuracy. The trial outcomes achieved from this proposed research work reveal the efficacy of the automatic Tamil script identification system. This proposed model includes segmentation, error detection/correction, and translation techniques that are also used to accomplish a better performance. From this research, the modified RAdam can execute well than the usual Adam optimizer with fewer fluctuations. The performance of the suggested model reveals that the model can detect the overlapped characters, which can also expose with the help of segmentation. Based on experimental research and

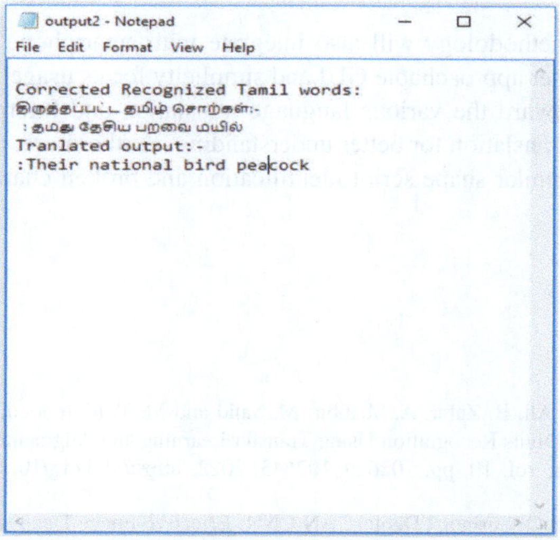

Fig. 14 (continued)

Table 2 Comparison of query image analysis

Query input	Correct Text	Before the error correction	After the error correction	After the translation
நமதுதேசியபறவைமயில் (image) (Namathuthesiyaparavaimayil)	Tamil: நமதுதேசியபறவைமயில் English: Our national bird is the peacock	Estimated Recognized Tamil words: கணிக்கப்பட்டதமிழ்சொற்கள்-: தமதுதேசியபறனவமயில் (Thamathudesiya paranavamayil)	Corrected Recognized Tamil words: திருத்தப்பட்டதமிழ்சொற்கள்-: தமதுதேசியபறவைமயில் (Thamathuthesiyaparavaimayil)	திருத்தப்பட்டதமிழ்சொற்கள்-: தமதுதேசியபறவைமயில் Translated output: Their national bird peacock

– Corrected character, - wrong character

result analysis, it has been successfully proved that deep learning is the most magnificent intelligent architecture to handle technical issues. Likewise, in this OHTT identification research, even the Inception-v3 model with the modified RAdam optimizer gives better performance. In future work, the study will concentrate on similar shape character recognition.

7.1 Future Scope

This proposed methodology will also integrate with smartphone applications that would provide user approachable GUI and simplicity for its usage. Future work will extend OHTT toward the various language translation and International Phonetic Alphabet (IPA) translation for better understanding. Furthermore, the algorithm will concentrate on similar shape script identification and broken characters to increase the accuracy rate.

References

1. A. Rasheed, N. Ali, B. Zafar, A. Shabbir, M. Sajid and M. T. Mahmood, "Handwritten Urdu Characters and Digits Recognition Using Transfer Learning and Augmentation WithAlexNet," in IEEE Access, vol. 10, pp. 102629–102645, 2022, https://doi.org/10.1109/ACCESS.2022.3208959
2. S. Gonwirat and O. Surinta, "DeblurGAN-CNN: Effective Image Denoising and Recognition for Noisy Handwritten Characters," in IEEE Access, vol. 10, pp. 90133–90148, 2022, https://doi.org/10.1109/ACCESS.2022.3201560
3. M. Shabir et al., "Real-Time Pashto Handwritten Character Recognition Using Salient Geometric and Spectral Features," in IEEE Access, vol. 9, pp. 160238–160248, 2021, https://doi.org/10.1109/ACCESS.2021.3123726.
4. R. Sahay and M. Coustaty, "An Enhanced Prototypical Network Architecture for Few-Shot Handwritten Urdu Character Recognition," in IEEE Access, vol. 11, pp. 33682–33696, 2023, https://doi.org/10.1109/ACCESS.2023.3263721.
5. N. Shaffi and F. Hajamohideen, "uTHCD: A New Benchmarking for Tamil Handwritten OCR," in IEEE Access, vol. 9, pp. 101469–101493, 2021, https://doi.org/10.1109/ACCESS.2021.3096823
6. I. Hussain, R. Ahmad, S. Muhammad, K. Ullah, H. Shah and A. Namoun, "PHTI: Pashto Handwritten Text Imagebase for Deep Learning Applications," in IEEE Access, vol. 10, pp. 113149–113157, 2022, https://doi.org/10.1109/ACCESS.2022.3216881.
7. X. Wu, Q. Chen, Y. Xiao, W. Li, X. Liu and B. Hu, "LCSegNet: An Efficient Semantic Segmentation Network for Large-Scale Complex Chinese Character Recognition," in IEEE Transactions on Multimedia, vol. 23, pp. 3427–3440, 2021, https://doi.org/10.1109/TMM.2020.3025696.
8. Y. Wang, Y. Yang, W. Ding and S. Li, "A Residual-Attention Offline Handwritten Chinese Text Recognition Based on Fully Convolutional Neural Networks," in IEEE Access, vol. 9, pp. 132301–132310, 2021, https://doi.org/10.1109/ACCESS.2021.3115606
9. N. Alrobah and S. Albahli, "A Hybrid Deep Model for Recognizing Arabic Handwritten Characters," in IEEE Access, vol. 9, pp. 87058–87069, 2021, https://doi.org/10.1109/ACCESS.2021.3087647.
10. S. Abirami, V. Essakiammal, and R. Baskaran, "Statistical features based character recognition for offline handwritten Tamil document images using HMM," International Journal of Computational Vision and Robotics, vol. 5, no. 4, p. 422, 2015, https://doi.org/10.1504/ijcvr.2015.072192.
11. M. Md Saufi, M. Afiq Zamanhuri, N. Mohammad, and Z. Ibrahim, "Deep Learning for Roman Handwritten Character Recognition," Indonesian Journal of Electrical Engineering and Computer Science, vol. 12, no. 2, p. 455, Nov. 2018, https://doi.org/10.11591/ijeecs.v12.i2.pp455-460

12. P. K. Sonawane and S. Shelke, "Handwritten Devanagari Character Classification using Deep Learning," 2018 International Conference on Information , Communication, Engineering and Technology (ICICET), Aug. 2018, https://doi.org/10.1109/icicet.2018.8533703.
13. S. Vijayalakshmi, K. R. Kavitha, B. Saravanan, R. Ajaybaskar, and M. Makesh, "Handwritten Character Recognition for Tamil Language Using Convolutional Recurrent Neural Network," Lecture Notes in Networks and Systems, pp. 369–384, 2022, https://doi.org/10.1007/978-981-19-1012-8_25.
14. H. Cai, X. Chen, J. Ling and Q. Xu, "Short-Term Load Forecasting Based on Radam Optimized CNN-BiLSTM-AE Hybrid Model," 2022 Power System and Green Energy Conference (PSGEC), Shanghai, China, 2022, pp. 626–631, https://doi.org/10.1109/PSGEC54663.2022.9881115.
15. M. Antony Robert Raj, S. Abirami, and S. M. Shyni, "Tamil Handwritten Character Recognition System using Statistical Algorithmic Approaches," Computer Speech and Language, vol. 78, p. 101448, Mar. 2023, https://doi.org/10.1016/j.csl.2022.101448.
16. G. P., B. G., K. V. and D. V., "Handwritten Tamil Character Recognition using Convolutional Neural Network," 2021 Sixth International Conference on Wireless Communications, Signal Processing and Networking (WiSPNET), Chennai, India, 2021, pp. 84–88, https://doi.org/10.1109/WiSPNET51692.2021.9419451.
17. D. Jagan, Mohan Reddy and A. Vishnuvardhan Reddy, "Recognition of handwritten characters using deep convolutional neural network", International Journal of Innovative Technology and Exploring Engineering (IJITEE), vol. 8, no. 6S4, pp. 314–317, 2019.
18. N. Altwaijry and I. Al-Turaiki, "Arabic handwriting recognition system using convolutional neural network," Neural Computing and Applications, vol. 33, no. 7, pp. 2249–2261, Jun. 2020, https://doi.org/10.1007/s00521-020-05070-8.
19. S. Attigeri, "Neural Network based Handwritten Character Recognition system," International Journal Of Engineering And Computer Science, vol. 7, no. 03, pp. 23761–23768, Mar. 2018, https://doi.org/10.18535/ijecs/v7i3.18.
20. A. James, M. J, and C. Saravanan, "Malayalam Handwritten Character Recognition Using AlexNet Based Architecture," Indonesian Journal of Electrical Engineering and Informatics (IJEEI), vol. 6, no. 4, Dec. 2018, https://doi.org/10.52549/ijeei.v6i4.518.
21. M. Soomro, M. A. Farooq, and R. H. Raza, "Performance Evaluation of Advanced Deep Learning Architectures for Offline Handwritten Character Recognition," 2017 International Conference on Frontiers of Information Technology (FIT), Dec. 2017, https://doi.org/10.1109/fit.2017.00071.
22. Y. Shima, Y. Nakashima, and M. Yasuda, "Handwritten Digits Recognition by Using CNN Alex-Net Pre-trained for Large-scale Object Image Dataset," Proceedings of the 3rd International Conference on Multimedia Systems and Signal Processing, Apr. 2018, https://doi.org/10.1145/3220162.3220163.
23. S. Naz et al., "Urdu Nastaliq recognition using convolutional–recursive deep learning," Neurocomputing, vol. 243, pp. 80–87, Jun. 2017, https://doi.org/10.1016/j.neucom.2017.02.081.
24. S. Naz et al., "Offline cursive Urdu-Nastaliq script recognition using multidimensional recurrent neural networks," Neurocomputing, vol. 177, pp. 228–241, Feb. 2016, https://doi.org/10.1016/j.neucom.2015.11.030.
25. N. Bi, J. Chen, and J. Tan, "The Handwritten Chinese Character Recognition Uses Convolutional Neural Networks with the GoogLeNet," International Journal of Pattern Recognition and Artificial Intelligence, vol. 33, no. 11, p. 1940016, Oct. 2019, https://doi.org/10.1142/s0218001419400160.
26. C. Szegedy, V. Vanhoucke, S. Ioffe, J. Shlens, and Z. Wojna, "Rethinking the Inception Architecture for Computer Vision," 2016 IEEE Conference on Computer Vision and Pattern Recognition (CVPR), Jun. 2016, https://doi.org/10.1109/cvpr.2016.308.

27. C. Lin, L. Li, W. Luo, K. C. P. Wang, and J. Guo, "Transfer Learning Based Traffic Sign Recognition Using Inception-v3 Model," PeriodicaPolytechnica Transportation Engineering, vol. 47, no. 3, pp. 242–250, Aug. 2018, https://doi.org/10.3311/pptr.11480.
28. M. N. AlJarrah, M. M. Zyout, and R. Duwairi, "Arabic Handwritten Characters Recognition Using Convolutional Neural Network," 2021 12th International Conference on Information and Communication Systems (ICICS), May 2021, https://doi.org/10.1109/icics52457.2021.9464596.
29. Y. Weng and C. Xia, "A New Deep Learning-Based Handwritten Character Recognition System on Mobile Computing Devices," Mobile Networks and Applications, vol. 25, no. 2, pp. 402–411, Mar. 2019, https://doi.org/10.1007/s11036-019-01243-5.
30. Q. Yue and C. Ma, "Deep Learning for Hyperspectral Data Classification through Exponential Momentum Deep Convolution Neural Networks," Journal of Sensors, vol. 2016, pp. 1–8, 2016, https://doi.org/10.1155/2016/3150632.
31. J. Chee and P. Li, "Understanding and Detecting Convergence for Stochastic Gradient Descent with Momentum," 2020 IEEE International Conference on Big Data (Big Data), Dec. 2020, https://doi.org/10.1109/bigdata50022.2020.9378129.
32. D. Valy, M. Verleysen, and K. Sok, "Line Segmentation Approach for Ancient Palm Leaf Manuscripts Using Competitive Learning Algorithm," 2016 15th International Conference on Frontiers in Handwriting Recognition (ICFHR), Oct. 2016, https://doi.org/10.1109/icfhr.2016.0032.
33. Y. Gao, X. Ding, and C. Liu, "A Multi-scale Text Line Segmentation Method in Freestyle Handwritten Documents," 2011 International Conference on Document Analysis and Recognition, Sep. 2011, https://doi.org/10.1109/icdar.2011.135.
34. P. Sahare and S. B. Dhok, "Multilingual Character Segmentation and Recognition Schemes for Indian Document Images," IEEE Access, vol. 6, pp. 10603–10617, 2018, https://doi.org/10.1109/access.2018.2795104.
35. G. D. Vignesh, A. Ramu, J. T. Raja, P. Ponmurugan, A. Haldorai, and G. Senthilkumar, "Sensor data fusion techniques in the construction of generalized VORONOI graph for on-line motion planning in robot navigation," International Journal of System Assurance Engineering and Management, Sep. 2022, https://doi.org/10.1007/s13198-022-01773-4.
36. A. Haldorai, S. Murugan, and A. Ramu, "Evolution, challenges, and application of intelligent ICT education: An overview," Computer Applications in Engineering Education, vol. 29, no. 3, pp. 562–571, Feb. 2020, https://doi.org/10.1002/cae.22217.
37. Matt Bowden, "A Review of Textual and Voice Processing Algorithms in the Field of Natural Language Processing", Journal of Computing and Natural Science, vol.3, no.4, pp. 194-203, October 2023. https://doi.org/10.53759//181X/JCNS/202303018.
38. P. Kumar, A. Kannan, and N. Goel, "Design and Implementation of NLP-based Spell Checker for the Tamil Language," Proceedings of 1st International Electronic Conference on Applied Sciences, Nov. 2020, https://doi.org/10.3390/asec2020-07636.
39. A. M. Al Trad and I. Abu Doush, "Improving post-processing optical character recognition documents with Arabic language using spelling error detection and correction," International Journal of Reasoning-based Intelligent Systems, vol. 8, no. 3/4, p. 91, 2016, https://doi.org/10.1504/ijris.2016.10003960.

Marine Vision-Based Situational Automatic Ship Detection Using Remote Sensing Images

1 Introduction

Nearly 70% of the Earth is covered by water. This includes ocean, lakes, and rivers. Due to the ferocious increase, marine surveillance with proper protection is necessary. When considering the global economy, countries should pay attention to maritime safety [1]. Marine safety/security is the main aim of Marine Surveillance System (MSS) [2]. MSS also monitors unwanted fishing, unwanted intrusion, and smuggling activity. Marine surveillance is a part of military and defense section. Additionally, it can be considered for marine pollution monitoring. But for this type of surveillance, the system needs a large number of manpower and materials. Yet, there is possibility of error with manual monitoring due to human fatigue. Hence, automatic MSS is desired [3, 4]. Already the world is shifting to automatic systems in various fields, such as the success seen in automatic road vehicle detection, automatic tumor detection, automatic emotion detection, and so on.

Specifically, automatic MSS is used for automatic ship detection in the deep sea. Generally, the position of a ship is determined by Global Positioning System (GPS), which contains the speed and location tracker of the ship. But these types of monitoring human experts are needed. One more thing is extra difficult; that is, moving ship detection [5, 6]. Tracking the ship in dim light is also challengeable. The expeditious development in communication area like data analysis, IoT (Internet of Things), big data, particularly Artificial Intelligence (AI), increase the demand of automation [7]. Already marine monitoring is a challengeable task, which needs an automatic type of surveillance system. In the current scenario, remote sensing area is growing quickly, because the remote sensing images have the ability to give high resolution image [8, 9]. The high resolution images give clear information. So only remote sensing is also considering the problems in automatic ship detection [10]. The proposed work also considers remote sensing data base for model training.

© European Alliance for Innovation 2024

A. Haldorai et al., *Artificial Intelligence for Sustainable Development*,
EAI/Springer Innovations in Communication and Computing,
https://doi.org/10.1007/978-3-031-53972-5_17

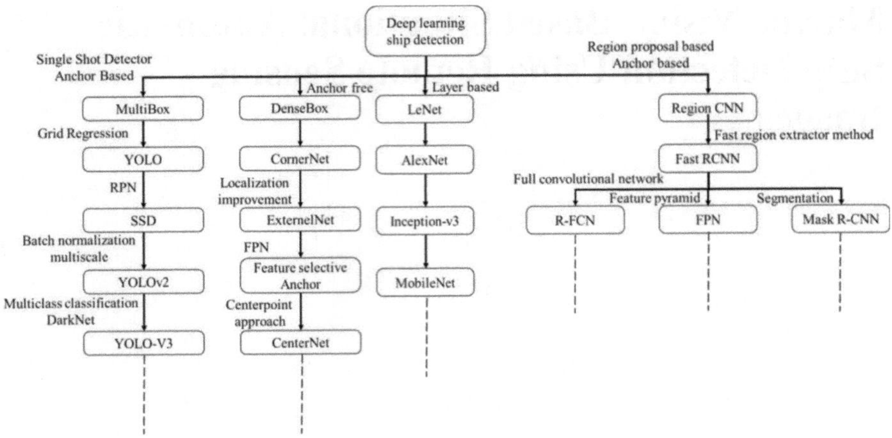

Fig. 1 Supporting algorithm for ship detection

Satellite images are one of the remote sensing types of images. Satellite images provide the appropriate information, since the image includes millions of pixels.

Object detection is the sub area of machine vision. To detect the ship, first it should deal with classification, which means it should identify the label from the input. Then the ship is detected by bounding box technique. During automatic ship detection, some problems will appear, such as (1) detection with complex backgrounds like shadow, cloud, mist, and waves, which will affect the quality of the data base. (2) confusion in shapes and color; for example, islands may look like a ship, and (3) resolution of the image also affects the accuracy. To overcome these issues in automatic ship detection, the deep learning model is used. Basically, the deep learning models use the traditional Convolutional Neural Network (CNN) variants for object detection. AlexNet, EfficientNet, R-CNN, LSTM are some of the variants of CNN. Some of the various detection algorithms are listed in Fig. 1.

Ship identification is necessary for the ship operation, ship crash avoidance, and danger prediction in the marine transport. The deep neural models and algorithms are generally considered for this automatic ship detection from the image or video-based input. The ship on the sea images is loaded and trained by the selected deep neural models and tested based on the trained model. The final tested model will categorize the ship from the maritime. The contribution of the proposed work is summarized as follows:

- In this study, the planned work introduced automatic ship detection that uses the MobileNet deep neural network-based learning without forgetting structure.
- The various permeances metrics are analyzed based on the trained model.
- The trained model is verified depending on various hyperparameters like learning rate, epochs, and image size.
- Then, the proposed work compares with various models and performance.

The chapter is planned in the following manner. Section 1 contains the introduction. The existing work is analyzed very deeply and clearly in Sect. 2, the literature survey section. Section 3 gives details about the planed model. Sections 4 and 5 analyze the performances, conclusion, and future work.

2 Literature Survey

The automatic and real-time identification of ship is necessary for the maritime security, because the maritime includes harbor surveillance, fishery management, marine pollution control, sea-based terrorist attack monitoring, and defense surveillance. There is always some interesting and useful research in this field. Specifically, the deep neural network-based ship identification is trendy.

Bingde Wang et al. [11] discussed a novel deep neural model for ship classification. Along with this deep neural network, K-means clustering was combined to improve the accuracy. The authors introduced Yolo-v4 architecture for ship classification. For this research work, the authors have collected private ship dataset. During the automatic ship detection, the illumination was the biggest challenge. S. Lei et al. [12] concentrate on this illumination problem and introduced YOLO-V3. Especially for this problem, own dataset is constructed by these researchers. The dataset is designed on various light setting. The attention mechanism is combined with YOLO-v3 for shift classification without illumination problem. Fine-grained recognition is another research section in ship identification. But there is less number of research work done on this particular section. J. Li and J. Tian [13] focused on this fine-grained recognition-based ship identification. They considered modified CNN model for automatic identification. Background filtering is considered for background removal. It is one of the successful models in large remote sensing image-based ship classification.

Most of the researchers use SAR images for ship classification. But it has less information. Instead of SAR image, Y. Chen et al. [14] used the optical images, since the optical images include detailed information, which is more suitable for feature selection. The researcher used the YOLO model with DarkNet residual block for ship classification. The ship was detected with the help of bounding box. M. Jianying et al. [15] use spaceborne rotating fan-beam radar for ship detection. They detect the ship based on the some of the parameters like doppler frequency, azimuth angle, length, and velocity of the ship. Y. Zhang et al. [16] proposed correlation filter algorithm, which has the ability to identify even the small ship. Particularly, they used kernelized correlation filters for automatic identification. The Faster R-CNN with dilated convolution technique was exclusively used by Wei et al. [17] on the HRSC2016 datasets. In this chapter dark channel prior technique is illustrated to enhance the image quality. S. Zhou and J. Yin [18] used Yolo model for ship classification on Ship7000 database. They used Yolo version three and five, and finally compared the experimental result with various performance metrics.

X. Leng et al. [19] focused the ship detection based on range compressed SAR data. They used CNN model on the range compressed SAR data for classification. Ship detection on or near shore atmosphere is difficult. So, Markov random fields and dual-attention network are considered by J. Yuan et al. [20]. The authors used two stage of ship detection for better result. They used GF-3 AIR-SARShip-1.0 dataset. Otsu algorithm was also used with the proposed algorithm to get high accuracy. Y. Bai et al. [21] used a novel technique, the combination of sea-sky-line inspection, Hough transform, and Otsu segmentation. The sea-sky-line inspection is used for detection of image interference. Single-stage detection is used for ship detection. Y. Zou et al. [22] considered a new method with MobilenetV2 model, which was trained by the Coco dataset. This research analyzed the performance in terms of Non-Maximum Suppression and recognition speed. The fusion of three-way decisions and multi-granularity feature was suggested by L. Ying et al. [23] for remote sensing image-based ship identification. The three-way decisions include two sections: (1) the grouping of deblurring and fuzzy classification and (2) extracting feature by the multi-granularity. Multi-granularity is used to solve issues related to different sizes. FGSD2021 and HRSC2016 were the different datasets utilized for this research.

Generally, transformer with encoder and decoder is a superior technique in detection problem. But for the ship detection, computational complexity and slow convergence speed are the main issues. So, Y. Zhang et al. [24] considered Efficient Ship Detection Transformer, which contains encoder/decoder and backbone. The backbone used the ResNet model. For the speed coverage, feature distillation technique was used. C. Zhang et al. [25] trained Centernet for ship detection. The Centernet was combined with saliency detection for high accuracy. The Yolo version seven was used by C. Yu and Y. Shin [26] for ship detection. SSDD was trained by this Yolo version. Y. Lei et al. [27] detect the ship with Google Earth Engine platform. This approach is well suited for practical application. N. Wang and F. Ma [28] introduced a new concept for ship identification. Edge computing technique is appropriate for moving object. So, the N. Wang and F. Ma used this edge computing to find moving ship. The dense connection DetNet, introduced by the K. Zhao et al. [29], was used particularly for SAR ship detection. This research focused on small size and complex background. Some of the recent research related to automatic ship detection based on deep learning model is listed in Table 1. They compared the proposed work with some other models like YOLOv3, Faster R-CNN model, and SSD.

3 Proposed Work

The proposed work includes two sections. They are (1) ship detection and (2) ship classification. The proposed structure is shown in Fig. 2. The loaded dataset is trained based on the MobileNet deep neural network-based learning without forgetting the structure. The detection of ship from an image is a crucial problem in real-

Table 1 Survey related to automatic ship detection

S.No	Paper	Method	Model	Dataset	Anchor free	Accuracy
1	Moosbauer, S [30]	Mask R-CNN	ResNet 101	SMD	No	0.875
2	Spraul, R [31]	RetinaNet	ResNet 50	SMD +MSCOCO	No	0.88
3	Betti, A [32]	Yolo-V3	DarkNet	Private	No	0.96
4	Nalamati, M [33]	Mask R-CNN	ResNet 101	MarDCT, IPATCH	No	0.96, 0.92
5	Shao, Z [34]	Faster R-CNN	ResNet-101	Seaship	No	0.92

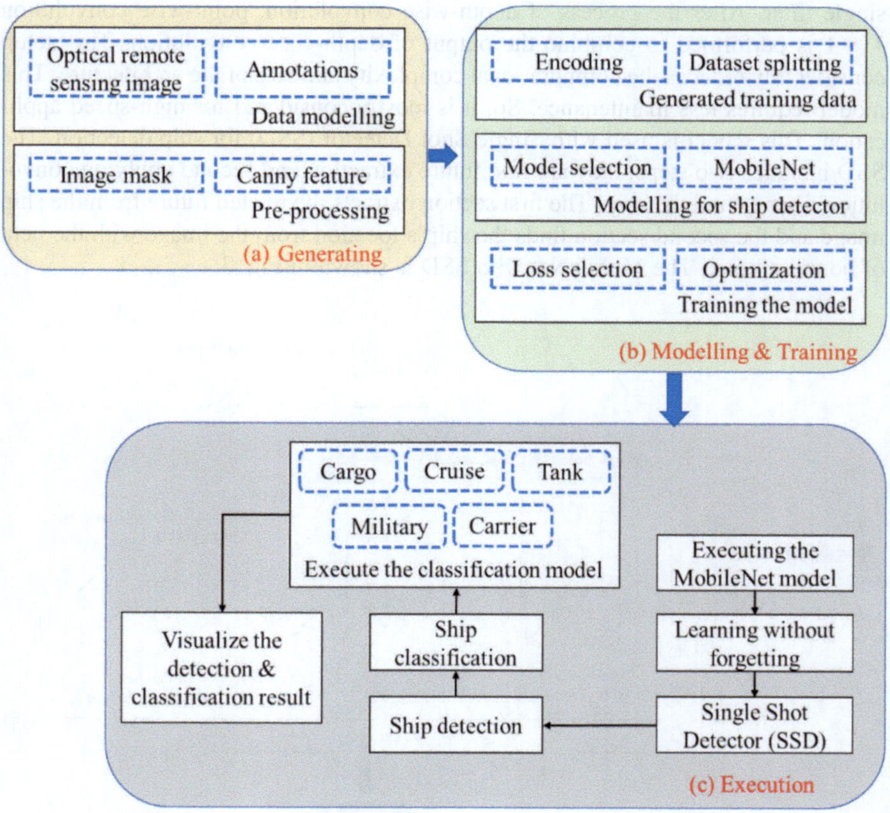

Fig. 2 Proposed system

time application. Since marine transport safety is necessary for country safety, this is an interesting and significant task. This will avoid unwanted ship accident, unknown outsiders' entry, and overfishing. Also, this is very much important for pollution monitoring and unlawful immigration. The proposed system selected the MobileNet model for the learning process.

The MobileNet architecture is a modest, at the same time effectual model. In recent times, most of the real-world practical applications consider MobileNet model for better efficiency. This is also familiar for light weight, because it performs depth-wise convolution and point-wise convolution rather than standard convolution. Depth-wise convolution means applying single convolution separately on each one of the color channels without convolution on combined manner. This process separates each channel into different layers. For this reason, only, the model utilizes single filter. After the process of depth-wise convolution, point-wise convolution 1×1 is performed to combine the output of depth-wise convolution. This setup considerably reduces the computational complexity and size of the architecture. This model requires less maintenance. So, it is mostly considered for high-speed application. This model is used with Single Shot Detector (SSD) for ship detection. The SSD includes two steps, they are first future extraction and second applying convolution filter to find the ship. The first section extracts the related future from the ship image and the second section finds the ship's location from the image with the help of bounding box. The MobileNet with SSD is shown in Fig. 3.

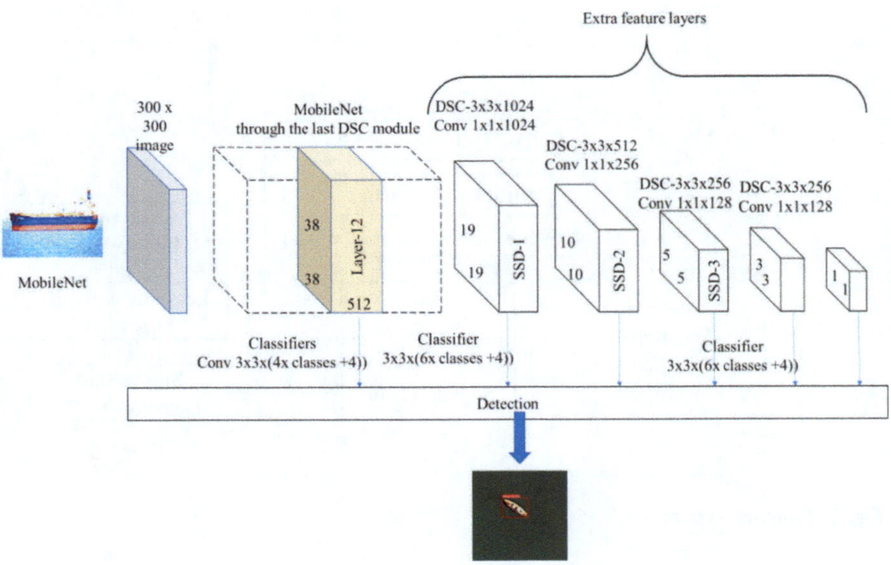

Fig. 3 MobileNet with SSD for ship detection

The general model training for computer vision task like object detection/classification takes long processing time based on model and layer structure. So always processing time is a notable point. Also, another one notable issue is small amount of data base. To solve these types of issues, deep learning moves to the new concept called Transfer learning. But, by using this transfer learning technique, there is a chance of missing the original ability of the selected model. The shared parameter of the pretrained model to the new task will not perform better future extraction always. Moreover, fine tuning may diminish the previous model's performance; for this reason the shared parameter is altered without varying task-specific parameter. So, retraining the model with new task will lose the original parameters and incapability to execute well on the new task. So, this proposed work suggested a new method "Learning Without Forgetting (LWF)" instead of common transfer learning practice. LWF is considered for new task learning, which will preserve the selected model's undamaged prior capabilities. The planned method learn the model with new samples while keeping its earlier skills. In this approach, the old network's skills are well maintained while the images from the new task are used to optimize the accuracy of the new task. Anyhow, the old task's samples and labels are needless for this method. The MobileNet deep neural network-based learning without forgetting structure is shown in Fig. 4.

The process of LWF is explained in detail below:

Step 1: Variable selection.

The shared parameter is mentioned as Ps, which is the model parameters restructured for original ship detection dataset. The task-specific parameter of the original database is mentioned as Po, and the new task-specific parameter is noted as Pn. The training images and labels are declared as Xn and Yn.

Step 2: Procedure.

The task-specific parameters of the ship detection database are calculated as Yo, which is given in Eq. 1. After this, add the nodes to the output layer for every label in the ship detection database. Initialize the Pn with random weights.

$$Y_o = \text{Pre} - \text{trainedMobileNet}\ (X_n, P_s, P_o) \tag{1}$$

Step 3: Training the model with ship detection dataset.

The $\widehat{Y_o}$ and $\widehat{Y_n}$ are calculated based on Eqs. (2) and (3). Find the loss function of base dataset and ship detection dataset, then update P_s, P_o, and P_n. Repeat the procedure until the convergence. The model learns new skills, though keeping its old skills working well, without using training data from the old tasks.

$$\widehat{Y_o} = \text{Pre} - \text{trainedMobileNet}\ \left(X_n, \widehat{P_s}, \widehat{P_o}\right) \tag{2}$$

$$\widehat{Y_n} = \text{Pre} - \text{trainedMobileNet}\ \left(X_n, \widehat{P_s}, \widehat{P_n}\right) \tag{3}$$

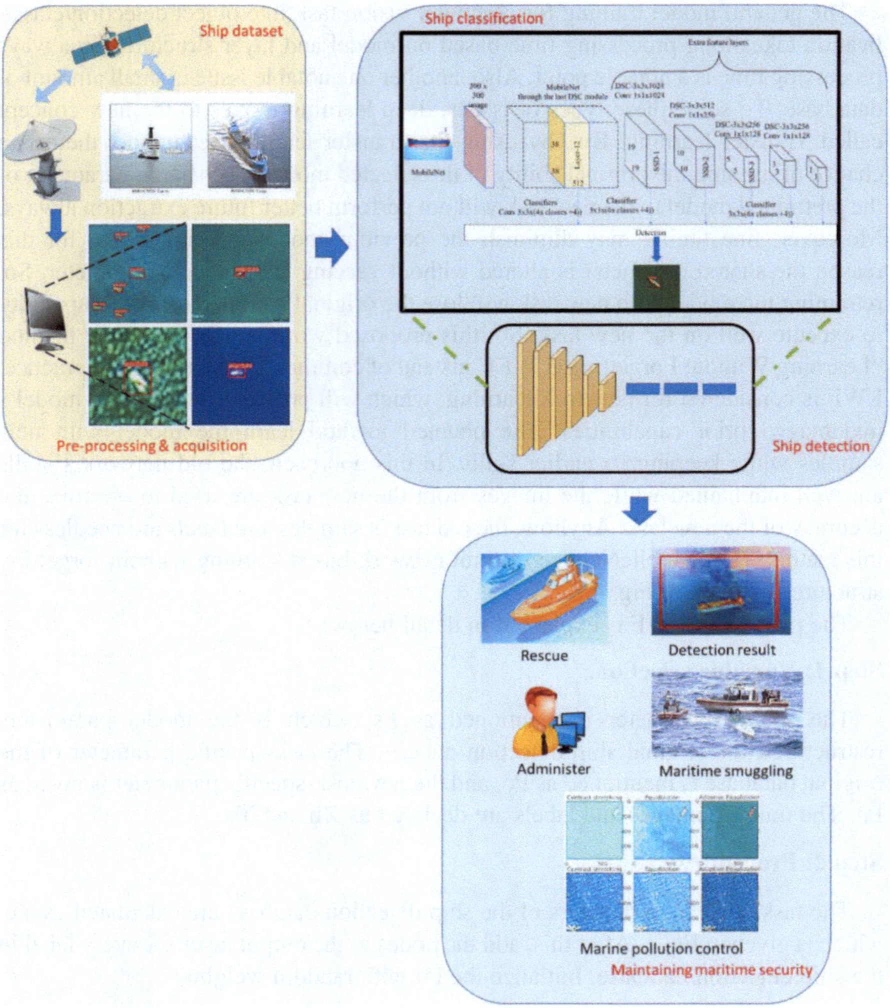

Fig. 4 MobileNet deep neural network-based marine surveillance system

4 Experimental Setup and Result Analysis

The ship detection section uses the dataset from Kaggle platform. The selected database includes 2800 images in two classes such as "ship" and "no ship." The label "ship" contains 700 images; the label "no ship" contains 2100 satellite-based images, which is displayed in Fig. 5. The pixel intensity of both labels is shown in Fig. 6.

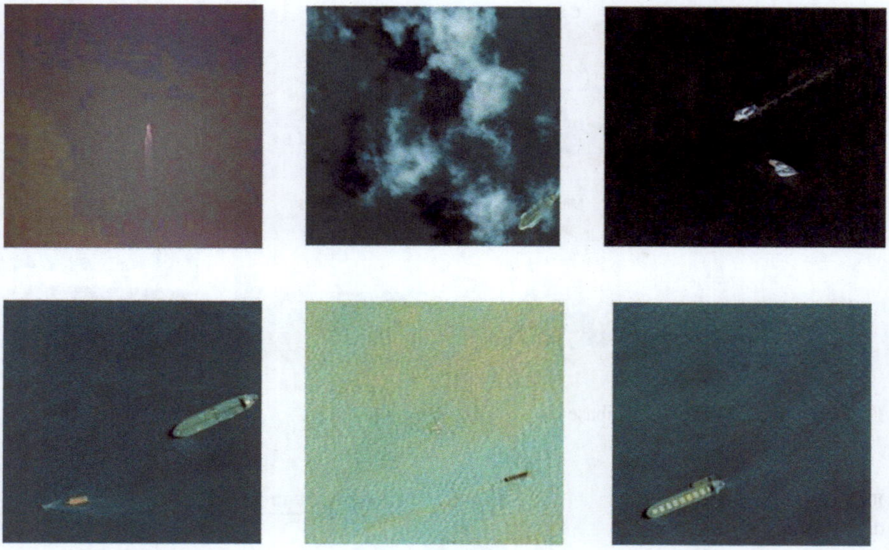

Fig. 5 Ship detection database samples

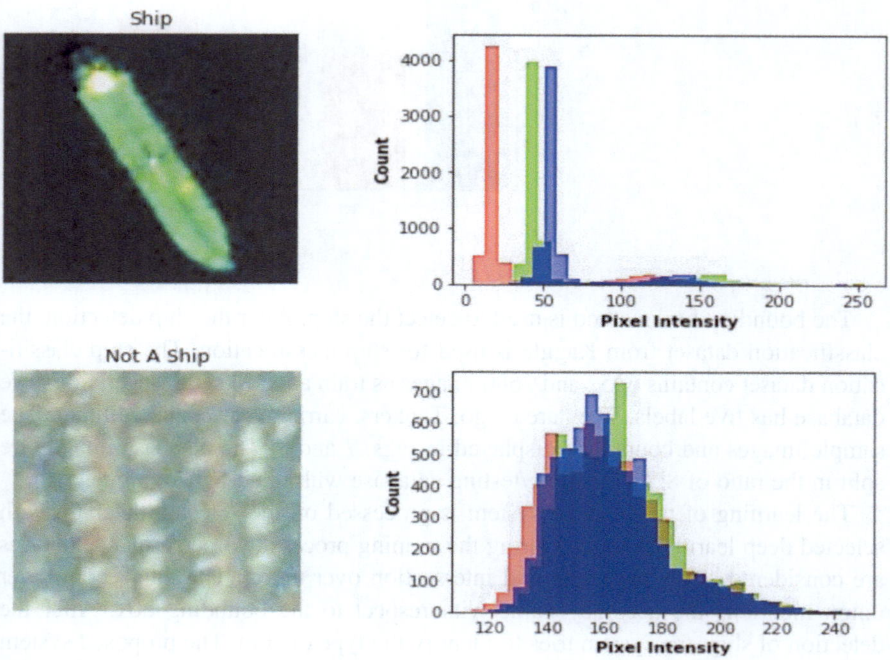

Fig. 6 Pixel intensity of database

Fig. 7 Ship classification database

Fig. 8 Ship category distribution

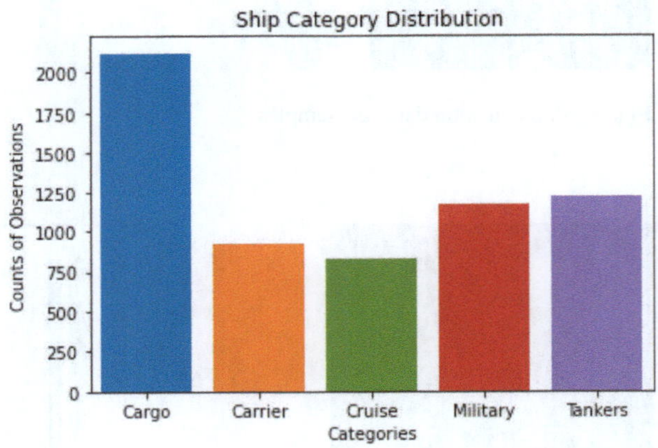

The bounding box method is used to detect the ship. After the ship detection, the classification dataset from Kaggle is used for ship classification. The ship classification dataset contains 6252 and 2680 images as train and test set, respectively. The database has five labels. They are cargo, Tankers, carrier, cruise, and Military. The sample images and counts are displayed in Figs. 7 and 8. The database images are split in the ratio of 8:2 as training/testing database with fixed hyperparameters.

The learning of the planned system is processed on the selected database with selected deep learning model. During the training process two performance metrics are considered. They are recall and intersection over union. The intersection over union mentions the accuracy value with respect to the bounding box. After the detection of ship, the system tries to identify the type of ship. The proposed system has the ability to identify the floating type of object only. Also, it has the ability to detect large number of ships with high accuracy. Generally, the bounding box method can give high IOU score. The ship detection is shown in Fig. 9.

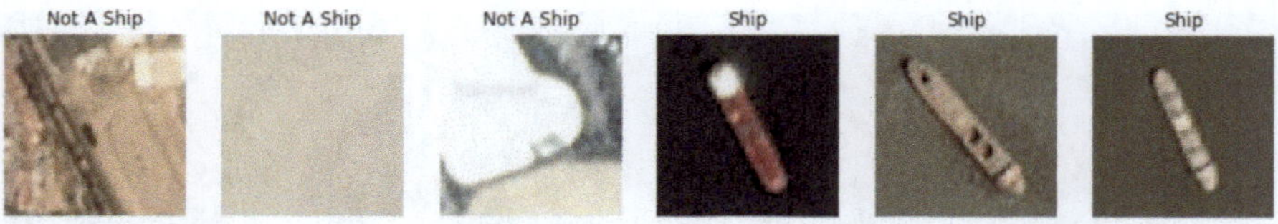

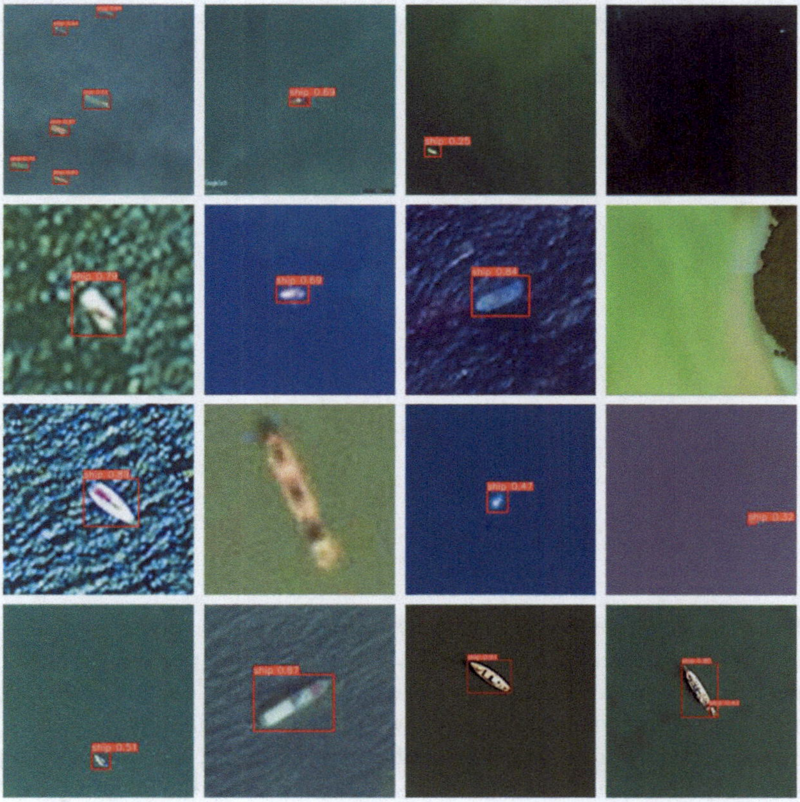

Fig. 9 Ship detection

Table 2 Ship detection accuracy for various optimizers

s.no	Optimizer	Batch size	Learning rate	Accuracy
1.	Adam	32	0.001	98.2%
2	Adagrad	32	0.001	93.4%
3	Adamax	32	0.001	89.67%

To reduce the loss function, different types of optimizers are considered for training. The proposed work checked for various optimizers. The accuracy value for various optimizer is listed in Table 2. Some other hyperparameters of the learning models are batch size, optimizer, learning rate, and so on. Batch size means total quantity of images processed for a single iteration. Epochs mean the entire image learning process. The smaller batch size produces the high accuracy than the higher

Fig. 10 Ship classification

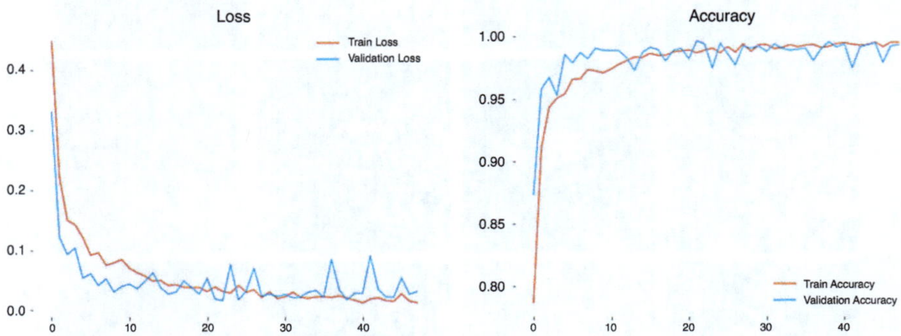

Fig. 11 Ship classification accuracy versus loss function

batch size. Based on the learning rate, the weight value will be tuned. Particularly, it will control the number of errors. Usually, smaller learning rate will give the high accuracy than the higher learning rate. The proposed system considers the 0.001 as learning rate value. Higher learning rate will make local optimization issues; at the same time small learning rate takes high processing time. The detected ship is classified with the trained model. The classification result is shown in Fig. 10. The learning accuracy and loss function are displayed in Fig. 11.

The ship images are loaded and preprocessed before the model training. Image masking, contrast sketching, equalization, adaptive equalization, Schar edge magnitude, and canny feature are some of the preprocessing techniques considered for this proposed work. The preprocessing images are displayed in Fig. 12. The future maps and intermediate layers are visualized in Fig. 13.

Each type of label is analyzed with the evaluation metrices like F1-score, Recall, and IOU. The classification accuracy is definite as the relation between properly classified images to the total quantity of images. F1-score is an inclusive calculation of classification process of the proposed model, which means that the ratio between precision and recall ratio. The precision ratio, defined as quantity of true positive, is divided by the predicted positive sample. Recall ratio is defined as the number of true positive divided by all positive samples. The performance analysis of each class is presented in Table 3. The confusion matrix is presented in Fig. 14.

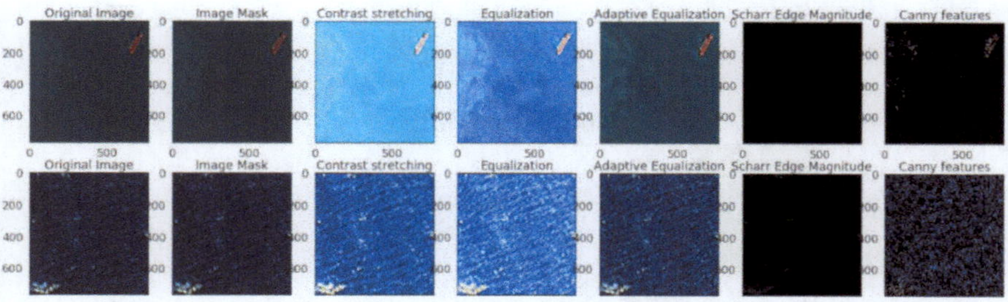

Fig. 12 Preprocessed image

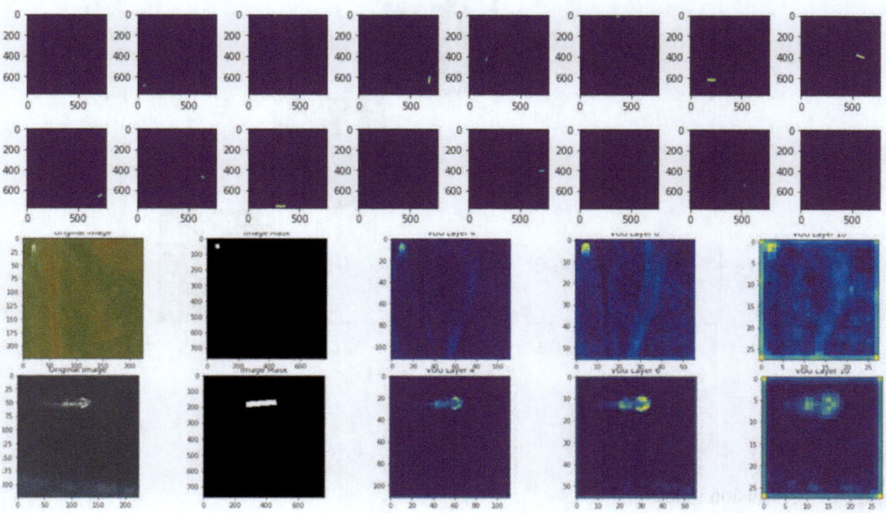

Fig. 13 Layer visualization

Table 3 Performance evaluation of each class

Class	Accuracy	Recall	Precision	IOU
Cargo	0.93	0.78	0.78	0.54
Military	0.98	0.92	0.92	0.7
Carrier	0.99	0.96	0.96	0.9
Cruise	0.98	0.92	0.92	0.7
Tankers	0.91	0.97	0.97	0.48

5 Conclusion

Maintaining flawless safety is the main aim of marine surveillance. The marine surveillance includes situational safety, fishery monitoring, illegal migration, and marine pollution monitoring and prevention of unwanted ship accident. Thus, marine safety is necessary for blue homeland security. Due to the improvements in

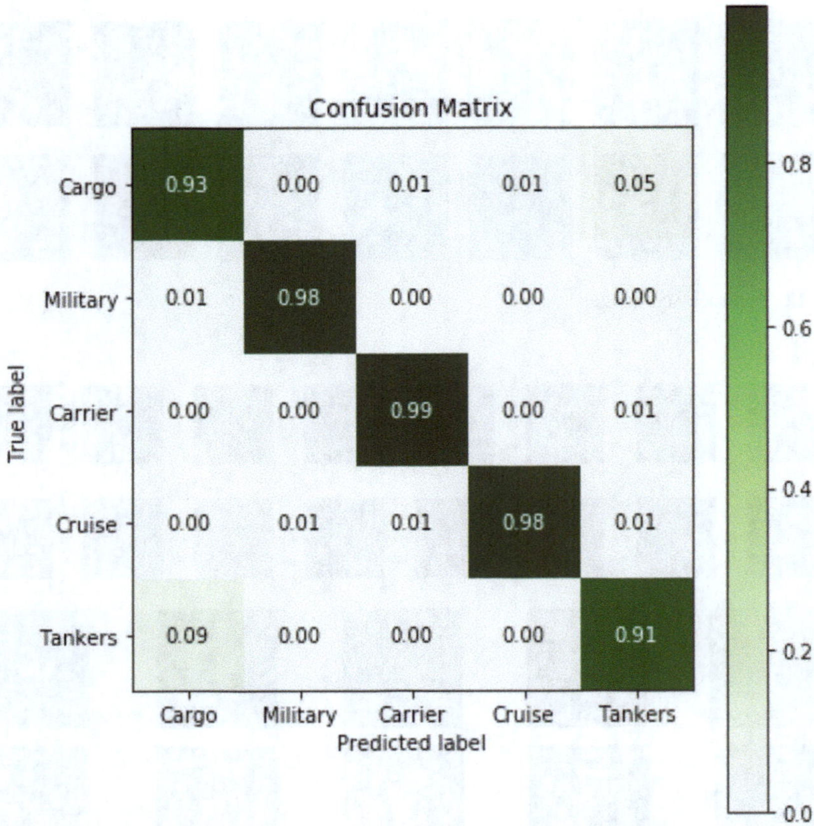

Fig. 14 Confusion matrix

wireless and artificial intelligence technology, automatic marine monitoring is also a noticeable research topic. So, the proposed work focused on automatic ship detection and classification. The planned model was trained by the MobileNet deep neural network-based learning without forgetting structure. The model achieved 98.1% accuracy rate. The proposed model has the ability to detect the ship that comes near the eyesight of humans. Due to the quality and quantity of the data samples, there are some limitations; some research gaps have occurred between real-time application and the project result. However, future work will concentrate on real-time ship detection with high-end deep learning models.

References

1. T. Zhang, Z. Zhang, H. Yang, W. Guo and Z. Yang, "Ship Detection of Polarimetric SAR Images Using a Nonlocal Spatial Information-Guided Method," in IEEE Geoscience and Remote Sensing Letters, vol. 19, pp. 1–5, 2022, Art no. 4513805, https://doi.org/10.1109/LGRS.2022.3205619.
2. M.-D. Li, X.-C. Cui and S.-W. Chen, "Adaptive Superpixel-Level CFAR Detector for SAR Inshore Dense Ship Detection," in IEEE Geoscience and Remote Sensing Letters, vol. 19, pp. 1–5, 2022, Art no. 4010405, https://doi.org/10.1109/LGRS.2021.3059253.
3. M. Zha, W. Qian, W. Yang and Y. Xu, "Multifeature Transformation and Fusion-Based Ship Detection With Small Targets and Complex Backgrounds," in IEEE Geoscience and Remote Sensing Letters, vol. 19, pp. 1–5, 2022, Art no. 4511405, https://doi.org/10.1109/LGRS.2022.3192559.
4. Y. Zhuang, Y. Liu, T. Zhang and H. Chen, "Contour Modeling Arbitrary-Oriented Ship Detection From Very High-Resolution Optical Remote Sensing Imagery," in IEEE Geoscience and Remote Sensing Letters, vol. 20, pp. 1–5, 2023, Art no. 6000805, https://doi.org/10.1109/LGRS.2023.3239016.
5. Bukuroshe Elira Epoka, "Literature Review of Qualitative Data with Natural Language Processing", Journal of Robotics Spectrum, vol.1, pp. 056–065, 2023. https://doi.org/10.53759/9852/JRS202301006.
6. K. Naveen Durai, R. Subha, and A. Haldorai, "Hybrid Invasive Weed Improved Grasshopper Optimization Algorithm for Cloud Load Balancing," Intelligent Automation Soft Computing, vol. 34, no. 1, pp. 467–483, 2022, https://doi.org/10.32604/iasc.2022.026020.
7. Xia Hao, "Examining Collaborative Business Process Modeling Techniques", Journal of Enterprise and Business Intelligence, vol.3, no.2, pp. 075–084, April 2023. https://doi.org/10.53759/5181/JEBI202303008.
8. Md. Z. Hussain, M. Ashraf, D. K. Singh, A. Haldorai, D. K. Mishra, and T. N. Shanavas, "Intelligent data post and read data system like to feed for IoT sensors," International Journal of System Assurance Engineering and Management, Jun. 2022, https://doi.org/10.1007/s13198-022-01683-5.
9. Y. Li, S. Zhang and W.-Q. Wang, "A Lightweight Faster R-CNN for Ship Detection in SAR Images," in IEEE Geoscience and Remote Sensing Letters, vol. 19, pp. 1–5, 2022, Art no. 4006105, https://doi.org/10.1109/LGRS.2020.3038901.
10. Q. Bai, G. Gao, X. Zhang, L. Yao and C. Zhang, "LSDNet: Lightweight CNN Model Driven by PNF for PolSAR Image Ship Detection," in IEEE Journal on Miniaturization for Air and Space Systems, vol. 3, no. 3, pp. 135–142, Sept. 2022, https://doi.org/10.1109/JMASS.2022.3203082.
11. B. Wang, B. Han and L. Yang, "Accurate Real-time Ship Target detection Using Yolov4," 2021 6th International Conference on Transportation Information and Safety (ICTIS), Wuhan, China, 2021, pp. 222–227, https://doi.org/10.1109/ICTIS54573.2021.9798495.
12. S. Lei, Y. Zhao and Y. Bai, "Ship detection based on deep learning under complex lighting," 2022 34th Chinese Control and Decision Conference (CCDC), Hefei, China, 2022, pp. 5694–5698, https://doi.org/10.1109/CCDC55256.2022.10033941.
13. J. Li, J. Tian, P. Gao and L. Li, "Ship Detection and Fine-Grained Recognition in Large-Format Remote Sensing Images Based on Convolutional Neural Network," IGARSS 2020—2020 IEEE International Geoscience and Remote Sensing Symposium, Waikoloa, HI, USA, 2020, pp. 2859–2862, https://doi.org/10.1109/IGARSS39084.2020.9323246.
14. Y. Chen, B. Zhang, Z. Li and Y. Qiao, "Ship Detection with Optical Image Based on Attention and Loss Improved YOLO," 2022 3rd International Conference on Pattern Recognition and Machine Learning (PRML), Chengdu, China, 2022, pp. 1–5, https://doi.org/10.1109/PRML56267.2022.9882217.
15. M. Jianying, Z. Di and Y. Risheng, "Ship detection using spaceborne rotating fan-beam radar," 2021 CIE International Conference on Radar (Radar), Haikou, Hainan, China, 2021, pp. 1301–1304, https://doi.org/10.1109/Radar53847.2021.10028518.

16. Y. Zhang, M. Xing and J. Zhang, "SAR Ship Targets Detection Based on Correlation Filters," 2021 CIE International Conference on Radar (Radar), Haikou, Hainan, China, 2021, pp. 1297–1300, https://doi.org/10.1109/Radar53847.2021.10028184.
17. S. Wei, H. Chen, X. Zhu and H. Zhang, "Ship Detection in Remote Sensing Image based on Faster R-CNN with Dilated Convolution," 2020 39th Chinese Control Conference (CCC), Shenyang, China, 2020, pp. 7148–7153, https://doi.org/10.23919/CCC50068.2020.9189467.
18. S. Zhou and J. Yin, "YOLO-Ship: A Visible Light Ship Detection Method," 2022 2nd International Conference on Consumer Electronics and Computer Engineering (ICCECE), Guangzhou, China, 2022, pp. 113–118, https://doi.org/10.1109/ICCECE54139.2022.9712768.
19. X. Leng, J. Wang, K. Ji and G. Kuang, "Ship Detection in Range-Compressed SAR Data," IGARSS 2022—2022 IEEE International Geoscience and Remote Sensing Symposium, Kuala Lumpur, Malaysia, 2022, pp. 2135–2138, https://doi.org/10.1109/IGARSS46834.2022.9884909.
20. J. Yuan, Z. Zhang and P. Zhang, "Ship Detection of SAR Image in Complex Nearshore Environment," 2022 17th International Conference on Control, Automation, Robotics and Vision (ICARCV), Singapore, Singapore, 2022, pp. 317–321, https://doi.org/10.1109/ICARCV57592.2022.10004222.
21. Y. Bai, S. Lei and L. Liu, "The Ship Target Detection Based on Sea-Sky-Line," 2021 6th International Conference on Automation, Control and Robotics Engineering (CACRE), Dalian, China, 2021, pp. 456–460, https://doi.org/10.1109/CACRE52464.2021.9501336.
22. Y. Zou, L. Zhao, S. Qin, M. Pan and Z. Li, "Ship target detection and identification based on SSD_MobilenetV2," 2020 IEEE 5th Information Technology and Mechatronics Engineering Conference (ITOEC), Chongqing, China, 2020, pp. 1676–1680, https://doi.org/10.1109/ITOEC49072.2020.9141734.
23. L. Ying, D. Miao, Z. Zhang and Q. Zhang, "Remote Sensing Ship Detection Based on Three-way Decisions and Multi-granularity Feature Fusion," 2022 IEEE 8th International Conference on Cloud Computing and Intelligent Systems (CCIS), Chengdu, China, 2022, pp. 234–238, https://doi.org/10.1109/CCIS57298.2022.10016364.
24. Y. Zhang, M. J. Er, W. Gao and J. Wu, "High Performance Ship Detection via Transformer and Feature Distillation," 2022 5th International Conference on Intelligent Autonomous Systems (ICoIAS), Dalian, China, 2022, pp. 31–36, https://doi.org/10.1109/ICoIAS56028.2022.9931223.
25. C. Zhang, P. Liu, H. Wang and Y. Jin, "Saliency-Based Centernet for Ship Detection in SAR Images," IGARSS 2022—2022 IEEE International Geoscience and Remote Sensing Symposium, Kuala Lumpur, Malaysia, 2022, pp. 1552–1555, https://doi.org/10.1109/IGARSS46834.2022.9883396.
26. C. Yu and Y. Shin, "A Deep Learning-Based SAR Ship Detection," 2023 International Conference on Artificial Intelligence in Information and Communication (ICAIIC), Bali, Indonesia, 2023, pp. 744–747, https://doi.org/10.1109/ICAIIC57133.2023.10067131.
27. Y. Lei, X. Leng and K. Ji, "Marine Ship Target Detection in SAR Image Based on Google Earth Engine," 2021 IEEE International Geoscience and Remote Sensing Symposium IGARSS, Brussels, Belgium, 2021, pp. 8574–8577, https://doi.org/10.1109/IGARSS47720.2021.9553711.
28. N. Wang and F. Ma, "Single ship target detection based on the concept of edge computing," 2021 6th International Conference on Transportation Information and Safety (ICTIS), Wuhan, China, 2021, pp. 559–564, https://doi.org/10.1109/ICTIS54573.2021.9798642.
29. K. Zhao, Y. Zhou and X. Chen, "A Dense Connection Based SAR Ship Detection network," 2020 IEEE 9th Joint International Information Technology and Artificial Intelligence Conference (ITAIC), Chongqing, China, 2020, pp. 669–673, https://doi.org/10.1109/ITAIC49862.2020.9338986.
30. Moosbauer, S.; Konig, D.; Jakel, J.; Teutsch, M. A benchmark for deep learning based object detection in maritime environments. In Proceedings of the IEEE Comput. Soc. Conf. Comput. Vis. Pattern Recognit. Work (CVPRW), Long Beach, CA, USA, 16–17 June 2019; pp. 916–925.

31. Spraul, R.; Sommer, L.; Arne, S. A comprehensive analysis of modern object detection methods for maritime vessel detection. Artificial Intelligence and Machine Learning in Defense Applications II. Int. Soc. Opt. Photonics 2020, 1154305.
32. Betti, A.; Michelozzi, B.; Bracci, A.; Masini, A. Real-Time target detection in maritime scenarios based on YOLOv3 model. arXiv 2020, arXiv:2003.00800.
33. Nalamati, M.; Sharma, N.; Saqib, M.; Blumenstein, M. Automated Monitoring in Maritime Video Surveillance System. In Proceedings of the 35th International Conference on Image and Vision Computing New Zealand (IVCNZ), Wellington, New Zealand, 25–27 November 2020; pp. 1–6.
34. Shao, Z.; Wu, W.; Wang, Z.; Du, W.; Li, C. SeaShips: A Large-Scale Precisely Annotated Dataset for Ship Detection. IEEE Trans. Multimed. 2018, 20, 2593–2604.

Enhancing Military Capability Through Artificial Intelligence: Trends, Opportunities, and Applications

1 Introduction

Utilizing cutting-edge technologies for military and defense objectives has become increasingly attractive to countries like the USA, China, and Russia in recent times. AI's rise has had a significant impact on many industries, leading some of them to integrate AI-powered solutions into their daily operations. The first AI plan was released by the US Department of Defense in 2019, which signaled the beginning of the integration of AI systems in studies, military, and defense projects. Artificial Intelligence has completely changed industries and has been a revolutionary force in the military and defense sectors for a considerable amount of time. It has therefore become an essential part of modern military and combat strategies. One of the most secure federal agencies dedicated to integrating AI into military and defense applications is the Department of Defense (DoD), which is well known for protecting sensitive and classified data. Wealthy countries are investing a lot of money in AI to bring new ideas to the military and defense industry [1].

The Department of Defense (DoD) has faced the difficulty of quickly and widely applying the most cutting-edge research and development breakthroughs available worldwide—a task that artificial intelligence is progressively helping to accomplish. Particularly in the field of generative AI, artificial intelligence has made significant strides in both capabilities and accessibility in the last year. For the military to maintain security and its technological edge, it is essential that it keep up with these developments. It will always be difficult for the military to fully utilize AI's promise to improve operational effectiveness as new applications and advancements in the technology arise. Since AI is always evolving, it is imperative that new approaches to incorporating it into military operations be investigated and implemented.

The market for artificial intelligence in military applications was estimated to be worth USD 7.93 billion globally in 2022 and is expected to grow to roughly USD 22.62 billion by 2032. This significant increase highlights the growing importance of

© European Alliance for Innovation 2024

A. Haldorai et al., *Artificial Intelligence for Sustainable Development*,
EAI/Springer Innovations in Communication and Computing,
https://doi.org/10.1007/978-3-031-53972-5_18

AI as a key component of contemporary warfare. With sophisticated computing and decision-making processes, artificial intelligence (AI) can effectively manage enormous amounts of military data and improve self-control, self-regulation, and self-actuation within military systems. As AI develops and becomes more sophisticated, its function in overseeing and streamlining military operations becomes more and more crucial [2].

The remaining part of the chapter is organized as follows: Section 2 elaborates the basic AI strategy for defense systems. Section 3 discusses the various AI algorithms supporting wargaming and Sect. 4 outlines the prevailing scope of AI for defense. Section 5 gives various AI factors required for defense application and Sect. 6 outlines the challenges and future scope of AI-based defense sector. Finally, Sect. 7 concludes the chapter.

2 Artificial Intelligence Strategy for Defense

AI has the extraordinary ability to imagine new frontiers in previously unheard-of ways. The emergence of new tools that facilitate decision-making in the public sector has been made possible by the combination of exponential data expansion, improved technical capabilities, higher processing capacity, and significant government funding in AI research and development. These defense and security agencies, be they border security, law enforcement, or the military, need to be prepared to handle a wide range of situations. This entails resolving ethical issues, integrating emerging technology, streamlining current procedures, operationalizing their data efficiently, and acquiring the necessary equipment to meet their goals. Artificial intelligence's rapid development has brought about a number of positive effects in a variety of industries, including transportation and healthcare. However, there are a lot of important concerns and debates when AI is used for military purposes. The potential of AI to transform military operations—including logistics, intelligence, surveillance, and, most significantly, weapons development—is becoming more widely recognized. But the current ambiguities and uncertainties surrounding AI and its military application have caused military strategists to engage in serious discussions. Although there is no denying AI's revolutionary potential in the military, there is also no denying the difficulties and ethical issues it raises, which have sparked a lively debate within the defense sector. Artificial Intelligence has the potential to completely change their operations and improve their ability to negotiate ever-changing, complex security environments [3]. Figure 1 illustrates the pipeline of AI strategy for defense application.

Consistent with advancements in other domains, artificial intelligence is driving military forces throughout the world to further incorporate AI into their combat systems. Command and control, intelligence, surveillance, logistics, healthcare, information warfare, cyber warfare, training and simulation, autonomous systems, and even deadly autonomous weapons are just a few of the military fields where artificial intelligence is now being used. This broad use of AI in the military

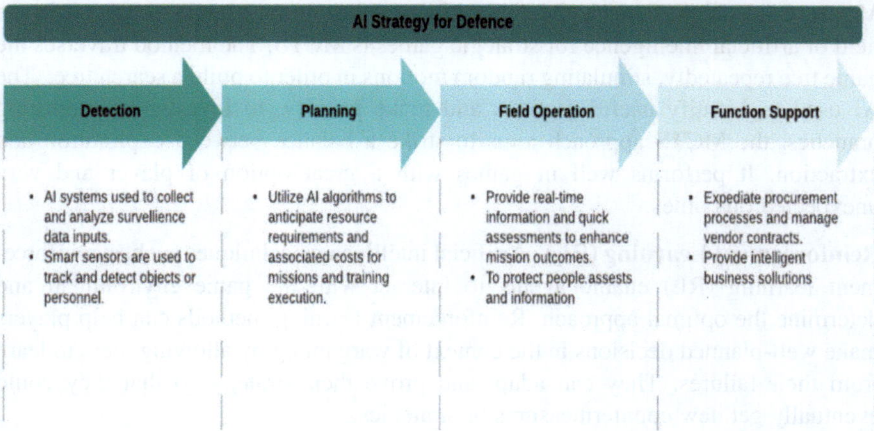

Fig. 1 Outline of AI strategy for defense

demonstrates the technology's ability to transform and optimize a wide range of military tasks, improving operational performance but also posing moral and geopolitical questions, especially in relation to autonomous weapon systems. AI is understood in the context of defense as a class of flexible technologies that have the capacity to enable robots to carry out tasks that traditionally require human or biological intelligence. Its contributions are manifold: it speeds up and improves the quality of decision-making and military operations; it strengthens the security and robustness of networks that are connected; it increases the size, endurance, reach, and efficacy of our military forces; and it protects people by automating jobs that are deemed "dull, dirty, and dangerous." The integration of AI into security operations demonstrates its potential to enhance security, efficiency, and strategic agility, while also optimizing military capabilities [4].

3 AI Algorithms for Wargaming

Artificial intelligence (AI) algorithms are essential in wargaming since they enable intelligent decision-making, simulate opponent behavior, optimize strategies, and improve overall game dynamics [5]. Below are a number of AI algorithms frequently employed in the context of wargaming:

Minimax Algorithm The minimax algorithm is a crucial method employed in wargaming to ascertain the most advantageous plays in a game that is played in turns. The algorithm works by considering every possible move and its potential outcomes, assuming the opponent would make the best possible countermove. The computer utilizes a function to evaluate possible moves selecting the one that maximizes the projected outcome for the player while minimizing the rivals.

Monte Carlo Tree Search (MCTS) The most common search technique in the field of artificial intelligence for strategic games is MCTS. The method traverses the game tree repeatedly, simulating random motions in order to build a search tree. The AI used to identify useful motions and make priority, to investigate interesting branches, the MCTS approach aims to strike a balance between exploration and extraction. It performs well in games with a great option of player and very unexpected outcomes.

Reinforcement Learning (RL) Artificial intelligence techniques such as reinforcement learning (RL) enable agents to interact with the game environment and determine the optimal approach. Reinforcement learning methods can help players make well-planned decisions in the context of wargaming by allowing them to learn from their failures. They can adapt and prove their strategy so that they could eventually get new countermeasures or strategies.

Genetic Algorithms (GA) GA is a computational optimization technique based on natural selection ideas. Genetic Algorithms (GA) can be used iteratively to improve tactics in wargaming. Selection crosses and mutation are examples of genetic agents employed over generations to gradually improve tactics. Different types of parameters are represented in a population, some individuals in each case.

Fuzzy Logic Fuzzy logic facilitates the logical expression and analysis of facts that are not more likely to be made. AI for wargaming might be used to replicate complex and unpredictable decision-making. Fuzzy logic may be used in circumstances where variables that are difficult to measure—like morale, uncertainty, or contextual factors—are present. By using vague concepts and linguistic elements, it provides an adaptable structure for assessment.

Neural Networks Wargaming is one of the many fields where neural networks, and more especially deep learning systems, have been effectively used. Through the study of input factors like game position as well historical data, or inputs from sensors, they may recognize patterns, recognize the behavior of their opponents, and forecast results. Neural networks are capable of making sophisticated decisions and adapting to changing game circumstances.

Particle Swarm Optimization (PSO) PSO is an optimization technique inspired by the social behavior seen in fish schools and bird flocks. PSO, in the context of wargaming, can be employed to optimize parameters or explore optimal methods across a vast solution space. It combines exploration and exploitation by iteratively updating a population of candidate solutions based on their individual and group performance.

4 Scope of AI for Defense Sector

It is anticipated that the military will continue to rely more and more on AI applications as they become an essential part of its operations. AI has advanced to the point where it can now manage activities that are getting more and more complicated, and in certain cases, it has almost completely replaced the need for human interaction [6–8]. Within the military, artificial intelligence has proven to be versatile in improving efficiency and capabilities, with applications ranging from data processing to battle simulations [11].

4.1 Platforms for Warfare

AI is being actively used by nations all over the world to develop cutting-edge weapon systems that can function in the air, space, land, and naval domains. The goal of this push to create AI-powered weapons is to make military operations more effective [17]. As a result, contemporary technology is encouraging increased cooperation and improvements in military prowess. AI-based weapons and systems stand out for their low maintenance requirements and decreased dependence on human interaction. AI is therefore at the forefront of developing autonomous, highly effective, and quickly deployed weaponry that can meet the needs of contemporary conflict. The transition to AI-powered weapons represents a significant development in military technology, characterized by greater efficiency and autonomy.

4.2 Training and Simulation

In any area related to the military or defense, training is unquestionably essential. Although it would be great, it isn't always possible to fully train cadets for every scenario they might face. A useful method for bridging this gap is revealed: simulation. Realistic models are created through the integration of computer science, software engineering, and system engineering via simulation. In order to familiarize oneself with the many fighting systems used in different military missions, troops might use these simulations. Sensor simulation systems have been widely used because of the thorough investigation of battle undertaken by the US Army and Navy. Notable corporations such as Orbital ATK, Leidos, SAIC, and AECOM have actively collaborated with the US Navy to enhance its training programs. The US Army also uses Millennium Engineering, Torch, CACI, and SAIC to improve its training programs. These collaborations emphasize the critical role of simulation in giving military personnel a safe and comprehensive learning environment, equipping them to handle the complexities of contemporary warfare [12].

4.3 Target Recognition

When it comes to complex military situations, artificial intelligence (AI) is now crucial for enhancing target detection via better equipment and processes. A thorough understanding of both the external environment and real-world activities may be achieved by the military with the help of artificial intelligence. The evaluation and analysis of a wide variety of reports, papers, and dispersed material leads to improved findings and outcomes. The military's target detecting systems are also greatly improved by artificial intelligence. These AI-powered gadgets make it much easier to foresee and assess the state of conflict. This is best shown by the DARPA-run TRACE program, which shows how effective machine learning algorithms are in locating and identifying targets. Synthetic Aperture Radar (SAR) data is used by TRACE, an automated target detection system, to enhance situational awareness and operational effectiveness in military environments.

4.4 Healthcare

The devastation caused by war has surely had a significant influence on the development of healthcare treatments. In the past, military personnel have often suffered burns, severe tissue damage, and catastrophic casualties that resulted in a large loss of blood. This gave rise to ideas like blood transfusions, which have saved many lives and are still relevant in modern medicine. Furthermore, artificial intelligence (AI) is now being used by the military to operate robotic surgical systems (RSS) and robotic ground platforms (RGPs). With these developments, there are a lot of potential to do remote surgical operations during crises and evacuations. The US military has set aside a significant amount of money to improve medical facilities on the front lines by acquiring RGPs, RSS, and other AI technology. AI systems are able to gather and evaluate a soldier's medical data during a combat because to this expenditure. One prime example is Watson from IBM, which is working with the US Veterans Administration to create a clinical reasoning prototype called EMRA (Electronic Medical Record Analyzer). This system automatically identifies and classifies health conditions according to their criticality by processing medical records using AI and machine learning algorithms. Such developments in AI-powered healthcare systems have the potential to completely change how healthcare is delivered, both in military and civilian contexts.

4.5 Cybersecurity

As with other traditional realms like land, sea, and air, cyberspace has definitely become a new theater of war. Now that they understand how crucial data protection is in the current day, military and defense companies are investing heavily in strengthening their networks. It is now more important to use AI and ML methods to improve data security and protect private data. AI and ML technologies are essential in preventing and stopping unauthorized breaches by using proactive measures to counter cyber threats. These systems provide a notable accuracy in detecting breaches and irregularities, which helps facilitate a prompt and efficient response to threats. Intelligent machine learning and artificial intelligence improve the army's defense capabilities. Further it demonstrates how dynamic and continually evolving the current war is and how cybersecurity is essential to maintaining national security.

4.6 Logistics

Logistics and transportation are essential for defense and military operations, in order to deploy soldiers, weapons equipment, and ammunition efficiently. Since businesses like Amazon began using artificial intelligence to enhance product delivery, the logistics industry has significantly advanced. AI is likely to improve military transportation by simplifying processes, cutting costs and requiring less human participation. By using this technology, we can identify any differences or irregularities in military fleets that may endanger our goals. One example is IBM's partnership with the US Army to use artificial intelligence in combat environments. One part of this collaborative project involved using artificial intelligence to pinpoint maintenance issues with the stryker, a military vehicle used to transport soldiers during conflict. By integrating AI with military mobility technology, military operations may be more efficient and effective overall.

4.7 Threat Monitoring

Data collection and analysis are crucial for military operations since they aid in many defensive duties, especially in monitoring potential threats and maintaining situational awareness. Autonomous systems are crucial in this scenario. These devices have the capability to operate autonomously or follow predetermined paths. Individuals serving in the military may find these gadgets beneficial because of their improved situational awareness and AI-driven danger identification capabilities. Such operations are more prevalent with drones equipped with artificial intelligence capabilities. One of its numerous advantages is its ability to detect danger, monitor border areas, and promptly call rescue teams. Drones equipped with artificial intelligence greatly improve military facilities' defenses and the safety of troops in battle. The military's capabilities and safety precautions are greatly enhanced by these technology developments in AI-driven threat monitoring and situation awareness systems.

4.8 Casualty Prediction and Support

Soldiers and medical professionals frequently have to make snap decisions in high stress situations. AI is a huge help to them, especially when one of their comrades needs medical attention. Using algorithms, this type of AI accesses large medical databases containing information on trauma cases, including diagnosis, vital signs, drugs, treatments, and their results. In order to provide indications, cautions, and treatment recommendations, the AI system integrates this medical data with information that has been manually entered. Even while AI is great at making data-driven, unbiased recommendations free from emotional bias, it still needs human oversight to function properly. In order to make informed judgments that consider these AI-generated recommendations, human operators need to be emotionally intelligent. AI is a useful tool that can quickly analyze data and provide humans with more information to make judgments, even though it is not equipped to make medical decisions. Especially in high-pressure and time-sensitive scenarios, this cooperative method makes use of the advantages of both AI and human decision-makers to guarantee the best possible results [16].

5 Key Factors for the Successful Use of AI in Defense

A branch of AI called "machine learning" involves using large datasets to train computer programs. By going through this process, the computer is prepared to learn from the data and then produce predictions or judgments based on what it has learned [9]. To guarantee the effective implementation of AI, certain essential conditions must be met:

(a) Data: The ability to gather relevant and reliable information from a variety of sources is vital in battle. Data cleaning, processing, and aggregation are crucial steps in ensuring the data is fit for use in AI systems. As an added requirement, the data must be compatible with the specific operating ecosystem. Findings from a desert location may not help algorithms designed for use at higher elevations. Customizing the data to fit the specific context and situations is essential for AI systems to work in military settings.
(b) Interoperability: When deploying AI systems in a military setting, interoperability must be the top priority. These systems must be able to quickly share data and interface with existing ones for AI to be used successfully. APIs, globally accepted data protocols, and other interoperability-promoting strategies are critical to reaching this objective. By taking these steps, many systems will be able to communicate and share data more easily.
(c) Power of Data Computation: AI systems require a significant amount of computer power to efficiently handle and analyze large amounts of data. Frequently, this necessitates the deployment of high-performance computer systems, cloud computing infrastructure, or other sophisticated computational resources. A further layer of intricacy occurs due to the requirement of evaluating a significant amount of data at the edges in close contact to the user or the location of action. Hence, it is crucial to possess a sufficient level of processing capability in order to enable prompt decision-making in a highly competitive battlespace.

(d) Security: Ensuring the security of AI systems used in military operations is of utmost importance in order to protect them from any cyberthreats and attacks. Currently, there is a danger to the data integrity of our AI technology. This presents a chance for malevolent actors to take advantage of these systems' flaws and use AI as a weapon. Proactive protection requires putting strong security measures in place, such as firewalls, encryption, and safe data storage. It is also advised to include AI technology into these procedures. To guarantee the resilience and dependability of AI systems employed in military environments, several precautions are crucial.

(e) Ethics: AI models have the potential to attain a degree of self-governance that raises serious moral and legal concerns about its use to military operations. Prioritizing upholding moral principles and keeping up with technological developments is essential. Efficient regulation of AI use in the military necessitates the formulation and adherence to clearly established ethical standards. To maintain an advantage over possible adversaries in this rapidly evolving context, it is imperative to ensure the ethical use of artificial intelligence and minimize unnecessary barriers to its implementation by implementing the following regulations.

(f) Expertise: The successful collaboration of a heterogeneous group is imperative for the advancement and use of AI in military contexts. Appropriate members of this team should have backgrounds in artificial intelligence, data science, cybersecurity, computer engineering, military activities, and ethics. Under certain circumstances, the military's existing resources may lack the requisite expertise. Hence, to ensure the availability of essential information and skills for the seamless incorporation of AI in military operations, it may be imperative to enlist civilian expertise and engage external specialists.

6 Challenges and Future Directions

In order to meet the changing needs of combat zones, modern warfare requires a high level of technical integration and strategic intelligence. An enormous amount of unstructured data is flooded into these spaces. Artificial intelligence has the capacity to provide real-time decision assistance, organize data, and classify event [13]. AI might offer useful alternatives that human cognition might find difficult to evaluate quickly owing to data volume and decision time limits, since information drives modern conflicts. International military powers are proactively creating a single platform to address situation-specific difficulties in space, on land, and at sea. They are doing this by combining disruptive technology with AI-enabled tools to achieve overall efficacy. Warfare is changing from a quantity-driven strategy that emphasizes damage and attrition to a knowledge-based paradigm that emphasizes effects and results [10].

The rise of a new doctrine of war is being driven by the combination of AI technology with conventional fighting skills. This idea places a strong emphasis on making quick, accurate decisions, deploying quickly, and undermining the enemy's ability and will to fight. It highlights the more general goal of impeding the enemy's capacity to wage war effectively, going beyond simply targeting opposing armaments and arsenals. This change emphasizes how combat is changing to become more strategic and knowledge-driven [14].

Without the underlying hardware and networked systems, AI cannot provide a complete answer. Improved seafloor submarine operations could result from AI's ingrained operational expertise. Submarines are able to recognize and react to qualitative changes in their surroundings with increasingly complex and advanced alternatives thanks to cognitive frameworks powered by AI. Underwater submarines are able to operate more efficiently and effectively because of the combination of AI with hardware, networks, and cognitive frameworks.

With AI's ability to increase productivity, reduce user workloads, and complete jobs faster than humans, the military can benefit greatly from its deployment. Improvements in AI's robustness, explainability, and capacities are the focus of ongoing research initiatives. Although the military cannot ignore this technology, it is imperative to understand that placing weak AI systems in contested environments and giving them decision-making authority could have disastrous results. Until such time as humans can offer the required oversight and accountability to guarantee the safe and ethical deployment of AI, humans will always have to make the crucial decisions.

7 Case Study: Image Analysis for Defense Intelligence

Project SPOTTER uses Machine Learning techniques to support and enhance imagery analysts' output, using automated detection and identification of objects within classified satellite imagery. Specialized graphics processing unit computing capabilities are being developed by the Centre for Intelligence Innovation. Through the utilization of Convolutional Neural Network techniques, they have effectively taught the program to identify particular objects of interest. In order to identify changes in activity, this permits automatic, continuous surveillance of sites around-the-clock. Prioritizing tasks with the help of this creative method improves picture intelligence reporting effectiveness and timeliness. An extensive investigation into several open-source algorithm frameworks, including YOLO, Mask R-CNN, and CenterNet, was carried out by the Centre for Intelligence Innovation [15].

The goal of this project was to create models that could function in a variety of habitats and biomes with a wide range of object kinds. A unique application architecture and interface were painstakingly created to maximize the benefit for Defense Intelligence. Scalability, modularity, and customization to satisfy the specific requirements of analysts are features of this architecture that guarantee an effective and appropriate solution for their jobs. Project SQUINTER, is primarily concerned with Synthetic Aperture Radar (SAR) satellite data as opposed to Electro-Optical data. To autonomously recognize and identify things of interest, it makes use of computer vision and machine learning algorithms. By providing analysts with a uniform user interface for effective and thorough data analysis, the SPOTTER application acts as a platform to host and exploit these models, streamlining these capabilities. In order to meet the unique requirements of Defense Intelligence, the Centre for Intelligence Innovation has accumulated a lot of experience in the application of AI and machine learning. Continued work will concentrate on improving algorithm performance, expanding the utility to several analytical teams, and widening the range of objects that can be detected. This exemplifies AI's disruptive potential since it offers analytical insights beyond what is currently possible given the constraints of few human resources. Defense Intelligence could undergo a revolution and broaden its scope of analysis with the continuous development of AI capabilities.

8 Conclusion

Future warfare will either undergo revolutionary or evolutionary changes as a result of AI, a topic of ongoing debate. Still, artificial intelligence is already having a significant effect on military equipment. Now, armed forces who do not adopt AI run the risk of falling behind in terms of technology. It's critical to realize that artificial intelligence is not a technology that can be easily plugged in and used in the armed forces. The Indian military must thoroughly reevaluate its workforce preparation, organizational structures, network integration, data management procedures, and organizational structure in order to fully utilize AI, even though certain simple applications may fall into this category. How to use technology to improve the overall effectiveness of our armed forces is currently the main challenge. This chapter has given a thorough review of the worldwide use of AI in a variety of military fields, stressing the requirements and current challenges as well as future directions.

References

1. Bhardwaj, "Application of Artificial Intelligence in Disaster Response," Artificial Intelligence, pp. 183–194, Oct. 2021, https://doi.org/10.1201/9781003140351-19.
2. Szabadföldi, "Artificial Intelligence in Military Application—Opportunities and Challenges," Land Forces Academy Review, vol. 26, no. 2, pp. 157–165, Jun. 2021, https://doi.org/10.2478/raft-2021-0022.
3. P. Jayakumar, S. N. Brohi, and N. Z. Jhanjhi, "Artificial Intelligence and Military Applications: Innovations, Cybersecurity Challenges and Open Research Areas," Aug. 2021, https://doi.org/10.20944/preprints202108.0047.v1.
4. U. S. Gaire, "Application of Artificial Intelligence in the Military: An Overview," Unity Journal, vol. 4, no. 01, pp. 161–174, Feb. 2023, https://doi.org/10.3126/unityj.v4i01.52237.
5. Robert Ahmed and Christna Ahler, "A Review on Background and Applications of Machine Learning in Materials Research", Journal of Computational Intelligence in Materials Science, vol.1, pp. 077–087, 2023. https://doi.org/10.53759/832X/JCIMS202301008.

6. J. T. Jacobsen and T. Liebetrau, "Artificial intelligence and military superiority," Artificial Intelligence and International Conflict in Cyberspace, pp. 135–156, Apr. 2023, https://doi.org/10.4324/9781003284093-8.
7. V.M. BURENOK, "Artificial Intelligence in Military Confrontations of the Future," Military Thought, vol. 30, no. 003, pp. 129–137, Sep. 2021, https://doi.org/10.21557/mth.70170498.
8. Neyigapula, B. S. (2023). Human-AI Collaborative Decision-making: A Cognitive Ergonomics Approach. https://doi.org/10.21203/rs.3.rs-3258718/v1
9. Rasch, R., Kott, A., & Forbus, K. D. (2003). Incorporating AI into military decision making: an experiment. IEEE Intelligent Systems, 18(4), 18–26. https://doi.org/10.1109/mis.2003.1217624
10. Nalin, A., & Tripodi, P. (2023). Future Warfare and Responsibility Management in the AI-based Military Decision-making Process. Journal of Advanced Military Studies, 14(1), 83–97. https://doi.org/10.21140/mcuj.20231401003
11. V. Kalpana, D. K. Mishra, K. Chanthirasekaran, A. Haldorai, Srigitha. S. Nath, and B. K. Saraswat, "On Reducing Energy Cost Consumption in Heterogeneous Cellular Networks using Optimal Time Constraint Algorithm," Nov. 2021, https://doi.org/10.21203/rs.3.rs-1090831/v1.
12. Uma Maheswari Arumugam, Suganthi Perumal, "Trust based Secure and Reliable Routing Protocol of Military Communication on MANETs", Journal of Machine and Computing, pp. 047–057, January 2023. https://doi.org/10.53759/7669/jmc202303006.
13. K. J, A. H, P. V, Y. S, and G. N, "A Study Of Dynamic Thresholds Power Detection Spectrum Sensing Techniques In CRN," 2023 9th International Conference on Advanced Computing and Communication Systems (ICACCS), Mar. 2023, https://doi.org/10.1109/icaccs57279.2023.10112923.
14. Garg, P. K., & Sharma, L. (2021). Artificial Intelligence: Challenges and Future Applications. Artificial Intelligence, 229–245. https://doi.org/10.1201/9781003140351-22
15. Jayakumar, P., Brohi, S. N., & Jhanjhi, N. Z. (2021). Artificial Intelligence and Military Applications: Innovations, Cybersecurity Challenges and Open Research Areas. https://doi.org/10.20944/preprints202108.0047.v1
16. St. Amant, R. (2023). Future trends and applications of AI to military challenges (Conference Presentation). Artificial Intelligence and Machine Learning for Multi-Domain Operations Applications V. https://doi.org/10.1117/12.2682188
17. Steff, R., & Abbasi, K. (2020). Artificial intelligence and the military balance of power. Emerging Technologies and International Security, 65–82. https://doi.org/10.4324/9780367808846-6

Crisis Assessment Through Satellite Footage Using Deep Learning Techniques for Efficient Disaster Response

1 Introduction

Humanitarian organizations are tasked with providing immediate, highly effective, and efficient assistance to those who are experiencing security challenges following a natural disaster. Decision-makers can gain important insights from damage assessment, which is the first assessment of the disaster's impact. Although satellite photography provides trustworthy and useful information for making well-informed decisions, the annotation process for this imagery is still a tedious and time-consuming undertaking. In the field of disaster risk reduction and management, AI is becoming more and more essential. Its uses range from developing hazard maps and forecasting extreme occurrences to detecting events in real time, giving situational awareness, and supporting decision-making processes [1]. AI refers to technologies that can mimic or even exceed human intelligence in particular activities. Within the field of artificial intelligence, machine learning (ML) comprises a range of techniques, such as supervised models (e.g., random forests or decision trees), unsupervised models (e.g., K-means), and reinforcement models (e.g., Markov decision process). Essentially, machine learning (ML) is the process of digesting data into algorithms that use that data to learn and provide predictions or classifications. AI methods open up new possibilities, especially in areas like post-processing forecast model outputs and preparing observational data. Technological developments in processors have further expanded these possibilities by enabling high-performance, parallel data processing [2].

Over the last decade cumulative impact of these disasters underscores the importance of a proper response, particularly in mitigating the loss of life, displacement, and other tragic consequences. Infrastructure damage emerges as a pivotal factor triggering these outcomes. Evaluating such damages, especially in the midst of a disaster or conflict, can be a time-consuming and resource-intensive process [24].

© European Alliance for Innovation 2024

A. Haldorai et al., *Artificial Intelligence for Sustainable Development*, EAI/Springer Innovations in Communication and Computing,
https://doi.org/10.1007/978-3-031-53972-5_19

Efficient response teams require precise information on locations requiring immediate attention. With the expanding market for geospatial data analytics and the integration of deep learning, there is potential to significantly reduce the consequences of these hazards.

This chapter explores how leveraging modern geospatial data can enhance awareness and response capabilities, establishing an automated pipeline for disaster management through the use of satellite imagery and deep learning for damage detection.

The remainder part of this chapter is as follows: Section 2 covers the various literature reviews. Section 3 gives an overview of disaster management. Types of disaster covered in Sect. 4 and its challenges in Sect. 5. The application of AI for damage detection from satellite images is given in Sect. 6. The proposed system methodology is outlined and experimental results in Sect. 7, in Sect 8, concludes the chapter.

2 Literature Review

Two main challenges remain concerning image classification for damage assessment: the availability of data resources and the timely gathering of accurate and actionable information. In this section, prevailing work addressing these challenges was reviewed, based on advanced approaches deep learning methods.

In [3], the author in this research utilizes machine learning techniques to automate the detection of building damage in satellite imagery. They conducted a comparative analysis of four distinct convolutional neural network models to identify damaged buildings in the context of the 2010 Haiti earthquake. Moreover, they evaluated the generalization capabilities of these models by training and testing them on various disaster events, aiming to quantify their performance in future disaster scenarios.

In work [4], the author proposed an innovative unsupervised deep learning method for detecting building damage after a disaster, capable of adapting to various damage types and geographic locations. Previous attempts in this area were hindered by a lack of high-quality training data. To address this, the author suggested employing a cutting-edge Anomaly Detecting Generative Adversarial Network (ADGAN) that relies solely on pre-event images showcasing undamaged buildings. This approach offers valuable support during the post-disaster response phase, as the model can be developed beforehand and swiftly deployed once the event occurs.

To ensure prompt and practical information for disaster response, a framework is introduced in [5], utilizing segmentation neural networks to identify affected areas and accessible roads in post-disaster situations. The study evaluates the impact of pretraining with ImageNet on aerial image segmentation and compares the performance of various popular segmentation models. The experimental findings demonstrate that pretraining on ImageNet typically enhances segmentation performance across multiple models.

In research [6], the author introduces and evaluates three multi-resolution CNN feature fusion approaches in comparison to two baseline mono-resolution methods for image classification of building damages. The findings reveal improved accuracy and localization capabilities when fusing multi-resolution feature maps, particularly when merging these maps and incorporating feature information from intermediate layers of networks at different resolution levels. In the case of satellite and aerial (unmanned) imagery, the accuracy improvements reached 2%, while the improvement in accuracy for the airborne (manned) imagery was minimal.

The author in [7] has proposed a unified model called RescueNet, designed to perform simultaneous building segmentation and assess damage levels for individual buildings. RescueNet is trained end-to-end and provides improved performance in building segmentation and damage classification compared to previous methods. The model is evaluated on the extensive and diverse xBD dataset, demonstrating its ability to generalize across different geographical regions and types of disasters.

In [8] the author presents a novel approach called the difference-aware attention network (D2ANet) for the simultaneous tasks of building localization and multilevel change detection using dual-temporal satellite imagery. In order to improve the model's ability to detect damage at different scales, a difference-attention module is made to take advantage of local correlations between the various degrees of alterations.

For the purpose of assessing building damage, the author in [9] introduces BDANet, a unique two-stage convolutional neural network. A U-Net is utilized in the first phase to locate building locations. The network assesses building damage in the following stage by using shared weights from the previous stage. A two-branch multiscale U-Net serves as the foundation of the second stage, which accepts different inputs for pre- and post-disaster photos. In order to obtain correlations between images, a cross-directional attention module is implemented.

3 Overview of Disaster Management

Disaster handling is a comprehensive approach designed to lessen the negative effects that natural and man-made disasters have on people, the environment, and infrastructure. This strategy includes a range of actions centered around readiness, quick action, recuperation, and prophylactics (Fig. 1).

Preparation is a crucial component of disaster management; it includes organizing, getting ready, and providing communities with the tools they need to handle calamities. This includes creating emergency response plans, carrying out exercises and drills, and educating the public on disaster preparedness measures. As soon as a disaster strikes, emergency services are called upon to conduct search and rescue operations, provide medical aid, and monitor the safety of those who are impacted. Activities during this stage include clearing debris, restoring infrastructure, and

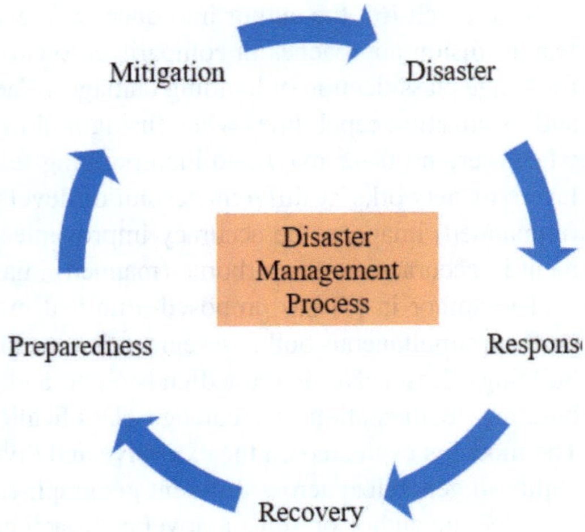

Fig. 1 Disaster management process

helping impacted people and communities start over. Long-term rehabilitation initiatives could focus on social and psychological effects as well as economic revival [10].

Mitigation is an ongoing effort to lessen the risks and vulnerabilities associated with calamities. Several steps must be taken in order to complete this process, such as developing early warning systems, strengthening infrastructure resilience, managing land use, and launching public awareness campaigns. These actions reduce the effects of approaching disasters and promote long-term resilience by reducing the basic risks. Strong coordination and collaboration among a variety of stakeholders—including governmental institutions, emergency services, nongovernmental organizations, communities, and individuals—are essential to the success of disaster management. In order to improve preparation, reaction, and recovery activities, it also depends on the integration of technology, data, and creative ideas. Adapting tactics and procedures based on knowledge from past experiences requires ongoing assessment, learning, and development.

4 Types of Catastrophes

Disasters range in kind and length, from short-lived interruptions to protracted destruction that lasts for days or even weeks. Figure 2 covers both natural and man-made events, and provides an overview of the several disaster kinds that could affect a community [11].

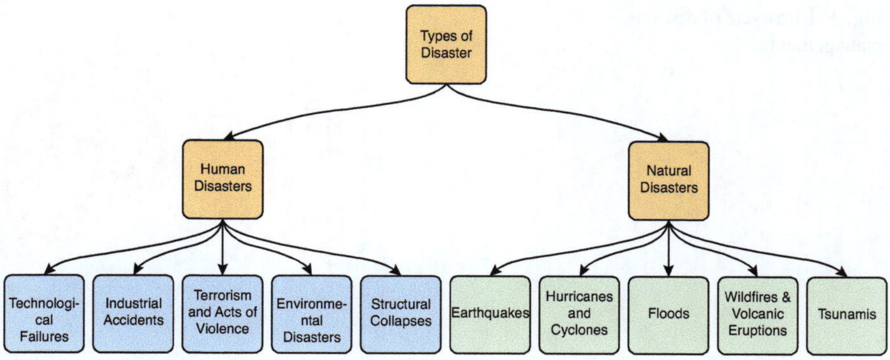

Fig. 2 Types of disasters

4.1 Human Catastrophes

These are incidents that cause significant destruction to infrastructure, the environment, or human life as a result of human activity, carelessness, or intentional intent. Human disasters include things like chemical spills, terrorism, industrial accidents, conflicts, and technical failures. Human disasters usually include a human component and can be prevented or mitigated by improved safety procedures, laws, and responsible behavior. Understanding the many types of human disasters is essential to developing preventive plans, improving safety standards, and fortifying emergency response capacities in order to mitigate their effects and protect infrastructure and people.

4.2 Natural Disasters

As opposed to human disasters, natural catastrophes are terrible events brought on by uncontrollably occurring natural processes and phenomena. These include natural disasters that result from geological, meteorological, or hydrological processes, such as earthquakes, hurricanes, floods, tsunamis, wildfires, volcanic eruptions, and droughts. Natural disasters cannot be stopped by people, but their effects can be lessened by forethought, early warning systems, and effective disaster management.

5 Challenges to the Use of AI for Disaster Management

Ultimately, the selection of an appropriate model architecture and the availability of high-quality data are the two main determinants of an AI model's efficacy for a given task. Instrumental networks (such as meteorological, hydrometeorological, and

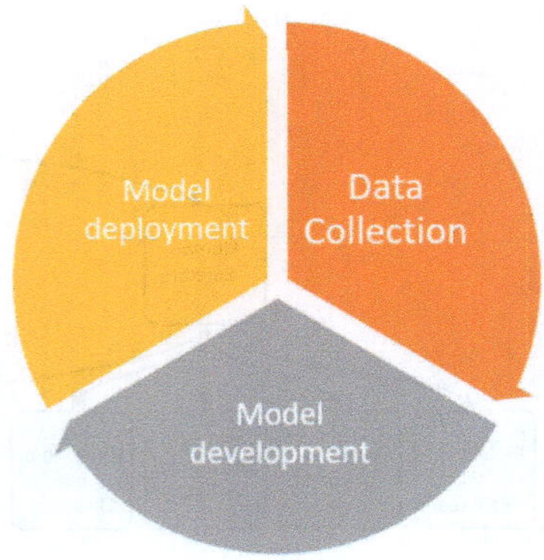

Fig. 3 Life cycle of disaster management

seismic stations), crowd-sourced data, and remote sensing technologies (such as satellites and drones) have all contributed to the tremendous expansion of Earth observational data. At the same time, model architectures are always being improved and refined [12].

Figure 3 depicts the basic lifecycle of disaster management, which includes data collection, model development, and deployment. Concerns about potential biases in training/testing datasets, the integration of new distributed AI technologies inside the data domain, and ethical considerations should all be taken into consideration during the data gathering and handling process.

1. To mitigate biases in training and testing datasets, it is crucial to guarantee accurate data sampling that includes an adequate representation of all important patterns for the given problem. For example, think about the difficulty of creating a representative dataset with uncommon severe occurrences in it. An inaccuracy in forecasting or bias in the results could result from a failure to supply relevant data.
2. Assuring objectivity of the dataset is only the first step. The subsequent phase involves incorporating novel distributed AI technologies into the data domain. Petabyte-scale volumes of data are being produced as a result of changes made to the design of space-based instrumentation, such as the placement of several tiny satellites, and the introduction of edge computing. Because AI depends on sophisticated machine learning calculations and data transmission, centralized data processing may provide difficulties. While machine learning algorithms usually work in a centralized way, necessitating data fusion on centralized servers, real-time catastrophe applications frequently require strong multinational relationships and data sharing. However, this centralized strategy may bring about privacy problems, limit transparency, erode trust, and make it more difficult to comply with legal requirements like the GDPR.
3. Another aspect of issues associated to data is ethical considerations. The ethical application of AI-driven technologies during the whole development and deployment process is at the center of them. Preventing socioeconomic biases from leaking into the system's models from the underlying data is the main goal. These

guidelines seek to lessen, if not completely eradicate, potential negative effects of artificial intelligence (AI), such as underrepresentation brought on by bias (technical or human). This way, the benefits of AI will be shared by all people, especially those who are more susceptible to the effects of natural disasters.

4. Issues with transparency and computing demands are most bothersome during the model construction phase, which follows data curation. During training, AI models frequently use complex designs that result in significant processing costs. 138 million trainable parameters, for example, are included in the VGG16 model that is used for picture categorization. Large and expensive computer resources are needed for training models of this magnitude, and these resources might not always be accessible.
5. It is crucial to make sure that an AI model produces outcomes that people can understand and find acceptable once it has been created. Since there isn't a human–machine interface that can be used everywhere that can explain the reasoning behind the AI model's judgments, achieving this can be challenging.

As a result, many researchers are committed to developing trustworthy AI solutions. This calls for a careful articulation of the problem and an awareness of the requirements and expected results of AI-based solutions. For the purpose of creating a suitable model and learning approach to deal with the current problem, this clarity is essential. Additionally, the selection and development of pertinent evaluation criteria are facilitated by a clearly specified problem scenario.

6 AI for Damage Detection from Satellite Images

Researchers have discovered that artificial intelligence (AI) has the ability to predict natural disasters. Through the use of large amounts of high-quality data, artificial intelligence (AI) can predict when natural disasters may occur. Machine learning algorithms can help solve various problems in remote sensing, such as hyperspectral image classification, semantic 3D reconstruction, and object detection. Deep learning can also be applied in terrain change detection. It is the process of spotting respective changes over time by observing the area of interest at different times. Such change detection can be applied in agriculture to observe fields' fertility over time, monitor deforestation, and city planning [13].

It's important to assess damage as soon as possible following a natural disaster, but doing so requires having post-disaster satellite imagery. When climatic conditions make it difficult to take images, the essential delay for this beginning usually varies from a few hours to many days. Post-disaster satellite imagery is quickly shared through partnerships between humanitarian organizations and Earth observation providers when situations allow. Pre-disaster image retrieval usually takes less time because the data is already in archives and just has to be shared and retrieved. The first priority after obtaining the satellite photographs is to quickly produce a preliminary report on damage assessment [14]. As shown in Fig. 4, this procedure consists of two main stages: mapping and data analytics.

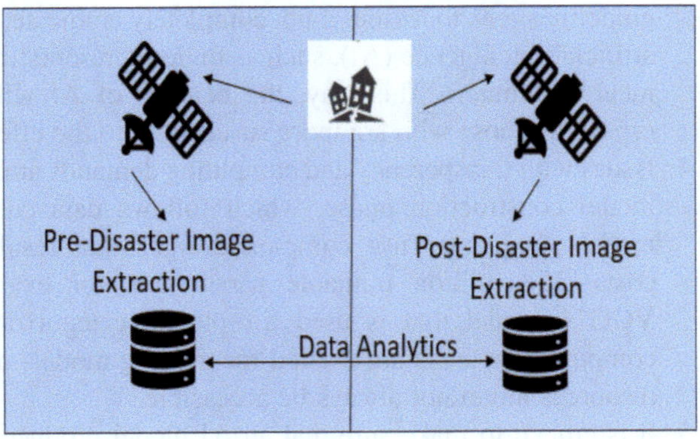

Fig. 4 Damage detection using AI

One of the fundamental tasks in the damage assessment process is mapping. This stage entails locating structures in satellite imagery and classifying them according to a damage rating that has been predetermined, paying special attention to buildings that have sustained damage [15]. When there has been significant destruction, the procedure includes locating damaged buildings over wide swaths of land. Field workers can use these maps directly to gather specific information, or they can undergo additional analysis to help in decision-making [16]. They provide important information, such as the number of damaged structures in a given area [17].

To aid in decision-making, the data analytics step incorporates extra demographic data with the original maps showing damaged buildings. In order to create a pertinent and customized reaction, this stage takes into account the particulars of the disaster event [18]. Demographic information, for instance, may indicate whether the calamity has affected a community that is particularly vulnerable, highlighting the critical need of offering food assistance [19]. The inherent inequalities across different crisis domains make it difficult to integrate machine learning methods into real-world humanitarian scenarios. These differences are caused by variables such as climate, season, damage kind, and geographic location. Because of the high variability of the situation, researchers have found that a model trained on supervised data from previous disaster events may not guarantee optimal performance for a new disaster event. Furthermore, restricting the amount of manually annotated labels for each new catastrophic occurrence is frequently necessary due to the urgency of model deployment [20].

7 System Methodology

Automated damage detection is a task that can also be approached as a supervised change detection problem [21]. Previous studies in this field show successful applications of image segmentation in order to solve this problem. Using a pair of

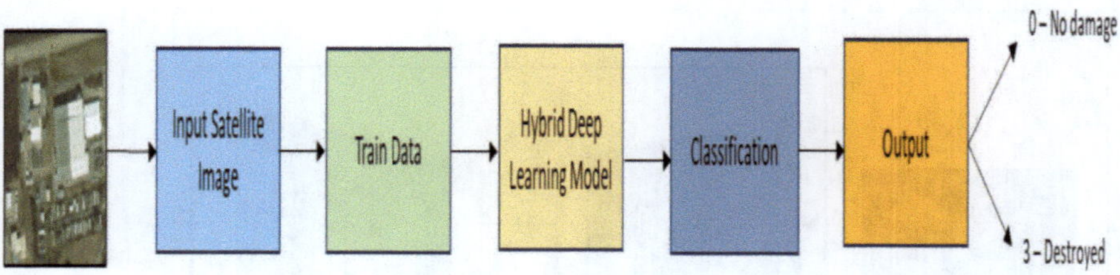

Fig. 5 Overview of the proposed system methodology

images that correspond to the same set of coordinates before and after a hazard (over a short period of time, assuming no structural patterns have not been changed), it is possible to identify image masks of structures in the pre-event image and then compare each mask with the post-event image. Images captured by satellites reveal priceless information about our planet [22]. Damage classification and building detection are the two separate tasks that make up the task of damage assessment. Many novel techniques have recently been introduced in the field of building detection [23]. The proposed method consists of four primary steps, as shown in Fig. 5 [24]. It starts with preparing the data by extracting training data for the task of interest (i.e., building footprints and manually labeling them as "Damaged" or "No Damage") after the satellite imagery has been pre-processed (atmospheric correction and orthorectification). This data is then separated into three categories: training, validation, and testing. Model fine-tuning follows, which modifies the classification layers of the existing architecture and trains the filter weights of the neural network. Finally, output accuracy from the fine-tuned classifier will be assessed using metrics widely available.

7.1 System Architecture

The primary objective of the damage detection involves detecting buildings in images taken prior to a disaster (pre-disaster), and then classifying each building in images taken after the disaster (post-disaster) as either damaged or undamaged. A novel training pipeline is built to classify the buildings' damage levels. In order to do this, the buildings in the post-disaster photos must be cropped off and resized to 128 by 128 by 3 dimensions, as many conventional neural networks support these input dimensions. Whether or not a building is present in the projected mask determines which cropping region is chosen (Fig. 6).

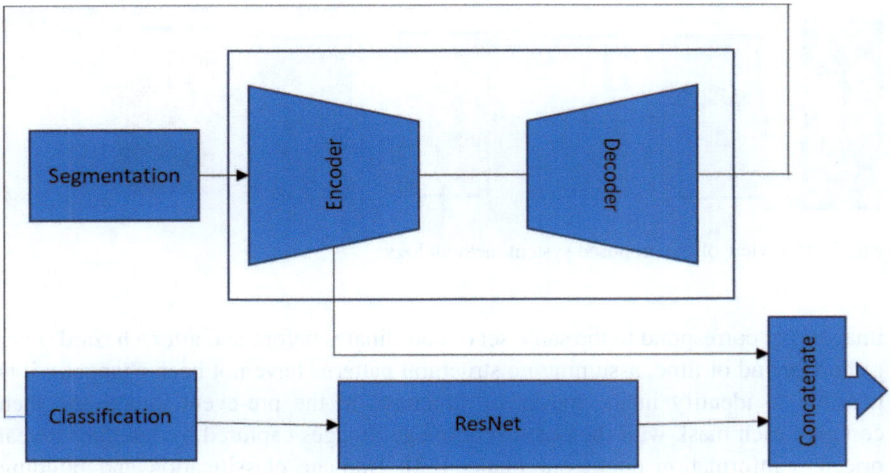

Fig. 6 Overall system architecture

8 Conclusion

The objective of this work was to apply neural network models to improve the ability to recognize and evaluate natural disaster-related damage in aerial photography. The strategy was building an efficient system by using historical data from past catastrophic situations. This system's fundamental component is a brand-new hybrid neural network model called UNet-ResNet. Disaster management is the main use for this model, which excels in classifying objects and structures within high-resolution multi-spectral satellite data. Superior picture processing and analysis capabilities of UNet-ResNet are demonstrated; this information is crucial for timely and well-informed decision-making during catastrophe reactions. It simplifies the damage assessment process by identifying objects and facilities correctly, guaranteeing focused and successful disaster management activities. The field of disaster management and response stands to gain greatly from this method, which combines the capabilities of neural networks with multispectral imaging.

References

1. Patel, A. K., & Jain, D. (2022). Disaster Risks and Management In India: A Critical Analysis of the Disaster Management Act. 5thWorld Congress on Disaster Management, 300–310. https://doi.org/10.4324/9781003341956-39
2. Xu, J. Z., Lu, W., Li, Z., Khaitan, P., & Zaytseva, V. (2019). Building damage detection in satellite imagery using convolutional neural networks. arXiv preprint arXiv:1910.06444.

3. Tilon, S., Nex, F., Kerle, N., & Vosselman, G. (2020). Post-disaster building damage detection from earth observation imagery using unsupervised and transferable anomaly detecting generative adversarial networks. Remote sensing, 12(24), 4193.
4. Gupta, A., Watson, S., & Yin, H. (2021). Deep learning-based aerial image segmentation with open data for disaster impact assessment. Neurocomputing, 439, 22-33.
5. Balakumar D and Rangaraj J, "A Prediction Model Based Energy Efficient Data Collection for Wireless Sensor Networks", Journal of Machine and Computing, vol. 3, no. 4, pp. 360–378, October 2023. https://doi.org/10.53759/7669/jmc202303031.
6. S. R and A. H, "Adaptive fuzzy logic inspired path longevity factor-based forecasting model reliable routing in MANETs," Sensors International, vol. 3, p. 100201, 2022, https://doi.org/10.1016/j.sintl.2022.100201.
7. R. Subha, A. Haldorai, and A. Ramu, "Artificial Intelligence Model for Software Reusability Prediction System," Intelligent Automation and Soft Computing, vol. 35, no. 3, pp. 2639–2654, 2023, https://doi.org/10.32604/iasc.2023.028153.
8. Ali-Khusein, "An Analysis of Multi Agent Systems Agent Based Programming", Journal of Computing and Natural Science, vol. 3, no. 4, pp. 182–193, October 2023. https://doi.org/10.53759//181X/JCNS/202303017.
9. Duarte, D. (2018). Nex F Kerle N Vosselman G. Multi-resolution feature fusion for image classification of building damages with convolutional neural networks Remote Sens, 10(1636), 10–3390.
10. Gupta, R., & Shah, M. (2021, January). Rescuenet: Joint building segmentation and damage assessment from satellite imagery. In 2020 25th International Conference on Pattern Recognition (ICPR) (pp. 4405–4411). IEEE.
11. Mei, J., Zheng, Y. B., & Cheng, M. M. (2023). D2ANet: Difference-aware attention network for multi-level change detection from satellite imagery. Computational Visual Media, 9(3), 563–579.
12. Shen, Y., Zhu, S., Yang, T., Chen, C., Pan, D., Chen, J., ... & Du, Q. (2021). Bdanet: Multiscale convolutional neural network with cross-directional attention for building damage assessment from satellite images. IEEE Transactions on Geoscience and Remote Sensing, 60, 1–14.
13. Gunturu, V. R. (2022). GIS, Remote Sensing and Drones for Disaster Risk Management. 5thWorld Congress on Disaster Management, 182–194. https://doi.org/10.4324/9781003341956-26
14. Mohamed Shaluf, I. (2007). An overview on disasters. Disaster Prevention and Management: An International Journal, 16(5), 687–703. https://doi.org/10.1108/09653560710837000
15. Gevaert, C. M., Carman, M., Rosman, B., Georgiadou, Y., & Soden, R. (2021). Fairness and accountability of AI in disaster risk management: Opportunities and challenges. Patterns, 2(11), 100363. https://doi.org/10.1016/j.patter.2021.100363
16. Nunavath, V., & Goodwin, M. (2019). The Use of Artificial Intelligence in Disaster Management - A Systematic Literature Review. 2019 International Conference on Information and Communication Technologies for Disaster Management (ICT-DM). https://doi.org/10.1109/ict-dm47966.2019.9032935
17. Subhashini, R., Thomas, J. J., Sivasangari, A., Mohana, P., Vigneshwari, S., & Asha, P. (2022). Artificial intelligence–based intelligent geospatial analysis in disaster management. Advances of Artificial Intelligence in a Green Energy Environment, 203–221. https://doi.org/10.1016/b978-0-323-89785-3.00006-2
18. Bouchard, I., Rancourt, M. È., Aloise, D., & Kalaitzis, F. (2022). On Transfer Learning for Building Damage Assessment from Satellite Imagery in Emergency Contexts. Remote Sensing, 14(11), 2532.
19. Dolce, M., & Goretti, A. (2015). Building damage assessment after the 2009 Abruzzi earthquake. Bulletin of Earthquake Engineering, 13(8), 2241–2264.
20. Deng, L., & Wang, Y. (2022). Post-disaster building damage assessment based on improved U-Net. Scientific Reports, 12(1), 15862.

21. Anandakumar Haldorai, "A Survey of Renewable Energy Sources and their Contribution to Sustainable Development", Journal of Enterprise and Business Intelligence, vol. 2, no. 4, pp. 211–222, October 2022. https://doi.org/10.53759/5181/JEBI202202021.
22. Toshihiro Endo, "Analysis of Conventional Feature Learning Algorithms and Advanced Deep Learning Models", Journal of Robotics Spectrum, vol. 1, pp. 001–012, 2023. https://doi.org/10.53759/9852/JRS202301001.
23. Cheng, C. S., Behzadan, A. H., & Noshadravan, A. (2021). Deep learning for post hurricane aerial damage assessment of buildings. Computer Aided Civil and Infrastructure Engineering, 36(6), 695–710.
24. https://eleks.com/research/deep-learning-for-damage-detection-using-satellite-images/

Climate Resilience Via Smart Technologies Over Natural Disaster

1 Introduction

Natural disasters are becoming more severe and devastating because of climate change, which has exacerbated humanitarian issues. Droughts, floods, disease outbreaks and earthquakes are becoming more frequent as a result of accelerating climate change. These events pose a serious threat to agricultural and economic infrastructure, putting millions of lives at risk. According to many studies, insufficient or delayed information dissemination, resource allocation, and risk mitigation are the major causes of problems in disaster management. It is important to be able to identify and predict the impact of natural disasters as soon as possible in a crisis. Modern technology can help a lot in this regard. Developing effective disaster management strategies requires collaboration between professionals from many disciplines, methodological diversity, and scientific data. Natural disaster research and education have received significant institutional funding from such prominent bodies such as the United Nations and the World Health Organization. An example of this is the United Nations International Disaster Risk Reduction Program (UNISDR), which was established in 1999 to coordinate and monitor programs aimed at reducing potential hazards. According to the UN ISDR, hazards include physical events, human activities, and any phenomena that has the potential to endanger lives, cause damage to the environment, or incite civil unrest. Examples include, among others, floods, hurricanes, droughts, epidemics, and nuclear disasters. On the other hand, when a community or population lacks the means to adequately handle the consequences posed by these risks, disasters can result as a potential outcome.

Given that a population is able to use the resources at their disposal, Fig. 1 illustrates the conceptual relationship between a hazard and a disaster. Accordingly, a hazard may be an occurrence that will happen regardless of human effort, as indicated by the ISDR in the aforementioned remark. However, with what will be

© European Alliance for Innovation 2024

A. Haldorai et al., *Artificial Intelligence for Sustainable Development*, EAI/Springer Innovations in Communication and Computing, https://doi.org/10.1007/978-3-031-53972-5_20

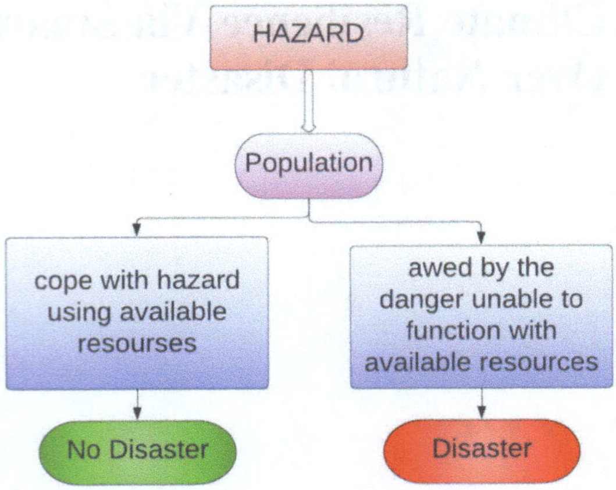

Fig. 1 Hazard and disaster

covered in more detail in this chapter, the impact of a disaster could be lessened or even prevented [3]. Prioritizing the integration of technology that can expertly handle the various stages and complexities of emergencies, extreme weather events, and disasters is crucial for mitigating losses. In order to enable the generation of preventative alarms, cloud platforms have a great potential for retrieving essential data in advance of disasters. Furthermore, these platforms are essential for carrying out rescue operations and efficiently prioritizing critical demands during the actual crisis. The addition of artificial intelligence (AI) increases this potential even further. Cloud platforms effectively manage the allocation of storage space thanks to their multi-tenant capabilities. These platforms enable communities and organizations to speed up recovery efforts after a disaster, allowing for a quick return to normal operations and enhancing readiness for any future events [2, 4].

Advanced geospatial and artificial intelligence technologies have made significant progress and offer a lot of potential in the field of crisis management [5]. The local geography, weather patterns, ecological elements, accessible resources, and logistical concerns are just a few of the factors that must be taken into account for effective disaster response planning. It is advised to take into account the societal impact while introducing management science and operations research principles to improve resilience in emergency relief [6]. A key player in disaster risk reduction (DRR) is machine learning (ML), a branch of artificial intelligence (AI). It makes contributions in a variety of areas, including hazard mapping, situational awareness, decision assistance, and more. Examples include anticipating extreme events and detecting events in real time. Utilizing advances in algorithms, computer power, and sizable datasets, the application of cutting-edge technologies like AI to disaster management has drawn more attention, much to how AI has impacted other sectors like finance and healthcare. Nevertheless, a number of academic studies have evaluated the performance of AI in disaster management [7].

India stands out due to significant differences between crisis response environments in various nations. As a result, it is crucial to identify and rank essential information for efficient crisis management during natural disasters [8]. The best ways to lessen the negative effects of disasters are through prevention, impact reduction, vulnerability mitigation, disaster readiness, and building resilience [9–11]. Examining the use of AI-based technologies in emergency response, disaster recovery, and catastrophe management situations has gained momentum [10]. Existing research, however, either lacks a thorough theoretical foundation or focuses primarily on particular stages like preparedness, response, recovery, or evaluation in pre- or post-disaster situations. In order to address these research gaps, we carried out a qualitative investigation and consulted industry professionals to create a solid framework. This framework includes the discovery, creation, and recommendation of a collaborative AI-based cloud platform connected with a spatial intelligent system to enhance disaster management skills.

The main contribution of this chapter is as follows:

1. In this chapter, the emerging technologies like AI, Cloud, and integration of AI-cloud are discussed.
2. The methods like Graphical Intelligent System (GIS) and Intelligent Decision Support System (IDSS) that can aid in disaster management have been examined and explained.
3. We have offered a study that allows us to examine how AI is helping in weather prediction.

The structure of this chapter is as follows. A survey on Emerging technology for disaster management and technique for the conceptual framework of AI and Cloud is covered in Sect. 2. In Sect. 3, methodology of GIS and IDSS is elaborated. Additionally, it introduces the study on AI for weather prediction in Sect. 4. The conclusion of the paper is done in Sect. 5.

2 Emerging Technologies for Disaster Management

Advances in AI have sparked significant innovations with far-reaching social repercussions. The rise in economic value of AI—predicted to reach $16 trillion by 2030 [19]—has made it a potent force for change. The current growth boom can be attributed to the quick development of high-throughput technology and the spread of widely available and rich data sources. The ubiquitous availability of high-bandwidth internet connections, the advancement of sensor technology, and the connectivity made possible by the Internet of Things (IoT) have all contributed to the development of this phenomenon. As a result, the AI community is shifting its attention to using its tools to fight climate change. The authors of the article "Tackling Climate Change with Machine Learning" [22] make a strong case for the usefulness of machine learning as a tool for reducing greenhouse gas emissions and supporting societal adaptation to changing climatic patterns. What makes this claim remarkable is ML's ability to provide significant improvements in performance and enhance public services. Additionally, it may have a substantial impact on "directing future development strategies" and "acquiring data for intelligent decision-making." A substantial obstacle still exists in the form of little interdisciplinary collaboration, despite the enthusiasm surrounding the use of AI to combat climate change. This deficiency prevents AI practitioners from using their methods to tackle climate change in an effective way [22].

2.1 Role of Technology for Disaster Management

Our daily lives include a lot of technology. We hardly ever get by without utilizing its advantages. This led to the development of the field of technical improvement, which has made steady progress over time and is constantly growing to keep up with the constantly changing requirements of the population. The age of technology has seen enormous technological advancements and quick expansion in terms of platforms and applications that streamline many parts of our daily lives.

Technology's contribution to the disaster management process is frequently overlooked. Figure 2 illustrates the Disaster Management Cycle (DMC) for the purpose of facilitating analysis. It may be used at any phase of the cycle, encompassing preparation, response, recovery, and mitigation.

- Preparation: Efficient use of technology is necessary for the creation and implementation of comprehensive emergency plans. It also has the ability to keep an eye out for potential threats in the area, such as weather patterns that might cause natural catastrophes, like a vigilant guardian.

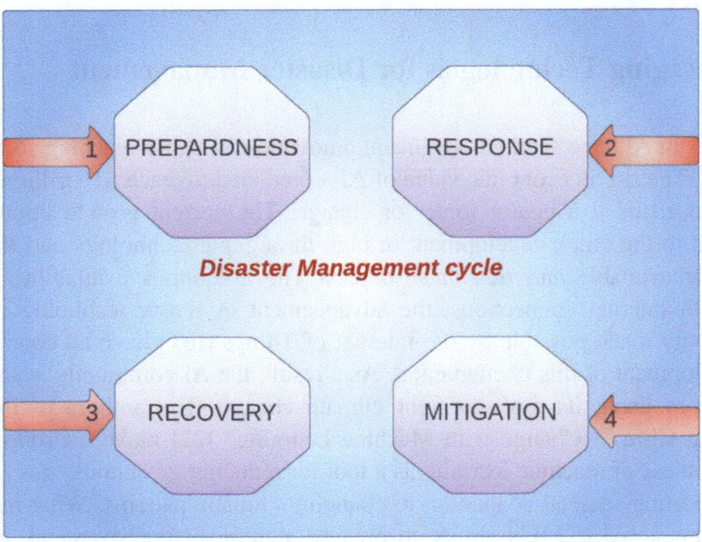

Fig. 2 Disaster management cycle

- Response: In an emergency, the use of technology can help plan and manage a response. Another potential use is the dissemination of information and assistance to those affected by disasters.
- Recovery: After a disaster, technology can make it easier to rebuild. It is a useful tool in damage assessment, construction planning, and coordination of relief work.
- Mitigation: This final phase of the cycle underscores the importance of having a comprehensive strategy. Mitigation, similar to preparation, involves implementing measures to reduce the likelihood of a disaster recurring. These safety precautions are crucial at all times, but their significance amplifies significantly in the aftermath of a tragedy, when a society or government may still be seriously impaired.

The DMC combines the activities to lessen the possibility of fatalities among people and property. Prior to a disaster, planning for mitigation entails creating laws to prevent future calamities and putting policies in place to handle emergencies. It also implies that you should plan ahead and make the most of your resources, as preparation is one of the DMC's most important stages for reducing disaster damage. Response and rescue operations are carried out during a disaster. First aid, humanitarian aid, and a damage assessment are the main objectives. Following the disaster, recovery efforts, including the overall community development assistance program, are launched with the goal of reestablishing the community's way of life and performing accurate damage assessments. Technology may assist us identify threats prior to they happen so that we can take action to prevent them from turning into a reality. Technology can also help us communicate with one another easily during an emergency so that we can work together to find solutions for problems related to catastrophic events such as earthquakes, hurricanes, floods, etc. [12]. The first thing to take into account when looking at technology in disaster management is that it can help us detect threats before they happen.

2.1.1 Artificial Intelligence

Due to its increasing economic and organizational significance in recent years, artificial intelligence (AI) has attracted a lot of attention [13]. The capabilities of AI lie in its ability to use specialized algorithms to digest information, which is useful in situations including emergencies, severe weather, and disaster relief efforts. If this use of capabilities is used, AI can transform various inputs into educated decisions and useful solutions [14]. This includes video, images, statistics, text, and music stored in cloud repositories. In some cases, the capabilities of artificial intelligence computers will match or exceed human intelligence. In this context, ML includes various AI subfields, such as supervised, unsupervised, and reinforcement learning. Machine learning uses algorithms to classify data or make predictions. The introduction of modern CPU components has opened up a whole new world of possibilities allowing for efficient parallel data processing. Artificial

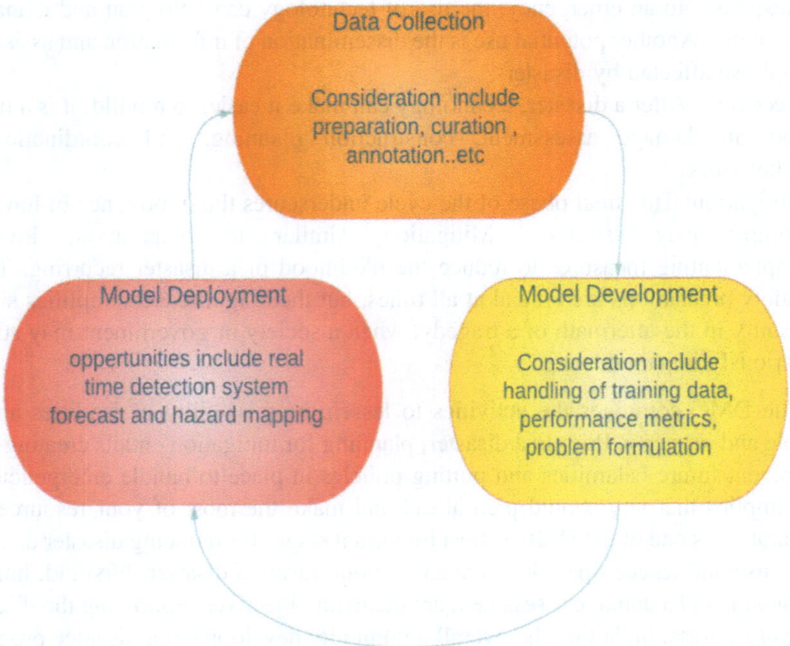

Fig. 3 AI lifecycle for disaster risk reduction

intelligence is great at helping people in crisis, especially when recycling using traditional hierarchical methods can take too long and leave spaces it is easier. Communication and cooperation between multiple stakeholders is enhanced by the process of artificial intelligence (AI) that distributes resources quickly and accurately [1].

Artificial intelligence has enormous promise, as seen in the area of predictive modeling, which uses past patterns to reduce fatalities. Using geological and seismic data, AI can forecast where and how big earthquakes will be [15, 16]. AI can also anticipate volcano eruptions, earthquake sites, and floods by analyzing rainfall data, population size, and flood simulations [17, 18]. As a result of taking the initiative, careful planning is carried out. Important steps make up the AI lifecycle for DRR, as shown in Fig. 3. Data processing and collection must adhere to certain ethical guidelines; it must also account for the potential for bias in training and testing datasets, as well as any newly introduced decentralized artificial intelligence, and for other factors. By allowing experts to accurately assess requirements and quickly transport aid to various places, AI greatly improves rescue operations.

Collaboration between public and commercial sector experts is essential for effective data application targeting and resource coordination. Drones equipped with artificial intelligence may also track the reach of severe weather, assess the impact of catastrophes, and determine the need for rescue efforts [58, 60]. There is great potential for continuous crisis scenario monitoring and preemptive response with the effective integration of drone and AI technology.

The study of satellite data is another capability that AI expands, allowing for the determination of hurricane and tornado intensity and paths. Natural language Processing (NLP) methods, such as "sentiment analysis," show how AI and social media are intersecting in a less contentious but nonetheless significant way. These techniques probe into social media conversations around "fake news" and advertising networks. An important aspect is the incorporation of AI into social media interfaces. Some social media-based organizations that contest the existence of global warming have benefited from the platform's popularity. Researchers have uncovered intricate relationships between numerous entities in this space by utilizing social network analysis methods [46, 48, 50]. The connections and shared interests between groups supporting Brexit and Trump, fossil fuel companies, and organizations that deny climate change have recently come to light [51, 53]. This new understanding highlights the strong connections that go beyond these seemingly unrelated fields. Similar tools and approaches could be used in a parallel setting to promote and track involvement in efforts connected to climate change [52, 54, 55]. The ability of AI to traverse and analyze enormous amounts of data can be used to support the movement for climate awareness and action, producing a significant beneficial impact.

Artificial intelligence, which refers to the cognitive abilities displayed by computer systems, stands out as a clear divergence from human intellect [15]. At its core, AI coordinates the convergence of several networked devices to mimic human actions. Computational activities essential to the construction of intelligent mechanisms are included in its scope. The trajectory of AI growth over the last 10 years has resulted in significant gains, especially in enhancing our capacity to predict disasters and enabling prompt response when necessary [16–18]. Search and rescue operations, the distribution of humanitarian goods as part of disaster planning, and crowdsourcing information systems are the domains where the effects of AI breakthroughs are most evident [15, 19]. Using a mix of robots, sensors, machine learning, drones, deep learning, and sophisticated algorithms, this study investigates how AI might be used to forecast catastrophes, aid rapid rescue attempts, and ease relief operations—all while recognizing the different shapes AI can take [20–23]. A relatively recent addition to the artificial intelligence landscape, machine learning differs from the more mature robotics discipline. Interestingly, machine learning algorithms may do jobs without explicit instructions by using patterns and conclusions found in data. They are classified as AI because of this, which allows them to complete tasks on their own [25]. A model that can identify changes in satellite data may be developed using convolutional neural networks (CNN), facilitating the coordination of relief activities and the identification of disaster-prone locations. With the introduction of artificial intelligence, disaster management as a whole undergoes a radical transformation. The impact of AI may also be seen in the development of multi-agent frameworks with real-time support in dynamic scenarios, the provision of data to disaster mitigation measures, and the promotion of

Table 1 Previous investigations using artificial intelligence for disaster management

SL. No	Analysis	AI algorithm	Input data
1	Damage mapping of post-earthquake	Artificial neural network	Satellite image
2	Landslide disaster exposure mapping	RFEs and NBT classifiers	Satellite spatial image and field survey data
3	Near real-time damage mapping	Convolutional neural networks	UAV images
4	Prediction of earthquake	Convolutional neural networks	3D point cloud
5	Plotting and flood vulnerability mapping	Artificial neural network	Satellite image
6	Landslide and flood disaster risk reduction	Convolutional neural networks	Satellite and spatial images
7	Classification of damaged building after earthquake	Convolutional neural networks	Satellite and UAV images
8	Disaster risk reduction through social media	CNN.SVM.RFS and GVN networks	Social media application and satellite image
9	Disaster assessment in coordinating relief	CNN and semantic segmentation model	Satellite image

knowledge exchange through ontologies [29–32]. The acquisition of information from many sources, its synthesis, and its effective application are necessary for effective decision-making in the field of disaster management [29, 34]. The multidimensional role of AI resonates as a crucial component in fostering efficacy and resilience within the challenging environment of catastrophe management.

Artificial intelligence is without a doubt the future severe disaster management since it significantly increases our ability to protect civilian populations in the case of calamities [40]. However, in order to ensure that the deployment of AI in hazard management is efficient and practical, governmental organizations must complete a development plan of fundamental requirements. Another choice is to use a variety or combination of deep learning algorithms to determine the forecast that is the quickest and most accurate [49, 50]. To predict future outcomes, especially in the case of natural disasters, enormous volumes of data are required, and high-performance algorithms are frequently crucial because of their big dense datasets with nonlinear properties [51]. Table 1 lists the earlier works on disaster prevention and the contribution of AI to effective disaster management techniques.

For the purpose of employing multiagency collaboration in resilient community construction, a new AI platform named MOBILISE was created. Catastrophe management organizations can collaborate on catastrophe mitigation thanks to MOBILISE's real-time information [60]. The MOBILISE platform is an intuitive API that enables agencies to upload and interactively analyze data on dangers, exposure, and vulnerability in order to develop an agreed-upon comprehension of their local risks and put catastrophe risk reduction measures into place. An Azure cloud service powers the information server for risks.

2.1.2 Cloud Technology

Pay-per-use cloud technology is also suitable for private as well as public companies, providing a flexible way to reach important stakeholders during emergencies with vital information [41, 42]. Beyond its function in fostering a sense of community, the combined influence of AI and cloud-based technologies is notably beneficial for businesses, giving them the ability to quickly and effectively create tactics that outperform conventional weather and catastrophe plans. The development of map overlays, which further improve situational awareness, is a key aspect of this synergy. The "infrastructure as a service," "platform as a service," and "software as a service" paradigms of cloud computing, which demonstrate its versatility, make it easy to generate, test, and update important insights. This dynamic approach guarantees that business processes are protected and continue without interruption, thereby strengthening security precautions and organizational resilience [56].

It is frequently necessary to access numerous remote places during disasters. Cloud technologies provide as a conduit for utilizing internet resources from anywhere in the world, encouraging cooperation among many stakeholders [45, 47]. The picture in Fig. 4 highlights how cloud computing employs parallel procedures that are similar to conventional methods and even predate the invention of this technology. Cloud technologies are steadily developing as the field of catastrophe management progresses, combining not only satellite and drone data

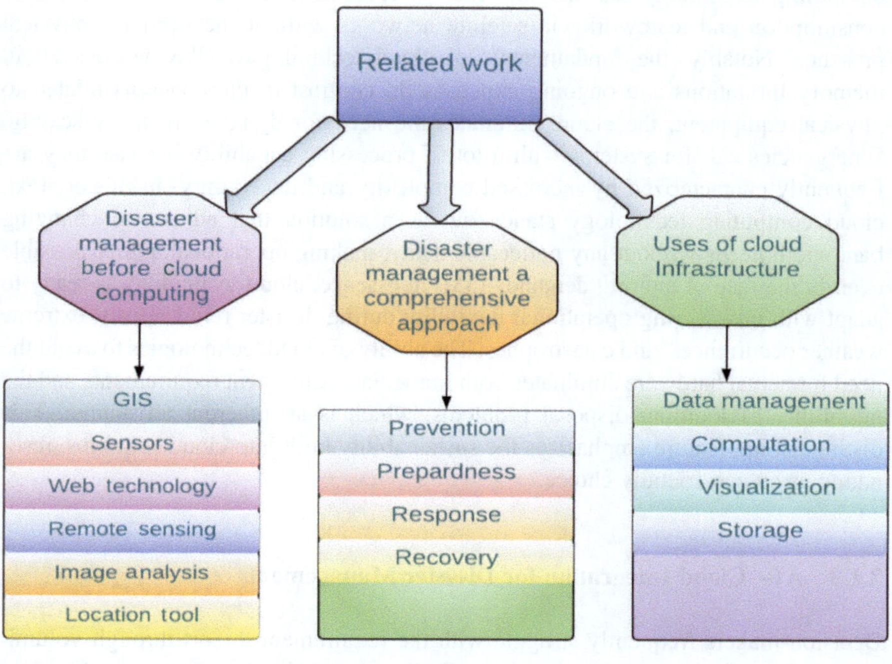

Fig. 4 Disaster management-related work with cloud computing

but also AI into the ecosystem. The addition of AI to cloud computing adds a level of accuracy to the trajectory predictions for hurricanes, earthquakes, wildfires, floods, and other weather disasters. This accuracy, when combined with accurate data, is essential for damage reduction, resource allocation, and the effectiveness of restoration strategies [63]. A proactive approach is developed through the use of AI in cloud computing, enabling early detection of probable changes in weather patterns. This quick identification enables authorities to assess the degree of physical damage quickly, expediting response operations [35–39]. Utilizing data from numerous social media platforms to understand changing circumstances is another example of how cloud computing and AI work together. Additionally, AI makes it easier for visual analytics to be presented via cloud platforms, which raises awareness [62]. Different stakeholders and several tenants safely store their data within the cloud architecture, with access restricted to those with the necessary permissions. Then, AI serves as a conduit to extract insights from this data, providing specific solutions to the problems presented by disasters. By incorporating the multi-tenancy concept, cloud platforms give government organizations access to a broad resource pool, making it easier to reach almost any place or device in an emergency. This interplay strengthens the capacity to manage crises successfully and boost's reaction capacities.

The cloud, which operates in an online virtual environment, acts as a digital safe haven where users can skillfully manage, store, and retrieve a variety of digital resources, including files, directories, programs, and applications, as needed. Cloud computing technology has the potential to revolutionize industries by enabling consumption and teamwork via satellite networks without the need for physical presence. Notably, the fundamentals of cloud technologies allay worries about memory limitations and ongoing expenses. In contrast to the concerns related to physical equipment, the cloud eliminates the need for dynamic memory scaling. Emergencies call for systems with a lot of processing capability because they are frequently characterized by increased complexity and uncertainty. In this context, cloud computing technology stands out as a solution that adjusts to changing bandwidth needs without any noticeable delay, making operational agility possible even in the face of altering demands [33]. In essence, cloud technology is ready to adapt with unwavering operational flexibility during disaster relief efforts, extreme weather occurrences, and catastrophes. The ability of cloud technologies to avoid the need for actual hardware eliminates both immediate equipment requirements and the possibility of looming disposal problems, which is an inherent advantage. This distinguishing feature emphasizes the sustainability built into cloud solutions, resonating as an eco-friendly choice.

2.1.3 AI—Cloud Integration for Disaster Management

Decision-makers frequently struggle with the requirement to sort through voluminous information in urgent situations. Different stakeholders approach this duty according to their unique knowledge and skills. However, the traditional strategy of independently processing enormous volumes of data results in bottlenecks because there isn't an integrated approach to tackling the intricacies inherent in extreme weather occurrences and disasters. As a result, it is crucial to incorporate cutting-edge technology like cloud computing and artificial intelligence to improve response times and maximize information processing effectiveness. A system or organization needs to have certain qualities in order to effectively manage emergencies [2]. For example, design principles that support organized frameworks are crucial for supporting efficient operations in emergency situations. To do this, responsibilities must be divided into manageable parts, and decisions must be well-informed and in accordance with the emergency operation's main goals [40].

The immediate, precise, and effective processing and distribution of information to stakeholders is essential in disaster and emergency situations [1]. The coordination of efforts between governmental and nongovernmental organizations, as well as the distribution of resources across urban and rural settings, are other factors that influence the effectiveness of the emergency management ecosystem. When disasters and extreme weather events occur, people and victims may be able to provide crucial real-time insights thanks to the disaster management ecosystem's skill in processing information.

Continuous data retrieval and preservation from cloud platforms play a crucial role in this dynamic environment because AI can access and analyze this data to train backup resources, including communities and specialized reaction teams. While a number of technologies have been investigated to improve information processing capabilities, cloud computing and artificial intelligence (AI) stand out as the best options due to their abilities to (i) improve organizational planning, (ii) process and integrate data insights, and (iii) facilitate remote accessibility. Additionally, the effectiveness of information processing abilities can be enhanced to fulfill the level of performance requirements unique to severe weather and extreme disaster circumstances.

3 Methods of GIS and IDSS

3.1 *Geographic Information System and Disaster Management*

The three main objectives of disaster management are the protection of life, the preservation of critical infrastructure, and the preservation of the environment. As part of the disaster management cycle, Geographic Information Systems (GIS) are integral to the stages of preparation, response, recovery, and mitigation, making them an indispensable tool in this endeavor. The integration of GIS-enabled data analysis is the defining feature of these stages, which together lead to the development of comprehensive disaster management plans. This approach uses a three-tier GIS framework designed to meet unique needs at each level. Getting information

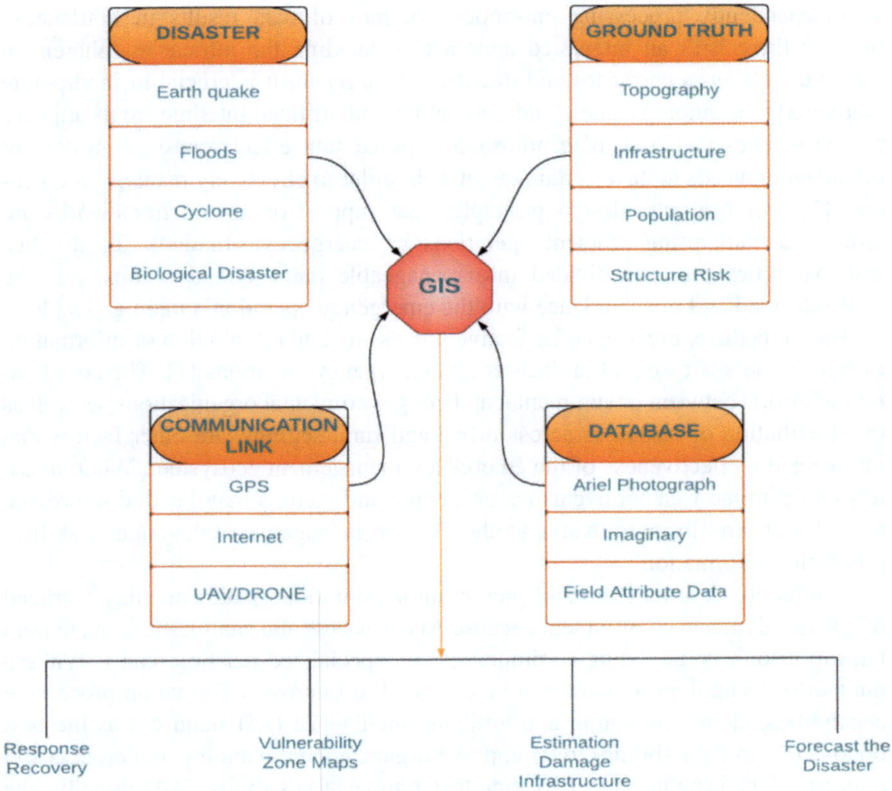

Fig. 5 Role of Geographic Information System

from prominent sources early is important to lay the foundation for later research. After careful measurement, spatial information is transformed into meaningful visual objects such as numbers, diagrams, charts, and numbers. Geographic databases (GDBs) govern the process of extracting valuable information from unstructured data [41]. Providing the user with well-designed and prepared information is the last step after data preparation. Statistical techniques are carefully applied throughout the delivery process to ensure proper dissemination of information. These GIS-supported geographic capabilities are powerful factors in improving the effectiveness of disaster management strategies. GIS is a reliable and efficient tool due to its excellent compatibility with spatial analysis [3]. Figure 5 is a graphic depiction that, among other things, shows how GIS is useful in many different contexts, how important it is, and how it can adapt to different problems.

A critical capability of Geographic Information Systems (GIS) is their ability to provide data in a tiered approach. Various characteristics may be captured at different granularities by using GIS's vast variety of display possibilities. Maps of flood-prone areas, architectural structures, landscapes, roads, medical facilities, schools, and populations at risk for vulnerability are all included in this. At the same time, satellite image technology makes it possible for researchers and scientists to "sense" the Earth through the field of remote sensing. Remote sensing, which makes use of high-resolution cameras, is an effective method for tracking flood dynamics and precisely measuring post-flood damages [42–45]. The complex combination of data needed for thorough disaster management is compiled from a variety of sources. The key is in the methodical gathering, preparation, and logical presentation of this material though [57]. The availability of essential information that is presented coherently during a crisis acquires critical importance in coordinating prompt responses and well-informed activities. The effects of disasters can be felt across the entire governmental spectrum or only in certain areas. Emergency responders frequently require vital information, ranging from drainage systems to electricity distribution. Through the use of GIS, data from several industries come together into a single, searchable repository that is made possible by computer-generated mapping, promoting easy information interchange [61, 62]. Being capable is extremely important, especially in light of the time restrictions that disasters frequently bring with them. Such situations demand hasty and well-informed decision-making due to their urgency. Without this centralized function, first responders would face the difficulty of locating map data spread across numerous administrative departments, which would be made even more difficult by the urgency of emergencies. Because of this, emergency personnel frequently have to make important decisions based on scant or inaccurate information, which emphasizes the need to provide them with excellent resources like GIS to deal with these difficult situations.

Maps, graphs, and tables are just a few of the dynamic graphical and visual outputs that geospatial analysis may produce, expanding the possibilities for in-depth analysis and forecasting flood alarms. The ability to act quickly in anticipation of looming tragedies is given to disaster agencies by this reservoir of geospatial insights, especially in favorable environmental conditions. A wide range of prevention and mitigation techniques are revealed through these illustrative maps, boosting the disaster management toolkit with a variety of proactive solutions [63]. By utilizing the strength of satellite images and remote sensing techniques, these maps stand out as crucial resources in the field of early warning systems. This instrumental fusion makes continual monitoring easier, which increases the effectiveness of activities for disaster preparedness. Aerial surveys cast their gaze over regions of interest, identifying flood-prone sites and establishing the groundwork for the best damage mitigation techniques in disaster zones. These surveys effortlessly integrate into this framework. Another distinguishing feature of GIS is the distinctive aspect of data provided in a layered style. By providing a broad visual panorama that includes flood-prone areas, architectural structures, landscapes, road networks, healthcare facilities, educational institutions, and populations vulnerable to vulnerabilities, GIS goes beyond conventional data representation. Satellite photography, a crucial component of remote sensing technology, enables researchers and scientists to "sense" Earth concurrently with these geographical breakthroughs. Remote sensing technology, which makes use of high-resolution cameras, efficiently monitors flood dynamics by anticipatorily estimating the level of damage before and after floods [24–28].

Geometric and mathematical analysis can be done using the GIS lens during each stage of catastrophe management. These reviews of previous flood incidents aid in predicting current patterns. Damage analysis is carried out using various levels [21]. The main contributors to flood dangers were found to be run-off generation, land use, and distance to channel [63]. The vulnerability map that was obtained shows that planning for flood mitigation is necessary in high-risk locations [25]. A technique for flood mitigation was presented by Puttinaovarat and Horkaew [26]. Two contributions formed the basis of the design. Input of data and central processing were first included in the designs, which produced a final flood report UI. Data intelligence was subsequently utilized to analyze the flood report using photographs that were remotely detected. As indicated in Fig. 6, the second phase included confirming the reported data by ensuring authentication with two factors using deep learning algorithms and reliable investigations.

Utilizing a GIS to manage disease in crisis areas following a disaster, especially a flood disaster, affected populations must deal with a number of new problems, and floods can cause a number of infectious diseases that can lead to viral infection. The second phase of the design of online flood information management systems [61] as shown in Fig. 6 involved verifying the reported data by establishing two-factor authentication with deep learning algorithms and trustworthy investigations. Puttinaovarat and Horkaew [56] presented a method for reducing flood damage. The design was built on the shoulders of two contributions. First, data input and central processing were incorporated into the designs, resulting in the creation of the final flood report UI. The flood report was then examined utilizing data intelligence using photographs that were remotely detected.

In order to better understand flood risks and risk assessment, numerous approaches and procedures have been created. The AHP approach, logistics regression, the analytic network method, statistical index, random forest, and flood zoning are a few of these methods. Geometric analysis and the assessment of flood damage using flood zoning maps are both prevalent in hydraulic science procedures. These maps show where a flood occurred and aid in determining the extent of the damage [15]. Flood zoning has grown to be an essential component of human security and safety. Additionally, it improves flood mitigation techniques and lessens flood-related damage [26]. To produce river maps, GIS is integrated with hydrologic engineering models (HEC-RAS). Many other states that are prone to flooding, including Warsaw, Columbia, and Dhaka, have effectively used these models [37, 38].

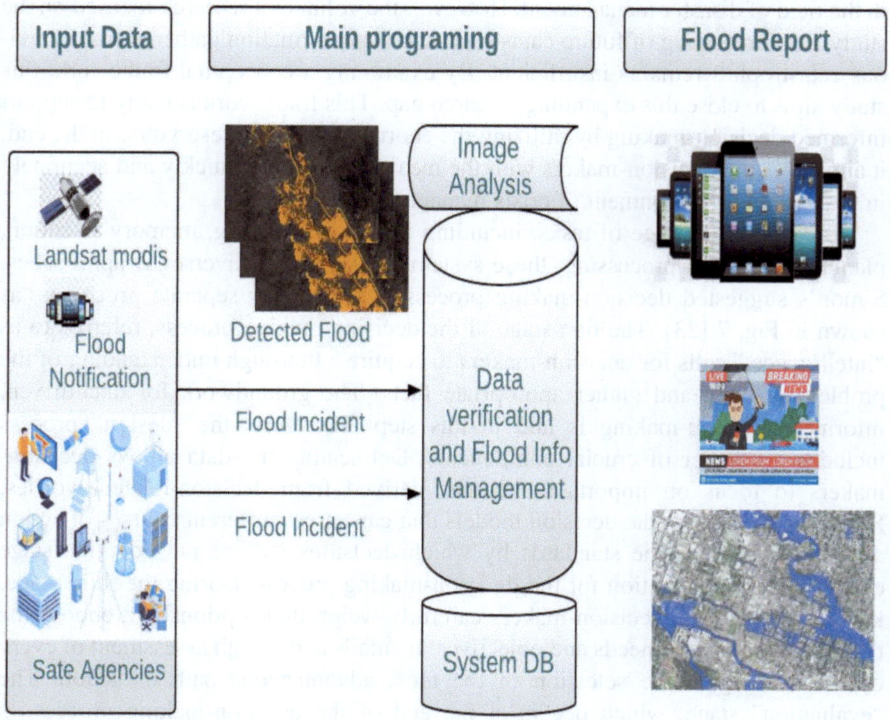

Fig. 6 Online flood information system design

3.2 *Intelligent Decision Support System for Disaster Management*

A crises' change continues to be considered a disaster until it is over. As a result, disaster response is an ongoing and dynamic process that requires continual decision-making based on prior experiences. This decision-making process has historically been hampered by a number of restrictions, most of which depend on the subjective insights that decision-makers share. The restrictions resulted from flaws in the information collection technology, making the decision-making process vulnerable to the inherently finite nature of the information and data provided. However, the modern environment has undergone a seismic change brought on by significant information technology breakthroughs. This revolutionary wave has brought us an era in which the amount of information available for wise decisions has increased exponentially. Notably, this has given decision-makers access to a multitude of information and insights. But it's important to understand that human cognitive abilities have intrinsic limitations, especially when it comes to crisis response, which necessitates quick decision-making. The advantages of combining a DSS with AI to create an Intelligent Decision Support System (IDSS) are clear-cut in the field of disaster management. However, the volume of research focused on the study and forecasting of future catastrophes using information gathered from previous catastrophes remains insufficient. By examining a conceptual framework, this study aims to close this expanding research gap. This framework is ready to support informed decision-making by utilizing the enormous big data reservoirs. In the end, it aims to equip decision-makers with the means to act more quickly and accurately in the dynamic environment of crisis management.

To carry out a range of tasks, including inference, learning, memory retention, planning, and data processing, these systems make use of diverse AI approaches. Simon's suggested decision-making process involves four separate processes, as shown in Fig. 7 [23]. The first stage of the decision-making process, referred to as "intelligence," calls for decision-makers to acquire a thorough understanding of the problem at hand and gather appropriate facts. The groundwork for data-driven, informed decision-making is laid in this step. Therefore, the "design" process includes a number of crucial components. Delineating the data allows decision-makers to focus on important elements derived from decision-related worries. Furthermore, they create decision models that can assess different courses of action and they determine the standards by which decisions will be judged. This stage establishes the foundation for the decision-making process. During the third stage, known as "choice," decision-makers carefully weigh their options and choose the best one that fits their needs and objectives. It entails a thorough assessment of every option, leading to the selection of the most advantageous path of action. The "evaluation" stage, which occurs at the end of the decision-making process, is when decision-makers thoroughly consider the implications and effects of the course of action they have chosen. This vital self-evaluation ensures that decisions are in keeping with the intended standards and aims.

While disaster recovery operations may benefit greatly from the use of DSSs, there are some disadvantages as well. One of the disadvantages is that quick and unfettered data transfer between people and organizations is impossible. This is because different organizations utilize different systems and technologies, or there may be incompatibilities between them. These disadvantages have been addressed, however, and managers now regularly use DSSs. The platform for integrating traffic, economic, geographic, and meteorological data was an intelligent disaster DSS (IDDSS).When there are floods, IDDSSs are utilized to manage the road networks. They give law enforcement the precise places to set up traffic management points (TMPs) during an emergency in order to avert dangerous trac scenarios.

4 Study on Extreme Weather and Prediction Via AI

The environment in the context of emergency circumstances has changed dramatically over the past few decades, resulting in a wider range of problems and a wider variety of viable solutions. Response and recovery tactics have historically been the main tools used for dealing with situations of emergency. The concept of readiness

Fig. 7 Four stages of decision making

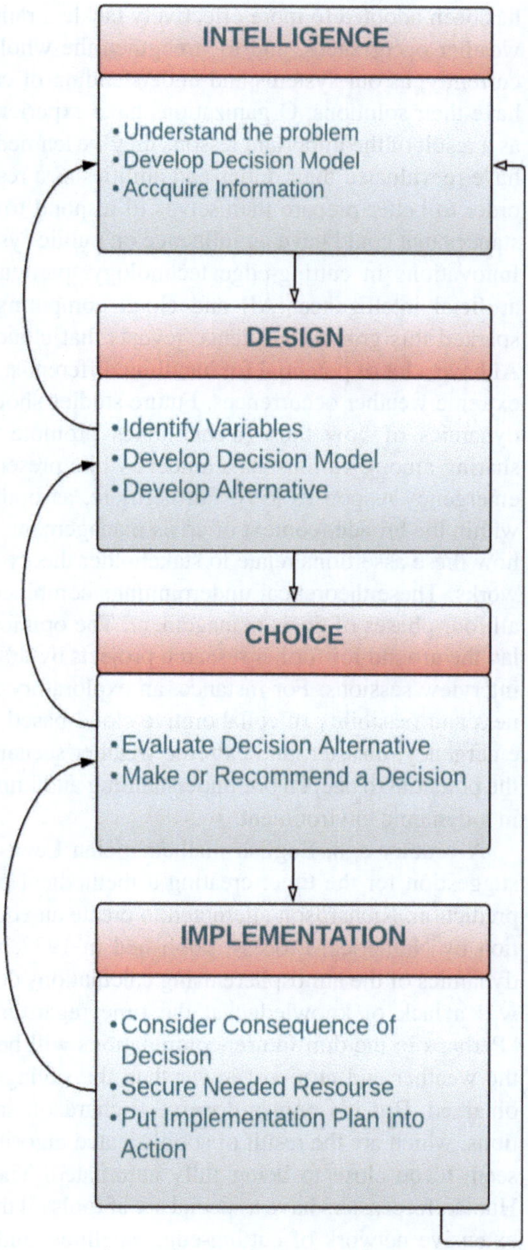

has been adopted to more effectively tackle a range of scenarios, including extreme weather occurrences, and to strengthen the whole disaster preparedness life cycle; currently, as our systems and understanding of catastrophes have advanced, so too have their solutions. Organizations have experienced a fundamental transformation as a result of the important lessons they've learned from previous emergencies. They have reevaluated their duties and abilities as a result of this introspective process in order to better prepare themselves to respond to unforeseen and emergent circumstances that could have an influence on public systems, enterprises, and ecosystems. Innovations in cutting-edge technology, particularly the increasing emphasis on artificial intelligence (AI) and cloud computing during the past 10 years, have sparked this growth. Evidence reveals that cloud-based collaboration systems and AI have a lot of potential for handling different stages of emergencies, disasters, and extreme weather occurrences. Future studies should, however, explore the complex dynamics of how these technologies promote communication and information-sharing among various stakeholders while preserving the vital capacity for prompt emergency responses [59]. Furthermore, as both of these ideas play crucial roles within the broader context of crisis management, it is important to take into account how these assertions relate to stakeholder theory and the use of contingency frameworks. These theoretical underpinnings combined can offer important insights into all four phases of crisis management. The opinions and conclusions presented here lay the ground for further research projects by drawing on data from semi-structured interview sessions. For instance, an exploratory study may look into the effectiveness and feasibility of collaborative cloud-based platforms and AI in the context of emergency, disaster, and extreme weather scenarios. Such research initiatives have the potential to deepen our understanding and improve our crisis management tactics in a dynamic environment.

A century ago, English mathematician Lewis Fry Richardson made a stunning suggestion for the time: creating a methodical procedure using maths for weather prediction. Richardson attempted to create an equation in his book Weather Prediction by Numerical Process, published in 1922, which he could use to resolve the dynamics of the atmosphere using calculations done by hand. It failed because there was a lack of knowledge at the time regarding the science of the atmosphere. "Perhaps in the dim future, computations will be able to evolve more quickly than the weather and at a cost lower than the savings to mankind from the knowledge obtained. But it's only a dream," Richardson said. Today's lightning-fast predictions, which are the result of sophisticated algorithms and extensive data collecting, seem to be close to being fully automated. They are not quite flawless, though. Human forecasters have a special set of tools all their own, despite the costly models, extensive network of cutting-edge satellites, and powerful computers. These forecasters continue to outperform the glitzy weather machines in the highest stakes situations thanks to experience—their capacity to observe and make connections where algorithms cannot. Now the situations have changed; emerging technologies are handling each and every sector in a tremendous way.

Advanced algorithms and extensive data collection have enabled today's lightning-fast forecasts, which seem to be close to becoming fully automated. But they're not yet flawless. Human forecasters have a special set of tools all their own, in spite of the pricey models, wide array of cutting-edge satellites, and massive computers. In scenarios with the highest stakes, these forecasters continue to outperform the flashy weather machines thanks to experience—their capacity to observe and make connections where computers cannot. Machine-learning algorithms generally try and optimize for one simple evaluation of how good its prediction explains by Niall Robinson, head of collaborations and product innovation at the Met Office and a co-author of the paper. However, there are numerous ways in which weather predictions might be good or disastrous. Maybe one forecast gets the proper amount of precipitation at the wrong area, while another gets the perfect combination of intensities at the wrong locations, and so on. In this research, we put a lot of effort into evaluating our algorithm using a variety of criteria.

Even roof kinds and conditions can be analyzed by AI algorithms to assess whether a roof can endure specific wind occurrences or if it has to be replaced. They are able to comprehend the state of a house before an occurrence of weather occurs by observing the actual damage by examining imagery captured throughout time. Utility businesses can better plan, prepare, and minimize risks and outages brought on by weather impacts and wildfires thanks to AI and ML models. These models take into account changes in network hardening as well as how infrastructure has reacted to previous storms. Utility workers can better position themselves thanks to this. AI models can estimate where and when wildfire conditions will occur in the case of wildfires to help assess the risk of a fire by ingesting wildfire factors. For example, whether to do work at a certain site on a particular day or the reaction times for fire trucks are some additional decision considerations that some utilities include to assist limit wildfire risks.

Transportation management can anticipate bad weather and proactively redirect a shipment if it is anticipated that it would affect a particular leg of a voyage. The output of the models can be used to choose the best places to buy the necessary items and the best routes to take in order to prevent further disruptions. To ensure that operations continue during inclement weather, appropriately equipped models can also initiate inventory transfers. Even though we are in the era of technology, the Field of Disaster management and weather prediction need more trustworthy techniques to manage disaster so that we can completely depend on an AI technology. Future research could conduct an exploratory study to examine the efficacy and suitability of AI and cloud-based collaboration tools in emergency, disaster, and extreme weather circumstances. In particular, a simulation-based study might be carried out to lay the groundwork for managing a genuine disaster and extreme weather situation through a cloud-based collaborative platform, and then compare the two approaches. Finally, since the function of artificial intelligence (AI) and cloud-based collaboration platforms can differ in different emergency scenarios, additional study may be undertaken to evaluate the research issue in the context of various catastrophes (floods, earthquakes, wildfires, cyclones, etc.).

5 Conclusion

This study examined a wide range of cloud computing, AI, and AI-cloud integrations in the context of disaster management. In doing so, it dismantled the current patterns that primarily concentrate on both proactive and reactive solutions for catastrophe mitigation. Geographic Information Systems (GIS) and IDSS have become effective instruments in this situation, providing new perspectives on disaster scenarios. A reliable support system for disaster response and decision-making is anticipated for the future, drawing on the possibilities of large data storage and artificial intelligence analysis. This conceptual framework holds out the possibility of providing decision-makers with the skills and knowledge necessary to make wise decisions. Such decisions could proactively avoid calamities, protect lives, and limit the destruction of property.

References

1. N. Bharosa, J. Lee, and M. Janssen, "Challenges and obstacles in sharing and coordinating information during multi-agency disaster response: Propositions from field exercises," Information Systems Frontiers, vol. 12, no. 1, pp. 49–65, May 2009, https://doi.org/10.1007/s10796-009-9174-z.
2. S. Gupta, S. Modgil, A. Kumar, U. Sivarajah, and Z. Irani, "Artificial intelligence and cloud-based Collaborative Platforms for Managing Disaster, extreme weather and emergency operations," International Journal of Production Economics, vol. 254, p. 108642, Dec. 2022, https://doi.org/10.1016/j.ijpe.2022.108642.
3. I. Ajzen, "The theory of planned behavior: Frequently asked questions," Human Behavior and Emerging Technologies, vol. 2, no. 4, pp. 314–324, Jun. 2020, https://doi.org/10.1002/hbe2.195.
4. S. Rehman, M. Sahana, H. Hong, H. Sajjad, and B. B. Ahmed, "A systematic review on approaches and methods used for flood vulnerability assessment: framework for future research," Natural Hazards, vol. 96, no. 2, pp. 975–998, Mar. 2019, https://doi.org/10.1007/s11069-018-03567-z.
5. X. Wang and H. Xie, "A Review on Applications of Remote Sensing and Geographic Information Systems (GIS) in Water Resources and Flood Risk Management," Water, vol. 10, no. 5, p. 608, May 2018, https://doi.org/10.3390/w10050608.
6. R. Pielke, "Tracking progress on the economic costs of disasters under the indicators of the sustainable development goals," Environmental Hazards, vol. 18, no. 1, pp. 1–6, Oct. 2018, https://doi.org/10.1080/17477891.2018.1540343.
7. R. I. Ogie, J. C. Rho, and R. J. Clarke, "Artificial Intelligence in Disaster Risk Communication: A Systematic Literature Review," 2018 5th International Conference on Information and Communication Technologies for Disaster Management (ICT-DM), Dec. 2018, https://doi.org/10.1109/ict-dm.2018.8636380.
8. V. Nunavath and M. Goodwin, "The Role of Artificial Intelligence in Social Media Big data Analytics for Disaster Management -Initial Results of a Systematic Literature Review," 2018 5th International Conference on Information and Communication Technologies for Disaster Management (ICT-DM), Dec. 2018, https://doi.org/10.1109/ict-dm.2018.8636388.

9. M. Abid et al., "Seismic Hazard Assessment of Shigo Kas Hydro-Power Project (Khyber Pakhtunkhwa, Pakistan)," Buildings, vol. 11, no. 8, p. 349, Aug. 2021, https://doi.org/10.3390/buildings11080349.
10. T.-H. Yang and W.-C. Liu, "A General Overview of the Risk-Reduction Strategies for Floods and Droughts," Sustainability, vol. 12, no. 7, p. 2687, Mar. 2020, https://doi.org/10.3390/su12072687.
11. S. Kamran Abid, N. Sulaiman, C. Shiau Wei, and U. Nazir, "Building Resilient Future: Information Technology and Disaster Management – A Malaysian Perspective," IOP Conference Series: Earth and Environmental Science, vol. 795, no. 1, p. 012026, Jun. 2021, https://doi.org/10.1088/1755-1315/795/1/012026.
12. R. Costache and D. Tien Bui, "Spatial prediction of flood potential using new ensembles of bivariate statistics and artificial intelligence: A case study at the Putna river catchment of Romania," Science of The Total Environment, vol. 691, pp. 1098–1118, Nov. 2019, https://doi.org/10.1016/j.scitotenv.2019.07.197.
13. R. R. Arinta and E. Andi W.R., "Natural Disaster Application on Big Data and Machine Learning: A Review," 2019 4th International Conference on Information Technology, Information Systems and Electrical Engineering (ICITISEE), Nov. 2019, https://doi.org/10.1109/icitisee48480.2019.9003984.
14. J. Chen, Q. Li, H. Wang, and M. Deng, "A Machine Learning Ensemble Approach Based on Random Forest and Radial Basis Function Neural Network for Risk Evaluation of Regional Flood Disaster: A Case Study of the Yangtze River Delta, China," International Journal of Environmental Research and Public Health, vol. 17, no. 1, p. 49, Dec. 2019, https://doi.org/10.3390/ijerph17010049.
15. A. Bhardwaj, "Application of Artificial Intelligence in Disaster Response," Artificial Intelligence, pp. 183–194, Oct. 2021, https://doi.org/10.1201/9781003140351-19.
16. Y. Han, "Gianluigi Negro (2017): 'The Internet in China. From Infrastructure to a Nascent Civil Society' (Palgrave Macmillan)," AI and SOCIETY, Dec. 2022, https://doi.org/10.1007/s00146-022-01623-8.
17. M. J. Canon, A. Satuito, and C. Sy, "Determining Disaster Risk Management Priorities through a Neural Network-Based Text Classifier," 2018 International Symposium on Computer, Consumer and Control (IS3C), Dec. 2018, https://doi.org/10.1109/is3c.2018.00067.
18. D. Fan and P. Shi, "Improvement of Dijkstra's algorithm and its application in route planning," 2010 Seventh International Conference on Fuzzy Systems and Knowledge Discovery, Aug. 2010, https://doi.org/10.1109/fskd.2010.5569452.
19. A. E. Akay, M. G. Wing, F. Sivrikaya, and D. Sakar, "A GIS-based decision support system for determining the shortest and safest route to forest fires: a case study in Mediterranean Region of Turkey," Environmental Monitoring and Assessment, vol. 184, no. 3, pp. 1391–1407, Apr. 2011, https://doi.org/10.1007/s10661-011-2049-z.
20. A. Correia, I. Severino, I. L. Nunes, and M. Simões-Marques, "Knowledge Management in the Development of an Intelligent System to Support Emergency Response," Advances in Human Factors and Systems Interaction, pp. 109–120, Jun. 2017, https://doi.org/10.1007/978-3-319-60366-7_11.
21. V. L. Sauter, "Decision Support Systems for Business Intelligence," Jan. 2011, https://doi.org/10.1002/9780470634431.
22. A. S. N. Alarifi, N. S. N. Alarifi, and S. Al-Humidan, "Earthquakes magnitude predication using artificial neural network in northern Red Sea area," Journal of King Saud University – Science, vol. 24, no. 4, pp. 301–313, Oct. 2012, https://doi.org/10.1016/j.jksus.2011.05.002.
23. "Prediction of Damage/Repair Rates in Water Distribution Systems Due to Seismic Events Using Artificial Neural Network," Intelligent Engineering Systems Through Artificial Neural Networks, Volume 17, pp. 149–154, 2007, https://doi.org/10.1115/1.802655.paper23.
24. "Introduction to Geographic Information Systems," Geographic Information Systems in Water Resources Engineering, pp. 33–44, Apr. 2016, https://doi.org/10.1201/9781420069143-8.

25. R. Der Sarkissian, J.-M. Zaninetti, and C. Abdallah, "The use of geospatial information as support for Disaster Risk Reduction; contextualization to Baalbek-Hermel Governorate/Lebanon," Applied Geography, vol. 111, p. 102075, Oct. 2019, https://doi.org/10.1016/j.apgeog.2019.102075.
26. M. Hoque, S. Tasfia, N. Ahmed, and B. Pradhan, "Assessing Spatial Flood Vulnerability at Kalapara Upazila in Bangladesh Using an Analytic Hierarchy Process," Sensors, vol. 19, no. 6, p. 1302, Mar. 2019, https://doi.org/10.3390/s19061302.
27. S. K. Sharma, S. K. Misra, and J. B. Singh, "The role of GIS-enabled mobile applications in disaster management: A case analysis of cyclone Gaja in India," International Journal of Information Management, vol. 51, p. 102030, Apr. 2020, https://doi.org/10.1016/j.ijinfomgt.2019.10.015.
28. S. K. Sharma, S. K. Misra, and J. B. Singh, "The role of GIS-enabled mobile applications in disaster management: A case analysis of cyclone Gaja in India," International Journal of Information Management, vol. 51, p. 102030, Apr. 2020, https://doi.org/10.1016/j.ijinfomgt.2019.10.015.
29. P. Fernandez, S. Mourato, and M. Moreira, "Social vulnerability assessment of flood risk using GIS-based multicriteria decision analysis. A case study of Vila Nova de Gaia (Portugal)," Geomatics, Natural Hazards and Risk, vol. 7, no. 4, pp. 1367–1389, Jun. 2015, https://doi.org/10.1080/19475705.2015.1052021.
30. A. G. Rumson and S. H. Hallett, "Innovations in the use of data facilitating insurance as a resilience mechanism for coastal flood risk," Science of The Total Environment, vol. 661, pp. 598–612, Apr. 2019, https://doi.org/10.1016/j.scitotenv.2019.01.114.
31. N. Ongdas, F. Akiyanova, Y. Karakulov, A. Muratbayeva, and N. Zinabdin, "Application of HEC-RAS (2D) for Flood Hazard Maps Generation for Yesil (Ishim) River in Kazakhstan," Water, vol. 12, no. 10, p. 2672, Sep. 2020, https://doi.org/10.3390/w12102672.
32. M. Vojtek and J. Vojteková, "Flood hazard and flood risk assessment at the local spatial scale: a case study," Geomatics, Natural Hazards and Risk, vol. 7, no. 6, pp. 1973–1992, Apr. 2016, https://doi.org/10.1080/19475705.2016.1166874.
33. I. Curebal, R. Efe, H. Ozdemir, A. Soykan, and S. Sönmez, "GIS-based approach for flood analysis: case study of Keçidere flash flood event (Turkey)," Geocarto International, vol. 31, no. 4, pp. 355–366, May 2015, https://doi.org/10.1080/10106049.2015.1047411.
34. V. Demir and O. Kisi, "Flood Hazard Mapping by Using Geographic Information System and Hydraulic Model: Mert River, Samsun, Turkey," Advances in Meteorology, vol. 2016, pp. 1–9, 2016, https://doi.org/10.1155/2016/4891015.
35. S. Puttinaovarat and P. Horkaew, "Internetworking flood disaster mitigation system based on remote sensing and mobile GIS," Geomatics, Natural Hazards and Risk, vol. 11, no. 1, pp. 1886–1911, Jan. 2020, https://doi.org/10.1080/19475705.2020.1815869.
36. D. Arruda and N. H. Madhavji, "The Role of Big Data Analytics in Corporate Decision-making," Proceedings of the 6th International Conference on Data Science, Technology and Applications, 2017, https://doi.org/10.5220/0006402300280037.
37. "Disaster Management," Encyclopedia of Big Data, pp. 400–400, 2022. https://doi.org/10.1007/978-3-319-32010-6_300090.
38. H. A. Simon, The new science of management decision. Harper & Brothers, 1960. https://doi.org/10.1037/13978-000.
39. G. Phillips-Wren, "Intelligent Decision Support Systems," Multicriteria Decision Aid and Artificial Intelligence, pp. 25–44, Feb. 2013, https://doi.org/10.1002/9781118522516.ch2.
40. A. Afuah and C. L. Tucci, "Value Capture and Crowdsourcing," Academy of Management Review, vol. 38, no. 3, pp. 457–460, Jul. 2013, https://doi.org/10.5465/amr.2012.0423.
41. P. Bhatt, A. J. Ahmad, and M. A. Roomi, "Social innovation with open source software: User engagement and development challenges in India," Technovation, vol. 52–53, pp. 28–39, Jun. 2016, https://doi.org/10.1016/j.technovation.2016.01.004.
42. F. Cappa, F. Rosso, and D. Hayes, "Monetary and Social Rewards for Crowdsourcing," Sustainability, vol. 11, no. 10, p. 2834, May 2019, https://doi.org/10.3390/su11102834.

43. K. R. Lakhani and R. G. Wolf, "Why Hackers Do What They Do: Understanding Motivation and Effort in Free/Open Source Software Projects," Perspectives on Free and Open Source Software, pp. 3–22, May 2005, https://doi.org/10.7551/mitpress/5326.003.0005.
44. L. Cazorla, C. Alcaraz, and J. Lopez, "A three-stage analysis of IDS for critical infrastructures," Computers and Security, vol. 55, pp. 235–250, Nov. 2015. https://doi.org/10.1016/j.cose.2015.07.005.
45. G. Grossel, A. Lyon, and M. Nunn, "Open-Source Intelligence Gathering and Open-Analysis Intelligence for Biosecurity," Invasive Species, pp. 84–92, Jun. 2017, https://doi.org/10.1017/9781139019606.005.
46. D. R. Hayes and F. Cappa, "Open-source intelligence for risk assessment," Business Horizons, vol. 61, no. 5, pp. 689–697, Sep. 2018, https://doi.org/10.1016/j.bushor.2018.02.001.
47. H. Darabi et al., "Development of a novel hybrid multi-boosting neural network model for spatial prediction of urban flood," Geocarto International, vol. 37, no. 19, pp. 5716–5741, May 2021, https://doi.org/10.1080/10106049.2021.1920629.
48. O. Rahmati et al., "Machine learning approaches for spatial modeling of agricultural droughts in the south-east region of Queensland Australia," Science of The Total Environment, vol. 699, p. 134230, Jan. 2020, https://doi.org/10.1016/j.scitotenv.2019.134230.
49. Z. Wu, Y. Zhou, H. Wang, and Z. Jiang, "Depth prediction of urban flood under different rainfall return periods based on deep learning and data warehouse," Science of The Total Environment, vol. 716, p. 137077, May 2020, https://doi.org/10.1016/j.scitotenv.2020.137077.
50. C. Campbell, S. Sands, C. Ferraro, H.-Y. (Jody) Tsao, and A. Mavrommatis, "From data to action: How marketers can leverage AI," Business Horizons, vol. 63, no. 2, pp. 227–243, Mar. 2020, https://doi.org/10.1016/j.bushor.2019.12.002.
51. Dae Geon Kim, "Artificial Intelligence and Agent based Modeling for Power System Engineering, Journal of Machine and Computing, vol.3, no.3, pp. 340–350, July 2023. https://doi.org/10.53759/7669/jmc202303029.
52. A. K. Dwivedi, U. Kaliyaperumal Subramanian, J. Kuruvilla, A. Thomas, D. Shanthi, and A. Haldorai, "Time-series data prediction problem analysis through multilayered intuitionistic fuzzy sets," Soft Computing, vol. 27, no. 3, pp. 1663–1671, Apr. 2022, https://doi.org/10.1007/s00500-022-07053-4.
53. Jain Emadi, "The Development of a Design Theory for Web Based Information Systems", Journal of Robotics Spectrum, vol.1, pp. 013–023, 2023. https://doi.org/10.53759/9852/JRS202301002.
54. V. L. Narla, R. Kachhoria, M. Arun, A. Haldorai, D. Vijendra Babu, and B. M. Jos, "IoT based energy efficient multipath power control for underwater sensor network," International Journal of System Assurance Engineering and Management, Apr. 2022, https://doi.org/10.1007/s13198-021-01560-7.
55. C. Cao, Y. Liu, O. Tang, and X. Gao, "A fuzzy bi-level optimization model for multi-period post-disaster relief distribution in sustainable humanitarian supply chains," International Journal of Production Economics, vol. 235, p. 108081, May 2021, https://doi.org/10.1016/j.ijpe.2021.108081.
56. G. Potutan, M. Arakida, and K. Suzuki, "Improving Disaster Recovery Governance: Lessons from the Two Ad-Hoc Recovery Agencies," Disaster Risk Reduction, pp. 63–86, Dec. 2021, https://doi.org/10.1007/978-981-16-7401-3_4.
57. T. Choi, S. Kumar, X. Yue, and H. Chan, "Disruptive Technologies and Operations Management in the Industry 4.0 Era and Beyond," Production and Operations Management, vol. 31, no. 1, pp. 9–31, Jan. 2022, https://doi.org/10.1111/poms.13622.
58. T.-M. Choi, C. K. Y. Lo, C. W. Y. Wong, and R. W. Y. Yee, "Green manufacturing and distribution in the fashion and apparel industries," International Journal of Production Economics, vol. 135, no. 2, p. 531, Feb. 2012, https://doi.org/10.1016/j.ijpe.2011.07.012.
59. C.-C. Chou, W.-C. Chiang, and A. Y. Chen, "Emergency medical response in mass casualty incidents considering the traffic congestions in proximity on-site and hospital delays," Transportation Research Part E: Logistics and Transportation Review, vol. 158, p. 102591, Feb. 2022, https://doi.org/10.1016/j.tre.2021.102591.

60. S. Chowdhury, A. Emelogu, M. Marufuzzaman, S. G. Nurre, and L. Bian, "Drones for disaster response and relief operations: A continuous approximation model," International Journal of Production Economics, vol. 188, pp. 167–184, Jun. 2017, https://doi.org/10.1016/j.ijpe.2017.03.024.
61. R. S. Cox and K.-M. E. Perry, "Like a Fish Out of Water: Reconsidering Disaster Recovery and the Role of Place and Social Capital in Community Disaster Resilience," American Journal of Community Psychology, vol. 48, no. 3–4, pp. 395–411, Feb. 2011, https://doi.org/10.1007/s10464-011-9427-0.
62. J. Davis, H.-G. Wolff, M. L. Forret, and S. E. Sullivan, "Networking via LinkedIn: An examination of usage and career benefits," Journal of Vocational Behavior, vol. 118, p. 103396, Apr. 2020, https://doi.org/10.1016/j.jvb.2020.103396.
63. Z. Piepiora and O. Belarga, "The local disaster management system: the crisis cluster in the face of natural disasters," Human and Social Sciences at the Common Conference, Oct. 2015, https://doi.org/10.18638/hassacc.2015.3.1.175.

Nature Inspired Optimizers and Their Importance for AI: An Inclusive Analysis

1 Introduction

Artificial intelligence (AI) is the learning of how to create machineries think and act like people. It has been able to do this since the creation of sophisticated artificial intelligence algorithms by building machines and robots that are used in various industries, including agriculture, healthcare, robotics, marketing, business analytics, and many more [1]. An algorithm typically takes some input and uses logic and mathematics to produce the result. In contrast, AI simultaneously uses a mixture of inputs/outputs to "study" the data and create outcomes when given fresh inputs. Artificial intelligence is a subfield of machine learning where we attempt to implement AI by learning the input data. Machine learning is only a neural network through three or additional layers, which is a subset of deep learning. This neural network's purpose is to imitate how the hominoid brain functions so that it can "learn" from a lot of data.

Deep learning, a form of machine learning, plays a crucial role in enabling computers to grasp concepts that humans often take for granted. This technique has found significant applications in various domains. For instance, self-driving cars rely on deep learning to recognize stop signs, differentiate pedestrians from streetlights, and make informed decisions on the road. Additionally, voice control functionality in consumer electronics, including speakerphones, tablets, TVs, and smartphones, heavily relies on deep learning algorithms. The recent surge in interest surrounding deep learning is well-founded, as it has demonstrated the ability to achieve previously unattainable results [2]. Through deep learning, computer models can learn to perform classification tasks directly from different types of input, such as images, text, or audio. This approach has led to remarkable advancements, with deep learning models frequently surpassing human capabilities in terms of accuracy. Training deep learning models involves utilizing extensive sets of labeled data and

© European Alliance for Innovation 2024
A. Haldorai et al., *Artificial Intelligence for Sustainable Development*, EAI/Springer Innovations in Communication and Computing,
https://doi.org/10.1007/978-3-031-53972-5_21

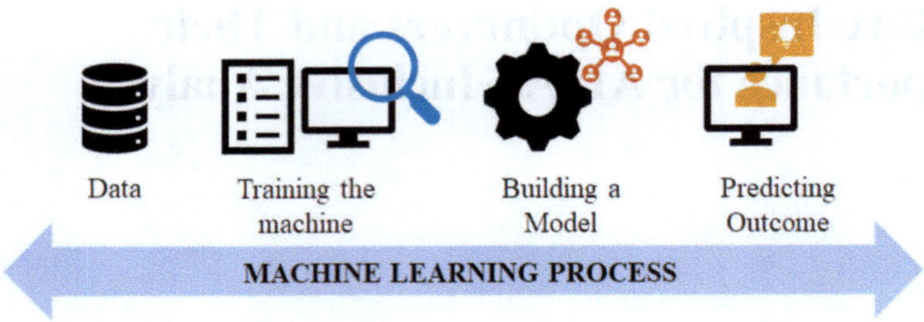

Fig. 1 Machine learning process

employing multilayer neural network architectures. This process allows the models to learn complex patterns and relationships within the data. Figure 1 provides an illustrative representation of the machine learning process.

Deep learning is a branch of machine learning used to perform difficult tasks such as text classification and speech recognition. Activation functions, inputs, outputs, hidden layers, loss functions, etc. are all components of a deep learning model [3]. All deep learning algorithms attempt to generalize the data they use and make predictions using data that has not yet been seen [4]. We need both an optimization technique and an algorithm that transforms sampled inputs into sampled outputs. In order to reduce the error in the mapping of inputs to outputs, optimization methods establish parameter values, or weights. This page contains comprehensive information on these deep learning optimizers and optimization methods. This section will examine the many optimizers that are frequently used to build deep learning models, their benefits and drawbacks, and the variables that affect the choice of optimizer for a certain application. Algorithms for optimization are essential for improving deep learning model performance [5]. The use of optimizers or optimization techniques has a major impact on these models' training speed and efficiency.

Minimizing the loss function is the goal of deep learning models, which need adjustments to the weights for each epoch during training. This is where the function or algorithm known as the optimizer is used. It makes it possible to alter the learning rate and weights of the neural network, which lowers total loss and raises accuracy. Due to the enormous number of parameters—which frequently approach millions—selecting the right weight for a deep learning model can be difficult [6]. Consequently, it becomes crucial to select an appropriate optimization algorithm tailored to your specific application. Thus, it is essential for data scientists to familiarize themselves with these machine learning techniques before delving into this subject matter. There exists a range of optimizers that can be employed to adjust the weights and learning rate in a machine learning model. However, determining the optimal optimizer depends on the unique requirements of the application at hand. Table 1 provides an overview of the fundamental optimizers examined within this context.

NIO (Nature-Inspired Optimization) emerges as a solution to address the aforementioned challenges, attracting significant attention from researchers. The appeal of NIO stems from its ability to leverage simulations of real-life phenomena, which

Table 1 The basic optimizers

S. no	Optimizer	Advantages	Disadvantages
1	Gradient Descent: This optimization technique leverages calculus to iteratively adjust values in order to reach a local minimum	Simple computation. Simple implementation. Easy to comprehend	1. Prone to getting stuck at local minima. 2. Convergence to the minima can be time-consuming for large datasets. 3. Demands significant memory resources
2	Stochastic Gradient Descent (SGD): SGD aims to update the model's parameters more frequently. Rather than utilizing the entire dataset in each iteration, SGD randomly selects batches of data for processing	1.Rapid updates of model parameters enable quicker convergence. 2.Demands lower memory requirements. 3.May discover new minima due to the random selection of data batches	1.Model parameters exhibit higher variance. 2. Requires a greater number of iterations to reach the local minima, leading to increased computation time
3	Stochastic Gradient Descent with Momentum: Momentum is introduced in SGD to address the issue of high variance and enhance the convergence process. It accelerates convergence toward the relevant direction while reducing fluctuations in the irrelevant direction	1.Mitigates oscillations and reduces high variance in parameter values. 2.Achieves faster convergence compared to standard gradient descent	1.Introduces an additional hyperparameter that requires careful and accurate manual selection
4	Mini-Batch Gradient Descent: In mini-batch gradient descent, the model parameters are updated after processing each batch of data. The dataset is divided into multiple smaller batches, and the parameters are updated iteratively after processing each batch	1.Updates the model parameters frequently, leading to reduced variance. 2. Requires a moderate amount of memory to handle the intermediate batch data	1. Selection of an optimal learning rate is necessary. 2. There is a possibility of getting trapped at local minima
5	Adagrad: Adagrad is a variant of gradient descent that incorporates adaptive learning rates. Unlike other gradient descent algorithms, it adjusts the learning rate for each iteration based on the parameter differences observed during training. If the parameters undergo significant changes, the learning rate is reduced proportionally, resulting in smaller updates	1. The learning rate adapts to the training parameters individually. 2. Eliminates the need for manual tuning of the learning rate. 3. Suitable for training on sparse data	1.Computationally expensive due to the requirement of calculating second-order derivatives. 2. The decreasing learning rate in Adagrad can result in slow training progress

(continued)

Table 1 (continued)

S. no	Optimizer	Advantages	Disadvantages
6	RMSProp: RMSProp is a widely used optimizer in the field of deep learning	1. RMSProp helps address the issue of fluctuating gradients during training	1. Manually defining the earning rate is necessary, as the suggested value does not universally apply to every application
7	AdaDelta differs from other optimization algorithms by limiting the accumulation of previously squared gradients to a fixed-size window, rather than accumulating all of them	1. Currently, there is no decay in the learning rate, and the training process continues without stopping	1. The process is computationally intensive or resource-intensive
8	Adam (Adaptive Moment Estimation) utilizes first and second-order momentums to optimize the learning process. The rationale behind Adam is to prevent excessively fast movement that could potentially overshoot the minimum. Instead, it aims to gradually reduce the velocity for a more cautious and precise search	1.The method exhibits exceptional speed and achieves rapid convergence. 2. It effectively addresses the issue of vanishing learning rates and mitigates high variance	1. The method is computationally expensive and requires significant computational resources

have proven instrumental in resolving various complex problems. By mimicking natural processes, scientific advancements have yielded some of the most remarkable inventions [7]. For instance, the development of drones drew inspiration from bird flight patterns, while deep neural networks were inspired by bee colony algorithms, and human neural networks have been discovered in nature. Nature's knack for simulating technological advancements has demonstrated great promise. In the realm of AI, approximately 120 algorithms and their variations have been described, offering effective and efficient solutions for optimization challenges. Many of these ideas have been influenced by natural phenomena, making NIO a captivating field that explores the potential of discovering new algorithms inspired by nature [8]. The promising nature of NIO has spurred numerous surveys, examining its applications and potential benefits.

Over the past three decades, several Non-Intrusive Optimization Algorithms (NIOAs) and their variations have been proposed. However, a significant gap exists due to the absence of a comprehensive and analysis of these various NIOAs. Consequently, researchers often select NIOAs for specific optimization problems based on their limited experiences. This study aims to address this gap by providing a thorough understanding of the concept of NIO and the role of optimizers in the training process. The primary objectives of this research are to bridge the aforementioned knowledge gap and enhance comprehension in these areas. In this chapter, the following contributions are made:

- We begin by summarizing the 11 common NIOAs' consistent formal description and comparing and contrasting them.
- Recognize the mathematical concepts such as population, inspiration, and others that underlie optimizers.
- Next, we thoroughly compare the routine of NIOAs and the key elements.
- Recognize how to modify optimizer settings to enhance training results and avoid overfitting.
- We present a detailed analysis of the issues and prospects facing the entire NIOAs field, guiding future NIOAs research.

The quest for a universal algorithm capable of addressing every point in every domain indefinitely is an insurmountable task. Instead, we present a valuable guide that can aid researchers in choosing suitable algorithms more effectively, depending on the specific application scenarios, thereby maximizing the utilization of diverse NIOAs. Through this study, we anticipate the emergence of more NIOAs tailored to novel problem domains in the future. Furthermore, we envision this work to serve as an exceptional manual and reference for fostering innovation and practical applications of NIOAs.

Section 2 explains the recent nature-inspired algorithms. The structure of nature-inspired algorithms is given in detail in Sect. 3. The next section, Sect. 4 gives the details about Algorithm classification and Sect. 5 presents challenges and conclusion.

2 Survey on Recent Nature-Inspired Optimization Algorithms

A new metaheuristic is suggested in the Abdollahzadeh et al. [9] study and is modeled after the way of life of African vultures. A system that mimics the hunting and navigating habits of African vultures is introduced: the African Vultures Optimization System (AVOA). Through a comparative analysis, the performance of AVOA is assessed using 36 widely used benchmark functions, demonstrating its superiority over several other methods. Eleven engineering design difficulties are then solved optimally by applying the algorithm, demonstrating the applicability of AVOA and its black-box nature. According to experimental results, AVOA performs better than other algorithms on 30 of the 30 benchmark functions and does well in most engineering case studies. The Wilcoxon rank-sum test, used for statistical analysis, shows that the AVOA algorithm significantly outperforms the competition at a 95% confidence level.

Remora Optimization Algorithm (ROA), novel bionics, naturally, and metaheuristic procedure, is proposed in this Jia et al. [10] research. The parasitic behavior of remora served as the primary source of inspiration for ROA. Different hosts update various locations: Remora feeds on the ectoparasites or wrecks of some huge hosts while avoiding natural predators, like in the case of enormous whales. For example, when a fast-moving swordfish is present, remora will follow the swordfish to a bait-rich location. Remora also makes some judgments in the context of these two updating mechanisms created on knowledge. If it choose to hunt, it keep learnt the host and does a global update. Remora does not alter the host if it eats around it; instead, it keeps doing local updates. For this reason the host in ROA can be adequately substituted; this method is more likely to offer a fresh concept for a memetic algorithm. The aforementioned dynamic mode and behavior are mathematically recreated, and 29 benchmark questions and 5 genuine engineering challenges are used to assess the ROA's validity. Ten additional natural heuristics are compared side by side. In assessment to other cutting-edge heuristic strategies, statistical data and comparisons demonstrate that ROA offers a very hopeful future and a great modest potential.

Abdollahzadeh et al. [11] conducted a study introducing the Artificial Gorilla Troops Optimizer (GTO), a novel metaheuristic algorithm inspired by the social intelligence observed in gorilla troops in the wild. This algorithm mathematically models the communal behavior of gorillas and incorporates new mechanisms for exploration and exploitation. To evaluate its performance, the GTO was applied to 7 engineering challenges and 52 commonly used benchmark functions. Statistical analysis was conducted using Friedman's test and Wilcoxon rank-sum tests, comparing the GTO with other existing metaheuristics. The results indicate that the GTO outperforms comparable algorithms on most benchmark functions, particularly for high-dimensional problems. These findings demonstrate that the GTO has the potential to produce superior results compared to other metaheuristics.

The study conducted by AL-kubaisy et al. [12] proposes the Red Colobuses Monkey (RCM) metaheuristic optimization algorithm, which emulates the behavior of red monkeys in nature. This algorithm can be effectively employed to tackle optimization problems. To evaluate the efficacy of the recommended algorithm, a set of 64 identified test functions commonly used in optimization were utilized as benchmarks. Both unconstrained and restricted test functions were employed in this assessment. As the algorithm progresses, the solutions have consistently improved, with higher-ranking outcomes based on the best answer found thus far. Furthermore, Dehghani and colleagues [13] introduce the Cat and Mouse-Based Optimizer (CMBO). Table 2 presents some of the newly introduced optimization algorithms for reference.

3 Structure of Nature-Inspired Optimizer

The 1980s and 1990s were the most exciting time for meta-heuristic algorithms. This algorithm is inspired by nature and is based on phenomena of physics, biology, animal behavior, and so on. For a long time, humans have used the direction of nature to find the most suitable answer for issues. Thus, during the most recent eras,

Table 2 Optimization algorithms

S. No	Optimization algorithm	Papers related to algorithms
1.	Artificial Gorilla Troops Optimizer	Solar Photo Voltaic Systems [14], Plant disease detection [15]
2.	Aquila Optimizer	Oil Production Forecasting [16], Wind energy assessment [17]
3.	Horse Herd Optimization Algorithm	Mathematical and Engineering Optimization for energy system [18], spam detection [19]
4.	Rat Swarm Optimizer	Feature selection [20], Energy system [21]
5.	Chameleon Swarm Algorithm	Engineering design [22], patch antenna design [23]
6.	Child Drawing Development Optimization	Mathematical and Engineering Optimization problems [24]
7.	Mayfly Optimization Algorithm	PEM Fuel Cell Parameter Identification [25]
8.	Coronavirus Optimization Algorithm	Deep Learning and Electricity Load Time Series Forecasting [26]
9.	Chimp Optimization Algorithm	Modified optimization technique [27], Large-Scale Spherical Vehicle Routing Problem [28]
10.	Bald Eagle Search	Photovoltaic Systems [29]

there has been an endeavor to develop in creating optimization algorithms inspired by nature. This meta-heuristic algorithm proved to be more efficient than heuristic algorithms in many applications. The algorithms listed in Table 3 above are often used by researchers in various fields. However, no single algorithm can be used to find the best solution for every optimization issue. Compared to other algorithms, some provide a better solution to a given problem. Therefore, academics still have a lot to learn about tracking for new meta-heuristic optimization algorithms. Stochastic and deterministic algorithms can be categorized as being inspired by nature. A random probability distribution can accurately depict anything that is stochastic. It alludes to a modeling strategy. An NIOA is referred to as stochastic or randomized if there is any indication of unpredictability in its general behavior. Deterministic systems are those in which the generation of outputs has not been affected by chance. The results may be anticipated or known in advance. Yes, there are always unknowns in experimental settings, but there may also be a range of expectations. The values are fixed numbers that cannot be changed to produce new solutions. We can say that it is simple to estimate the values.

The majority of NIOAs are stochastic because they obtain solutions from the search space at random. Others are deterministic because they are in a more regulated environment. A local optimum may be the outcome. According to research, NIOAs exhibit typical behaviors. This information can assist scientists better understand how to find new types. It becomes vital to recognize the similarities between them in order to advance in this field. Figure 2 illustrates how the standard procedure for the majority of NIOAs is broken down into four fundamental components.

Table 3 Meta-heuristic algorithms with advantages

S. no	Meta-heuristic algorithms	Source of inspiration	Comment	Advantage
1	Genetic algorithms (GA) [30]	Biological evolution based on Charles Darwin's natural selection theory	Most popular evolutionary algorithms with a diverse range of applications	Achieving excellent results in optimizing both constrained and unconstrained design problems
2	Simulated Annealing (SA) [31]	Metal annealing processing	Have the ability to avoid stuck in local optima	Exceptional utilization of the search space for unimodal functions, resulting in superior exploration and optimal solutions
3	Differential evolution (DE) [32]	Development of genetic algorithms	The evolutionary algorithm based on vector	Demonstrating remarkable adaptability in tackling optimization problems with a combination of reliability and efficiency
4	Ant colony optimization (ACO) [33]	Foraging behavior of social ants	Famous meta-heuristic algorithm	Efficiently navigating non-convex and non-separable functions by effectively avoiding local minima and achieving rapid convergence
5	Particle swarm optimization (PSO) [34]	Fish and bird schooling	Applied to almost every area in optimization, computational intelligence	Successfully converging to the global optimum and generating high-quality approximations of Pareto optimal solutions for multi-objective problems
6	Bee algorithms (BA) [35]	Foraging behavior of bees	It introduces many sub-algorithms	Delivering exceptional efficiency in optimal reactive power dispatch and exhibiting strong exploration abilities for multi-modal functions
7	Honey-bee mating optimization (HBMO) [36]	Based on BA	Applied to reservoir modeling and clustering	The approach eliminates the need for arbitrary values during population initialization, ensuring a more robust and reliable optimization process
8	Artificial Bee Colony (ABC) [37]	Based on BA	For numerical function optimization	Well suited for continuous nonlinear optimization problems, with a tendency to converge early while maintaining optimal performance

(continued)

Table 3 (continued)

S. no	Meta-heuristic algorithms	Source of inspiration	Comment	Advantage
9	Harmony Search (HS) [38]	Related to the improvisation process of a musician	Prominent meta-heuristic algorithm	Exhibiting robustness in handling constrained optimization problems, ensuring reliable and consistent results even in the presence of constraints
10	Firefly Algorithm (FA) [39]	Flashing patterns and behavior of fireflies	Nature-based algorithm	Highly effective in solving global optimization problems with continuous variables and boundary constraints, delivering accurate and optimal solutions
11	Cuckoo search (CS) [40]	Brood parasitism of cuckoo species	Latest nature-inspired meta-heuristic algorithms	Ideal for continuous nonlinear optimization problems, showcasing a propensity for early convergence while maintaining exceptional performance throughout the process

Fig. 2 Process of NIOs

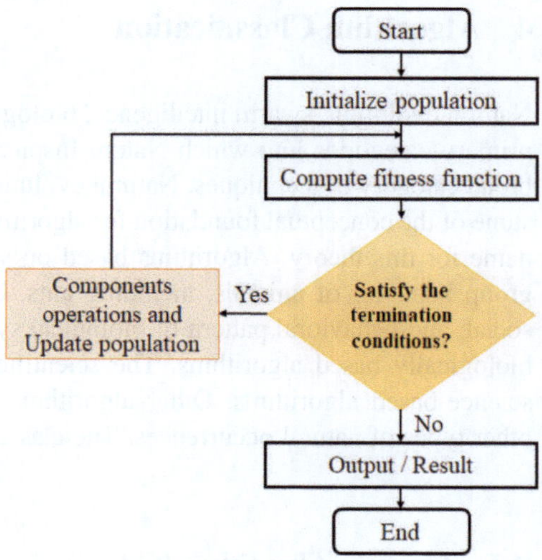

First, the population's characteristics are determined. The initialization of each NIOA's parameters occurs before the optimization procedure is launched. To ensure that it covers the largest feasible portion of the solution space, the initial population is produced using random methods. When the optimization process starts, the parameter values are fixed, but they may vary later on. Based on the optimization's criteria, the population and its size are specified. Later in the method, the parameters often go through an iterative process, but these conditions are established when the parameters are started for the first time. The parameters used to initialize the algorithm at step one provide it shape and performance control. If the outcomes fall short of expectations, the initial set of parameters can be changed. It will be simpler to alter these parameters if they are modularized. The initialization of the population's parameters is initially calculated in step two. The iteration starts and goes on. This is the initial indicator of whether the parameters created will achieve objectives. Later in the flow, the terminating condition will be satisfied.

The second stage's output is represented by stage three. This stage is typically in charge of carrying out the primary duty and using operators and variables to change values across the population's parameters while ensuring system control. The system constantly modifies and updates when the parameter values are changed, improving its efforts to accomplish the goals. For instance, there will be specifications for the Artificial Bee Colony's desertion criterion, employed bee count, observer, and scout bee counts. The variables will be food source selection and food source position updating. This will change when the algorithm runs its course. The algorithm finishes its work and produces a result in step four. This might or might not be satisfactory. By changing the settings and values and rerunning the program, the cycle can begin again.

4 Algorithm Classification

Natural evolution, swarm intelligence, biology, science, and other fields are the five primary categories into which Nature-Inspired Algorithms (NIAs) fall. NIAs are a broad category of techniques. Natural evolution theories' tenets serve as the cornerstone of the conceptual foundation for algorithms based on them. "Darwinism" is the name for this theory. Algorithms based on swarm intelligence are inspired by the group behavior of animals, including ants, bats, bees, cuckoos, and fireflies. The social, and behavioral pattern of biological systems serves as the inspiration for the biologically based algorithms. The scientific notions serve as the foundation for science-based algorithms. Other algorithms are those that draw inspiration from other types of natural occurrences. The classification of NIAs is shown in Fig. 3.

4.1 Genetic Algorithm (GA)

The GA concept is built upon Darwin's theory of biological evolution known as "Survival of the Fittest." This theory forms the basis for GA, where unfit individuals are eliminated, and only the most adapted individuals survive and pass on their traits to the next generation. Holland's proposed GA algorithm relies on three fundamental mechanisms [41]. The encoding scheme used in the GA algorithm is determined by

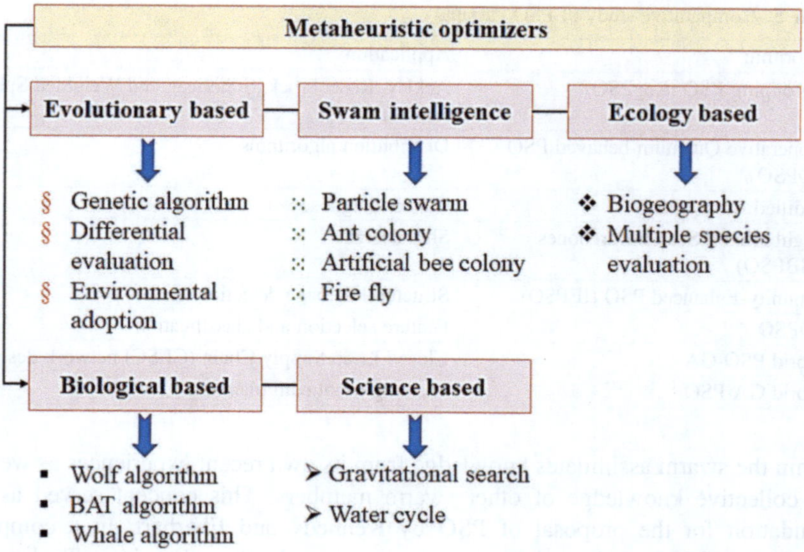

Fig. 3 NIA algorithm classification

Table 4 Comparative study of GA variants

Algorithm	Application
Non-dominated Sorting GA II	Multi-objective test problems
Real-coded GA	Brain images segmentation
Binary-Real Coded GA	Unit Commitment (UC) problem
Hybrid GA/PSO	Procedure preparation and arrangement
Simplified GA	Software energy approximation
Elitism GA (EGA)	Extreme Knowledge

the specific challenges faced by individuals [42]. Selection, crossover, and mutation operators come in various forms. echolocation behavior of bats. The mutation operator involves randomly changing one or more genes [43]. The choice of these three operators depends on the encoding scheme employed. Table 4 provides an overview of the diverse applications of GA in various domains.

4.2 *Particle Swarm Optimization (PSO)*

The PSO draws inspiration from the coordinated behavior observed in flocks of birds or schools of fish, where a swarm represents the collective group and each individual member is referred to as a particle. PSO was designed to emulate the cooperative dynamics exhibited by these swarms, which involve synchronized movement, abrupt changes in direction, dispersal, regrouping, and more. In PSO, every particle

Table 5 Comparative study of PSO variants

Algorithm	Application
Regrouping PSO (RegPSO)	Ackley, Rosenbrock, Spherical, and Weighted Sphere function
Cooperative Quantum-behaved PSO (CQPSO)	Distribution algorithms
Modified Binary PSO	Cancer diagnoses
Neighborhood Search Barebones (NSBPSO)	Ship design
Immunity-Enhanced PSO (IEPSO)	Structural damage detection
MOPSO	Feature selection and classification
Hybrid PSO-GA	Closed-Loop Supply Chain (CLSC) network design
Hybrid GA-PSO	Welding robot path planning

within the swarm assimilates knowledge from its own recent experiences as well as the collective knowledge of other swarm members. This concept served as the foundation for the proposal of PSO by Kennedy and Eberhart. In a computer simulation of PSO, each particle possesses its own velocity and position. To discover high-quality solutions, each particle remembers its own best position (referred to as "pbest") and shares the globally best position (referred to as "gbest") with other particles in the swarm. The new position of a particle depends on the distance between its current position and both its pbest and gbest. Equations 1 and 2 are employed to calculate the new velocity and position of particles. These equations have been refined through various developmental stages to improve the performance and effectiveness of PSO

$$\begin{aligned} \text{new\_velocity} = \text{current\_velocity} + c_1 r_1 \times (\text{pbest} - \text{current\_position}) \\ + c_2 r_2 \times (\text{gbest} - \text{current\_position}) \end{aligned} \tag{1}$$

$$\text{new\_position} = \text{current\_position} + \text{new\_velocity} \tag{2}$$

where, r_1, r_2 are two random variables with range [0,1]; c_1, c_2 are learning factors. Table 5. Gives the comparative study of PSO variants.

4.3 *Artificial Bee Colony (ABC)*

A new swarm-based algorithm was proposed by Karaboga as a result of the honey bees' intelligent foraging activity. ABC is the name for this. ABC uses the concepts of self-organization and labor division to achieve the swarm's intelligent behavior. Positive feedback, negative feedback, fluctuation, and various interactions are all parts of self-organizing. The division of labor entails the use of specialists for each task. Bees perform three fundamental functions throughout each search cycle. Employed bees identify food sources and gauge nectar value in the first step. The worth of nectar is discussed with bystanders by a hired bee in the second stage, after which onlookers choose a good food source. Scout bees search the search region for new food sources in the third step. The new velocity and new position of particles are calculated using Eqs. 3 and 4.

$$\begin{aligned}\text{new\_position} = \text{current\_position} + \text{rand} \\ \times (\text{current\_position} - \text{current\_position\_k})\end{aligned} \tag{3}$$

$$\text{probability} = \frac{\text{fitness}}{\text{total\_fitness}} \tag{4}$$

where rand mention the random number between [−1,1] and random number index chosen between 1 and n. The position of the ith member on iteration tin is mentioned in Eq. (5), where, L_d and U_d are the upper bounds. The forager's search is presented in Eq. (6)

$$x_{i,d}(t) = L_d + \text{rand }(0,1) * (U_d - L_d) \tag{5}$$

$$\vartheta_{i,d}(t+1) = x_{i,d}(t) + \varphi\big(x_{i,d}(t) - x_{j,d}(t)\big) \tag{6}$$

Here, $0 < i, j \leq M$, $I \neq j$, φ is the random number in (0,1). The food sources are given in Eqs. (7), (8). Table 6 presents the various variants of ABC algorithm.

$$p_i = \frac{fit_i}{\sum_{i=1}^{N} fit_i} \tag{7}$$

$$fit_i = \begin{cases} \dfrac{1}{1+f(i)}, f(i) \geq 0 \\ 1+abs\ (f(i)), \text{otherwise} \end{cases} \tag{8}$$

4.4 *Firefly Algorithm (FA)*

Fireflies' intelligent flashing behavior has served as a valuable source of inspiration for the development of novel algorithms. Each firefly exhibits a distinct pattern of flashing, which serves as a signaling and communication mechanism among them, as well as a means to attract prey and function as a protective warning system. Usually, fireflies use distinct flashing patterns to entice one another in order to mate. The firefly' attraction, their distance from one another, and the degree of absorption in the surrounding media are some of the variables that affect the intensity of their flashing light. Yang proposes a novel algorithm called the Firefly Algorithm (FA) after finding a relationship between the objective function and the firefly's flashing behavior. An overview of the many FA algorithm versions is given in Table 7.

Table 6 Various variants of ABC algorithm

Algorithm	Application
Chaotic ABC (CABC)	Rosenbrock, Griewangk, and Rastrigin function
Cooperative ABC (CABC)	Sphere, Rosenbrock, Griewank, Rastrigin, Ackley, and Schwefel function
Hybrid GA-ABC	Proportional Integral (PI) speed controller of Permanent Magnet Synchronous Motor (PMSM)
Discrete ABC (DABC)	Blocking Flow Shop (BFS) scheduling problem
Constrained ABC	13 linear, nonlinear, and quadratic test functions
Hybrid ABC-SPSO	CEC05 benchmark functions
Multi-objective ABC (MOABC)	Unconstrained and constrained test problems
Binary ABC (DisABC)	Uncapacitated Facility Location Problem (UFLP)
ABC with Levy Flight distribution (LFABC)	Steel space frame design problem
Covariance guided ABC (M-CABC)	Portfolio optimization

Table 7 Variants of FA algorithm

Algorithm	Application
Levy Flights FA (LFA)	Michalewicz, Rosenbrock, De Jong, Schwefel, Ackley, Rastrigin, Easom, Griewank, Yang, and Shubert function
Gaussian FA (GD-	Sphere, Rosenbrock, Rastrigin
Elitist FA	Seven benchmark problems
Binary Real Coded FA (BRCFF)	Unit Commitment Problem (UCP)
Parallel FA	Graphical Processing Unit (GPU) implementation
Hybrid Evolutionary Firefly Algorithm (HEFA)	Parameters optimization in a complex and nonlinear problem
Hybrid ACO–FA	Several benchmark problems of unconstrained optimization
Chaos FA (CFA)	Sphere and Griewank function
Multi-objective FA (MOFA)	Design optimization benchmarks in industrial engineering
Discrete Firefly Algorithm with the Standard GA	Capacitated Facility Location Problem (CFLP)
Hybrid Discrete Firefly Algorithm (HDFA)	Flexible Job Shop Scheduling Problem (FJSP)

The luminance and attraction degree is defined in (9), (10), and (11). The new position is updated in (12).

$$I = I_0 * \exp\left(-\gamma * r_{i,j}\right) \tag{9}$$

$$r_{i,j} = \sqrt{\sum_{k=1}^{D} \left(x_{i,k} - x_{i,k}\right)^2} \tag{10}$$

$$\beta(r_{i,j}) = \beta_0 * \exp\left(-\gamma * r_{i,j}{}^2\right) \tag{11}$$

$$x_i(t+1) = x_i(t) + \beta(r_{i,j}) * \left(x_j(t) - x_i(t)\right) + \text{step} * \varepsilon_i \tag{12}$$

4.5 *Bat Algorithm*

Yang introduced BA algorithm, which is based on echo sounding habit of bats. The frequency range of sound wave is presented in Eq. (13), where Freqmin and Freqmax represent the lower and upper bounds of frequency range. The sound intensity and new position updates given in Eqs. (14) and (15)

$$\text{Freq}_i = \text{Freq}_{\min} + (\text{Freq}_{\max} - \text{Freq}_{\min}) * \beta_i \tag{13}$$

$$v_i(t+1) = v_i(t) + \left(x_i(t) - p_g(t)\right) * \text{Freq}_{\text{i}} \tag{14}$$

$$x_i(t+1) = x_i(t) + v_i(t+1) \tag{15}$$

The new solution generated by bat is expressed in Eq. (16). β is the randum number between [0,1]. In the Eq. (16), $\varepsilon$is the random number between [−1,1]. The updated values are given in Eqs. (17) and (18); α, γ are the contant values. r_i^0 is the initialization value of r in the BA algorithm.

$$x_{\text{new}} = x_{\text{old}} + \varepsilon * \overline{A} \tag{16}$$

$$A_i(t+1) = \alpha * A_i(t) \tag{17}$$

$$r_i(t+1) = r_i^0 * [1 - \exp(-\gamma * t)] \tag{18}$$

Table 8 presents the various NIO from the year 2016 to 2023. The comparison between various nature inspired algorithm is presented in the Fig. 4.

5 Challenges and Future Direction of NIOA

Having a benchmark test that aligns with the tuning toolset of NIOAs and is user-friendly is highly desirable. This is because the process of automatically tuning parameters can be computationally expensive, particularly when dealing with real-world application problems. Given that most NIOAs function in a black-box manner without a strong mathematical foundation, researchers are constantly introducing supposedly "novel" algorithms and claiming superior optimization results compared to other NIOAs. However, there is a scarcity of discussions regarding the research visualization platforms of NIOAs, except for an initial, straightforward attempt.

Table 8 List of NIO from 2016 to 2023

Year	Algorithm	Inspiration
2016	Whale optimization	Hunting behavior of humpback whales
	Crow search	Food hiding behavior of crows
2017	Grasshopper	Foraging and swarming behavior of grasshoppers
	Salp Swarm	Salps swarm behavior when navigating and foraging
2018	Emperor Penguin Optimizer	Huddling behavior of emperor penguins
	Butterfly optimization	Foraging and mating behavior of butterflies
2019	Spider Monkey Optimization	Foraging behavior of spider monkeys
	Sailfish Optimizer	Hunting strategy of sailfishes
	Squirrel search	Foraging behavior of southern flying squirrels and locomotion
2020	Corona virus optimization	Coronavirus susceptible infectious removed model
	Black Widow Optimization	Mating behavior of black widow spiders
2021	Horse herd optimization	Social performances of horses at different ages
	Jellyfish optimizer	Foraging behavior of jellyfish in the ocean
2022	Artificial hummingbird	Special flight skills and intelligent foraging strategies
	Mountain Gazelle Optimizer	Inspired by the social life and hierarchy of wild mountain gazelles
2023	Nut cracker	Based on the nutcracker's intelligent behaviors

Fig. 4 Comparison between various nature inspired algorithm

Another significant challenge is the ability of NIOAs to handle continual optimization problems in demanding scenarios. Real-world optimization environments can be uncertain, while the problems themselves may be multi-objective. Additionally, the fitness evaluations can be noisy, imprecise, and time-consuming, and the fitness functions can occasionally exhibit nondeterministic behavior. The complexity of real-world situations poses a significant hurdle for NIOAs. Despite some efforts being made to address these aforementioned challenges, finding a satisfactory solution remains a daunting task.

We might be interested to hear how you can help if we decide to conduct additional research on these exciting subjects. To identify potential study areas, we must consider the ensuing problems. Operations at the NIOA can be very complicated. There are still challenging problems to resolve despite the NIOAs' widespread and effective use in numerous application domains. The first is a dearth of investigating the underlying theories and instruments of NIOAs. Only a small number of already existing benchmark tests are applicable to particular NIOA optimization issues, including automatic parameter tuning. Because of this, there has been insufficient theoretical research, which has made it challenging to discern between the traits of various NIOAs.

It is desirable to have a benchmark test that fits the NIOAs' tuning toolset and is simple to use because an automatic tuning process for parameters is typically c expensive. Since practically all NIOAs operate in the black-box mode and lack a strong mathematical foundation, researchers are constantly proposing supposedly "novel" algorithms and claiming that their optimizers find better solutions than those of other NIOAs. The NIOA's research visualization platforms are another issue. Our research shows that talks on this topic are scarce, with the exception of an early, straightforward attempt. The second is that in challenging situations, NIOAs struggle to solve continual optimization issues. The optimization environments in real-world environments can be dynamic and indeterminate. The fitness evaluations can also be noisy, imprecise, and time-consuming, and the fitness functions can occasionally be nondeterministic. For NIOAs, the complexity of real-world situations presents a significant difficulty. Even though some efforts have been made to address the aforementioned difficulties, finding a solution is still challenging.

NIOAs naturally have a parallel and distributed design, but combinations of parallel and distributed technologies, such as GPU-based hardware, robot swarms, and cloud platforms, receive less attention. On the one hand, it is actually not viable to have a single NIOA that can address all application-related issues. On the other hand, some NIOAs may work better for particular issues than others. The simulations carried out in NIOAs are very intricate. This is due to the complexity of real-world contexts and the potential for large-scale, multidimensional optimization problems with numerous variables and operators. Real-world conditions might change, and optimization environments can also be unpredictable and dynamic. It can require finances and time for development.

6 Conclusion

The domain of Nature-Inspired Optimization Algorithms (NIOAs) has demonstrated efficacy in tackling complex issues that conventional optimization techniques find difficult to tackle. Numerous NIOAs have been proposed in the last few decades, demonstrating the significance and promise of this sector. But even with this advancement, there are still a number of obstacles to be solved. One strategy for addressing these issues is to group related issues together and integrate NIOAs with other pertinent domains. By combining concepts and methods, scientists can investigate new areas for optimization and broaden their field of study. Since many optimization issues in the real world are quite complicated, effective and targeted NIOAs that are designed to meet these particular challenges are necessary. Additionally, a thorough examination of the theoretical underpinnings of the current NIOAs is required. This assessment will improve our understanding of these algorithms generally and assist identify areas that need more work. Furthermore, the development of a user-friendly algorithm toolbox and benchmark test suite is essential to support research and real-world applications. This suite should cover a number of topics, such as the previously mentioned complexity and automatic parameter modification. It is possible for researchers to efficiently compare and assess the effectiveness of various NIOAs by offering standardized techniques and resources. To sum up, NIOAs face a number of hurdles that must be addressed. These challenges include interdisciplinary collaboration, the creation of specialized algorithms for challenging issues, in-depth theoretical evaluations, benchmark test suites, and approachable algorithm toolboxes.

References

1. M. Nazar, M. M. Alam, E. Yafi and M. M. Su'ud, "A Systematic Review of Human–Computer Interaction and Explainable Artificial Intelligence in Healthcare With Artificial Intelligence Techniques," in *IEEE Access*, vol. 9, pp. 153316–153348, 2021, https://doi.org/10.1109/ACCESS.2021.3127881.
2. S. A. Hassan, S. Akbar, A. Rehman, T. Saba, H. Kolivand and S. A. Bahaj, "Recent Developments in Detection of Central Serous Retinopathy Through Imaging and Artificial Intelligence Techniques–A Review," in *IEEE Access*, vol. 9, pp. 168731–168748, 2021, https://doi.org/10.1109/ACCESS.2021.3108395.
3. F. Cristovao *et al.*, "Investigating Deep Learning Based Breast Cancer Subtyping Using Pan-Cancer and Multi-Omic Data," in *IEEE/ACM Transactions on Computational Biology and Bioinformatics*, vol. 19, no. 1, pp. 121–134, 1 Jan.-Feb. 2022, https://doi.org/10.1109/TCBB.2020.3042309.
4. P. Khan *et al.*, "Machine Learning and Deep Learning Approaches for Brain Disease Diagnosis: Principles and Recent Advances," in *IEEE Access*, vol. 9, pp. 37622–37655, 2021, https://doi.org/10.1109/ACCESS.2021.3062484.
5. H. Singh, S. Sharma, M. Khurana, M. Kaur and H. -N. Lee, "Binary Drone Squadron Optimization Approaches for Feature Selection," in *IEEE Access*, vol. 10, pp. 87099–87114, 2022, https://doi.org/10.1109/ACCESS.2022.3192836.

6. D. Gut, Z. Tabor, M. Szymkowski, M. Rozynek, I. Kucybała and W. Wojciechowski, "Benchmarking of Deep Architectures for Segmentation of Medical Images," in *IEEE Transactions on Medical Imaging*, vol. 41, no. 11, pp. 3231–3241, Nov. 2022, https://doi.org/10.1109/TMI.2022.3180435.
7. Siva Kumar, S. Raja, N. Pritha, Havaldar Raviraj, An adaptive transformer model for anomaly detection in wireless sensor networks in real-time, Measurement: Sensors, Volume 25, 2023,100625. https://doi.org/10.1016/j.measen.2022.100625
8. B. Merikhi and M. R. Soleymani, "Automatic Data Clustering Framework Using Nature-Inspired Binary Optimization Algorithms," in *IEEE Access*, vol. 9, pp. 93703–93722, 2021, https://doi.org/10.1109/ACCESS.2021.3091397.
9. B. Abdollahzadeh, F. S. Gharehchopogh, and S. Mirjalili, "African vultures optimization algorithm: A new nature-inspired metaheuristic algorithm for global optimization problems," Computers & Industrial Engineering, vol. 158, p. 107408, Aug. 2021, https://doi.org/10.1016/j.cie.2021.107408.
10. H. Jia, X. Peng, and C. Lang, "Remora optimization algorithm," Expert Systems with Applications, vol. 185, p. 115665, Dec. 2021, https://doi.org/10.1016/j.eswa.2021.115665.
11. M. T, S. Upadhyay, R. Beaulah Jeyavathana and A. Gopatoti, "Big Data Analytics with Deep Learning based Intracranial Haemorrhage Diagnosis and Classification Model," 2022 International Conference on Augmented Intelligence and Sustainable Systems (ICAISS), Trichy, India, 2022, pp. 877–883, https://doi.org/10.1109/ICAISS55157.2022.10010826.
12. W. J. AL-kubaisy, M. Yousif, B. Al-Khateeb, M. Mahmood, and D.-N. Le, "The Red Colobuses Monkey: A New Nature–Inspired Metaheuristic Optimization Algorithm," International Journal of Computational Intelligence Systems, vol. 14, no. 1, p. 1108, 2021, https://doi.org/10.2991/ijcis.d.210301.004.
13. M. Dehghani, Š. Hubálovský, and P. Trojovský, "Cat and Mouse Based Optimizer: A New Nature-Inspired Optimization Algorithm," Sensors, vol. 21, no. 15, p. 5214, Jul. 2021, https://doi.org/10.3390/s21155214.
14. A. R. Ginidi, A. M. Shaheen, R. A. El-Sehiemy, and E. Elattar, "Supply demand optimization algorithm for parameter extraction of various solar cell models," Energy Reports, vol. 7, pp. 5772–5794, Nov. 2021, https://doi.org/10.1016/j.egyr.2021.08.188.
15. Parthasaradi, V., Kavitha, M., Sridevi, A. et al. Novel rare-earth Eu and La co-doped ZnO nanoparticles synthesized via co-precipitation method: optical, electrical, and magnetic properties. J Mater Sci: Mater Electron (2022). https://doi.org/10.1007/s10854-022-09272-9
16. M. A. A. Al-qaness, A. A. Ewees, H. Fan, A. M. AlRassas, and M. Abd Elaziz, "Modified aquila optimizer for forecasting oil production," Geo-spatial Information Science, vol. 25, no. 4, pp. 519–535, May 2022, https://doi.org/10.1080/10095020.2022.2068385.
17. A. A. Abou El-Ela, R. A. El-Sehiemy, A. M. Shaheen, and A. S. Shalaby, "Aquila Optimization Algorithm for Wind Energy Potential Assessment Relying on Weibull Parameters Estimation," Wind, vol. 2, no. 4, pp. 617–635, Sep. 2022, https://doi.org/10.3390/wind2040033.
18. S. Sarwar, M. A. Hafeez, M. Y. Javed, A. B. Asghar, and K. Ejsmont, "A Horse Herd Optimization Algorithm (HOA)-Based MPPT Technique under Partial and Complex Partial Shading Conditions," Energies, vol. 15, no. 5, p. 1880, Mar. 2022, https://doi.org/10.3390/en15051880.
19. S. S. C, B. V and B. L. R, "High-Performance Computing Based on Residue Number System: A Review," 2023 9th International Conference on Advanced Computing and Communication Systems (ICACCS), Coimbatore, India, 2023, pp. 639–647, https://doi.org/10.1109/ICACCS57279.2023.10112959.
20. M. A. Awadallah, M. A. Al-Betar, M. S. Braik, A. I. Hammouri, I. A. Doush, and R. A. Zitar, "An enhanced binary Rat Swarm Optimizer based on local-best concepts of PSO and collaborative crossover operators for feature selection," Computers in Biology and Medicine, vol. 147, p. 105675, Aug. 2022, https://doi.org/10.1016/j.compbiomed.2022.105675.

21. K. K. Mohammed, S. Mekhilef, and S. Buyamin, "Improved Rat Swarm Optimizer Algorithm-based MPPT under partially shaded conditions and load variation for PV systems," IEEE Transactions on Sustainable Energy, pp. 1–13, 2022, https://doi.org/10.1109/tste.2022.3233112.
22. V. Ranganayaki, P. S. Ramesh, K. Rammohan, R. B. Lincy and A. Deepak, "Machine Learning Approaches on Pedestrian Detection in an autonomous vehicle," 2023 Fifth International Conference on Electrical, Computer and Communication Technologies (ICECCT), Erode, India, 2023, pp. 1–8, https://doi.org/10.1109/ICECCT56650.2023.10179836.
23. A. Uluslu, "Chameleon Swarm Algorithm Assisted Optimization of U-Slot Patch Antenna for Quad-Band Applications," IEEE Access, vol. 10, pp. 74152–74163, 2022, https://doi.org/10.1109/access.2022.3190378.
24. D. Pysal, S. J. Abdulkadir, S. R. Mohd Shukri, and H. Alhussian, "Classification of children's drawing strategies on touch-screen of seriation objects using a novel deep learning hybrid model," Alexandria Engineering Journal, vol. 60, no. 1, pp. 115–129, Feb. 2021, https://doi.org/10.1016/j.aej.2020.06.019.
25. J. Gupta, P. Nijhawan, and S. Ganguli, "Parameter Estimation of Fuel Cell Using Chaotic Mayflies Optimization Algorithm," Advanced Theory and Simulations, vol. 4, no. 12, p. 2100183, Nov. 2021, https://doi.org/10.1002/adts.202100183.
26. X. Li, Y. Wang, G. Ma, X. Chen, Q. Shen, and B. Yang, "Electric load forecasting based on Long-Short-Term-Memory network via simplex optimizer during COVID-19," Energy Reports, vol. 8, pp. 1–12, Sep. 2022, https://doi.org/10.1016/j.egyr.2022.03.051.
27. H. Jia, K. Sun, W. Zhang, and X. Leng, "An enhanced chimp optimization algorithm for continuous optimization domains," Complex & Intelligent Systems, vol. 8, no. 1, pp. 65–82, Apr. 2021, https://doi.org/10.1007/s40747-021-00346-5.
28. Y. Xiang, Y. Zhou, H. Huang, and Q. Luo, "An Improved Chimp-Inspired Optimization Algorithm for Large-Scale Spherical Vehicle Routing Problem with Time Windows," Biomimetics, vol. 7, no. 4, p. 241, Dec. 2022, https://doi.org/10.3390/biomimetics7040241.
29. S. S. C, B. L. R and D. S, "Design and Analysis of CNN based Residue Number System for Performance Enhancement," 2023 Third International Conference on Artificial Intelligence and Smart Energy (ICAIS), Coimbatore, India, 2023, pp. 1182-1187, https://doi.org/10.1109/ICAIS56108.2023.10073805.
30. S. Ai and Y. Wang, "Application of Improved Genetic Algorithms in Structural Optimization Design," Information and Management Engineering, pp. 480–487, 2011, https://doi.org/10.1007/978-3-642-24097-3_72.
31. S. Kirkpatrick, C. D. Gelatt, and M. P. Vecchi, "Optimization by Simulated Annealing," Science, vol. 220, no. 4598, pp. 671–680, May 1983, https://doi.org/10.1126/science.220.4598.671.
32. G. Y. -, "An Improved Differential Evolution Algorithm for Multi-objective Optimization Problems," International Journal of Advancements in Computing Technology, vol. 3, no. 9, pp. 106–113, Oct. 2011, https://doi.org/10.4156/ijact.vol3.issue9.14.
33. M. Dorigo and C. Blum, "Ant colony optimization theory: A survey," Theoretical Computer Science, vol. 344, no. 2–3, pp. 243–278, Nov. 2005, https://doi.org/10.1016/j.tcs.2005.05.020.
34. I. C. Trelea, "The particle swarm optimization algorithm: convergence analysis and parameter selection," Information Processing Letters, vol. 85, no. 6, pp. 317–325, Mar. 2003, https://doi.org/10.1016/s0020-0190(02)00447-7.
35. B. Yuce, M. Packianather, E. Mastrocinque, D. Pham, and A. Lambiase, "Honey Bees Inspired Optimization Method: The Bees Algorithm," Insects, vol. 4, no. 4, pp. 646–662, Nov. 2013, https://doi.org/10.3390/insects4040646.
36. O. B. Haddad, A. Afshar, and M. A. Mariño, "Honey-bee mating optimization (HBMO) algorithm in deriving optimal operation rules for reservoirs," Journal of Hydroinformatics, vol. 10, no. 3, pp. 257–264, May 2008, https://doi.org/10.2166/hydro.2008.018.

37. A. Baykasolu, L. Oumlzbakr, and P. Tapk, "Artificial Bee Colony Algorithm and Its Application to Generalized Assignment Problem," Swarm Intelligence, Focus on Ant and Particle Swarm Optimization, Dec. 2007, https://doi.org/10.5772/5101.
38. J. H. Kim, "Harmony Search Algorithm: A Unique Music-inspired Algorithm," Procedia Engineering, vol. 154, pp. 1401–1405, 2016, https://doi.org/10.1016/j.proeng.2016.07.510.
39. N. F. Johari, A. M. Zain, M. H. Noorfa, and A. Udin, "Firefly Algorithm for Optimization Problem," Applied Mechanics and Materials, vol. 421, pp. 512–517, Sep. 2013, https://doi.org/10.4028/www.scientific.net/amm.421.512.
40. K. Chandrasekaran and S. P. Simon, "Multi-objective scheduling problem: Hybrid approach using fuzzy assisted cuckoo search algorithm," Swarm and Evolutionary Computation, vol. 5, pp. 1–16, Aug. 2012, https://doi.org/10.1016/j.swevo.2012.01.001.
41. R. Subha, A. Haldorai, and A. Ramu, "An Optimal Approach to Enhance Context Aware Description Administration Service for Cloud Robots in a Deep Learning Environment," Wireless Personal Communications, vol. 117, no. 4, pp. 3343–3358, Feb. 2021, https://doi.org/10.1007/s11277-021-08073-3.
42. Fauziya Njeru, "A Review of Artificial Intelligence and its Application in Business", Journal of Enterprise and Business Intelligence, vol.3, no.1, pp. 044–053, January 2023. https://doi.org/10.53759/5181/JEBI202303005.
43. S. Ayub, N. Singh, Md. Z. Hussain, M. Ashraf, D. K. Singh, and A. Haldorai, "Hybrid approach to implement multi robotic navigation system using neural network, fuzzy logic, and bioinspired optimization methodologies," Computational Intelligence, vol. 39, no. 4, pp. 592–606, Sep. 2022, https://doi.org/10.1111/coin.12547.

Harnessing Intelligent AI to Elevate Business Modeling: A Perspective

1 Introduction

Artificial intelligence (AI) and machine learning (ML) are ubiquitous in modern culture, impacting many aspects of life, especially process automation and commercial decision-making. Decision-making processes are now streamlined and expedited by the typical practice of integrating ML prediction models within automated workflows. The benefits of forecasting scenarios and learning from past data are clear, but it's important to highlight that ML approaches haven't always followed the same guidelines on interpretability, audibility, and transparency as their counterparts in process automation. Concerns concerning the capacity to comprehend and closely examine the decision-making processes within ML models are raised by the dependence on these sophisticated learning algorithms [1].

Establishing a harmonious equilibrium between the potential of machine learning to augment automation and guaranteeing transparency is imperative in cultivating reliance and comprehension in the application of these technologies, especially given their growing significance in molding company environments and procedures. Evaluating, understanding, debugging, and benchmarking ML and AI models is a fundamental challenge, especially when these technologies are used in processes that have a direct impact on people's lives and corporate choices. According to the 2022 PwC AI Survey, more than thirty-five percent of companies and business executives support the use of cutting-edge AI technology. Artificial intelligence has a significant impact on businesses, changing their operational paradigms and creating new opportunities for growth. By virtue of its ability to manage large datasets, AI makes a substantial contribution to improving key performance indicators. These metrics include revenue generation, productivity, business growth, digital transformation, and overall operational efficiency [2]. With companies adopting cutting-edge AI solutions more and more, it is critical to have a thorough grasp of these technologies

© European Alliance for Innovation 2024

A. Haldorai et al., *Artificial Intelligence for Sustainable Development*,
EAI/Springer Innovations in Communication and Computing,
https://doi.org/10.1007/978-3-031-53972-5_22

in order to harness their revolutionary potential in an ethical and responsible manner, assuring beneficial outcomes for corporate strategy and societal effect.

Artificial intelligence has effortlessly permeated almost every aspect of our lives, from vital medical devices to driverless cars. Application areas include workflow optimization, human error reduction, deep analytics, informed decision-making, and 24/7 functionality. Improving our life' efficiency is the main reason artificial intelligence was developed. The worldwide big data market is expected to grow to an estimated $103 billion by 2027, making it clear that traditional human-led analysis devoid of technological assistance would not be able to keep up. The amount of data that companies are using to find possibilities is growing at an accelerated rate [3]. Artificial intelligence can analyze information faster than humans can because of its deep-learning algorithms, which enable quick and intelligent decision-making. The increasing dependence on insights derived from data makes AI an indispensable tool for managing the intricacies of today's data-laden world.

The rise of AI has enabled businesses to produce previously inconceivable new products and services. The ability of AI to build new business models is one of the most significant implications of AI on business. AI-powered technologies are changing how companies work and causing fundamental shifts in various industries.

1.1 Leveraging AI to Enhance Business Modeling

AI has become a game-changer for businesses across industries, and with the most recent advances, it is not just a nice to have. Going forward, AI will be a must-have to ensure the survival of many businesses by enhancing following business requirements as outlined in Fig. 1:

1.1.1 Better Decision-Making

One of the essential advantages of AI is its capacity to evaluate data and deliver insights that a human eye may not detect. AI algorithms can quickly identify patterns and trends and provide actionable insights in real time. This can assist organizations in making informed decisions.

1.1.2 Enhanced Customer Experience

AI optimizes the experience of customers. Chatbots, for example, are gaining popularity in customer service. Chatbots can understand and answer client inquiries fast and accurately by employing natural language processing (NLP). AI customizes the client experience. Businesses can create individualized communications and product recommendations by evaluating client data, increasing the possibility of a sale.

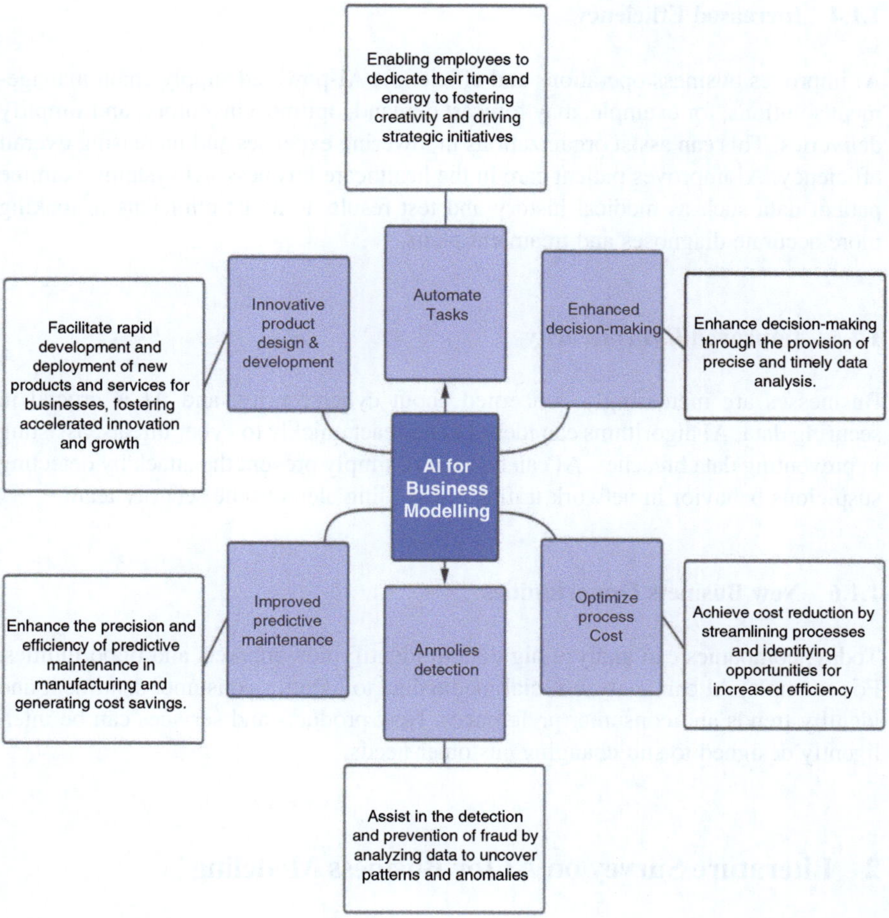

Fig. 1 AI and business modeling overview

1.1.3 Automation of Business Processes

AI-driven automation tools automate mundane tasks and free human resources. RPA (Robotic Process Automation) tools now perform much of the tedious work. Further, with advancements in AI, experts predict a new era of automation where AI rapidly automates routine cognitive activities. Machine algorithms have become core to many industries, including retail, hospitality, and financial services. These algorithms not only perform highly effectively, but they can work continuously for 24 h.

1.1.4 Increased Efficiency

AI improves business operations and efficiency. AI-powered supply chain management solutions, for example, may forecast demand, optimize inventory, and simplify deliveries. This can assist organizations in lowering expenses and increasing overall efficiency. AI improves patient care in the healthcare business. AI systems examine patient data such as medical history and test results to assist clinicians in making more accurate diagnoses and treatment plans.

1.1.5 Improved Cybersecurity

Businesses are increasingly concerned about cybersecurity, and AI is crucial in securing data. AI algorithms can identify and react quickly to cyber threats, assisting in preventing data breaches. AI can be used to simply prevent the attack by detecting suspicious behavior in network traffic and sending alerts to the security team.

1.1.6 New Business Opportunities

Today, companies can analyze big data to identify new markets and opportunities. For example, AI can analyze social media data to identify consumer sentiment and identify trends and consumer preferences. New products and services can be intelligently designed to suit changing customer needs.

2 Literature Survey on AI for Business Modeling

Businesses can uncover hidden trends, identify opportunities for expansion and improvement, optimize company procedures, and, above all, make educated decisions that drive their success by leveraging business intelligence. Within the field of business intelligence, machine learning and artificial intelligence provide enormous and potentially revolutionary benefits. With the use of these technologies, businesses may anticipate trends, automate repetitive activities, and gain a competitive advantage in the fast-paced business world of today. They can also extract insightful knowledge from large databases. Business intelligence and AI work together to create a proactive, data-driven approach that helps organizations negotiate complexity, seize opportunities, and keep improving as they strive for greatness.

The study in [4] explores about how established manufacturing companies leverage artificial intelligence (AI) to facilitate innovation in their business models within industrial ecosystems. Employing a qualitative approach, authors examine four major global manufacturing incumbents currently reshaping their business models through AI integration. Over 30 semi-structured, in-depth interviews were conducted with key strategic personnel to gain insights into the successful implementation of AI and the consequent transformation of their business models. The primary contribution of our research underscores the importance of aligning AI-driven business model innovation with ecosystem innovation.

The work in [5] seeks to enhance comprehension of artificial intelligence (AI) and its role as a catalyst for business model innovation. Given the expanding array of published materials on the subject, we conducted a comprehensive literature review to synthesize current knowledge in the realm of AI-driven business model innovation. The outcomes are elucidated in the form of a roadmap designed to offer guidance in the application of AI to organizational operations. Our findings propose a four-step approach for effective AI implementation: (1) grasp the fundamentals of AI and the organizational capabilities required for digital transformation; (2) comprehend the existing business model, recognize potential for innovation, and understand the role within the business ecosystem; (3) cultivate and refine the capabilities necessary for successful AI implementation; and (4) secure organizational acceptance while nurturing internal competencies.

The work in [6] focuses on observations and insights related to the application of AI solutions in business. While emphasizing the importance of identifying a specific problem to be solved, the chapter emphasizes that the desired benefit of using AI in solving a well-defined problem in or solely finding solutions for anomaly detection, event prediction, and process optimization, AI can address these challenges effectively. The chapter begins by breaking down the research landscape and identifying strategies for integrating AI into business processes. Additionally, it explores potential business opportunities for AI and identifies the benefits companies can gain by integrating AI into their business processes.

This part of the review in [7] provides a brief introduction to artificial intelligence (AI), addresses the ongoing challenges in AI development, and provides transformative impacts on business systematically is clear. Through an examination of two companies that revolutionized their business models using AI, the study highlights the potential impact of this technology. Additionally, it delves into the strategies through which executives can foster an innovative culture centered around AI, thereby reframing the approach to AI-driven business model innovation. Successfully leveraging AI can empower companies to usher in disruptive innovation, reshaping not only their business models and processes but also potentially transforming the global competitive landscape.

The primary objective of the study in [8] was to pinpoint the application of machine learning (ML) analysis in Malaysia's upstream capital projects, with the aim of enhancing business decisions through the utilization of statistical models and ML algorithms. By integrating ML algorithms and statistical models, the study anticipates an improvement in business decision-making processes, leading to increased efficiency, productivity, accelerated monetization, and a reduction in risk and returns. In essence, the incorporation of a diverse set of analytical elements is expected to provide superior decision support for stakeholders and company owners, offering valuable insights before critical business decisions are made.

According to the survey, prediction models are being employed in a wide range of commercial domains as technology advances. Therefore, this area of study requires collaboration with technologies like artificial intelligence, data science, and machine learning.

3 Empowering AI Breakthroughs and Tools for Business Modeling

There have been several notable innovations in the field of AI for business modeling as shown in Fig. 2. Here are some key advancements:

Explainable AI: Traditionally, AI models were seen as "black boxes" that provided accurate predictions but lacked clarity. With the rise of interpretable AI, efforts are being made to develop models that can provide insight into the reasoning behind those predictions. This will lead to better understanding and confidence in the AI-driven business model.

Reinforcement Learning: A subcategory of machine learning, reinforcement learning, has gained attention for its ability to modify business processes through trial and error. AI can provide the agent with a reward system, it can learn how to make decisions, and take actions that maximize long-term goals, resulting in more efficient business models.

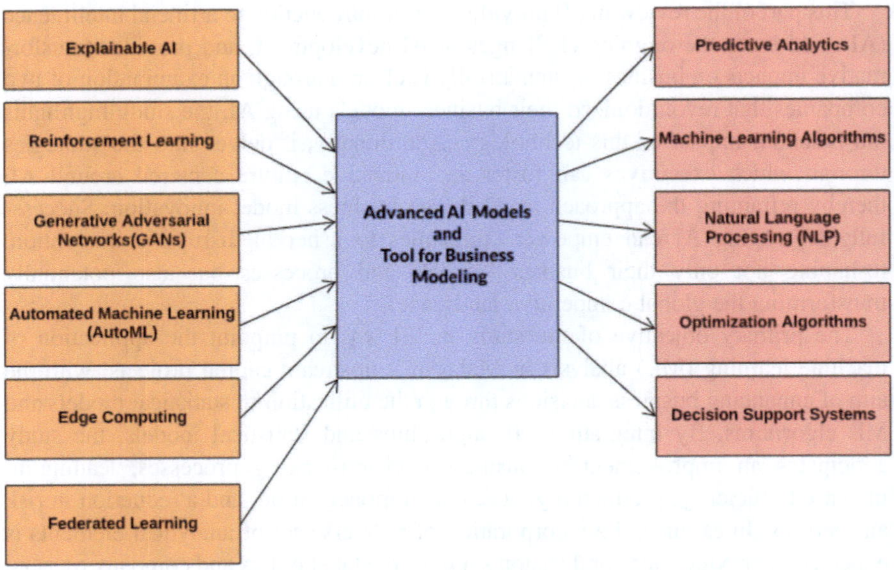

Fig. 2 AI Models and tools for business modeling

Generative Adversarial Networks (GANs): GANs have proved to be relevant in generating synthetic data similar to real-world data. This innovation is valuable for business modeling, as it allows organizations to create simulated scenarios, test hypotheses, and generate large datasets for training AI models, all without having to collect large amounts of data.

Deep Learning Architectures: A subset of machine learning has made great strides in deep learning, especially in the field of motor skills. New technologies such as convolutional neural networks (CNNs) and recurrent neural networks (RNNs) have improved the ability to process and analyze complex data types such as images, videos, and sequences, and serve value for business model services of such data..

Automated Machine Learning (AutoML): AutoML is a rapidly growing field that aims to automate the process of building and deploying machine learning models. It simplifies the complexities of model selection, feature engineering and hyperparameter tuning, enabling nonexperts to take advantage of the age of AI for performance modeling with little technical knowledge.

Edge Computing for AI: Edge computing requires data processing and the use of AI models on local machines or edge servers, rather than relying solely on cloud-based solutions This innovation has enabled tasks to be executed in real time analysis and decision-making on the stream, and reduces latency and response to AI -driven business models Increased.

Federated Learning: This type of learning enables multiple companies or organizations to train in collaboration with AI models, while keeping their data localized and secure. This new workflow is especially important when multiple stakeholders need to provide their data without compromising confidentiality or data privacy.

These innovations in AI for business modeling provide organizations with exciting opportunities to leverage state-of-the-art technologies and processes, enabling them to make more accurate forecasts, optimize their processes, and gain a competitive edge in the marketplace.

3.1 Business Modeling with AI-Driven Tools

AI tools for business modeling refer to various technologies and techniques that use artificial intelligence to enhance and optimize different aspects of business modeling These tools use AI algorithms, machine learning and data analytics to deliver valuable insights, predictive capabilities and automation to support decision-making and improve overall business processes.

Here are some of the common AI tools:

Predictive Analytics: AI algorithms are needed to identify patterns and trends in historical data. Managers can use this skill to adjust their models in response to accurate predictions of future outcomes. Organizations can improve a number of facets of their operations, including decision-making processes and strategic planning, by utilizing the insights obtained from the analysis of historical data. AI-enabled iterative cycles of research and optimization help firms remain flexible in ever-changing contexts while also increasing prediction accuracy. In the end, businesses are able to proactively alter their plans and remain ahead of the curve in a constantly changing landscape by employing AI algorithms to extract valuable information from past data.

Machine Learning Algorithms: To aid in different practices, machine learning (ML) systems refer to learning from data and adapting to changing business models. Key contributors to this change are customer segmentation, price optimization, and demand forecasting. The way machine learning systems identify patterns and trends in a corporate environment is strengthened by the constant use of new data. These algorithms can take into account past data in demand forecasting, for example, and change forecasts in response to new trends. The ability of ML to analyze multiple factors including consumer behavior and market dynamics is beneficial for pricing efficiency. Additionally, with subtle trends in data, machine learning algorithms are making inroads into customer segmentation, allowing companies to tailor their strategies to target specific customer segments.

Natural Language Processing (NLP): NLP enables analysis and interpretation of unstructured data from sources such as social media posts and customer surveys. This capability makes it possible to extract important insights from textual data, which is very helpful for business modeling. Because NLP algorithms are designed to understand the nuances of human language, they can be used to sift through large amounts of unstructured data to find patterns, extract important information, and identify emotions, and professionals can use this to hear their customers better, understanding thoughts, feelings, and preferences. Companies can access critical information with NLP to help make informed decisions. This includes identifying new trends, often seeking public opinion on the brand, improving products and services in response to customer feedback. NLP in particular is a powerful tool in the company's toolbox if it makes it possible to turn raw data into useful insights that can inform strategic choices.

Optimization Algorithms: Supply chain optimization, production scheduling, resource allocation, and other complex business problems all benefit greatly from the use of AI-powered optimization algorithms. With artificial intelligence these sophisticated algorithms can take over decision centers robust and identify effective and efficient solutions. In the presence of AI driven optimization algorithms are a particularly valuable asset to find the optimal solution.

Intelligent Decision Support Systems: Using artificial intelligence approaches, intelligent decision support systems provide real-time insights, recommendations, and ideas to help decision-makers make well-informed decisions during the business modeling process. These systems use AI to scan large, complicated datasets for patterns and produce insightful reports that help decision-makers make sense of challenging situations. Through the use of sophisticated algorithms, machine learning, and data analytics, these systems are able to quickly process enormous amounts of data and offer relevant and current advice to decision-makers. Intelligent decision support systems facilitate rapid and actionable information delivery, which improves decision-making in a variety of contexts, including corporate strategic planning, forecasting market trends, and operations including challenges.

These AI tools provide more accurate, efficient, and agile business models, enabling organizations to make data-driven decisions, change their strategies, and better adapt to changing market dynamics.

4 AI Framework for Business Modeling

AI process modeling typically involves a systematic approach that uses artificial intelligence and tools in developing and optimizing business processes. Figure 3 illustrates the AI process modeling process, including financial statements, customer profiles, and unstructured data (e.g., social media posts, customer feedback). The data is then preprocessed and cleaned to ensure quality and readiness for analysis [9].

The process of scrutinizing data to understand its characteristics, identify trends, and uncover insights is known as Exploratory Data Analysis, or EDA. This process analyzes data structures and searches for important features using various statistical and graphical tools. Plots, graphs, and charts are examples of visualization tools that help present complex information in a logical way. Researchers and data scientists can use this visual search to explore correlations, outliers, and probabilities in a dataset. Exploratory data analysis helps formulate hypotheses and identify key variables for business modeling. Researchers can make informed decisions about feature engineering, model selection, and hypothesis testing, as well as other steps in the research process, by using EDA to gain a deeper understanding of the data. Data

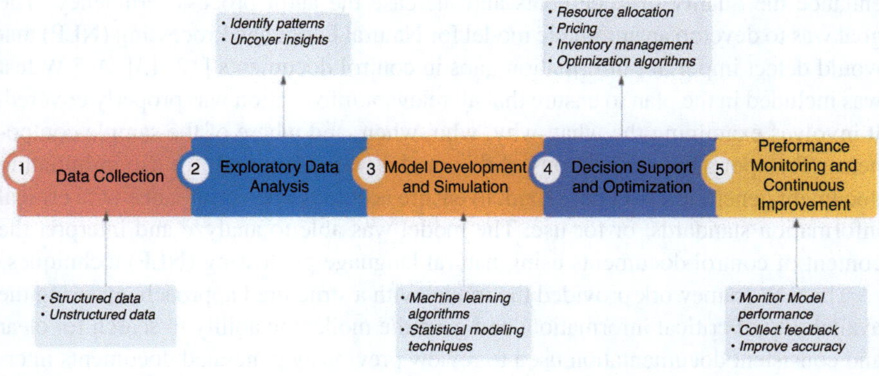

Fig. 3 AI process flow for business modeling

professionals can identify nuances in a data system and actively prepare for complex analytical tasks by completing an EDA, which is an important first step.

Subsequently, various AI techniques are used to model mode-Ls representing performance dynamics and capturing related relationships. This can use machine learning algorithms, statistical modeling techniques, or optimization techniques based on the specific requirements of the business problem. Models are trained and validated with data prepared. Once the model is developed, simulations are constructed, and integrated into a coherent system that simulates the operating environment. This allows situations to be tested and strategies or decisions to be tested. A model simulates the behavior of a project under different conditions and provides insight into the possible outcomes of different actions [10].

In addition, Decision Support and Optimization enables decision support by leveraging evolving models. It helps decision-makers evaluate different strategies, determine the impact of decisions on performance, and identify the best course of action. Optimization algorithms can be used to determine optimal solutions for allocation, pricing, inventory, or other aspects of business.

Finally, they allow for continuous performance management and improvement using AI systems to monitor the performance of business processes and collect feedback data This data is used to continuously develop models that are updated and refined, ensuring accuracy and relevance over time. Continuous research and development help adapt to changing business dynamics and improve overall AI-driven business models [11].

From the analysis, we can see that the AI framework for business modeling can vary depending on organizational objectives, tasks, and data availability The framework described above uses the keywords of Provides a detailed description of the use of AI for business modeling.

Case Study: NLP was adopted by the Global Bank to actively identify potential gaps in information.

By using an "instant proofreader" for document writers, Global Bank aims to enhance the quality of documents and increase the audit process' efficiency. The goal was to develop an automatic model for Natural Language Processing (NLP) that would detect important information gaps in control documents [12, 13]. A 5 W test was included in the plan to ensure that all relevant information was properly covered. It involves examining the what, why, who, when, and where of the sample components. Two potential applications of this model were identified: for use in initiating document generation at Global Bank to ensure accuracy and compliance with critical information standards, or for use. The model was able to analyze and interpret the content of control documents using natural language processing (NLP) techniques.

The 5 W framework provided the model with a structured approach, ensuring the availability of critical information, and gave the model the ability to search for clear and consistent documentation used to review previously generated documents internally to amplify any potential mirror errors. This activity demonstrated the potential of using state-of-the-art NLP technology to improve the efficiency and accuracy of the document review process, ensuring that critical information is consistently recorded and collected on corporate documents [14].

Global Bank built a proof-of-concept based on a pretrained Bidirectional Encoder Representations from Transformers (BERT) model (a technique for NLP). It used the model for named-entity recognition, attempting to identify terms that represented each of the 5Ws. To recognize Global Bank-specific terminology, the model would need to be fine-tuned, and because of new audits coming in and planned rollouts of the capability to other departments, the model would need to be retrained multiple times. However, BERT was a large, complex model and would require significant compute resources to retrain. Additionally, Global Bank was limited to using on-prem hardware for security reasons [15].

The solution was two-pronged. First, Global Bank built a new model that was easier to retrain. It partnered with IBM to build this model, bootstrapped from a prior engagement, and put in an on-prem implementation using IBM Watson® Studio. Second, Global Bank set out to increase the data available for this new model. Previously, it had built a tagging system in Python directly connected to their internal audit platform, and the system enabled auditors to make new annotations as they worked on audits. The company augmented this system with the original BERT model, which provided auditors instant feedback, and made this data available for the new IBM model.

The system provides three key benefits. First, by providing immediate feedback on what's missing from the description of controls, it raises the completeness and accuracy of documentation at the time of writing, which also helps new joiners get up to speed much more quickly. Second, there is improved consistency of write-ups against company standards. Third, the improved quality reduces the need for back-and-forth between auditors and control owners, which enables each to be more productive. For Global Bank, this improved efficiency resulted in substantial advantages. In addition to saving time, the expedited review process freed up resources worth an estimated 30,000 h. After these hours were recovered, they could be used for other assurance-related tasks that would not have been possible without the increased productivity provided by the NLP model. The system's success shows how valuable it is for streamlining processes, increasing output, and empowering businesses to strategically spend resources for more thorough assurance initiatives.

5 Conclusion

Many different industries and businesses are interested in AI due to the rapid advances in technology and the ongoing study in this field. Even still, little economic benefits have come from a lack of understanding about how to use AI technologies effectively. Through an exploration of current research in the AI area, this chapter seeks to address and improve the knowledge of AI implementation. Researchers in this field are guided by two main goals: (i) to investigate how AI might enable task automation through technological developments, and (ii) to offer advice on both existing and advanced algorithms that are customized for particular commercial applications. To summarize, the goal of this work is to enhance the overall comprehension of AI implementation by highlighting the automation of tasks through technological developments and providing recommendations for the selection of algorithms that are tailored to particular business needs. By utilizing these insights, companies may fully utilize AI and convert technological advancements into noticeable and beneficial outcomes.

References

1. W. Reim, J. Åström, and O. Eriksson, "Implementation of Artificial Intelligence (AI): A Roadmap for Business Model Innovation," AI, vol. 1, no. 2, pp. 180–191, May 2020, https://doi.org/10.3390/ai1020011.
2. B. Moustaid and M. Fakir, "Implementation of Business Intelligence For Sales Management," IAES International Journal of Artificial Intelligence (IJ-AI), vol. 5, no. 1, p. 22, Mar. 2016, https://doi.org/10.11591/ijai.v5.i1.pp22-34.
3. M. Stefik, "The AI business: Commercial uses of artificial intelligence," Artificial Intelligence, vol. 28, no. 3, pp. 345–348, May 1986, https://doi.org/10.1016/0004-3702(86)90057-3.
4. Burström, T., Parida, V., Lahti, T., & Wincent, J. (2021). AI-enabled business-model innovation and transformation in industrial ecosystems: A framework, model and outline for further research. Journal of Business Research, 127, 85–95.
5. Kraus, M., Feuerriegel, S., &Oztekin, A. (2020). Deep learning in business analytics and operations research: Models, applications and managerial implications. European Journal of Operational Research, 281(3), 628–641.
6. Akerkar, R. (2019). Employing AI in Business. In: Artificial Intelligence for Business. Springer Briefs in Business. Springer, Cham. https://doi.org/10.1007/978-3-319-97436-1_6.
7. D. Ramkumar, M. Ashraf, K. Sathesh Kumar, Md. Z. Hussain, A. Haldorai, and D. K. Mishra, "Defining multiple geometrical areas with modeling of elementary geometrical volumes in robot-environment interaction," International Journal of System Assurance Engineering and Management, Jun. 2022, https://doi.org/10.1007/s13198-022-01708-z.
8. Francisco Pedro, "A Review of Data Mining, Big Data Analytics, and Machine Learning Approaches", Journal of Computing and Natural Science, vol.3, no.4, pp. 169–181, October 2023. https://doi.org/10.53759//181X/JCNS/202303016.
9. R. Subha and A. Haldorai, "An Efficient Identification of Security Threats in Requirement Engineering Methodology," Computational Intelligence and Neuroscience, vol. 2022, pp. 1–14, Aug. 2022, https://doi.org/10.1155/2022/1872079.
10. Lee, J.; Suh, T.; Roy, D.; Baucus, M. Emerging Technology and Business Model Innovation: The Case of Artificial Intelligence. J. Open Innov. Technol. Mark. Complex. 2019, 5, 44.
11. M. S. A. Rahman, N. A. A. Jamaludin, Z. Zainol and T. M. T. Sembok, "Machine Learning Algorithm Model for Improving Business Decisions Making in Upstream Oil & Gas," 2021 International Congress of Advanced Technology and Engineering (ICOTEN), Taiz, Yemen, 2021, pp. 1–5, https://doi.org/10.1109/ICOTEN52080.2021.9493499.
12. Metelskaia, O. Ignatyeva, S. Denef, and T. Samsonowa, "A business model template for AI solutions," Proceedings of the International Conference on Intelligent Science and Technology, Jun. 2018, https://doi.org/10.1145/3233740.3233750.
13. S. Mishra and P. Mishra, "AI business models and its impact on business strategic framework," International Journal of Financial Engineering, vol. 10, no. 02, Feb. 2023, https://doi.org/10.1142/s2424786323500019.
14. W. Spiess-Knafl, "AI-first impact business models," Artificial Intelligence and Blockchain for Social Impact, pp. 40–62, Oct. 2022, https://doi.org/10.4324/9781003218913-3.
15. W. Pfau and P. Rimpp, "AI-Enhanced Business Models for Digital Entrepreneurship," Future of Business and Finance, pp. 121–140, Nov. 2020, https://doi.org/10.1007/978-3-030-53914-6_7.

Automatic Human Activity Detection Using Novel Deep Learning Architecture

1 Introduction

Human Activity Detection (HAD) using information from video signals or multichannel time series data has made great strides in human motion classification. The HAD is an important field dedicated to recognizing routine behaviors people just do it themselves. It's increasingly important, especially in wearable sensor technology, because it can segment and track many daily activities that aren't related to physical activity, such as eating, drinking, brushing teeth, and even if they will experience irregular sleep. They basically consist of two categories: basic behaviors and phenomenal behaviors. Simple activities, including activities such as jogging, walking, and running, are defined by body movements and postures. On the other hand, complex actions are straightforward actions that serve a specific purpose [1, 2]. A meal, for example, includes the basic act of sitting as well as the purpose for which he eats. However, because human activities exhibit substantial interclass and intra-class variation, difficulties arise in accurately identifying these activities The classification process is complicated by the fact that different individuals may perform comparable tasks in various ways [3, 4].

Human Activity Detection (HAD) uses highly accessible devices such as cell phones and video cameras that are excellent at collecting data on human movement, and claim a wide range of applications when electronic devices and their functionality continue to fuel the growth of Artificial Intelligence (AI) and play a transformative role in unlocking deep insights of precision detection and interpretation. The rapidly expanding acquisition devices, powerful AI techniques, and applications all under one roof thanks to this HAD and AI a combined therefore. Rapid advancement in AI technology, by creating new opportunities for the application of these devices in various sectors such as industry, medicine and education drives the development

© European Alliance for Innovation 2024

A. Haldorai et al., *Artificial Intelligence for Sustainable Development*,
EAI/Springer Innovations in Communication and Computing,
https://doi.org/10.1007/978-3-031-53972-5_23

of HAD technology. This dynamic interaction between AI models and HAD devices allows for applications beyond models based on single images or short sequences. They have been able to take testing in more complex ways [5].

A subset of supervised learning, human activity detection using labeled data, presents a challenge in multivariate time series classifications. Many function recognition techniques that have been studied in the past include advanced deep learning techniques such as ANN, CNN, RNN, LSTM, as well as more traditional algorithms such as SVM, Random Forest, XGBOst Traditional systems with sometimes extensive feature engineering and time-consuming human feature extraction calls. In contrast, deep learning methods are able to automatically find relevant attributes in raw sensor data, making them ideal for classifying complex human activities. DL, applied to the HAD domain, has greatly simplified the task of extracting useful attributes from raw sensor data. An important development is the use of convolutional neural networks (CNN) for HAD. This broadened the utility of transfer learning methods. Knowledge is reapplied by transfer learning using a recognition model trained on one dataset in another test dataset. This shared information facilitates the implementation of pretrained models and accelerates training by applying acquired properties to relevant domains.

Convolutional neural networks (CNNs) have been shown to be very successful in solving the problems associated with human activity recognition (HAR), especially when it comes to recognizing common actions and gestures. CNNs provide realistic solutions an end-to-end combination of feature extraction and classification that is simpler than traditional methods. The CNNs' feature extractors use nonlinear transformations to learn directly from the raw data, producing representations of human action-classes that are more discriminative. Traditional handcrafted characteristics, on the other hand, are difficult to compute and scale efficiently. CNNs get over this restriction by using a variety of filters and pooling techniques, which enables them to extract basic and complicated human actions in a hierarchical manner while also learning their nonlinear and temporal correlations. To effectively process temporal data, these CNN architectures frequently use convolution and pooling operations along the time axis.

The general HAD system is mentioned in Fig. 1. HAD typically consists of four primary stages: dataset collecting, data preprocessing, AI-based activity recognition, and HAR administration user interface. Each level presents several implementation strategies, giving the HAD system a range of choices. As a result, choosing the application domain, type of data gathering equipment, and AI algorithms for activity detection can be more difficult. The remainder of the chapter is organized as follows: The related research on HAD using multichannel time series data will be covered in Sect. 2. CNN will be introduced as a method for HAD in Sect. 3, along with the innovative CNN architectures. The datasets utilized will be discussed in Sect. 4 along with implementation information. It will also include discussions based on the experiments that were conducted as well as the experimental outcomes. The study's findings will be presented in Sect. 5 to close.

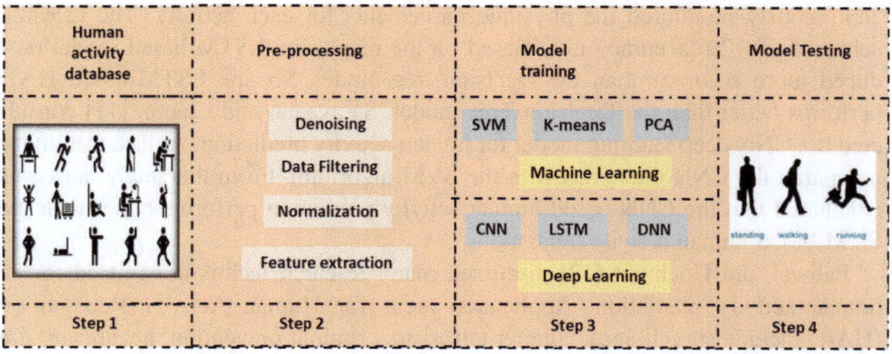

Fig. 1 General architecture of HAD

2 Litrature Survey

Many researchers explore many algorithms and techniques on the HAD. Some of the works related to HAD are discussed in this section. Yu et al. [6], considered the HAD problem since the multisensory technology achieved the high success with maximum reliability. This research collected data from the various sensing systems and the collected data was put on the neuromorphic computing system to detect the human activity. The information was composed with the help of inertial measurement techniques. Based on this technology, each of the human activity had individual activity matrix, respectively. The Hopfield NN system worked based on the Hebbian learning law, which was responsible for the finding of the activity feature pattern. Thus, the proposed method achieved 95.4% as output rate. Dutta and Zielinska [7] detected the human activity to improve the performance of the robot assistance. Every human activity is accountable for successive actions. So, these actions are restricted and shaped by the motion trajectory. The input samples are collected by the various sensors and states are distinct with the training process. In the training phase, the input samples are processed, and structured database was generated for prediction. This research considered CAD 60, CAD-120, WUT-17, and WUT-18 databases for action prediction. And finally, all the results are compared with previous models.

William et al. [8] identified the human action by analyzing smartphone sample data. These data are collected by the accelerometer and gyroscopic sensors. This research is mostly used in the healthcare domains. The suitable features are generated, since accuracy of the system is mostly depending on these selected features. Park et al. [9] considered the deep learning architectures for HAD. The dataset was selected from the sensors. The research used to encode regression technique with the deep learning models for best accuracy rate. The regression technique used to study the temporal dependency from sequence information. Charan et al. [10] suggested LSTM with Decision Tree algorithms for HAD. The study made a clear analysis about the HAD based on the various machine learning techniques. The model continuously monitored the physique movements for each activity. The research achieved 90.12% accuracy rate. Based on the result, the LSTM-based model produced more accuracy than the Decision tree model. So, the LSTM-based HAD performs better than the Decision Tree model. Saravanan and Charan [11] considered the CNN deep learning model for human activity prediction. At last, the author compared the CNN outcomes with the SVM algorithm. From this analysis, it was concluded that the CNN-based human activity prediction performs better than the SVM-based human activity prediction.

Paliwal and Gochhait [12] discussed many research techniques related to the human activity prediction. Application areas for Human Activity Recognition (HAR) include surveillance, counter terrorism, criminal prevention, healthcare, life tracking, and help. Beyond its beneficial effects on well-being, the capacity to recognize human activity has several useful applications. The recognition of human actions is a topic of great interest in the fields of computer vision and machine learning. Pose estimation and categorization algorithms for picture and video inputs have made it possible to collect and preserve data on several facets of human movement in a variety of uncontrolled contexts. Human activities can be divided into three categories according to their nature because they have a built-in hierarchical structure. For instance, common interior activities include standing, sitting, conversing, and strolling. There are also specialized tasks like those done in factories or kitchens. Nevertheless, the variety of approaches used by people poses a significant obstacle to activity identification.

It is projected that this technology would primarily use the Internet of Things (IoT) for applications in healthcare and eldercare. The research report uses a hierarchy of analytical approaches that is clearly outlined to solve this issue and discusses numerous tactics that have been investigated in earlier works in this field. Despite substantial progress in this important area, there is always opportunity for improvement, especially when using cutting-edge classification algorithms on a variety of problem sets. Pose estimation and classification algorithms help in effectively assessing data collection and interpreting human behaviors to address these issues. The study also examines how users interact with video and image inputs.

Charan et al. [13] used a novel machine learning techniques like K-nearest neighbor and Navie Bayes techniques for human pose detection. The training samples are collected from the internet. The performance of K-nearest neighbor is better than the Navie Bayes techniques. Nagpal and Gupta [14] consider human activity detection on the healthcare domain. This research discussed the importance of the human activity detection in the medical. Also, this research discussed about the various novel algorithms and provided detailed review about HAD. Kumar and Prasad [15] used accelerometers and Gyroscopes for data collection. The author used deep LSTM algorithm for human poster prediction. The model can easily predict the human activity without the help of the domain expertise. Ghosh et al. [16] detected the human activity like walking, running, singing, and so on. To recognize these activities the author used the Logistic Regression and Random Forest algorithm on the publicly available dataset.

Moola and Hossain [17] surveyed the research papers related to HAD. Due to its numerous applications in areas including security, entertainment, ambient-assisted living, health management, and monitoring, machine learning research primarily focuses on the identification of human behaviors. Through their studies on human activity recognition, researchers' keen interest in comprehending daily human activities is clear. As a result, this paper clarifies the basic design of Human Activity Recognition (HAR) systems and presents an explanation of their key elements. While other deep learning models have shown respectable accuracy, recent research has primarily moved toward using Convolutional Neural Networks (CNNs) for HAR tasks, according to an analysis of the present landscape in accelerometer-based human activity detection. Based on the sort of machine learning used (conventional or deep learning) and mode of process (online or offline), this survey divides the research into two main areas. This research scrutinizes prediction performance, applied algorithms, activity categories, and the sensor equipment by examining 48 distinct investigations. In conclusion, the study contrasts the difficulties and problems of applying deep learning to traditional machine learning techniques for recognizing human movements based on accelerometer sensor data, while also taking into account the online versus offline execution scenarios.

Zamichos et al. [18] used RNN-based LSTM model for detection of human activity detection. An individual's past actions are listed together with their demographic characteristics in their activity profile. This manuscript describes the concept and provides a three-pronged evaluation strategy. For this, it is necessary to (a) compare its results to those of a Markov Chain model, (b) evaluate the effect of the number of activities within a sequence on prediction accuracy, and (c) evaluate the inclusion of characteristics as network inputs, both in conjunction with historical sequences and separately. A ground-breaking method appears in response to the restrictions present in existing human activity recognition projects, which mainly rely on manually collected characteristics and some rudimentary machine learning models. The approach provides a novel multitask layered neural network model built by the Wang and Peng [19], which has the ability of deep neural networks to find characteristics on their own in the setting of smart homes. In order to fully utilize the potential of both recurrent neural networks and convolutional neural networks for seamless feature extraction, this innovative model combines their most recent innovations.

Krishnaprabha and Raju [20] utilized CNN for human activity detection. The activity recognition model that has been created is specifically intended for identifying older citizens' actions inside of their houses. The model may also distinguish between people's actions inside a densely populated location when permitted to do so. Sharma et al. [21] used the LSTM architecture for human activity prediction, which achieved 92.1% on the HAR UCI Machine Learning Repository. Saroja et al. [22] discussed various techniques on HAD. Within the developing field of computer vision-driven technologies, human posture estimation offers significant advantages. Human–computer interactions can be significantly improved by using applications of human pose estimation [23]. This extensive survey will include a number of publications that evaluate performance and address occlusion problems [24]. In this research, we investigate different approaches used for human pose estimation with the goal of improving the approach and learning more about the drawbacks of each methodology [25].

3 Proposed Work

The impressive technological advances in the field of artificial intelligence, particularly in deep learning and machine learning techniques, must be emphasized. The ability to extract complex hidden insights into identifiable and accurate interpretation is fundamentally changing as a result of this development. As a result, it becomes urgently important to comprehend the crucial role performed by these cutting-edge paradigms that are quickly changing the landscape of HAD devices. This highlights the need for comprehensive research that also examines the changing interactions between AI and HAD devices. With the integration of devices and the accompanying application domains inside the specialized AI context, this review seeks to clarify the complex HAD framework. The HAD system is presented in the Fig. 2.

The four main steps of the HAD process are data collection, preprocessing, model training, and performance assessment. The first HAD device selection is based on the particular application. For instance, cameras are used as the data collecting tool in surveillance scenarios involving numerous people. On the other hand, sensors are the best data source for tasks that include tracking a person's daily activities. Although cameras can be used, they demand a lot of processing power and present privacy issues. In the second stage of data cleaning, filters (low-pass or high-pass) are used to reduce noise or improve pictures. The segmentation of these revised data into regions and boundaries follows. As handy references, our group has written specialized monographs that describe segmentation paradigms. The divided data serves as the basis for training the model. In stage 3, either deep learning (DL) or machine learning (ML) methods are used to train the HAD model. While DL frameworks are best for automatic feature extraction, ML-based approaches are

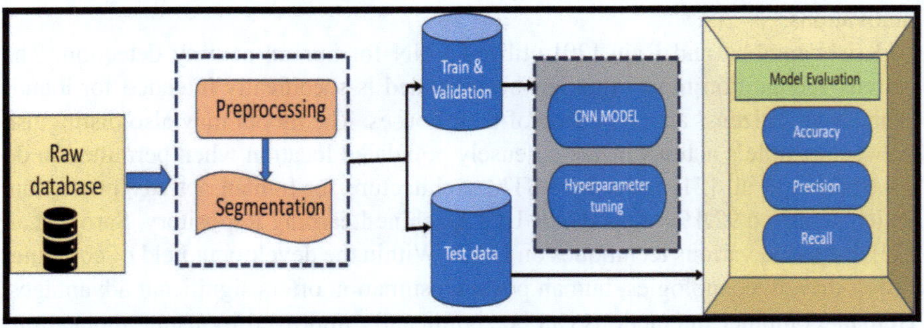

Fig. 2 HAD system design

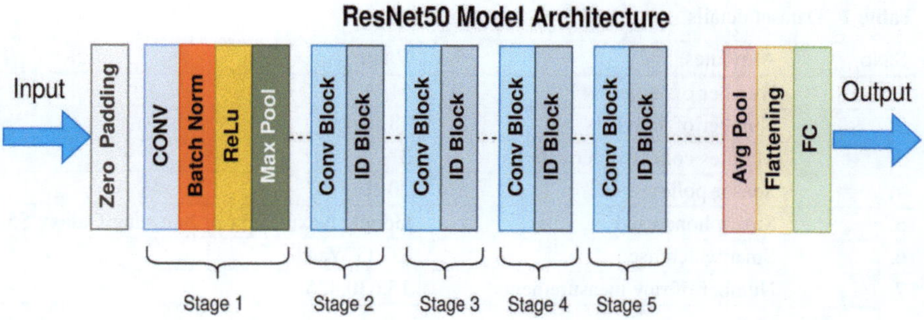

Fig. 3 ResNet-50 architecture

appropriate for manually created feature utilization. Additionally, DL benefits include the exploration of large datasets, transfer learning models for reuse of information, and hybrid models with the ability to recognize and learn spatial and temporal characteristics. The HAD model is ready for application or prediction once stage 3 is finished. The application of the model to real-world data, which differs depending on physical parameters including age, physique, and task execution style, proves to be stage 4's most challenging stage. The performance of an effective HAD model is independent of these physical characteristics.

The model used the Fast ResNet-50 deep learning architecture for HAD. A well-known subset of neural networks primarily used for processing visual data, CNNs, ResNet-50 is a pretrained deep learning model created for image categorization. ResNet-50, which consists of 50 layers, was trained using a dataset of one million photos from the ImageNet database that were divided into 1000 categories. This model is impressive in that it includes more than 23 million trainable parameters, highlighting the sophisticated design that boosts its effectiveness in image recognition tasks. The term Residual Network, often known as ResNet, stands for a certain Residual Neural Network topology. To clarify, there are five phases in the ResNet-50 variation, each of which contains a residual block. Each residual block consists of three layers with 1x1 and 3x3 convolutions in each layer. Fundamentally, leftover blocks are a simple concept. The output from each layer acts as the input for the following layer in conventional neural networks. The output of each layer in a network with residual blocks, however, not only connects to the next layer but also directly with layers that are two to three layers away, creating what are known as identity connections. The ResNet-50 architecture is shown in Fig. 3.

4 Experimental Setup and Result Analysis

The research process includes a number of steps. These steps involve collecting or acquiring data related to the detection of human activity using sensors, preprocessing the data, segmenting the raw sensor data using an appropriate sliding window

Table 1 Dataset details

S.No	Attributes	Value
1.	Number of subjects	51
2.	Number of activities	18
3.	Minutes collected per activity	3
4.	Sensor polling rate	20 Hz
5.	Smartphone used	Google Nexus 5/5x or Samsung Galaxy S5
6.	Smartwatch used	LG G Watch
7.	Number of raw measurements	15,630,426

technique, and breaking the dataset into train, validation, and test subsets. After that, models are built using various deep learning methods, the hyperparameters are adjusted, and the models' performance is evaluated using a variety of metrics. These stages lead to supervised categorization, which makes it possible to recognize human behaviors from sensor-gathered data. This identification can be helpful in remotely tracking the activities of elderly or severely ill patients based on their physical movements. Through the UCI Machine Learning Repository, the data used in this study has already been gathered and is available to the general audience. This dataset differs from others in that it classifies a wide range of activities. These tasks go beyond simple movement and fall under a variety of categories, including eating spaghetti, eating a sandwich, folding laundry and cleaning teeth. Comparatively, to datasets that only concentrate on movement or exercise patterns, this broader focus makes its possible applications more diverse. Fifty-one different people made up the cohort from which this dataset was created. Notably, information like age and gender is kept private. Table 1 provides a concise summary of the key properties of the data as well as the methodology used to obtain the data.

As a result of the dataset's hierarchical flat file format, the first preparation step entails combining these flat files to create a coherent data frame. Due to a lack of synchronization in the data obtained from the smartwatch and smartphone, separate data frames had to be made for each device. The data from the accelerometer and gyroscope sensors will be integrated into a single data frame for each individual device. In order to make this easier, the participant data has been combined into separate data frames for each type of sensor (gyroscope and accelerometer), and they are arranged according to the timestamp column. Then, distinct datasets for the smartwatch and smartphone are created using these various sensor data frames. The dataset description is shown in Fig. 4.

A mitigation technique has been developed for model development to handle noise existing in the dataset, which may result from unintentional oscillations caused by human activity. In this case, a max-pooling layer is added in between Convolutional Neural Network (CNN) layers. This helps to lessen the effect of such noise while the model is learning. The resilient volume of the dataset allows for the removal of rows with missing values reported as 0 or NaN. Notably, the 18 different human behaviors under consideration are evenly dispersed in terms of data points across both the smartphone and smartwatch datasets. Due to the balanced

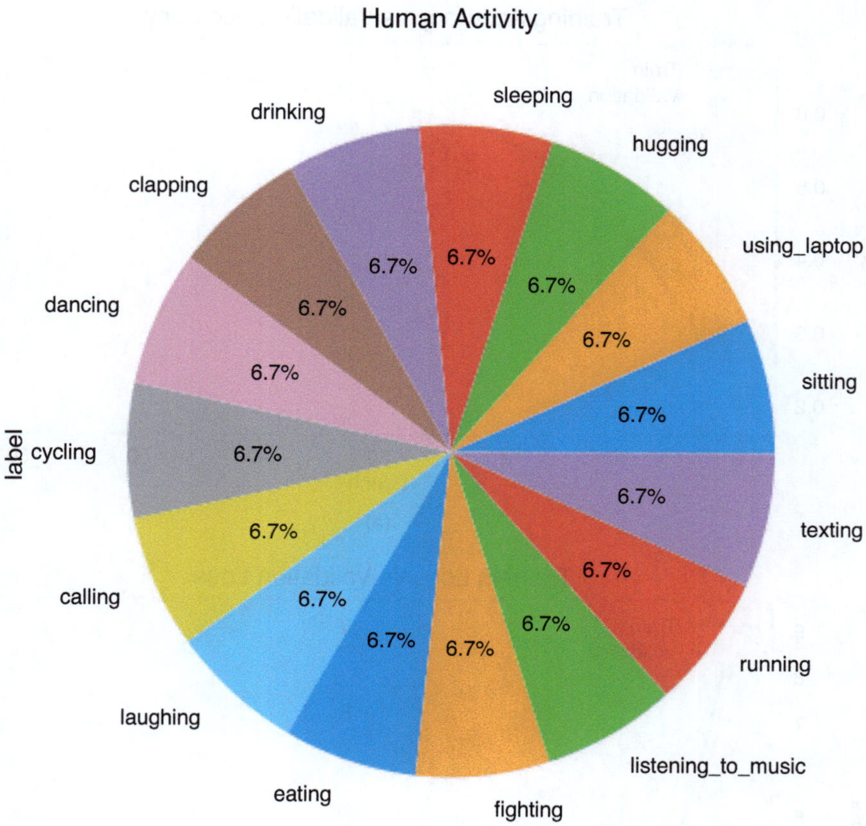

Fig. 4 Dataset description

distribution, there is no need to solve any class imbalance problem in the layers outside the analysis. Using a sliding window method, the data is split into blocks, where each block is 10 s long and overlaps 50% with the previous block This split uses 50 steps, in which the data needs to be transformed well for time intervals and attributes This approach is necessary to establish a valid and significant data structure. The original dataset is divided into three distinct subsets: training, validation, and testing, in accordance with the study objective of building models that can cope with input from different users meet; this section has a ratio of 60:20:20, in that order. The deep learning models used in this study are trained and optimized with training and validation subsets to facilitate the tuning of hyperparameters and after that, the models are tested on a specified subset of tests, and culminate in the establishment of significant outcomes.

Using deep learning models like VGG, Inception, ResNet, and EfficientNet, we started exploring Human Activity Detection (HAD). These models were tested using data from our Human Activity Recognition (HAR) dataset. The results shown in

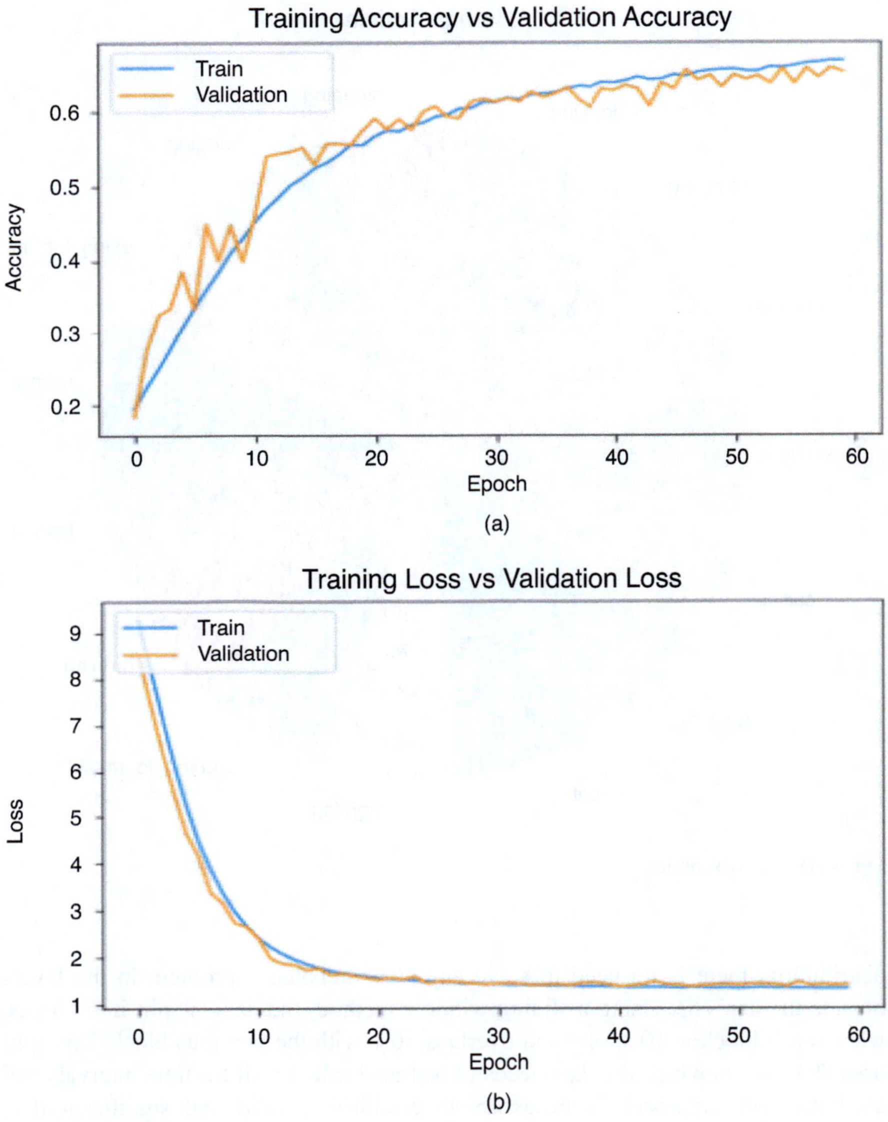

Fig. 5 The loss and accuracy function

Fig. 5 show how well each different model performed in classification. The results of the confusion matrices show that the ResNet-50 algorithm performs better than the other models in terms of classification accuracy, with a model accuracy score of 92.13%. It is worth noting that the ResNet-50 algorithm demonstrates high accuracy by reducing classification errors It is important to keep in mind, however, that this success occurs with greater sampling loss than with any other model. Moreover, the

Fig. 6 Human activity detection

ResNet-50 model necessitates more processing time compared to the Inception models and VGG. However, the ResNet-50 model consistently maintains its excellent classification accuracy for both simple and complex activities. The ResNet-50 model thus stands out as the best option because of its remarkable effectiveness in identifying machine-related activities. Some of the image classifications of human activity is displayed in Fig. 6.

5 Conclusion

The proposed method considerably improves the performance of Human Activity Detection by utilizing the mask-enhanced ResNet-50 architecture's capabilities for human body estimation. The results of our testing demonstrate that our suggested HAD system performs better in terms of classification accuracy than traditional sensor-based approaches. Our HAD system exhibits higher classification results in comparison to conventional sensor-based HAD systems, which are prone to sensor errors leading to mediocre outcomes in complicated activities. It is adept at both simple and complex user motions, constantly maintaining high classification accuracy for a variety of user actions. Our thorough examination of the results of HAD classification highlights the contribution of High-Intensity Training machines to activity recognition. Our machine-based HAD system is very effective when camera-based input devices are used. Through rigorous experimentation and analysis, the validity of our HIT machine-based HAR system for activity recognition is demonstrated. In our upcoming work, we want to improve classification performance by combining sensor data with image data for more accurate activity recognition. Additionally, in order to assess the performance of our HAR datasets against industry standards, we intend to interact with widely known public datasets like the UCI-Human Activity Recognition using Smartphones dataset.

References

1. S. Park, H. O. Lee, Y. M. Hwang, S. -K. Ko and B. -T. Lee, "Enhanced Prediction Model for Human Activity Using an End-to-End Approach," in IEEE Internet of Things Journal, vol. 10, no. 7, pp. 6031–6041, 1 April1, 2023, https://doi.org/10.1109/JIOT.2022.3223674.
2. Y. Tang, L. Zhang, H. Wu, J. He and A. Song, "Dual-Branch Interactive Networks on Multichannel Time Series for Human Activity Recognition," in IEEE Journal of Biomedical and Health Informatics, vol. 26, no. 10, pp. 5223–5234, Oct. 2022, https://doi.org/10.1109/JBHI.2022.3193148.
3. F. Duan, T. Zhu, J. Wang, L. Chen, H. Ning and Y. Wan, "A Multitask Deep Learning Approach for Sensor-Based Human Activity Recognition and Segmentation," in IEEE Transactions on Instrumentation and Measurement, vol. 72, pp. 1–12, 2023, Art no. 2514012, https://doi.org/10.1109/TIM.2023.3273673.
4. W. Kong, L. He and H. Wang, "Exploratory Data Analysis of Human Activity Recognition Based on Smart Phone," in IEEE Access, vol. 9, pp. 73355–73364, 2021, https://doi.org/10.1109/ACCESS.2021.3079434.
5. R. Mojarad, F. Attal, A. Chibani and Y. Amirat, "Automatic Classification Error Detection and Correction for Robust Human Activity Recognition," in IEEE Robotics and Automation Letters, vol. 5, no. 2, pp. 2208–2215, April 2020, https://doi.org/10.1109/LRA.2020.2970667.
6. Z. Yu, Adnan Zahid, Ahmad Taha., "An Intelligent Implementation of Multi-Sensing Data Fusion With Neuromorphic Computing for Human Activity Recognition," in IEEE Internet of Things Journal, vol. 10, no. 2, pp. 1124–1133, 15 Jan.15, 2023, https://doi.org/10.1109/JIOT.2022.3204581.
7. V. Dutta and T. Zielinska, "Prognosing Human Activity Using Actions Forecast and Structured Database," in IEEE Access, vol. 8, pp. 6098–6116, 2020, https://doi.org/10.1109/ACCESS.2020.2963933.
8. P. William, G. R. Lanke, D. Bordoloi, A. Shrivastava, A. P. Srivastavaa and S. V. Deshmukh, "Assessment of Human Activity Recognition based on Impact of Feature Extraction Prediction Accuracy," 2023 4th International Conference on Intelligent Engineering and Management (ICIEM), London, United Kingdom, 2023, pp. 1–6, https://doi.org/10.1109/ICIEM59379.2023.10166247.
9. S. Park, H. O. Lee, Y. M. Hwang, S. -K. Ko and B. -T. Lee, "Enhanced Prediction Model for Human Activity Using an End-to-End Approach," in IEEE Internet of Things Journal, vol. 10, no. 7, pp. 6031–6041, 1 April1, 2023, https://doi.org/10.1109/JIOT.2022.3223674.
10. S. Charan, M. S. Saravanan. and R. Surendran., "Prediction of Sufficient Accuracy for Human Activity Recognition using Novel Long Short Term Memory in Compared with Decision Tree," 2023 International Conference on Sustainable Computing and Data Communication Systems (ICSCDS), Erode, India, 2023, pp. 609–614, https://doi.org/10.1109/ICSCDS56580.2023.10104587.
11. M. S. Saravanan and S. Charan, "Prediction of Insufficient Accuracy for Human Activity Recognition using Convolutional Neural Network in Compared with Support Vector Machine," 2022 5th International Conference on Contemporary Computing and Informatics (IC3I), Uttar Pradesh, India, 2022, pp. 1915–1919, https://doi.org/10.1109/IC3I56241.2022.10072905.

12. Paliwal and S. Gochhait, "Classification of Machine Learning and Power Requirements for HAI (Human Activity Identification)," 2023 International Conference on Inventive Computation Technologies (ICICT), Lalitpur, Nepal, 2023, pp. 199–204, https://doi.org/10.1109/ICICT57646.2023.10134470.
13. S. Charan, S. M.S. and S. R, "Prediction of Insufficient Accuracy for Human Activity Recognition with Limited Range of Age using K-Nearest Neighbor," 2023 Second International Conference on Electronics and Renewable Systems (ICEARS), Tuticorin, India, 2023, pp. 926–930, https://doi.org/10.1109/ICEARS56392.2023.10085062.
14. D. Nagpal and S. Gupta, "Human Activity Recognition and Prediction: Overview and Research Gaps," 2023 IEEE 8th International Conference for Convergence in Technology (I2CT), Lonavla, India, 2023, pp. 1–5, https://doi.org/10.1109/I2CT57861.2023.10126458.
15. T. Kumar and B. Prasad, "Prediction of Human Physical Activities With Consumer Smartphone Sensor Data," 2021 IEEE Bombay Section Signature Conference (IBSSC), Gwalior, India, 2021, pp. 1–6, https://doi.org/10.1109/IBSSC53889.2021.9673366.
16. N. S. Ghosh, R. Majumdar, B. Giri and A. Ghosh, "Detection of Human Activity by Widget," 2020 8th International Conference on Reliability, Infocom Technologies and Optimization (Trends and Future Directions) (ICRITO), Noida, India, 2020, pp. 1330–1334, https://doi.org/10.1109/ICRITO48877.2020.9197982.
17. R. Moola and A. Hossain, "Human Activity Recognition using Deep Learning," 2022 URSI Regional Conference on Radio Science (USRI-RCRS), Indore, India, 2022, pp. 1–4, https://doi.org/10.23919/URSI-RCRS56822.2022.10118525.
18. Zamichos, M. Tsourma, S. Papadopoulos, A. Drosou and D. Tzovaras, "User Profile-aware Daily Activity Prediction," 2023 24th International Conference on Digital Signal Processing (DSP), Rhodes (Rodos), Greece, 2023, pp. 1–5, https://doi.org/10.1109/DSP58604.2023.10167947.
19. C. Wang and Z. Peng, "Deep Learning Model for Human Activity Recognition and Prediction in Smart Homes," 2020 International Conference on Intelligent Transportation, Big Data & Smart City (ICITBS), Vientiane, Laos, 2020, pp. 741–744, https://doi.org/10.1109/ICITBS49701.2020.00163.
20. K. K. Krishnaprabha and C. K. Raju, "Predicting Human Activity from Mobile Sensor Data Using CNN Architecture," 2020 Advanced Computing and Communication Technologies for High Performance Applications (ACCTHPA), Cochin, India, 2020, pp. 206–210, https://doi.org/10.1109/ACCTHPA49271.2020.9213225.
21. T. Sharma, P. Kavipriya and U. Kiruthika, "Towards Improving Human Activity Recognition Using Artificial Neural Network," 2021 Emerging Trends in Industry 4.0 (ETI 4.0), Raigarh, India, 2021, pp. 1–5, https://doi.org/10.1109/ETI4.051663.2021.9619361.
22. M. N. Saroja, K. R. Baskaran and P. Priyanka, "Human pose estimation approaches for human activity recognition," 2021 International Conference on Advancements in Electrical, Electronics, Communication, Computing and Automation (ICAECA), Coimbatore, India, 2021, pp. 1–4, https://doi.org/10.1109/ICAECA52838.2021.9675787.
23. S. Deepak, H. Anandakumar, S. Pavithra, V. Keerthika, and K. Nandhini, "Performance Analysis of Star Topology for Small Networks Using Riverbed," 2022 8th International Conference on Advanced Computing and Communication Systems (ICACCS), Mar. 2022, https://doi.org/10.1109/icaccs54159.2022.9785012.
24. C P Shirley, Kantilal Rane, Kolli Himantha Rao, B Bradley Bright, Prashant Agrawal and Neelam Rawat, "Machine learning and sensor-Based Multi-Robot System with Voice Recognition for Assisting the Visually Impaired, Journal of Machine and Computing, vol.3, no.3, pp. 206–215, July 2023. https://doi.org/10.53759/7669/jmc202303019.
25. A. Haldorai, Ravishankar C. V, Q. S. Mahdi, and P. Devasudha, "Application of AI/ML in Network-Slicing-Based Infrastructure of the Next-Generation Wireless Networking Systems," 2023 Fifth International Conference on Electrical, Computer and Communication Technologies (ICECCT), Feb. 2023, https://doi.org/10.1109/icecct56650.2023.10179649.

Part V
Wildlife Monitoring and Management Using AI

Vision Transformer-Based Forest Fire Classification: Wild Life Management System

1 Introduction

Forest is the home for much rich variety of species such as animals and plants. Based on this forest environment, human's natural life system is moving [1]. Also, this forest environment is only the reason behind stable climate. In the world level four billion hectares of aquatic zone, which means 29% of the globe is considered to be forest area. Nearly one billion people depend on forest life for sustenance, minerals medicines, materials, and fuel. One of the main advantages is global warming reduction, because the trees from the forest absorb and emits the carbon dioxide and oxygen, respectively. The maintenance of oxygen level is the basic need for social survival. So, forest preservation is important for future generation to maintain the diversity. But the forest area is slowly reducing day by day due to many reasons. Among the many facts, forest fire is the primary cause. Usually, forest fire is unrestrained burning of trees in the forest area. Recently, due to temperature variation the number of forest fire cases expanded. Forest fire affects resources of nature, flora/fauna, and environment very badly. This threatens the life of living things, which produce greenhouse gases madly [2]. So, timely detection of forest fire is necessary to stay away from unfathomable damage.

The forest fire occurs by reason of man-made or natural reason. Mostly, cigarette and electric sparks are man-made reasons. At the same time, temperature also will cause fires. Coniferous trees in the forest easily catch fire. This forest fire may lead to loss of lives, along with economic loss. This may lead to unhealthy living situations and increase the rate of carbon dioxide in the air [3, 4]. Therefore, early stage of fire detection is vital to reduce the loss. Researchers all over the world and nature lovers are trying to find a solution for early fire detection. Habitual human supervision is difficult and it is not possible to supervise the forest area manually as it is time-consuming and also because the forests are spread over a large area. For this reason, every government is trying to produce Fire Forecasting System for successful forest

© European Alliance for Innovation 2024
A. Haldorai et al., *Artificial Intelligence for Sustainable Development*, EAI/Springer Innovations in Communication and Computing,
https://doi.org/10.1007/978-3-031-53972-5_24

fire management and administration. As a result, there is a need for automatic forest fire observation and monitoring system to prevent the forest ruin, since in the recent years there have been large numbers of forest fire accidents.

With the goal of automatic forest fire monitoring, many environment researchers produced many techniques such as IoT (Internet of Things) based, Artificial Intelligence (AI), and computer vision, since human-based monitoring is unreliable. IoT-based forest fire monitoring system will produce large quantity of database, which is always not possible manually. So, AI-based techniques are introduced [5, 6]. AI-based machine/deep learning models has the ability to do effective intelligent surveillance. Deep learning is the fundamental algorithm of computer vision with AI technology. Deep learning-based forest fire management has some of the following advantages such as less cheap data storage and GPU with high performance. Basically, the deep learning model requires high quantity of dataset; but nowadays modified technology reduces these types issues.

Continuous surveillance is required, only then action can be taken at the correct time and the forest be saved from big disaster and loss. Deep learning is the one of the promising continuous surveillance technologies. Deep Learning Models (DLM) extend the hands toward countless concerns in the modern world. E-commerce, Nature Language Processing, web search, object identification, data analysis, and audio to text conversion. DLM is defined as very deep structure, hierarchical-based machine learning models [7–9]. In DLM, the model contains many layers with neurons depending on the problem requirement. With the help of these layers, the neurons learned very deeply about the nature of data base. The DLM are considered for various applications such as image classification and detection. The proposed work also planned to use DLM model based on this image classification.

The identification of unique characteristics and features from the image is denoted as image classification. Hence, image classification is considered as the most important research in digital image analysis. Image classification means classification of images from the predefined labels. Manual classification might be difficult task, particularly during the image classification with massive image. So, if the process is automatic, then the process is easy. Due to this automatic requirement, the computer vision combines the image processing with DLM. This success only extends the need of DLM in face identification, automatic car, detection, and so on. Usually, the DLM model includes the following steps during the image classification such as image preprocessing, extracting unique features, classification, and detection [10]. The DLM model changes the training process to intelligent and independent process. So, the proposed automatic forest fire detection also considers DLM to achieve the high performance. The following are key contributions of the planned work:

- Proposing a novel and real-time DLM-based forest fire detection system, since the DLM model is producing best accuracy.
- Presenting the high efficiency classification model with Vision Transformer Model.

- Suggesting automatic and accurate forest fire forecasting system, because manual supervision is difficult.
- Evaluating the planned Vision Transformer Model with various performance metrics for forest fire data base.

The structure of the planned work is as follows. Section 1, presents the basic concepts of the DLM model, needs of forest forecasting system, and the key contribution. The related works are discussed in Sect. 2. Section 3 gives the detail about the proposed Vision Transformer Model, process, and working flow. The processing steps are explained in Sect. 4, as experimental setup. The results are analyzed in the Sect. 5, and finally the proposed research is concluded in Sect. 6.

2 Literature Review

In recent days, many researchers are presenting and suggesting the many forest fire systems. Fire is an undefined object with irregular shape. So, Martin et al. [11] consider the color feature to detect the fire. Most of the work related to fire detection is considered as surveillance scenario. Some of the researchers introduced conventional and cutting-edge-based fire detection. To reduce the damage from the fire accident, the fire is detected from a smoky area. The detection of smoke also help in finding the fire at an early stage. So, sensor and vision-based smoke detection model was suggested by many researchers. Basically, this sensor-vision-based approach includes the five sub-model. They are (a) smoke based (b) light based (c) gas based (d) temperature, and (e) composite model [12, 13]. But mostly, smoke and temperature sensors are considered. But the lack of less detection coverage area and speed [14], image-based surveillance system introduced by the researchers. Some of the fire detection methods are presented in Table 1.

In the image-based fire detection, the researchers considered color features to classify the fire image [15]. The fire-based image includes smoke content in the

Table 1 Fire detection method

S. no	Method	Techniques	Comments
1	IoT	Smoke Sensor Temperature sensor	Less detection ranges Less detection speed
2	Image processing	Color feature based Dynamic texture Shape	False color in images Difficult to understand mist and smoke
3	Video processing	Video image based	More expensive Difficult to analyze the video image
4	Computer vision	Machine learning Deep learning Deep learning with Transfer learning	Need high amount of data base Over- or underfit issues may occur Local optimum

image. But identification of smoke and mist is difficult from the image. Generally, image processing includes some other characteristics also like shape, dynamic structure with color for fire detection [16, 17]. Next, some researchers have introduced video processing-based fire detection. Zhao et al. [18] introduced fire surveillance system with the help of video processing technology. In the next scenario, the researchers were started to consider computer vision technologies like machine or DLM algorithms. The papers [19–21] introduced machine learning model to detect the forest fire. Machine learning-based Fire classification from image always depends on manual feature extraction. For manual feature extraction deep knowledge or expert is needed. But the DLM is different from machine learning in the terms of feature extraction because it will extract the feature automatically. Generally, Convolutional Neural Network (CNN) is the advanced and always noteworthy DLM algorithm. There are lots of CNN models such as AlexNet, ResNet, MobileNet, VGG [22], Inception [23], YOLO, GAN, and so on. These types of DLM techniques are also applied on fire classification problems.

The various CNN techniques for fire classification are deeply analyzed and discussed in paper [24]. Usually, the DLM techniques need large parameter space. So, transfer learning technique is introduced to solve issues like large processing time and the need for large parameter space. In the transfer learning-based training, there is no need of scratch training. Transfer learning means knowledge sharing between two tasks. Similarly, some other advanced DLM algorithms, in particular, Faster Resolution CNN, YOLO-V3 was used by the Li et al. Among these three models, YOLO-v3 performed well than other models. The YOLO-V3 model brings out 83.5% accuracy rate. A time efficient fire detection model is presented by the Mahmoud et al. [25]. In this paper the author suggested transfer learning algorithm with DLM for better reduction of processing time. Bari et al. [26] exploited some of the DLM models like Inception V3 and MobileNet V2 for fire classification system. The researcher Cheng et al. [27] introduced a new concept Fast Regional CNN to identify the smoke. With this Fast Regional CNN, a novel selective search method merged to find the exact location of smoke. Based on the results and discussion, this smoke detection model generates the best classification rate. Some of the papers are reviewed in Table 2.

The author Govil presented with his research team [40] the Inception-V3 model to detect smoke and non-smoke areas. His proposed model generated 91% accuracy rate for smoke detection system. Similarly, Park et al. [41] offered many DLM algorithms like VGG16, DenseNet, ResNet50 for detection of flame, smoking area, and forest fire; these models give the accuracy rate as 96.4%, 98.1%, and 96.6%, respectively. The researchers Qin et al. [42] recommended a fire detection system-based DLM algorithm. The main aim of this paper is finding the exact location of the fire. The suggested models are depth-wise separable DLM. In this paper fire image is categorized with the help of YOLO-v3 model. The research starts the training with the CNN model to reduce the training timings. Then this is followed with advanced YOLO-v3 model, for the concern of high accuracy. One of the main advantages of this paper is reduction in identification time. A novel outline was suggested by Jeon et al. [43] for fire image classification. In this paper, the author launched multiscale

Table 2 Review paper related to fire detection

S.no	Research	Method	Accuracy	Recall	Precision
1	Alexandrov et al. [28]	Yolo-v2 Faster RCNN	98.3% 95.9%	98.3% 95.9%	100% 100%
2	Zhang et al. [29]	SVM-pool5 CNN-pool5	95.6% 97.3%	– –	76.2% 84.8%
3	Jiao et al. [30]	Yolo-v3	–	78%	84%
4	Li et al. [31]	Yolo-edge	–	51.2%	78.5%
5	Lee et al. [32]	Alexnet VGG13 GoogleNet	94.8% 86.2% 96.2%	–	–
6	Kaabi et al. [33]	DBN	95%	–	–
7	Zhao et al. [34]	Fire-Net	98%	98.8%	–
8	Chen et al. [35]	CNN	86%	98%	–
9	Cao et al. [36]	Abi-LSTM	97.8%	97.5%	–
10	Sousa et al. [37]	Inception-v3	93.6%	93.1%	94.1%
11	Tang et al. [38]	ForestResNet	92%		
12	Sun et al. [39]	CNN	94.1%	–	–

classification with high accuracy. Particularly, the characteristics of the fire image were carefully selected with the help of densely stacked CNN layers. In the modified version, the CNN layer is combined with feature-squeeze block, which is considered for spatial and channel-based feature map compression.

For the power-constrained system, a DL-based model was planned by Vinicius et al. [44]. In this work, cost complexity is reduced by eradicating some unwanted filters. Even with the deletion of unwanted filters, the model maintains the best performance. The author Dampage et al. [45] introduced a fire detection system with the help of sensors. With these wireless sensors, machine learning algorithm also combined to get high-performance rate. A fire was automatically detected by the Dogan et al. [46] in the research. The author used ResNet models for automatic fire detection. Thus, many of the research papers proved to be good with satisfactory classification rate in the area of fire detection. From these researchers, one thing is noticeable, that is, the DLM is well suited for image-based identification problems.

Based on the investigation above, it is understandable that DLM holds great promise for fire recognition and can contribute to the development of a robust system capable of drastically minimizing humanoid and economic injury due to fires. According to the evaluation of literature, while detection of forest fire/smoke from pictures has been studied, no work has been done on the ViTM, when trained models are utilized for new tasks with fire/smoke images. Furthermore, significant limitations in the use of DLM for fire and smoke discovery remain, such as quicker parameter learning and training/validation, hyperparameter tuning, and transfer learning over novel data bases. Although a few trials used transfer learning to speed up training, none of the studies cited above did. Although transfer learning was used in a few experiments to speed up the training procedure, still something is lacking in automatic classification. Since the main aim of the proposed research is high accuracy, the planned model is going to consider the novel DLM for automatic forest fire detection.

3 Forest Fire Detection System

Forest fire always depends on climate and atmospheric conditions. So only the early identification of forest fire detection is a challenge and demands research in mountain areas. This research considers the novel ViTM model for automatic identification of forest fire. In recent times, the proposed ViTM is one of the emerging technologies, which appeared instead of traditional CNN DLM. The ViTM act as substitute for CNN in the research area of image classification. This is considered as advanced image classification model, which is four times better than traditional CNN image classification in terms of complexity and accuracy. ChatGPT is one of the notable and successful Transformer-based technique. The ChatGPT used the combination of pretrained transformer with self-attention technique; which is mainly used in NLP applications.

3.1 Vision Transformer Model

Even though CNN has been the king of computer vision field for several years, the novel ViTM is also slowly entering into the computer vision area by the success of ChatGPT. ViTM too revealed the capabilities and skills in computer vision applications like image classification, object detection, and segmentation.

The ViTM was presented by Neil Houlsby [47] and his Google Research Brain Team members at ICLR 2021. The model was pretrained based on ImageNet and ImageNet-21 k data bases. The Vi-TM evolved from transformer architecture. Basically, the transformer architecture was designed particularly for NLP. In the image classification problem, the transformer architecture applied straight on the sequence of patches of an image. The transformer architecture, which was introduced in 2017 [48], is a combination of RNN and self-attention mechanism. The transformer model is shown in Fig. 1. Attention mechanism means the transformer has the ability to access past information. Generally, the transformer model is nonsequential; but the RNN only accept the sequential form of data. During the processing of data, RNN accepts only word; at the same time the transformer architecture accepts all the words at a single moment. RNN accesses the past information until some point only, after which the data is lost. But the accessing process of the transformer is opposite. It has no limit to access the past data. Additionally, the transformer architecture considers the positional embeddings to store data. These points are the main reasons to change the transformer to next level than RNN.

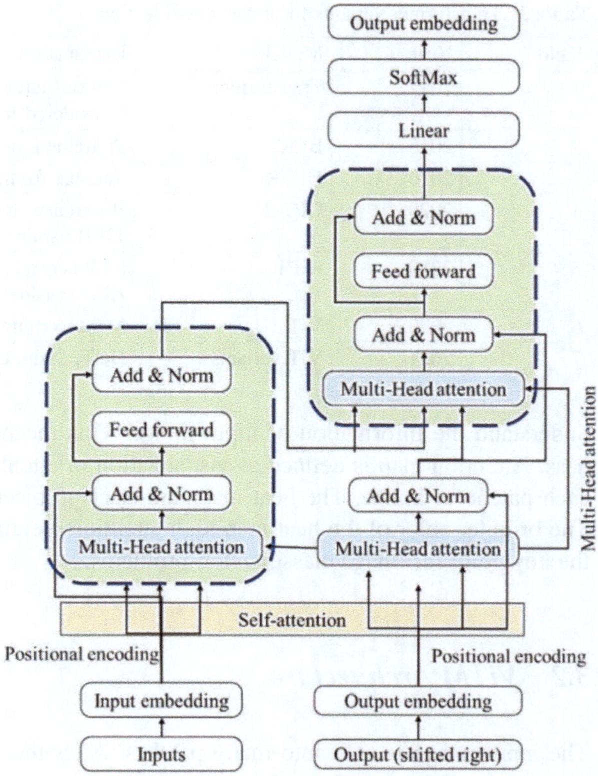

Fig. 1 Transformer model

The ViTM learn the structure of the input image autonomously and by its own without any dependency, where the input image is considered as sequence. These sequences or patches are flattened as single vector form by connecting all pixel channels. Since the ViTM considers the self-attention mechanism, every pixel of the input image should attend and compare by other pixels. Due to this scenario, the computational cost value becomes double during the pixel value increment. So, this is not suitable for big resolution image. To resolve this issue, the splitting concept was considered during the ViT architecture design, i.e., separation of image into many patches. For this separation, ViT consider two methods: (1) the image is resized (H x W x C) into two-dimensional (number of patches x size of the patch2. Channel) size flattened patches. Next, these flattened patches were converted to linear layer. (2) converting the image by convolutional layer; where the size of the kernel and stride is equal to the size of the patch. After the conversion, the flattened layer will be available. Some of the transformer model is presented in Table 3.

The transformer model is made up of several self-attention layers. So, this self-attention is the main module in transformer model. This mechanism is used for capturing the contextual data from input. This allows the model to analyze the all-region or patches of input to calculate the weighted sum by the comparison between patches. So, it is easy to recognize the relevant features, which help us to

Table 3 Transformer variants for image classification

S.No	Year	Model	Explanation
1	2017	Transformer	Use self-attention mechanism Considered for NLP application
2	2018	BERT	Pretrained model used in NLP
3	2020	DETR	Suitable for high-level vision object detection
4	2020	GPT-3	Big transformer model 170B parameter
5	2020	iGPT	NLP is original application Also consider for pretraining
6	2020	ViT	Most recommendable for visual recognition
7	2021	ViT variants	DeiT, Swin, CSWin are introduced

understand the information of input image. This mechanism increases the robustness. Attention map is defined as visualization of attention weight considered from each patche of image. The heat maps are easily projected with this attention map. The brighter color of the heat map high attention weight. Attention map analysis is the key point for image classification problems.

3.2 ViTM Architecture

The input image is split into many patches. After the separation of patches, these patches are flattened. From these flattened patches generate the low-dimensional linear embeddings. In the next process, the positional embedding is added. The patches are considered as input to ViTM. After the pretraining process, fine-tuning will be considered. There is some difference between ViTM and traditional CNN. Different from CNN, the ViTM get global representation from the shallow structure layers. The skip type of connection is additionally powerful than CNN model. The ViTM has the potential to recall more the past information than CNN. Also, it allows the learning with high quality intermediate representation data. The structure of ViTM is given in Fig. 2. When compared to ViT, optimization is easy for CNN. For the transformer, CNN performance as a front end. Generally, to increase the accuracy the image is processed by 3*3 convolution with stride value 2. The pixels of the image are transformed to future map; these feature maps are converted to tokens or patches. These patches are given as input to transformer. After the processing with self-attention technique, output patches are generated. At the same time, output patches are compared with future map.

The ViTM architecture contains three main blocks such as layer norm, MSP (Multi-head Attention Network), and finally MLP (Multilayer Perceptron). The layer norm helps in the training process, which means that it helps the model to accept the variations during the learning process. MSP is accountable for creating attention maps from image patches, to focus the important region of the input image. The MLP is the classification layer with Gaussian Error Linear Unit, which has MLP

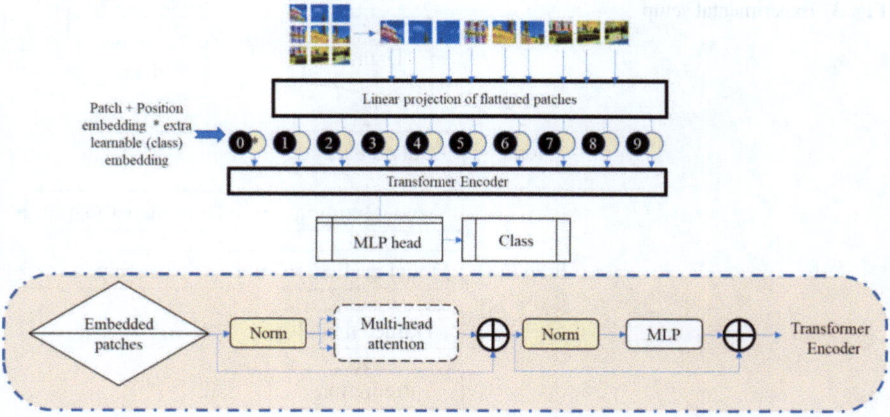

Fig. 2 ViTM architecture

head as the final block. The ViT includes many layers; each layer is a transformer block. Here, self-attention mechanism is applied on MHA (Multi-Head Attention). Each transformer block introduces batch size, num_patches, and projection_dim as tensor. These tensors are processed by softmax classifier to give the output with probability value. The output of the transformer is reshaped by the flatten layer. This flattened layer acts as the classifier head. If the projection of the patches is big, then the model uses the average pooling layer for dimension reduction. For the best result the model should train with more epochs, resized input, large quantity of transformer layer, correct patch, and projection size. Additionally, learning rate, model choice, optimizer type, and weight decay are also considered for quality output. Transfer learning with fine tune process is well suggestable for large database with high resolution images.

4 Experimental Setup

The experimental setup for the forest fire detection is shown in Fig. 3. The setup includes training and validation process. The dataset is taken from Kaggle platform. The images are reshaped and data augmentation techniques like shifting, scaling, cropping, flipping, and padding are applied on the reshaped image. After the data augmentation, the data base is divided into training/validation and test data bases. The images from the data base are converted to patches by the convolutional layer. After the patch conversion, the patches are flattened. In the next procedure, the flattened image is processed with vector embedding and position embedding techniques. Now this is given to transformer section as input. The quantity of transformer depends on the types of ViTM architecture. The ViTM has three types of model; they are ViTM-base, ViTM-large, and ViTM-huge. The ViTM-base includes totally

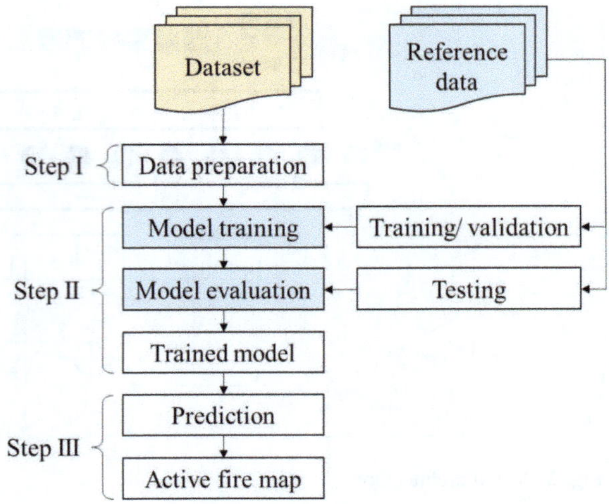

Fig. 3 Experimental setup

12 layers with 86 M parametrs and 768 hidden size value. The ViTM-large contains 24 layers with 307 M parameters and 1024 hidden size. Similarly, the ViTM-huge has 32 layers with 632 M parameter and 1280 hidden size. In the proposed work, the model chose the ViTM-Base model for training process.

The size of the transformer output is based on the ViTM type. Generally, the model is fine-tuned based on the application. The selected ViTM is trained on ImageNet-21 K database. So, the pretrained model is selected and fine-tuned. The model has 21,843 classes. But we have only two classes. In the fine-tuning process the front layers are frozen layers and the classification layers only act as unfreeze layers. So, during the fine-tuning, there is no need of scratch training. Scratch training means training the layers from start layer itself. But in the fine tuning, only the unfreeze layers trained. This process reduces the processing time. The top layers have the ability to extract the general features such as edge, corner, and so on. But the final layers only extract the features related to data base and selected problems. So, during training, unfrozen layer and modified classification layer only trained. The parameters like learning rate, epochs, patches, optimizer type and loss function are selected based on our problem. The selected parameters are listed in Table 4. The fine-tuned model for forest fire detection is shown in Fig. 4.

The Weight-Binary-Cross-Entropy-Dice (WBCED) loss function is considered for loss analysis. The WBCED is mentioned in Eq. 1. The WBCED is the combination of dice loss and weighted cross-entropy loss. The dice loss is mentioned in Eq. 2. In Eq. 3 is mentioned the weighted cross-entropy loss. The weighted function is shown in Eq. 5.

$$\mathrm{Loss}_{\mathrm{WBCED}} = \mathrm{Loss}_{\text{weighted binay cross entropy}} + \mathrm{Loss}_{\mathrm{Dice}} \tag{1}$$

Table 4 Parameters list

S. no	Parameters	Value
1	Image size	224 × 224
2	Learning rate	0.001
3	Number of epochs	100
4	Batch size	32
5	Projection dimension	64
6	Number of heads	4
7	Transformer layer	8
8	Weight decay	0.0001
9	Patch size	6 × 6
10	Loss function	Weight-binary-cross-entropy-dice loss (WBCED)
11	Optimizer	Adam

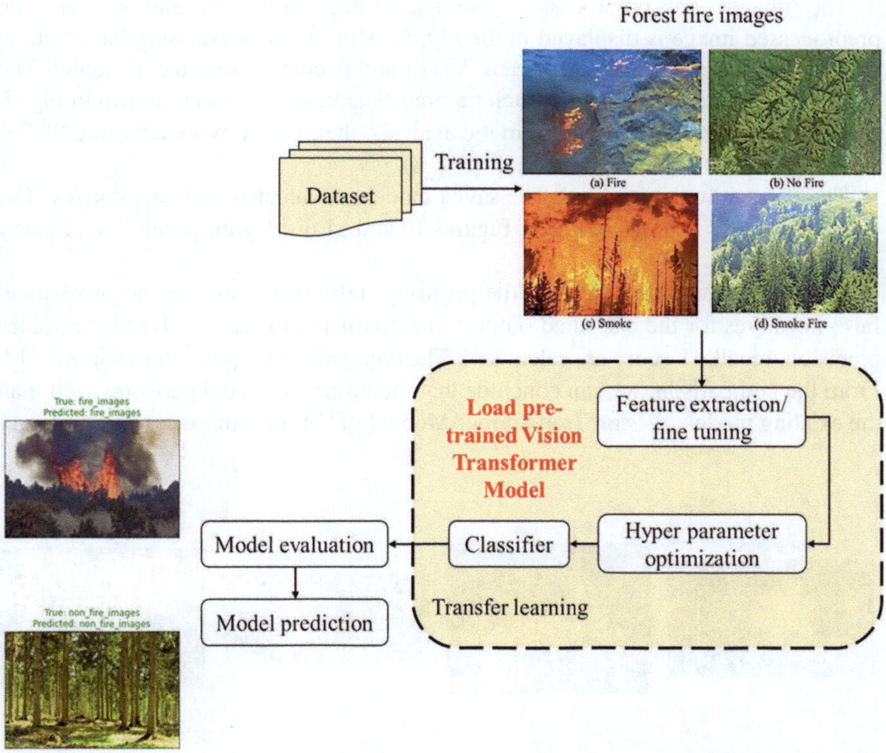

Fig. 4 Fine-tuned forest fire detection

$$\text{Loss}_{\text{Dice}} = 1 - \frac{2\sum y\, x\, p}{\sum y + \sum p} \tag{2}$$

$$\text{Loss}_{\text{weighted binary cross entropy}} = -w\,(y\,\log\,(p) + (1-y)\,\log\,(1-p)) \tag{3}$$

$$\text{Loss}_{\text{binary cross entropy}} = -(y \log(p) + (1-y)\log(1-p)) \tag{4}$$

$$w = \frac{\sum_{i=1}^{s}\sum_{j=1}^{t} x_{i,j}}{e^{-5x|\text{Mask}-0.5|} / \sum e^{-5x|\text{Mask}-0.5|}} \tag{5}$$

5 Result and Discussion

The dataset image is selected from Kaggle. The sample image is shown in Fig. 4. The data base contains two classes, they are fire and no fire. The sample image is reshaped and converted to patches, which is shown in Fig. 5.

The images are preprocessed like sharpening, cropping, and so on. The preprocessed image is displayed in the Fig. 6. After the preprocessing the database is trained by the selected model. Here ViTM architecture is selected as model. The model was fine-tuned without scratch training. The trained model is shown in Fig. 7, in terms of accuracy and loss. From the analysis, the planned work achieved 98.7% accuracy rate for forest fire detection (Fig. 8).

The trained model is saved. The saved model is evaluated with test images. The evaluated result is shown in Fig. 9. Figures 10 and 11 display the prediction of query image.

The error is analyzed by the True positives, false positives, true negatives, and false negatives for the classified output. The confusion matrix, analyzed the values precision, recall, f1-score are calculated. The comparison graph is shown in Fig. 12. From the comparison, we can conclude that the proposed model performs well than the existing models. Vision Transformer Model (ViTM) accomplishes extraordinary

Fig. 5 Sample data base

Fig. 6 Patch images

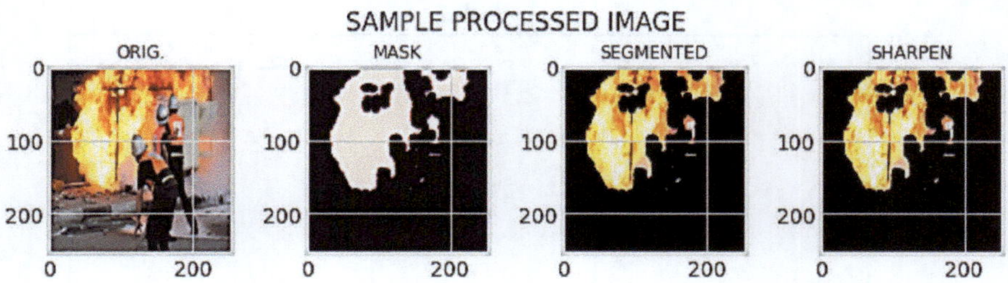

Fig. 7 Preprocessed image

outcomes compared to some other existing convolutional neural networks (CNN) models.

6 Conclusion

The proposed work focused on Vision Transformer Model architecture-based forest fire identification at early stage. This chapter concentrated on the vision-based image analysis. The main aim of this research is avoiding the big disaster. The performance of the proposed approach for classifying Fire and No-Fire classes was evaluated on different performance metrics and compared with other CNN models. The proposed

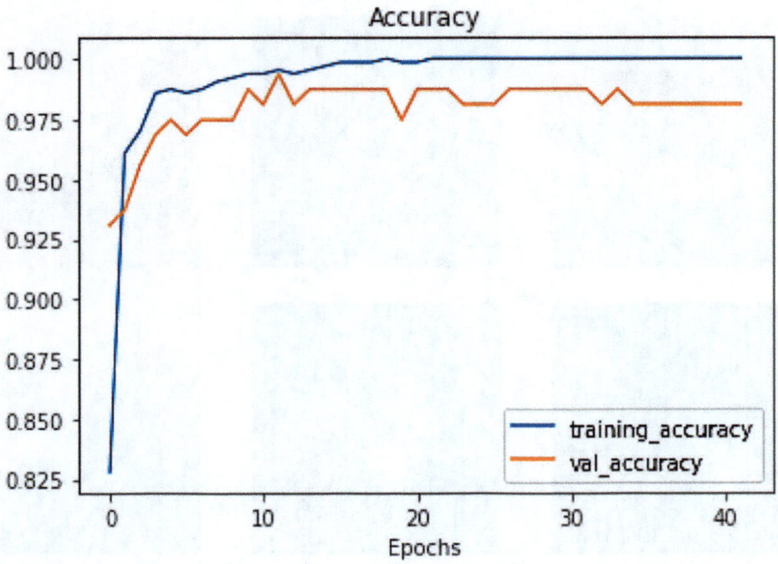

Fig. 8 Training analysis

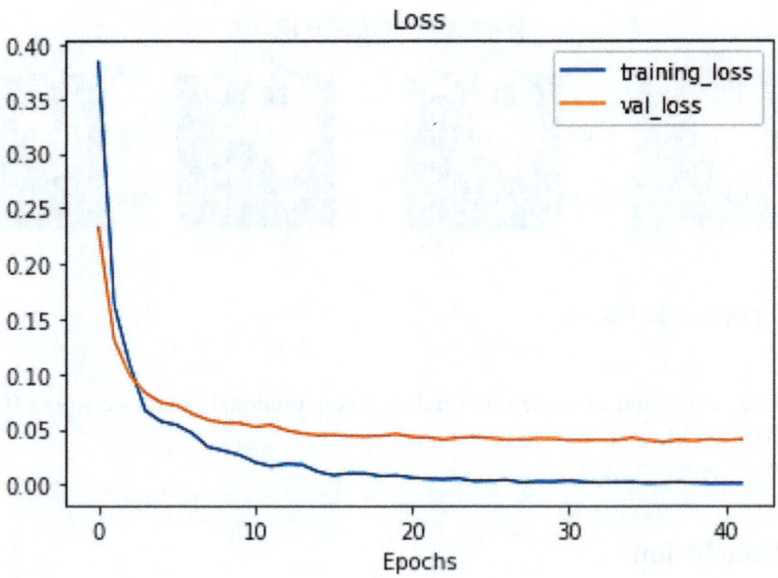

Fig. 9 Evaluated analysis

Fig. 10 Test image analysis

Fig. 11 Query image analysis

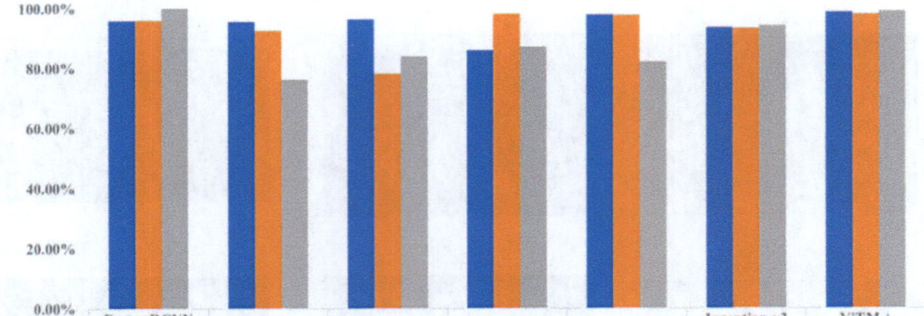

	Faster RCNN [28]	SVM-pool5 [29]	Yolo-v3 [30]	CNN [35]	Abi-LSTM [36]	Inception-v3 [37]	ViTM + WBCED loss
Accuracy	95.90%	95.60%	96.30%	86%	97.80%	93.60%	98.70%
Recall	95.90%	92.50%	78%	98%	97.50%	93.10%	97.80%
Precision	100%	76.20%	84%	87%	82%	94.10%	98.90%

Fig. 12 Comparison analysis

approach achieved 98.7% accuracy, 1.3% error rate, 97.8% recall, and 97.9% precision in classifying the Fire and No-Fire images. The results of the proposed approach showed superiority over other CNN models such as Faster RCNN, CNN, Abi-LSTM, and Inception-v3. The outcomes of the proposed approach are promising for the classification problem considering forest fire detection datasets. Future work will increase the spatial resolution of the images in the forest fire detection dataset.

References

1. X. Chen et al., "Wildland Fire Detection and Monitoring Using a Drone-Collected RGB/IR Image Dataset," in IEEE Access, vol. 10, pp. 121301–121317, 2022, https://doi.org/10.1109/ACCESS.2022.3222805.
2. D. Q. Tran, M. Park, Y. Jeon, J. Bak and S. Park, "Forest-Fire Response System Using Deep-Learning-Based Approaches With CCTV Images and Weather Data," in IEEE Access, vol. 10, pp. 66061–66071, 2022, https://doi.org/10.1109/ACCESS.2022.3184707.
3. H. Yar, T. Hussain, M. Agarwal, Z. A. Khan, S. K. Gupta and S. W. Baik, "Optimized Dual Fire Attention Network and Medium-Scale Fire Classification Benchmark," in IEEE Transactions on Image Processing, vol. 31, pp. 6331–6343, 2022, https://doi.org/10.1109/TIP.2022.3207006.
4. J. S. Almeida, C. Huang, F. G. Nogueira, S. Bhatia and V. H. C. de Albuquerque, "EdgeFireSmoke: A Novel Lightweight CNN Model for Real-Time Video Fire–Smoke Detection," in IEEE Transactions on Industrial Informatics, vol. 18, no. 11, pp. 7889–7898, Nov. 2022, https://doi.org/10.1109/TII.2021.3138752.
5. F. H. Panahi, F. H. Panahi and T. Ohtsuki, "A Reinforcement Learning-Based Fire Warning and Suppression System Using Unmanned Aerial Vehicles," in IEEE Transactions on Instrumentation and Measurement, vol. 72, pp. 1–16, 2023, Art no. 5500216, https://doi.org/10.1109/TIM.2022.3227558.
6. W. Bo, J. Liu, X. Fan, T. Tjahjadi, Q. Ye and L. Fu, "BASNet: Burned Area Segmentation Network for Real-Time Detection of Damage Maps in Remote Sensing Images," in IEEE

Transactions on Geoscience and Remote Sensing, vol. 60, pp. 1–13, 2022, Art no. 5627913, https://doi.org/10.1109/TGRS.2022.3197647.
7. L. Li et al., "Estimation of Ground Water Level (GWL) for Tropical Peatland Forest Using Machine Learning," in IEEE Access, vol. 10, pp. 126180–126187, 2022, https://doi.org/10.1109/ACCESS.2022.3225906.
8. M. Amanullah, S. Thanga Ramya, M. Sudha, V. P. Gladis Pushparathi, A. Haldorai, and B. Pant, "Data sampling approach using heuristic Learning Vector Quantization (LVQ) classifier for software defect prediction," Journal of Intelligent Fuzzy Systems, vol. 44, no. 3, pp. 3867–3876, Mar. 2023, https://doi.org/10.3233/jifs-220480.
9. A. Haldorai and K. K, "An Analysis of Software Defined Networks and Possibilities of Network Attacks," Journal of Machine and Computing, pp. 42–52, Jan. 2022, https://doi.org/10.53759/7669/jmc202202006.
10. J. Shi, W. Wang, Y. Gao and N. Yu, "Optimal Placement and Intelligent Smoke Detection Algorithm for Wildfire-Monitoring Cameras," in IEEE Access, vol. 8, pp. 72326–72339, 2020, https://doi.org/10.1109/ACCESS.2020.2987991.
11. K. M. Keerthi, "Optical Flow Estimation for Efficient Flame Detection in Videos," Proceedings of the International Conference on Emerging Trends in Engineering & Technology (ICETET-2015), 2015, https://doi.org/10.3850/978-981-09-5346-1_cse-535.
12. V. E. Sathishkumar, J. Cho, M. Subramanian, and O. S. Naren, "Forest fire and smoke detection using deep learning-based learning without forgetting," Fire Ecology, vol. 19, no. 1, Feb. 2023, https://doi.org/10.1186/s42408-022-00165-0.
13. B. Kizilkaya, E. Ever, H. Y. Yatbaz, and A. Yazici, "An Effective Forest Fire Detection Framework Using Heterogeneous Wireless Multimedia Sensor Networks," ACM Transactions on Multimedia Computing, Communications, and Applications, vol. 18, no. 2, pp. 1–21, Feb. 2022, https://doi.org/10.1145/3473037.
14. M. Park and B. C. Ko, "Two-Step Real-Time Night-Time Fire Detection in an Urban Environment Using Static ELASTIC-YOLOv3 and Temporal Fire-Tube," Sensors, vol. 20, no. 8, p. 2202, Apr. 2020, https://doi.org/10.3390/s20082202.
15. Y. Hu et al., "Fast forest fire smoke detection using MVMNet," Knowledge-Based Systems, vol. 241, p. 108219, Apr. 2022, https://doi.org/10.1016/j.knosys.2022.108219.
16. S. Yang, S. Zhang, X. Chen, J. Li, E. Li, and W. Chen, "A Fire Detection Method based on Computer Vision," 2022 IEEE 2nd International Conference on Electronic Technology, Communication and Information (ICETCI), May 2022, https://doi.org/10.1109/icetci55101.2022.9832265.
17. S. T. Seydi, V. Saeidi, B. Kalantar, N. Ueda, and A. A. Halin, "Fire-Net: A Deep Learning Framework for Active Forest Fire Detection," Journal of Sensors, vol. 2022, pp. 1–14, Feb. 2022, https://doi.org/10.1155/2022/8044390.
18. L. Zhao, L. Zhi, C. Zhao, and W. Zheng, "Fire-YOLO: A Small Target Object Detection Method for Fire Inspection," Sustainability, vol. 14, no. 9, p. 4930, Apr. 2022, https://doi.org/10.3390/su14094930.
19. F. Abid, "A Survey of Machine Learning Algorithms Based Forest Fires Prediction and Detection Systems," Fire Technology, vol. 57, no. 2, pp. 559–590, Nov. 2020, https://doi.org/10.1007/s10694-020-01056-z.
20. B. C. Ko, K.-H. Cheong, and J.-Y. Nam, "Fire detection based on vision sensor and support vector machines," Fire Safety Journal, vol. 44, no. 3, pp. 322–329, Apr. 2009, https://doi.org/10.1016/j.firesaf.2008.07.006.
21. S. G. Kong, D. Jin, S. Li, and H. Kim, "Fast fire flame detection in surveillance video using logistic regression and temporal smoothing," Fire Safety Journal, vol. 79, pp. 37–43, Jan. 2016, https://doi.org/10.1016/j.firesaf.2015.11.015.
22. J. Kim, J. K. Lee, and K. M. Lee, "Accurate Image Super-Resolution Using Very Deep Convolutional Networks," 2016 IEEE Conference on Computer Vision and Pattern Recognition (CVPR), Jun. 2016, https://doi.org/10.1109/cvpr.2016.182.

23. C. Szegedy et al., "Going deeper with convolutions," 2015 IEEE Conference on Computer Vision and Pattern Recognition (CVPR), Jun. 2015, https://doi.org/10.1109/cvpr.2015.7298594.
24. P. Li and W. Zhao, "Image fire detection algorithms based on convolutional neural networks," Case Studies in Thermal Engineering, vol. 19, p. 100625, Jun. 2020, https://doi.org/10.1016/j.csite.2020.100625.
25. H. A. Hosni Mahmoud, A. H. Alharbi, and N. S. Alghamdi, "Time-Efficient Fire Detection Convolutional Neural Network Coupled with Transfer Learning," Intelligent Automation & Soft Computing, vol. 31, no. 3, pp. 1393–1403, 2022, https://doi.org/10.32604/iasc.2022.020629.
26. A. Bari, T. Saini, and A. Kumar, "Fire Detection Using Deep Transfer Learning on Surveillance Videos," 2021 Third International Conference on Intelligent Communication Technologies and Virtual Mobile Networks (ICICV), Feb. 2021, https://doi.org/10.1109/icicv50876.2021.9388485.
27. X. Cheng, "Research on Application of the Feature Transfer Method Based on Fast R-CNN in Smoke Image Recognition," Advances in Multimedia, vol. 2021, pp. 1–7, Nov. 2021, https://doi.org/10.1155/2021/6147860.
28. D. Alexandrov, E. Pertseva, I. Berman, I. Pantiukhin, and A. Kapitonov, "Analysis of Machine Learning Methods for Wildfire Security Monitoring with an Unmanned Aerial Vehicles," 2019 24th Conference of Open Innovations Association (FRUCT), Apr. 2019, https://doi.org/10.23919/fruct.2019.8711917.
29. J. Zhang, L. Ye, and Y. Lai, "Stock Price Prediction Using CNN-BiLSTM-Attention Model," Mathematics, vol. 11, no. 9, p. 1985, Apr. 2023, https://doi.org/10.3390/math11091985.
30. Jiao, Z.; Zhang, Y.; Mu, L.; Xin, J.; Jiao, S.; Liu, H.; Liu, D. A YOLOv3-based Learning Strategy for Real-time UAV-based Forest Fire Detection. In Proceedings of the Chinese Control and Decision Conference (CCDC), Hefei, China, 22–24 August 2020; IEEE: New York, NY, USA, 2020; pp. 4963–4967.
31. Li, W.; Yu, Z. A Lightweight Convolutional Neural Network Flame Detection Algorithm. In Proceedings of the IEEE International Conference on Electronics Information and Emergency Communication (ICEIEC), Beijing, China, 18–20 June 2021; IEEE: New York, NY, USA, 2021; pp. 83–86.
32. S. Lee and Y. J. Jung, "Hint-Based Image Colorization Based on Hierarchical Vision Transformer," Sensors, vol. 22, no. 19, p. 7419, Sep. 2022, https://doi.org/10.3390/s22197419.
33. Kaabi, R.; Sayadi, M.; Bouchouicha, M.; Fnaiech, F.; Moreau, E.; Ginoux, J.M. Early smoke detection of forest wildfire video using deep belief network. In Proceedings of the International Conference on Advanced Technologies for Signal and Image Processing (ATSIP), Sousse, Tunisia, 21–24 March 2018; IEEE: Piscataway, NJ, USA, 2018; pp. 1–6.
34. Y. Zhao, J. Ma, X. Li, and J. Zhang, "Saliency Detection and Deep Learning-Based Wildfire Identification in UAV Imagery," Sensors, vol. 18, no. 3, p. 712, Feb. 2018, https://doi.org/10.3390/s18030712.
35. Chen, Y.; Zhang, Y.; Xin, J.; Wang, G.; Mu, L.; Yi, Y.; Liu, H.; Liu, D. UAV Image-based Forest Fire Detection Approach Using Convolutional Neural Network. In Proceedings of the IEEE Conference on Industrial Electronics and Applications (ICIEA), Xi'an, China, 19–21 June 2019; IEEE: Piscataway, NJ, USA, 2019; pp. 2118–2123.
36. Cao, Y.; Yang, F.; Tang, Q.; Lu, X. An Attention Enhanced Bidirectional LSTM for Early Forest Fire Smoke Recognition. IEEE Access 2019, 7, 154732–154742.
37. Sousa, M.J.; Moutinho, A.; Almeida, M. Wildfire detection using transfer learning on augmented datasets. Expert Syst. Appl. 2019, 142, 112975.
38. Tang, Y.; Feng, H.; Chen, J.; Chen, Y. ForestResNet: A Deep Learning Algorithm for Forest Image Classification. J. Physics Conf.Ser. 2021, 2024, 012053.
39. S. S. C, B. L. R and D. S, "Design and Analysis of CNN based Residue Number System for Performance Enhancement," 2023 Third International Conference on Artificial Intelligence and

Smart Energy (ICAIS), Coimbatore, India, 2023, pp. 1182–1187, https://doi.org/10.1109/ICAIS56108.2023.10073805.
40. Govil, K.;Welch, M.L.; Ball, J.T.; Pennypacker, C.R. Preliminary Results from aWildfire Detection System Using Deep Learning on Remote Camera Images. Remote Sens. 2020, 12, 166.
41. Park, M.; Tran, D.Q.; Lee, S.; Park, S. Multilabel Image Classification with Deep Transfer Learning for Decision Support on Wildfire Response. Remote Sens. 2021, 13, 3985.
42. Qin, YY., Cao, JT. & Ji, XF. Fire Detection Method Based on Depthwise Separable Convolution and YOLOv3. Int. J. Autom. Comput. 18, 300–310 (2021). https://doi.org/10.1007/s11633-020-1269-5
43. Jeon, M., H.-S. Choi, J. Lee, and M. Kang. 2021. Multi-scale prediction for fire detection using convolutional neural network. Fire Technology 57 (5): 2533–2551.
44. P. V. A. B. de Venâncio, A. C. Lisboa, and A. V. Barbosa, "An automatic fire detection system based on deep convolutional neural networks for low-power, resource-constrained devices," Neural Computing and Applications, vol. 34, no. 18, pp. 15349–15368, Jun. 2022, https://doi.org/10.1007/s00521-022-07467-z.
45. Dampage, U., L. Bandaranayake, R. Wanasinghe, K. Kottahachchi, and B. Jayasanka. 2022. Forest fire detection system using wireless sensor networks and machine learning. Scientific reports 12 (1): 1–11.
46. R. Dogan and M. E. Karsligil, "Fire detection using color and motion features in video sequences," 2010 IEEE 18th Signal Processing and Communications Applications Conference, Apr. 2010, https://doi.org/10.1109/siu.2010.5651727.
47. Alexey Dosovitskiy, Lucas Beyer, Alexander Kolesnikov, Dirk Weissenborn, Xiaohua Zhai, Thomas Unterthiner, Mostafa Dehghani, Matthias Minderer, Georg Heigold, Sylvain Gelly, Jakob Uszkoreit, Neil Houlsby , "An Image is Worth 16*16 Words: Transformers for Image Recognition at Scale" ICLR 2021.
48. Ashish Vaswani, Noam Shazeer, Niki Parmar, Jakob Uszkoreit, Llion Jones, Aidan N. Gomez, Lukasz Kaiser, Illia Polosukhin , "Attention Is All You Need", 31st Conference on Neural Information Processing Systems (NIPS 2017), Long Beach, CA, USA.

A Modified AI Model for Automatic and Precision Monitoring System of Wildlife in Forest Areas

1 Introduction

Surveillance of the wild life is important for preserving the ecosystem. The surveillance of wild life includes identification of territories and populations of wild animals and birds. In the earth, forest and wildlife are interrelated. The forest contains high volume of land enclosed by trees, insects, and different varieties of flora and fauna [1]. The forest gets disturbed by human beings also; simultaneously the humans are also attacked by wild animals. The properties and crops of humans may get damaged by wild animals. When considering "life," it is common for both. Hence, a procedure is needed to protect the life of humans, as well as wild animals and birds. So, wild life monitoring is the main role of the wild life preservation system [2]. This evaluates the unwanted differences in territory, illegal hunting, risky attack of humans and wild animals, health of the wild life members. For the ecological balance, proper maintenance of the wild life is obligatory. This is also needed for ecological and economic improvement of the earth. The importance of the wild life is presented in Fig. 1.

The threats of the wild life are listed in Fig. 2. The behavior, activities and living places of wild animals and birds are carefully monitored by various monitoring methodology. Initially, the presence of wild animals is identified by the noise pattern of the wild animal. But the presence of animals without noise is difficult to identify. Nighttime identification of wild animals is also a problematic one. In recent time, digital photography is presenting a nonintrusive and active method for wild life surveillance. Even though this is effective, the involving people is dangerous [3, 4]; this process needs technicians and experts. So, the researchers started to use automatic monitoring system, which includes motion detection sensor and solar-based camera. The Naturalist use camera for the standard data collection as there is less disturbance for the wild life [5]. But it is a time-consuming process to align and label the photos. The weather conditions like wind, rain, and temperature will trigger the

© European Alliance for Innovation 2024

A. Haldorai et al., *Artificial Intelligence for Sustainable Development*,
EAI/Springer Innovations in Communication and Computing,
https://doi.org/10.1007/978-3-031-53972-5_25

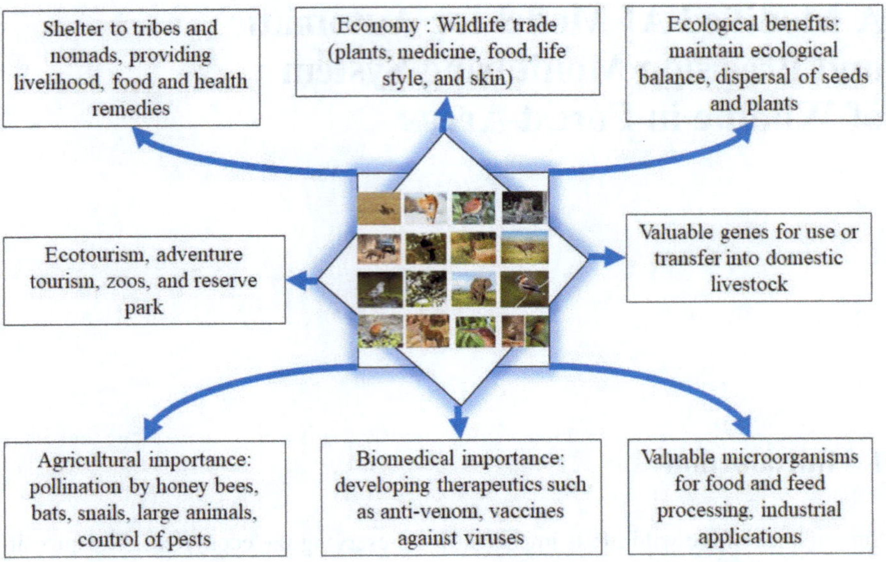

Fig. 1 Importance of the wild life

Fig. 2 The threats of the wild life

data collection. The rapid improvement and success of this system motivate the researchers for further improvement.

Due to the high number of wildlife species, normal manual monitoring and identification of the irregular changes are difficult. The video-based data collection includes high quantity of data, analyzing and processing of these types of data is too time consuming and costly. So, the researchers considered the image processing technique for wild life monitoring. Since, the success of the computer vision methodology in automation projects, which is combined with image processing technology for automatic wild life monitoring [6, 7]. The computer vision-based deep learning technique automatically monitors and identifies the living place, population, and location of the wild life member. The Convolutional Neural Networks (CNN) model is used for automatic detection and classification of wild life animal and birds. The deep learning model extracts relevant feature automatically, and produces the best performance with less error rate. So, the deep learning models are considered for object detection/classification [8–10]. The CNN has the richer structure, which has the capability to extract the pattern and information easily from the given input image [35]. Many of the deep learning models proved that it is one of the promising methods for wild life monitoring. Region-based CNN (RNN), You Only Look Once (YOLO), and Single Shot Detector (SSD) are the some of the notable deep learning models [36]. The innovations of the proposed system are listed below:

- Design an automated wild life monitoring system, which will monitor the location, and habits of the wild animal and birds.
- After the detection process classify the detected image, which helps to identify the species.
- For the proposed system, ROLO model is implemented on the wild life database.

Section 1 gives details about the importance of wild life and the challenges. Section 2 discusses the existing works related to wild life maintenance. The proposed model is presented in detail in Sect. 3. The experimental setup and results are analyzed in the Sects. 3.1 and 3.2, respectively. The conclusion is presented in Sect. 4.

2 Literature Review

The researchers are presenting many works related to wild life animal and birds monitoring. This section is going to discuss the previous existing works deeply. Sheela and Muthumanickam [11] proposed a deep learning model for wild animal detection in the agriculture area. The animals like elephant, boars, monkeys etc., and birds interrupt the farm lands, which may make great damage on the cultivation areas. The intrusion of animals and birds causes loss to agriculturalist. The animals sometime attack the farmers also. Humans do not have the ability and time to monitor their farmlands to save them from the wild animals and birds. These issues are overcome by the modified CNN models, which professionally spot the presence of animals and birds' interruption in the crop arena. This research combines the modified CNN model with Raspberry Pi and GSM models. Madhumathi et al. [12] used an innovative algorithm for wild animal tracking. The authors used YOLO-V5 model to perform the efficient detection of wild animals, which surmount the problems like human attack and farm land loss. The proposed model successfully performed the animal trespass in real time. The animal trespass was designed for human safety in high wild life population areas. It alerted the presence of animals. In this method, after the tracking of the wild animal, the humans are alerted with bullet firing sound.

Another important problem is traffic, accident of vehicle, and death of wild animals due to the intrusion of wild animals on roads, particularly near forest areas. During the travel, clean observation and manual detection of wild animals is not possible. It requires costly systems to decrease this human–wildlife conflicts. So, S. K. L and Edison [13] introduced a modified CNN model to cut down the unwanted accidents of vehicle and wild animals. In this method, the point and depth wise CNN models were combined, which was tested on the IWildCam database. The proposed model gave the noted accuracy rate than other models. Improvements and growth of the smart cities, over population, and high development of the human civilization reduce the territories of the wild life. So unwanted entry of wild animals into human living areas happens and threatens the life of humans, and wild animals too. So timely detection of wild animals is necessary. Ma and Yang [14] introduced a deep learning model for timely wild animal detection. They used YOLO-V5 for speedy identification of animal presence. The proposed model combined with weighted Bidirectional Feature Pyramid Network and Effective Channel Attention for better performance.

Similarly, the identification of birds is also necessary. Schiano et al. [15] introduced a novel deep learning model for pigeons tracking because the pigeons bring ailments to humans and damage to buildings and construction areas. This proposed method suggests an automatic system using drones for tracking pigeons without causing harm to them. The location and activities are monitored by the neural network model and deter the presence of pigeons on the top of buildings through a fully automatic mode. Alagarsamy et al. [16] presented an algorithm for wild animal monitoring systems in outdoors areas. Since the forest land covers a variety of spices, manual inspection and tracking is difficult. So, the proposed model track and categorize the animals based on the deep learning model. This helps in unwanted wild animal accidents. This work alerts the road traveler about the presence, location, and activities of wild animals. Su et al. [17] used YOLOx model for wild life monitoring, which is verified on the PASCAL VOC database.

Since the deep learning shows high involvements and achievements in object detection areas, the Endris et al. [18] considered the YOLO-V3 architecture for the wild life tracking. The proposed model combined with GUI for mobile application. They performed on their own dataset, which contains six labels such as red fox, bale mountain vervet, Somali wild ass, gelada baboon, black maned lion, and walia ibex. The wild life tracking also includes monitoring the extinct level of wild animals and birds. Weather changes and illegal hunting are the main reasons for the extinction of wild animals and birds. This also effects biodiversity. So, identification of threats, reasons of that threats, selecting the solutions, and tracking of the extinct wild species are obligatory. By considering this concept, Huang and Basanta [19] suggested a hybrid deep learning model Inception-ResNet-v2 with transfer learning technique to identify the extinct birds. After the detection, the model classifies the birds from the 29 varieties of the birds from Taiwan.

Table 1 Literature review

S.No	Paper	Model	Accuracy (%)
1.	Jiang et al. [26]	BCNet4	89
2.	Yin and You [27]	VGG16	88
3.	Miao et al. [28]	ResNet	88.1
4.	Chen et al. [29]	Novel 7 L CNN	90.32
5.	Lai et al. [30]	Xception	91.1
6.	Hu and You [31]	ResNet18	92
7.	Song [32]	Improved CNN	85
8.	El Abbadi [33]	Deep CNN	97.1
9.	Fujimori [34]	DeepLabCut	91

Ibraheam et al. [20] combined the deep learning model YOLO-V2 with the embedded system for wild life tracking. They used high speed processing model, which is compared with some other models like YOLO-V3 and YOLO-V4. Rajan et al. [21] concentrated on the road accidents of the wild animals; they used IoT models for the wild life tracking. This model avoids the unwanted entry of the wild animals into the road areas. The model includes PIR and IR sensors for motion variation; MEMS sensor for pressure differences; and the ultrasonic sensor for location prediction. They are necessary in road and railway tracking. The authors used NODE MCU for hardware implementation. All the information is stored and controlled by the IoT cloud. Yang and Song [22] used YOLO-V3 model for bird monitoring system. This work concentrates on the accuracy and stability of the detection of the fine-grained birds. The model YOLO-V3 with depth-wise convolution and point wise convolution. Focal loss was added as the loss function for fine-grained birds tracking.

Hadji et al. [23] proposed a machine learning model for birds' classification on Wetlands or humid zones. By reason of weather change, the counts of the birds are cautiously reducing. Also, it threatened the life of the wild birds. They used the advanced future extraction technique with machine learning algorithm for birds' classification. Zhao [24] used YOLO-V4 deep learning model to detect the behavior of birds. The authors used the CSPdarknet for future extraction. The proposed model continuously monitors the behavior and motion of the birds. Similarly, Zhao [25] used Convolutional Block Attention Module with YOLO-V5 architecture for high precision bird detection. The model achieves high accuracy rate. Some of the other works are presented in Table 1.

3 Proposed Model

In recent times, the research of image classification is highly solved with the help of computer vision and advanced deep learning methodology. The success of image classification moved and concentrated on some other areas such as image generation, visual story narration, detection, and tracking. In the above challenging research, the proposed work considers the tracking problem. The modern technology tracks the object through detection methodologies. These types of recent tracking system comprise various subsystems like future extractor, classifier, and detector. From these components, future extractor is the main component, since the appropriate features can progress along the tracking rate. The hand-crafted feature selection is difficult. So, the modern technology considers deep learning-based tracking with automatic future extraction.

Generally, the object tracking is used in surveillance application, activity tracking, game analyzing, and visual traffic monitoring. The result of tracking system is effected by the illumination condition, object distortions, over speed, and blur. The first main problem is identity switches, where the system can't identify the exact suitable object after the crossing. In the motion blur problem, the target is unclear due to the movement of the camera. This changes the entire look of the target. The view point difference and scale changes also may affect the detection accuracy. In background clutters, it is not easy to identify the target from the background, since the color of the background near the target is equivalent to target. Similarly, low resolution and illumination variation also reduce the target detection accuracy.

The object detection algorithms are mostly performed under the two models. They are (1) motion model and (2) visual appearance model. In the motion model-based object detection, the quality tracker observes the dynamic changes of the target. This motion model-based object detection speculates the potential position. But it doesn't have the ability to predict unexpected changes. The second, the visual appearance-based model, will track the target with respect to the appearance by detecting discriminative model from the background. Usually the target detection/tracking consists of four modules. They are (1) target initialization (2) appearance modeling (3) motion estimation, and (4) target positioning. In the first, target initialization module, the object is identified by bounding box. The bounding box identifies the location of the target. The next step is appearance modeling, where the visual features related to target like scale changes and illumination changes are learned by the learning algorithm. In the following module, the motion estimation is concentrated for zone identification. The target positioning module presents the possible region and fixes the correct location of the target.

Generally, many tracking algorithms are considered for object detection. The detection-based tracking system use tracker for target identification. At the same time, detection-free tracking method identifies the object without the help of tracker. The single object tracking system has the ability to track the single object only, even though the background contains multiple objects. But the multiple object detector can identify multiple objects at the same time. The online tracking system sense the target during the process and run time itself; the offline tracking sense the object after the process and run time. The different types of tracking systems are presented in Fig. 3. There are a lot of traditional tracking algorithms followed by researchers. The deep learning-based tracking systems are now improved in terms of precision. This research planned to use one of the hybrid deep learning ROLO model for the forest surveillance system.

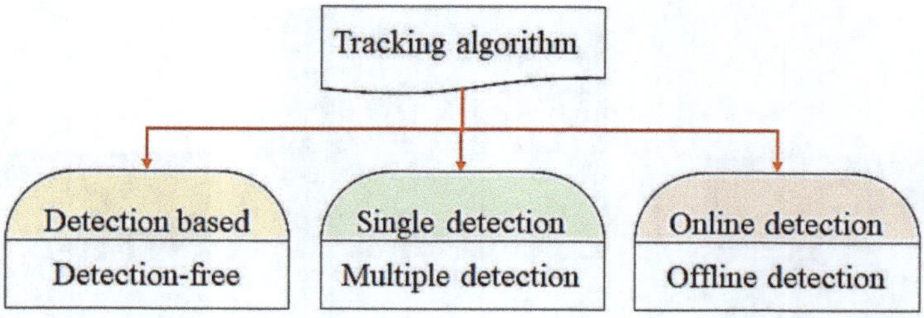

Fig. 3 The different types of tracking system

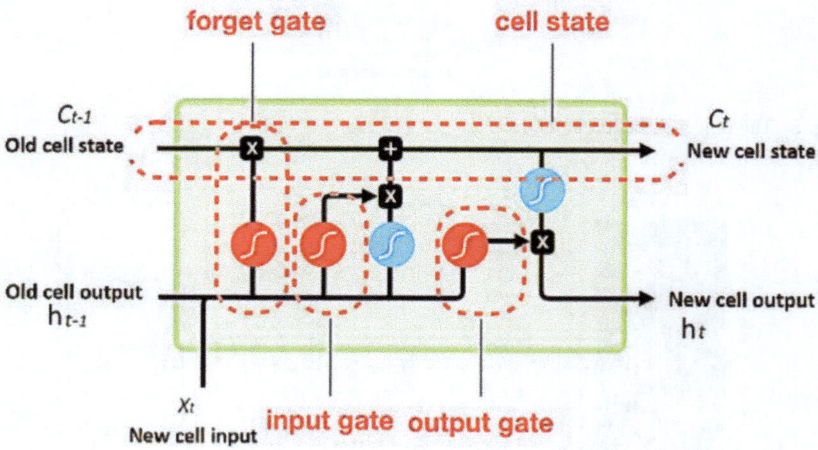

Fig. 4 LSTM

3.1 The ROLO Model

The ROLO is a short term for the Recurrent YOLO (You Only Look Once), which is the combination of the Recurrent LSTM (long short-term memory) and YOLO architecture. The ROLO model is considered for target detection and tracking system. The ROLO model has the ability to convert a sequence of the high-level characteristics into coordinates to track the object, since the LSTM model performs the regression temporally and spatially. The high-level features are combined and partially managed by the YOLO model. But the ROLO is the solo object tracker system.

The LSTM is one type of RNN model that has a complex structure than the traditional Recurrent model. This model regulates the system by the learn or forget process with various types of input sources. The LSTM contains cell memory (cell state) and gates, they are responsible for the information flow and regulation of information flow. The cell memory encodes the data. The LSTM is shown in Fig. 4.

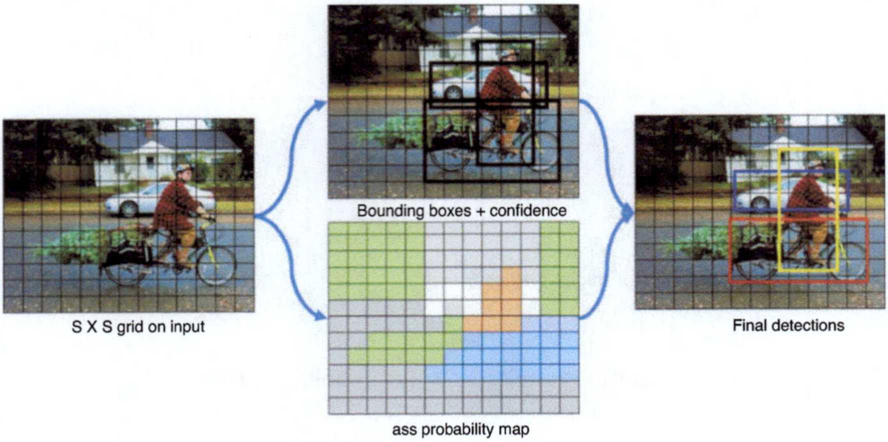

Fig. 5 The object detection in YOLO

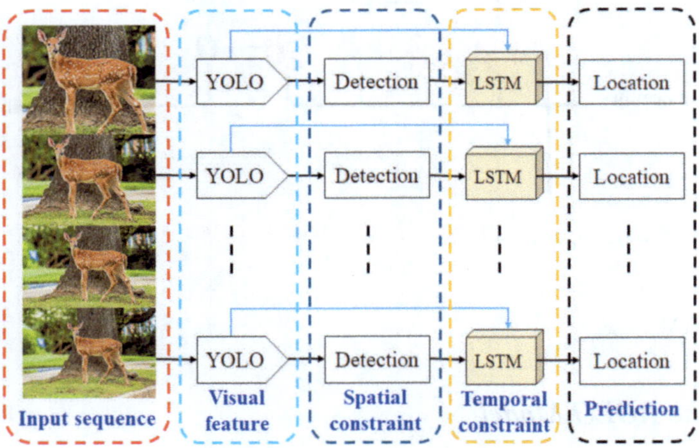

Fig. 6 Object identification

When comparing with other detection models like Fully CNN, the YOLO performs better. It identifies the object by predicting on different regions in a single image itself. This model splits the image into various sub-grids, every grid generates two bounding box; based on these grids and bounding box the object is identified. Always the bounding box is greater than the grid size. The object detection in YOLO is displayed in Fig. 5.

The ROLO is a single item detector, which identifies the object based on the tracking algorithms. The ROLO use the LSTM for the identification of target's trajectory and YOLO for target detection. Generally, the LSTM is well appropriate for pattern identification with fast tracking ability. Figure 6 presents the ROLO model for object identification. The raw input is given to the YOLO architecture, which gives the feature vectors as output. This output is passed to the LSTM model, where, the selected features and bounding box coordinate. The LSTM gives the coordinates of the tracked object. The inference identifies the location of the object. The ROLO have the ability to track location with spatial-temporal features, even in the blur condition. In the first step, the YOLO extract the suitable and robust features with location inferences. In the next step, LSTM performs the classification.

The LSTM model overcome the vanishing gradient problem than the traditional RNN model. The vanishing gradient problem is also mentioned as back-propagation error. The memory cell is mentioned as mt. This memory cell is performed as accumulator. Each input is controlled and regularized by the accumulator and input gates. The input gate is noted as int. The forget gate and the output gate is mentioned as fot and out. Thus, the LSTM model contains many self-parameterized gates. The parameter updates of the LSTM model are given in the Eqs. (1–5).

$$in_t = \sigma(W_{xi}x_t + W_{hin}h_{t-1} + b_{in}) \tag{1}$$

$$fo_t = \sigma\left(W_{xfo}x_t + W_{hfo}h_{t-1} + b_{fo}\right) \tag{2}$$

$$ou_t = \sigma(W_{xou}x_t + W_{hou}h_{t-1} + b_{ou}) \tag{3}$$

$$g_t = \sigma(W_{xm}x_t + W_{hc}h_{t-1} + b_m) \tag{4}$$

$$h_t = o_t \odot \varnothing(c_t) \tag{5}$$

where σ is the sigmoid function, the value of the σ is mentioned as in Eq. 5. In the YOLO model intersection-over-union (IOU) is calculated for the target identification. The objective function of the ROLO is defined in the Eq. (6). In this equation, H_{target} mention the heatmap vector function of the target and H_{pred} denotes the heatmap prediction.

$$L_{\text{MSE}} = \frac{1}{n}\sum_{i=1}^{n} \left\|H_{\text{target}} - H_{\text{pred}}\right\|_2^2 \tag{6}$$

3.2 *Experimental Setup and Result Discussion*

The experimental setup of the proposed work contains three phases: (1) preprocessing phase (2) YOLO for feature selection, and (3) LSTM for object identification. First, the required data and pretrained models are downloaded from the suitable platform. The ROLO model is based on the YOLO model, since the YOLO is common object detection system. The YOLO model is one of the pretrained models, which trained based on the IMAGENET, stored in Keras. The selected YOLO model is fine-tuned based on our dataset.

Fig. 7 Data samples

The YOLO model selects the features based on our dataset. The selected features are passed to LSTM for tracking purpose. The input gives mostly raw images, which means that noisy images. So, the input images undergo the preprocessing techniques like noise removal, resizing, and array conversion. The dataset samples are displayed in Fig. 7. Figure 8 presents the segments of the input images for easy processing. The dataset samples are separated as training, validation, and testing in the ratio of 8:2 for learning process. The dataset samples are segregated for the purpose of using the whole data samples in a useful manner. The training samples are used for learning and the validation samples are considered for regulating and fine-tuning the important values of the hyper-parameters by the method of trial and error. The efficiency of the proposed model highly depends on the suitable hyper parameter. The testing images are utilized for the verification purpose. Since the proposed model is a deep learning-based method, there is no need of manual feature extraction. The model itself has the facility of automatic feature extraction.

The proposed model uses the Adam optimizer for the weight modification, since the Adam is one of the successful optimizers in deep learning field. The Adam optimizer uses the 0.9 value for momentum and 0.0005 as weight decay with the learning rate of 0.001 to make the model to converge. This research planned to use ROLO model, which is the combination of YOLO and LSTM for forest surveillance system. After the process, the result of the suggested system is compared with other

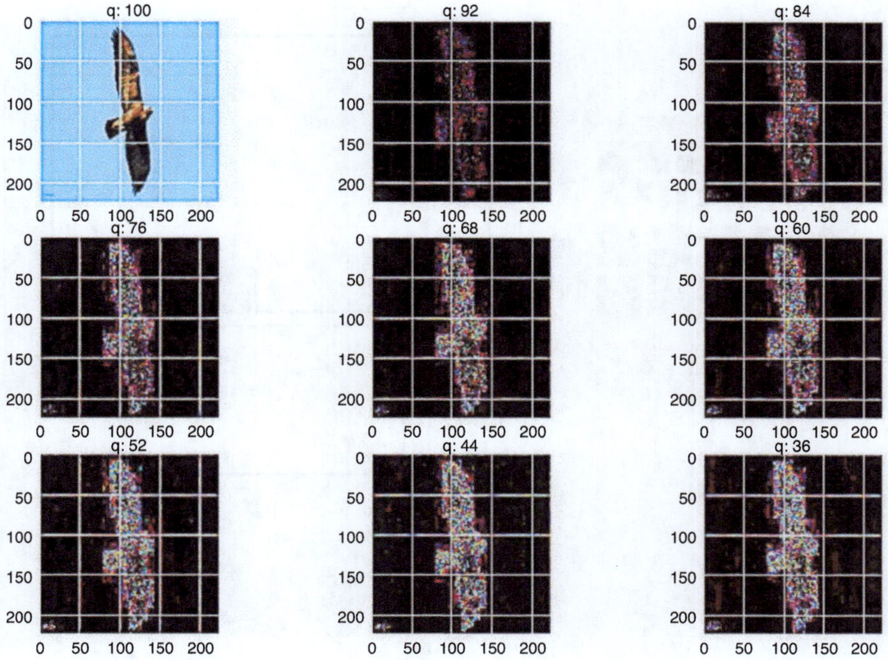

Fig. 8 Data sample segmentation

models such as YOLO, LSTM, and FastRCNN. The experimental setup for the forest surveillance system is given in Fig. 9.

The suggested hybrid model YOLO+LSTM system produces significant improvements in the way of accuracy value during training and validation process. The planned model produces 94.5% accuracy. In Fig. 10a and b, we plotted the accuracy and loss curve for the suggested forest surveillance system. This figure explains and analyzes the performance and loss value of the system. Here, the model uses the mean square error loss as the loss function for weight updates. Thus the Fig. 10 explains the performance analysis.

The YOLO model uses the rectangle bounding box for the object tracking in the sample image. The proposed model uses the bounding box method for the prediction in the proficient way. Figure 11 explains the bounding box method for the forest surveillance system. After the detection, the animals and birds are classified with the help of classification algorithm. The classified output is mentioned in Fig. 12.

The proposed model results are compared with some other models. The comparison is given in Table 2. In the comparison table, some of the parameters like sensitivity, specificity, accuracy, and precision are discussed. There are lots of performance analysis metrics like accuracy, sensitivity, precision, and confusion matrix for result analysis. The confusion matrix includes some performance metrics such as True Positive, True Negative, False Positive, and False Negative. The

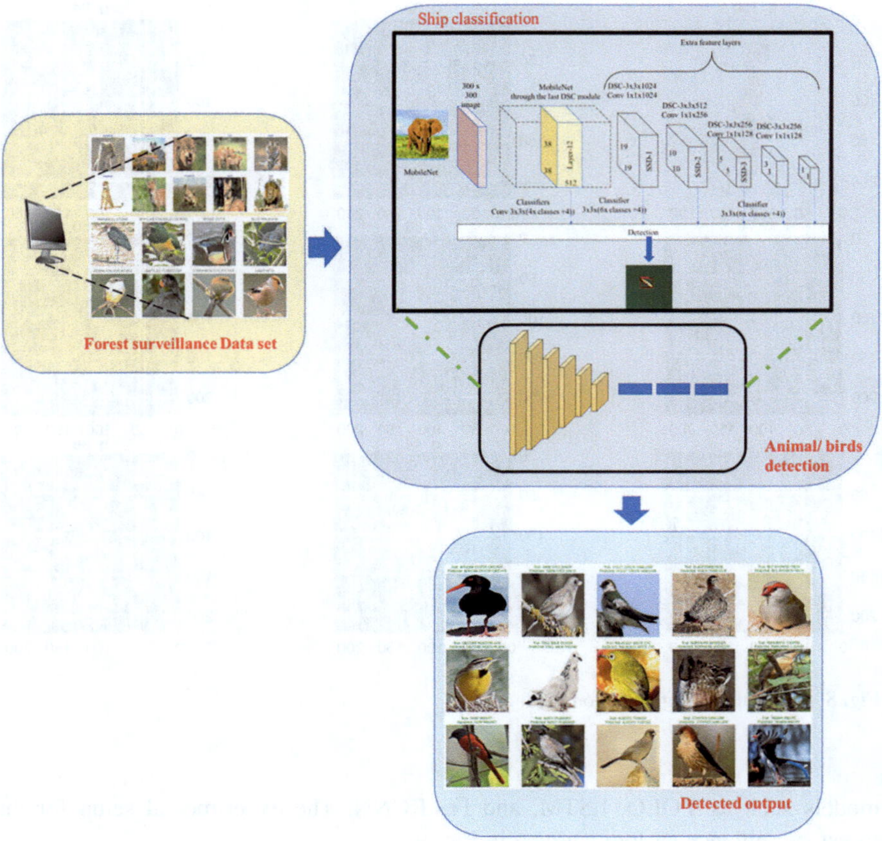

Fig. 9 Experimental setup

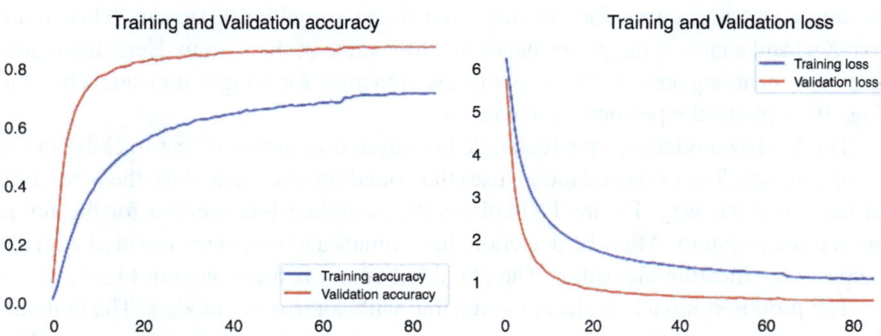

Fig. 10 (**a**) accuracy (**b**) loss

Fig. 11 Forest surveillance detection using bounding box

correctly classified true value is mentioned as True Positive. Similarly, the correctly classified false value is noted as True negative. If the true value is classified with false value, it is called as False positive. The Accuracy, sensitivity, specificity, and precision are mentioned in the following Eqs. (7–10).

$$\text{Accuracy} = (\text{TP} + \text{TN})/(\text{TP} + \text{TN} + \text{FP} + \text{FN}) \tag{7}$$

$$\text{Precision} = \text{TP}/(\text{TP} + \text{FP}) \tag{8}$$

$$\text{Sensitivity} = (\text{TP})/(\text{TP} + \text{FN}) \tag{9}$$

$$\text{Specificity} = \text{TN}/(\text{TN} + \text{FP}) \tag{10}$$

The word accuracy informs us of the proportion of correct classifications among all possible classifications. The precision is mentioned as how many of the things that were tagged as favorable are really positive. The sensitivity is expressed as how many of the things classified as favorable are actually favorable. The term specificity is expressed as how many of the actual negative cases were classified as such. The F1 score value is expressed in Eq. 11. Since the F1 score is a weighted average of

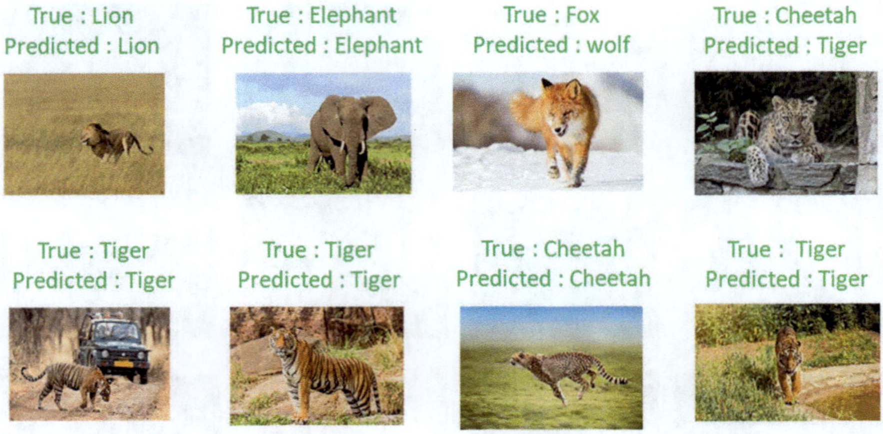

Fig. 12 Classification result

Table 2 Comparison Table

S.No	Method	Accuracy	Sensitivity	Specificity	F1-score	Precision
1.	CNN	76.29	76.3	62.52	75.7	77.2
2	Inception-v3	82.3	80.6	79.6	80.22	81.93
3	LSTM	85.4	85.94	83.14	84.4	83.4
4	YOLO-v3	89.7	88.2	88.7	87.45	88.9
5	Proposed (ROLO)	94.5	94.1	93.77	94.1	93.9

Precision and Recall, FP and FN are given equal weight. Compared to "Accuracy," this measure is quite helpful.

$$\text{F1 score} = 2^{*}(\text{precision}^{*}\text{Recall})/(\text{Precision} + \text{Recall}) \tag{11}$$

4 Conclusion

The chapter suggested the hybrid deep learning ROLO model, which is the combination of YOLO and LSTM architectures for the forest surveillance. This hybrid deep learning model is introduced to save the animals and birds from the human stalking and human from the unexpected attack of the animals. The proposed system introduced an advanced algorithm for forest surveillance, which is evaluated by the various performance metrics. The investigational outcomes demonstrate the enhanced performance of the proposed system. The suggested system presents the 94.5% accuracy rate for classification and 75.8% mean Average Precision (mAP). Thus, the planned hybrid deep learning ROLO model outperforms earlier approaches and produces superior outcomes with lesser computation time.

References

1. A. H M, N. E, P. Kambli, D. P and N. M, "Identification of Harmful animal detection using Image Processing Technique," 2022 Fourth International Conference on Cognitive Computing and Information Processing (CCIP), Bengaluru, India, 2022, pp. 1–5, https://doi.org/10.1109/CCIP57447.2022.10058651.
2. R. Gandhi, A. Gupta, A. K. Yadav and S. Rathee, "A Novel Approach of Object Detection using Deep Learning for Animal Safety," 2022 12th International Conference on Cloud Computing, Data Science & Engineering (Confluence), Noida, India, 2022, pp. 573–577, https://doi.org/10.1109/Confluence52989.2022.9734225.
3. F. Williams, L. I. Kuncheva, J. J. Rodríguez and S. L. Hennessey, "Combination of Object Tracking and Object Detection for Animal Recognition," 2022 IEEE 5th International Conference on Image Processing Applications and Systems (IPAS), Genova, Italy, 2022, pp. 1–6, https://doi.org/10.1109/IPAS55744.2022.10053017.
4. P. T. Sheeba, "Deep learning based Rice plant Disease Identification with Transfer learning," 2022 International Conference on Computing, Communication, and Intelligent Systems (ICCCIS), Greater Noida, India, 2022, pp. 697–702, https://doi.org/10.1109/ICCCIS56430.2022.10037638.
5. N. Mamat, M. F. Othman and F. Yakub, "Animal Intrusion Detection in Farming Area using YOLOv5 Approach," 2022 22nd International Conference on Control, Automation and Systems (ICCAS), Jeju, Korea, Republic of, 2022, pp. 1–5, https://doi.org/10.23919/ICCAS55662.2022.10003780.
6. S. Santhanam, S. S. B, S. S. Panigrahi, S. K. Kashyap and B. K. Duriseti, "Animal Detection for Road safety using Deep Learning," 2021 International Conference on Computational Intelligence and Computing Applications (ICCICA), Nagpur, India, 2021, pp. 1–5, https://doi.org/10.1109/ICCICA52458.2021.9697287.
7. K. Khatri, A. C. S and J. M. D'Souza, "Detection of Animals in Thermal Imagery for Surveillance using GAN and Object Detection Framework," 2022 International Conference for Advancement in Technology (ICONAT), Goa, India, 2022, pp. 1–6, https://doi.org/10.1109/ICONAT53423.2022.9725883.
8. J. J. Daniel Raj, C. N. Sangeetha, S. Ghorai, S. Das, Manish and S. Ahmed, "Wild Animals Intrusion Detection for Safe Commuting in Forest Corridors using AI Techniques," 2023 3rd International Conference on Innovative Practices in Technology and Management (ICIPTM), Uttar Pradesh, India, 2023, pp. 1–4, https://doi.org/10.1109/ICIPTM57143.2023.10117831.
9. B. Meenakshi, A. B. SR Y, H. N C, K. L and A. J, "Animal Intrusion Detection and Ranging system using YOLOv4 and LoRa," 2022 International Conference on Power, Energy, Control and Transmission Systems (ICPECTS), Chennai, India, 2022, pp. 1–6, https://doi.org/10.1109/ICPECTS56089.2022.10047729.
10. S. S. C, B. L. R and D. S, "Design and Analysis of CNN based Residue Number System for Performance Enhancement," 2023 Third International Conference on Artificial Intelligence and Smart Energy (ICAIS), Coimbatore, India, 2023, pp. 1182–1187, https://doi.org/10.1109/ICAIS56108.2023.10073805.
11. S. M. S, T. Sheela and T. Muthumanickam, "Development of Animal-Detection System using Modified CNN Algorithm," 2022 International Conference on Augmented Intelligence and Sustainable Systems (ICAISS), Trichy, India, 2022, pp. 105-109, doi: 10.1109/ICAISS55157.2022.10011014.

12. C. S. Madhumathi, V. Naveen, N. Akshay, M. S. Kumar and M. M. Aslam, "Advanced Wild Animal Detection and Alert System using YOLO V5 Model," 2023 7th International Conference on Trends in Electronics and Informatics (ICOEI), Tirunelveli, India, 2023, pp. 365–371, https://doi.org/10.1109/ICOEI56765.2023.10126065.
13. S. K. L and A. Edison, "Wild Animal Detection using Deep learning," 2022 IEEE 19th India Council International Conference (INDICON), Kochi, India, 2022, pp. 1–5, https://doi.org/10.1109/INDICON56171.2022.10039799.
14. D. Ma and J. Yang, "YOLO-Animal: An efficient wildlife detection network based on improved YOLOv5," 2022 International Conference on Image Processing, Computer Vision and Machine Learning (ICICML), Xi'an, China, 2022, pp. 464-468, https://doi.org/10.1109/ICICML57342.2022.10009855.
15. F. Schiano, D. Natter, D. Zambrano and D. Floreano, "Autonomous Detection and Deterrence of Pigeons on Buildings by Drones," in IEEE Access, vol. 10, pp. 1745–1755, 2022, https://doi.org/10.1109/ACCESS.2021.3137031.
16. S. Alagarsamy, R. Rajasekar, S. K. M. Krishna, K. P. Chand, K. Nikhil and K. V. P. Teja, "Detection of Animals for Road Safety Using Deep Learning Method," 2023 International Conference on Computer Communication and Informatics (ICCCI), Coimbatore, India, 2023, pp. 1–5, https://doi.org/10.1109/ICCCI56745.2023.10128493.
17. M. T, S. Upadhyay, R. Beaulah Jeyavathana and A. Gopatoti, "Big Data Analytics with Deep Learning based Intracranial Haemorrhage Diagnosis and Classification Model," 2022 International Conference on Augmented Intelligence and Sustainable Systems (ICAISS), Trichy, India, 2022, pp. 877–883, https://doi.org/10.1109/ICAISS55157.2022.10010826.
18. A. Endris, Zakria, J. Deng, M. Ahmednasir, J. Mohammed and N. Kewyu, "Towards Automatic Ethiopian Endemic Animals Detection on Android Using Deep Learning," 2021 4th International Conference on Pattern Recognition and Artificial Intelligence (PRAI), Yibin, China, 2021, pp. 463–468, https://doi.org/10.1109/PRAI53619.2021.9550798.
19. Y. -P. Huang and H. Basanta, "Recognition of Endemic Bird Species Using Deep Learning Models," in IEEE Access, vol. 9, pp. 102975–102984, 2021, https://doi.org/10.1109/ACCESS.2021.3098532.
20. M. Ibraheam, K. F. Li and F. Gebali, "An Accurate and Fast Animal Species Detection System for Embedded Devices," in IEEE Access, vol. 11, pp. 23462–23473, 2023, https://doi.org/10.1109/ACCESS.2023.3252499.
21. T. K. Rajan, S. Karthick, S. Nirmal, S. N. Kumar and S. Senthilmurugan, "IoT Based Remote Surveillance For Animal Tracking Near Railway Tracks," 2023 International Conference on Networking and Communications (ICNWC), Chennai, India, 2023, pp. 1–7, https://doi.org/10.1109/ICNWC57852.2023.10127346.
22. K. Yang and Z. Song, "Deep Learning-Based Object Detection Improvement for Fine-Grained Birds," in IEEE Access, vol. 9, pp. 67901–67915, 2021, https://doi.org/10.1109/ACCESS.2021.3076429.
23. O. Hadji, O. Kadri, M. Maimour, E. Rondeau and A. Benyahia, "Region of Interest and Redundancy Problem in Migratory Birds Wild Life Surveillance," 2022 International Conference on Advanced Aspects of Software Engineering (ICAASE), Constantine, Algeria, 2022, pp. 1–8, https://doi.org/10.1109/ICAASE56196.2022.9931576.
24. Zhao, T.; Yi, X.; Zeng, Z.; Feng, T. MobileNet-Yolo based wildlife detection model: A case study in Yunnan Tongbiguan Nature Reserve, China. J. Intell. Fuzzy Syst. 2021, 41, 2171–2181.
25. H. Zhao, D. Cai, Z. Liang and Y. Wang, "An Improved Method for Farm Birds Detection Based on YOLOv5s," 2022 4th International Conference on Machine Learning, Big Data and Business Intelligence (MLBDBI), Shanghai, China, 2022, pp. 183–187, https://doi.org/10.1109/MLBDBI58171.2022.00043.
26. Jiang, B.; Huang, W.; Tu, W.; Yang, C. An Animal Classification Based on Light Convolutional Network Neural Network. In Proceedings of the International Conference on Intelligent Computing and Its Emerging Applications (ICEA), Tainan, Taiwan, 30 August–1 September 2019; pp. 45–50

27. Yin, Z.; You, F. Animal Image Recognition Based on Convolutional Neural Network. In Proceedings of the 4th International Conference on Electronic Information Technology and Computer Engineering, Xiamen, China, 6–8 November 2020; Association for Computing Machinery: New York, NY, USA, 2020; pp. 545–549
28. Miao, Z.; Gaynor, K.M.; Wang, J.; Liu, Z.; Muellerklein, O.; Norouzzadeh, M.S.; McInturff, A.; Bowie, R.C.K.; Nathan, R.; Yu, S.X.; et al. Insights and Approaches Using Deep Learning to Classify Wildlife. Sci. Rep. 2019, 9, 8137.
29. Chen, R.; Little, R.; Mihaylova, L.; Delahay, R.; Cox, R. Wildlife Surveillance Using Deep Learning Methods. Ecol. Evol. 2019, 9, 9453–9466.
30. Lai, K.; Tu, X.; Yanushkevich, S. Dog Identification Using Soft Biometrics and Neural Networks. In Proceedings of the International Joint Conference on Neural Networks (IJCNN), Budapest, Hungary, 14–19 July 2019; pp. 1–8.
31. Hu, M.; You, F. Research on Animal Image Classification Based on Transfer Learning. In Proceedings of the 4th International Conference on Electronic Information Technology and Computer Engineering, Xiamen, China, 6–8 November 2020; Association for Computing Machinery: New York, NY, USA, 2020; pp. 756–761
32. Z. Song, W. Gong, C. Li and T. T. Toe, "Animals Image Classification Method Based on Improved Convolutional Neural Network," 2022 IEEE 10th Joint International Information Technology and Artificial Intelligence Conference (ITAIC), Chongqing, China, 2022, pp. 1997–2001, https://doi.org/10.1109/ITAIC54216.2022.9836676.
33. N. K. El Abbadi and E. M. T. A. Alsaadi, "An Automated Vertebrate Animals Classification Using Deep Convolution Neural Networks," 2020 International Conference on Computer Science and Software Engineering (CSASE), Duhok, Iraq, 2020, pp. 72–77, https://doi.org/10.1109/CSASE48920.2020.9142070.
34. S. Fujimori, T. Ishikawa and H. Watanabe, "Animal Behavior Classification Using DeepLabCut," 2020 IEEE 9th Global Conference on Consumer Electronics (GCCE), Kobe, Japan, 2020, pp. 254–257, https://doi.org/10.1109/GCCE50665.2020.9291715.
35. G. Gokilakrishnan, P. A. Varthnan, D. V. Kumar, Ram. Subbiah, and A. H, "Modeling and Performance Evaluation for Intelligent Internet of Intelligence Things," 2023 9th International Conference on Advanced Computing and Communication Systems (ICACCS), Mar. 2023, https://doi.org/10.1109/icaccs57279.2023.10112692.
36. A. Haldorai, Ravishankar C. V, Q. S. Mahdi, and G. J. Nehru, "Advanced Communication in Cyber Physical System Infrastructure, Protocols, and Challenges," 2023 Fifth International Conference on Electrical, Computer and Communication Technologies (ICECCT), Feb. 2023, https://doi.org/10.1109/icecct56650.2023.10179710.

Index

© European Alliance for Innovation 2024

A. Haldorai et al., *Artificial Intelligence for Sustainable Development*,
EAI/Springer Innovations in Communication and Computing,
https://doi.org/10.1007/978-3-031-53972-5

I

K

L

M

N

O

P

X

Y

Z

MIX
Papier aus verantwortungsvollen Quellen
Paper from responsible sources
FSC® C105338

If you have any concerns about our products,
you can contact us on
ProductSafety@springernature.com

In case Publisher is established outside the EU,
the EU authorized representative is:
Springer Nature Customer Service Center GmbH
Europaplatz 3, 69115 Heidelberg, Germany

Printed by Libri Plureos GmbH
in Hamburg, Germany